Advanced Unsaturated Soil Mechanics and Engineering

T0092919

Also available from Taylor & Francis

Ground Improvement 2nd edition

K Kirsch et al.   Hb: ISBN 978–0–415–27455–5

Geotechnical Modelling

D Muir Wood   Hb: ISBN 978–0–415–34304–6
Pb: ISBN 978–0–419–23730–3

Cone Penetration Testing in Geotechnical Practice

T Lunne et al.   Hb: ISBN 978–0–419–23750–1

Soil Liquefaction

M Jefferies et al.   Hb: ISBN 978–0–419–16170–7

Introductory Geotechnical Engineering

HY Fang et al.   Hb: ISBN 978–0–415–30401–6
Pb: ISBN 978–0–415–30402–3

Information and ordering details

For price availability and ordering visit our website
www.tandfbuiltenvironment.com
Alternatively our books are available from all good bookshops.

Advanced Unsaturated Soil Mechanics and Engineering

Charles W.W. Ng and
Bruce Menzies

CRC Press
Taylor & Francis Group
Boca Raton London New York

CRC Press is an imprint of the
Taylor & Francis Group, an **informa** business
A TAYLOR & FRANCIS BOOK

CRC Press
Taylor & Francis Group
6000 Broken Sound Parkway NW, Suite 300
Boca Raton, FL 33487-2742

First issued in paperback 2019

© 2007 Charles W.W. Ng and Bruce Menzies
CRC Press is an imprint of Taylor & Francis Group, an Informa business

Typeset in Sabon by
Integra Software Services Pvt. Ltd, Pondicherry, India

No claim to original U.S. Government works

ISBN-13: 978-0-415-43679-3 (hbk)
ISBN-13: 978-0-367-38803-4 (pbk)

British Library Cataloguing in Publication Data
A catalogue record for this book is available from the British Library

Library of Congress Cataloging in Publication Data
Ng, C. W. W.
Unsaturated soil mechanics and engineering / Charles W.W. Ng and Bruce Menzies.
p. cm.
Includes bibliographical references and index.
ISBN 978-0-415-43679-3 (hardback : alk. paper) --
ISBN 978-0-203-93972-7 (ebook) 1. Soil mechanics. 2. Swelling soils.
3. Zone of aeration. 4. Soil moisture. I. Menzies, Bruce Keith. II. Title.
TA710.N475 2007
624.1′5136--dc22
2007008742

Visit the Taylor & Francis Web site at
http://www.taylorandfrancis.com

and the CRC Press Web site at
http://www.crcpress.com

Dedicated To Professor Delwyn G. Fredlund, Unsaturated Soil Mechanics Pioneer

Delwyn G. Fredlund O.C., Ph.D., F.E.I.C., P.Eng.
Senior Geotechnical Engineering Specialist, Golder Associates, 2005–
Professor Emeritus, University of Saskatchewan, from 2000
K.P. Chao Chair, Zhejiang University, 2007–2009
Adjunct Professor, Hong Kong University of Science and Technology, 2001–
Adjunct Professor, University of Alberta, 2002–
Honorary Professor, University of British Columbia, 2002–
Research Professor, Arizona State University, 2004–

If you learn to think the way unsaturated soil behaves, then you will make wise engineering decisions.
(Del Fredlund's advice to students)

Cover Frontispiece

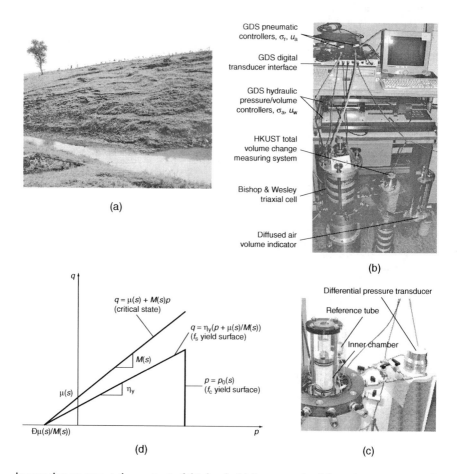

(a)

GDS pneumatic controllers, σ_r, u_a

GDS digital transducer interface

GDS hydraulic pressure/volume controllers, σ_a, u_w

HKUST total volume change measuring system

Bishop & Wesley triaxial cell

Diffused air volume indicator

(b)

$q = \mu(s) + M(s)p$
(critical state)

$q = \eta_y(p + \mu(s)/M(s))$
(f_s yield surface)

$M(s)$

η_y

$p = p_0(s)$
(f_c yield surface)

$\mu(s)$

$Đ\mu(s)/M(s))$

(d)

Differential pressure transducer

Reference tube

Inner chamber

(c)

Images that represent the content of this book: (a) Retrogressive failures in an unsaturated expansive soil slope (Ng et al., 2003); (b) HKUST–GDS computer controlled triaxial stress-path apparatus equipped with double wall total volume measuring system; (c) HKUST measurement of total volume change (double wall system) using a differential pressure transducer (Ng et al., 2002); (d) Yield surface in $q-p$ space at constant suction (Chiu and Ng, 2003).

Book overview

Civil engineers build on or in the earth's surface. Most of the earth's land surface comprises notoriously hazardous geomaterials called 'unsaturated soils'. These soils are a hazard to earth structures and earth-supported structures because on wetting, by rain or other means, they can expand or collapse with serious consequences for cost and safety. This book puts the mechanics and engineering of unsaturated soils into a logical framework for civil engineering analysis and design. It also explains the laboratory and field testing and research that are the logical basis of this modern approach to safe construction in these hazardous geomaterials.

This state-of-the-art book is written in four parts as follows:

1. Physical and flow characteristics of unsaturated soils;
2. Collapse, swelling, shear strength and stiffness of unsaturated soils;
3. State-dependent elasto-plastic modelling of unsaturated soils;
4. Field trials and numerical studies in slope engineering of unsaturated soils.

This book is unique because:

- it illustrates the importance of state-dependent soil–water characteristic curves;
- highlights modern soil testing of unsaturated soil behaviour, including accurate measurement of total volume changes and the measurement of anisotropic soil stiffness at very small strains;
- introduces an advanced state-dependent elasto-plastic constitutive model for both saturated and unsaturated soil;
- demonstrates the power of numerical analysis which is at the heart of modern soil mechanics;
- studies and simulates the behaviour of loose fills from unsaturated to saturated states, explains the difference between strain-softening and static liquefaction, and describes real applications in unsaturated soil slope engineering;
- includes purpose-designed field trials to capture the effects of two independent stress variables, and reports comprehensive measurements of soil suction, water contents, stress changes and ground deformations in bare slopes;
- introduces a new conjunctive surface and subsurface transient flow model for realistically analysing rainfall infiltration in unsaturated soil slopes, and illustrates the importance of the flow model in slope engineering.

Contents

Authors' biographical details   xiii
Preface   xiv
Acknowledgements   xvii

PART I
**Physical and flow characteristics of
unsaturated soils**   I
   Part 1 Frontispiece 2
   Chapter synopses 3

1 Basic physics, phases and stress state variables   5
   Sources 5
   Introduction (Fredlund, 1996) 5
   Four-phase materials (Fredlund and Rahardjo, 1993) 20
   Cavitation (Young, 1989; Marinho, 1995;
     Marinho and de Sousa Pinto, 1997) 24

2 Measurement and control of suction: methods and
   applications   31
   Sources 31
   Theory of soil suction (Fredlund and Rahardjo, 1993) 32
   The capillary model 32
   Suction control and measurement methods
     (Ng and Chen, 2005, 2006; Ng et al., 2007a) 36
   Suction measuring devices (Fredlund and Rahardjo, 1993;
     Ridley and Wray, 1995; Feng and Fredlund, 2003) 63
   Case study: comparisons of in situ suction
     measurements (Harrison and Blight, 2000) 87

3   Flow laws, seepage and state-dependent soil–water
     characteristics   93
     Sources 93
     *Flow laws for water and air (Fredlund and Rahardjo,
       1993) 94*
     *Soil–water characteristic curves and water permeability
       functions (Fredlund et al., 2001a, 2002) 122*
     *The state-dependent soil–water characteristic curve:
       a laboratory investigation (Ng and Pang, 2000a,b) 147*
     *Generalised triaxial apparatus for determination of the
       state-dependent soil–water characteristic curve
       (Ng et al., 2001a) 166*
     *Measurements of the state-dependent soil–water characteristic
       curve in a centrifuge (Khanzode et al., 2002) 168*
     *Steady-state and transient flows (Fredlund and
       Rahardjo, 1993) 174*
     *Analytical analysis of rainfall infiltration mechanism
       in unsaturated soils (Zhan and Ng, 2004) 194*

**PART 2
Collapse, swelling, strength and stiffness
of unsaturated soils**   **221**
     Part 2 Frontispiece 222
     Chapter synopses 224

4   Collapse and swelling caused by wetting   227
     Sources 227
     Overview 227
     *Collapsible soils (Bell and Culshaw, 2001; Houston,
       1995; Fredlund, 1996) 228*
     Concept of virgin and non-virgin fills (Ng et al., 1998) 240
     Estimations of foundation settlements (Houston, 1995) 249
     *Mitigation measures against collapsible soils (Houston,
       1995) 250*
     *Expansive soils (Bell and Culshaw, 2001; Tripathy et al.,
       2002) 252*
     *Design of foundations on expansive swelling
       soils (Jimenez-Salas, 1995) 273*

5   Measurement of shear strength and shear behaviour of
     unsaturated soils   279
     Sources 279
     *Introduction to shear strength (Fredlund and
       Rahardjo, 1993) 280*

*A new simple system for measuring volume changes of
    unsaturated soils in the triaxial cell
    (Ng et al., 2002a) 287*
*Comparisons of axis-translation and osmotic techniques
    for shear testing of unsaturated soils (Ng et al., 2007) 294*
*Extended Mohr–Coulomb failure
    criterion (Fredlund and Rahardjo, 1993) 307*
*The relationship of the unsaturated soil shear strength to
    the soil–water characteristic curve (Fredlund et al., 1996) 313*
*Effects of soil suction on dilatancy of an
    unsaturated soil (Ng and Zhou, 2005) 326*
*Behaviour of a loosely compacted unsaturated
    volcanic soil (Ng and Chiu, 2001) 335*
*Laboratory study of a loose saturated and unsaturated
    decomposed granitic soil (Ng and Chiu, 2003a) 358*

6   **Measurement of shear stiffness**   380
    Sources 380
    Introduction (Ng et al., 2000b) 380
    *Effects of initial water contents on small strain
        shear stiffness of sands (Qian et al., 1993) 382*
    *Effects of suction on elastic shear modulus of
        quartz silt and decomposed granite (silty sand) 394*
    *Anisotropic shear stiffness of completely
        decomposed tuff (clayey silt) (Ng and Yung, 2007) 408*

PART 3
**State-dependent elasto-plastic modelling
of unsaturated soil**   **431**
    Part 3 Frontispiece 432
    Chapter synopsis 433

7   **A state-dependent elasto-plastic critical state-based
    constitutive model**   435
    Sources 435
    Introduction 435
    Mathematical formulations 436
    Determination of model parameters 447
    *Comparisons between model predictions and experimental
        results for decomposed granite 457*
    *Comparisons between model predictions and experimental
        results for decomposed volcanic soil 463*
    Summary 466

PART 4
Field trials and numerical studies in
slope engineering of unsaturated soils   469
Part 4 Frontispiece 470
Chapter synopses 471

8   Instrumentation and performance: A case study in
    slope engineering   473
    Source 473
    The South-to-North Water Transfer Project,
    China (Ng et al., 2003) 473

9   Engineering applications for slope stability   501
    Sources 501
    Methods of slope stability analysis in unsaturated
    soils (Fredlund and Rahardjo, 1993) 502
    Three-dimensional numerical parametric study of rainfall
    infiltration into an unsaturated soil slope
    (Ng et al., 2001b) 515
    Influence of state-dependent soil–water characteristic curve
    and wetting and drying on slope stability (Ng and Pang,
    2000b) 537
    Effects of state-dependent soil–water characteristic
    curves and damming on slope stability
    (Ng and Lai, 2004) 547
    Fundamentals of recompaction of unsaturated
    loose fill slopes (Ng and Zhan, 2001) 559
    Effects of surface cover and impeding layers
    on slope stability (Ng and Pang, 1998a) 567
    Numerical experiments of soil nails in loose fill slopes
    subjected to rainfall infiltration effects
    (Cheuk et al., 2005) 575
    Effects of conjunctive modelling on
    slope stability (Tung et al., 2004) 599

Appendix   611
Notation   618
References   621
Author Index   648
Subject Index   654

Authors' biographical details

Charles Ng graduated from Hong Kong Polytechnic in 1986. Following periods at University of Southampton, the University of Hong Kong, Ove Arup & Partners, University of Bristol, Mott MacDonald Group and Cambridge University, he moved to the Hong Kong University of Science and Technology (HKUST). He is Associate Editor of the Canadian Geotechnical Journal and International Editorial Panel Member for Geotechnical Engineering, Proceedings of the Institution of Civil Engineers, UK and has been Editorial Board Member of the Journal of Geotechnical and Geo-environmental Engineering, ASCE. He established the unsaturated soil engineering research group at HKUST in 1995 and has taught post-graduate courses on unsaturated soil mechanics since 1996. He has published widely on the behaviour of unsaturated soils and slopes, on applications of centrifuge modelling and on the soil-structure engineering of piles, barrettes, tunnels and deep excavations. In 2005, he was elected as Overseas Fellow at Churchill College, Cambridge University. He is Professor of Civil Engineering at HKUST where he is also Director of the Geotechnical Centrifuge Facility and Associate Dean of Engineering.

Bruce Menzies graduated from the University of Auckland in 1962. After periods at Kingston Reynolds & Thom, Soil Mechanics Ltd, Kingston University and University College London, he became lecturer in Soil Mechanics at the University of Surrey. His publications include papers on progressive failure of foundations and on the performance of buried pipelines. In 1979, he founded Geotechnical Digital Systems Ltd (GDS) that has supplied computer controlled soil and rock testing systems to over 500 universities and research establishments. He invented the digital pressure controller used for pressure regulation and volume change measurement in soil and rock mechanics laboratories. Over 2000 of the GDS pressure controllers have been sold. At GDS, he has participated in the development of unsaturated soil testing systems. He is a winner of the British Geotechnical Association Prize 2002. In 2003, he was Visiting Scholar at Hong Kong University of Science and Technology. He is Chairman of GDS.

Preface

Civil engineers build on or in the earth's surface. Most of the earth's land surface comprises notoriously hazardous geomaterials called 'unsaturated soils'. These soils are a hazard to earth structures and earth-supported structures because on wetting, by rain or other means, they can expand or collapse with serious consequences for cost and safety. This book puts the mechanics and engineering of unsaturated soils into a logical framework for civil engineering analysis and design. It also explains the laboratory and field testing and research that are the logical basis of this modern approach to safe construction in these hazardous geomaterials.

A search of the Institution of Civil Engineers' library web site returns a total of 462 titles of books, conference proceedings and articles on 'soil mechanics'. A search for 'unsaturated soil mechanics', however, returns only 66 titles, or just over 14 per cent of the total. Of these only two of the titles are text books. This is in spite of the fact that most of the soils on our planet's land surface are unsaturated. The small proportion of unsaturated soil titles is hardly surprising, however. This is because post-Terzaghi soil mechanics had its origins in northern Europe and North America where soils involved in civil engineering construction are mainly saturated or near-saturated. So naturally the science developed on the (simplified) assumption of saturation.

With the rapid development, however, of China, India, Central and South America, and Africa (covering regions where foundation soils are rarely saturated near the surface), geotechnical engineers can no longer ignore the complication of unsaturated soils and the challenges they present. Accordingly, the international geotechnical community has responded. Since the 7th International Conference on Expansive Soils held in Dallas, USA, in 1992, there have been over 10 major conferences or workshops up to and including the 3rd International Conference on Unsaturated Soils (UNSAT, 2002) held in Recife, Brazil, in 2002, plus the 4th International Conference on Unsaturated Soils held in Carefree, Arizona in 2006 (the international conferences are convened at 4 yearly intervals). These conferences have brought into the public domain the results of research conducted mainly at universities. But the teaching at universities of unsaturated soil mechanics and engineering is still relatively rare and when covered is confined mainly to post graduate courses including short courses for continuing professional development. Part of the problem of teaching the subject is the lack of up-to-date textbooks. There are two books as mentioned above. There is the classic text *Soil Mechanics for Unsaturated Soils* by Fredlund and Rahardjo (1993) which starts from the concepts of agronomy and soil science and moves on to unsaturated soil mechanics, based mainly on the Mohr–Coulomb failure criterion, and the problems of unsaturated soil slopes. Also, there is the more recent text *Unsaturated Soil Mechanics* by Lu and Likos (2004) which concentrates more on flow in unsaturated soils. Both books are excellent and are major aids to the teaching of unsaturated soil mechanics.

Based on our teaching, research and test equipment development, particularly at Hong Kong University of Science and Technology and at GDS Instruments, we felt, however, that there was a need for a third text book – one where modern soil testing of unsaturated soil behaviour, including measurement of the state-dependent soil–water characteristic curve (SDSWCC) and anisotropic soil stiffness, is highlighted; where constitutive modelling, which is at the heart of modern soil mechanics, is central and its' power in numerical analysis demonstrated; and where real applications in unsaturated soil slope engineering field trials are described. Accordingly, we wrote this state-of-the-art book which is in four parts as follows:

- Part 1: Physical and flow characteristics of unsaturated soils;
- Part 2: Collapse, swelling, strength and stiffness of unsaturated soils;
- Part 3: State-dependent elasto-plastic modelling of unsaturated soils;
- Part 4: Field trials and numerical studies in slope engineering of unsaturated soils.

Each part is prefaced with a frontispiece of images that represent the content of that part. The images are then followed by synopses of each chapter in that part. These synopses give overviews of each chapter and explain how they enhance geotechnical knowledge. The reader can immediately judge from this the benefit to be gained by looking into a particular part or chapter.

This state-of-the-art book is unique because it illustrates the importance of state-dependent soil–water characteristic curves and how to measure them, describes how to measure small-strain anisotropic soil stiffness, develops state-dependent elasto-plastic constitutive modelling as well as conjunctive modelling of rain infiltration, and studies the behaviour of loose fills from unsaturated to saturated states. A major part of the book addresses the hugely challenging geotechnical problem of unsaturated soil slope stability.

This state-of-the-art book was written at Churchill College, Cambridge where Charles Ng was Overseas Fellow in 2005. The book is dedicated to Prof. Delwyn G. Fredlund, unsaturated soil mechanics pioneer.

Charles W.W. Ng and Bruce Menzies
Churchill College, Cambridge, 2005

Acknowledgements

We acknowledge permissions from publishers to make verbatim extracts from their published works as follows:
John Wiley & Sons, Inc. from

- Fredlund, D.G. and Rahardjo, H. (1993). *Soil Mechanics for Unsaturated Soils*. John Wiley & Sons, Inc, New York, 517 p.
  Reprinted with permission of John Wiley & Sons, Inc. to republish abstract from pages 3, 20, 24–25, 66–69, 71, 77, 85–87, 105–123, 134, 150–153, 159–160, 171–173, 175–177, 225–226, 228–229, 237, 247–248, 321–326, 331–332 and 380 from Fredlund/Soil Mechanics for Unsaturated Soils/047185008X.

The American Society for Civil Engineers from

- Ng, C.W.W and Chiu, C.F. (2003). Laboratory study of loose saturated and unsaturated decomposed granitic soil. *J. Geotech. Geoenviron. Eng., ASCE.* 129(6), 550–559.
- Ng, C.W.W. and Chiu, A.C.F. (2001). Behaviour of a loosely compacted unsaturated volcanic soil. *J. Geotech. Geoenviron. Eng., ASCE,* December, 1027–1036 (Figures 1–14; Tables 1–3).
- Ng, C.W.W. and Chiu, A.C.F. (2003). Laboratory study of loose saturated and unsaturated decomposed granitic soil. *J. Geotech. Geoenviron. Eng., ASCE,* June, 550–559 (Figures 1–10; Tables 1–3).
- Ng, C.W.W. and Pang, Y.W. (2000b). Influence of stress state on soil–water characteristics and slope stability. *J. Geotech. Environ. Eng., ASCE,* 26(2), 157–166 (Figures 5–12; Tables 3–4).
- Ng, C.W.W., Pun, W.K. and Pang, R.P.L. (2000b). Small strain stiffness of natural granitic saprolite in Hong Kong. *J. Geotech. Geoeviron. Eng., ASCE,* 819–833 (Figure 1).
- Qian, X., Gray, D.H. and Woods, R.D. (1993). Voids and granulometry: effects on shear modulus of unsaturated sands. *J. Geotech. Eng., ASCE,* 119(2), 295–314 (Figures 3–8, 11–13, 17–19; Table 1).

- Zhan, L.T. and Ng, C.W.W. (2004). Analytical analysis of rainfall infiltration mechanism in unsaturated soils. *Int. J. Geomech., ASCE,* 4(4), 273–284 (Figures 1–15; Table 1).

Taylor & Francis Books UK, reproduced by permission from

- Cabarkapa, Z., Cuccovillo, T. and Gunn, M. (1999). Some aspects of the pre-failure behaviour of unsaturated soil. *Proc. Conf. Pre-failure Deformation Characteristics of Geomaterials,* Jamiolkowski, Lancellotta and Lo Presti, eds, Vol. 1, Balkema, Rotterdam, 159–165.
- Fredlund, D.G. (2002). Use of soil–water characteristic curves in the implementation of unsaturated soil mechanics. *Proc. 3rd Int. Conf. Unsat. Soils,* Recife, Brazil, Balkema, Rotterdam.
- Fredlund, D.G., Fredlund, M.D. and Zakerzadeh, N. (2001b). Predicting the permeability function for unsaturated soils. *Proc. Int. Symp. Suction, Swelling, Permeability and Structured Clays,* IS-Shizuoka, Shizuoka, Japan.
- Fredlund, D.G., Rahardjo, H., Leong, E.C. and Ng, C.W.W. (2001a). Suggestions and recommendations for the interpretation of soil–water characteristic curves. *Proc. 14th Southeast Asian Geotech. Conf.,* Hong Kong, Vol. 1, Balkema, Rotterdam, pp. 503–508.
- Harrison, B.A. and Blight, G.E. (2000). A comparison of in situ soil suction measurements. *Proc. Conf. Unsaturated Soils for Asia,* Singapore, H. Rahardjo, D.G. Toll and, E.C. Leong, eds, Balkema, Rotterdam, pp. 281–284.
- Houston, S.L. (1995). Foundations and pavements on unsaturated soils – Part one: Collapsible soils. *Proc. Conf. Unsaturated Soils,* Paris, Alonso and Delage, eds, Balkema, Rotterdam, pp. 1421–1439.
- Jiménez-Salas, J.A. (1995). Foundations and pavements on unsaturated soils – Part two: Expansive clays. *Proc. Conf. Unsaturated Soils,* Paris, Alonso and Delage, eds, Balkema, Rotterdam, pp. 1441–1464.
- Mancuso, C., Vassallo, R. and d'Onofrio, A. (2000). Soil behaviour in suction controlled cyclic and dynamic torsional shear tests. *Proc. Conf. Unsaturated Soils for Asia,* Rahardjo, Toll and Leong, eds, Balkema, Rotterdam, pp. 539–544.
- Marhino, F.A.M. and de Sousa Pinto, C. (1997). Soil suction measurement using a tensiometer. *Proc. Conf. Recent Developments Soil Pavement Mech.,* Almeida, eds, pp. 249–254.
- Marinho, F.A.M. (1995). Cavitation and the direct measurement of soil suction. *Proc. Conf. Unsaturated Soils,* Paris, Alonso and Delage, eds, Balkema, Rotterdam, pp. 623–630.
- Ng, C.W.W. and Pang, Y.W. (1998). Role of surface cover and impeding layer on slope stability in unsaturated soils. *Proc. 13th Southeast Asian Geotech. Conf.,* Taipei, Taiwan, ROC, pp. 135–140.

- Ng, C.W.W. and Zhan, L.T. (2001). Fundamentals of re-compaction of unsaturated loose fill slopes. *Proc. Int. Conf. Landslides – Causes, Impacts and Countermeasures*. Davos, Switzerland, pp. 557–564.
- Ng, C.W.W., Chiu, C.F. and Shen, C.K. (1998). Effects of wetting history on the volumetric deformations of unsaturated loose fill. *Proc. 13th Southeast Asian Geotech. Conf.*, Taipei, Taiwan, ROC, pp. 141–146.
- Ng, C.W.W., Wang, B. and Gong, B.W. (2001a). A new triaxial apparatus for studying stress effects on soil–water characteristics of unsaturated soils. *Proc. 15th Conf. of Int. Soc. Soil Mech. and Geotech. Engrg*, Istanbul, Balkema, Rotterdam.
- Ridley, A.M. and Wray, W.K. (1995). Suction measurement: A review of current theory and practices. *Proc. Conf. Unsaturated Soils*, Paris, Alonso and Delage, eds, Balkema, Rotterdam, pp. 1293–1322.

Elsevier, reprinted with permission from

- Cheuk, C.Y., Ng, C.W.W. and Sun, H.W. (2005). Numerical experiments of soil nails in loose fill slopes subjected to rainfall infiltration effects. *Comput. Geotech.*, Elsevier (Figures 1–15; Tables 1–3).
- Ng, C.W.W. and Shi, Q. (1998). A numerical investigation of the stability of unsaturated slopes subject to transient seepage. *Comput. Geotech.*, Vol. 22, 1–28. (No Figures and Tables).
- Hillel, D. (1998). *Introduction to Environmental Soil Physics*. Academic Press, San Diego, CA, USA, p. 364.

Thomas Telford Ltd from

- Bell, F.G. and Culshaw, M.G. (2001). Problem soils: A review from a British perspective. *Proc. Symp. Problematic Soils*, Nottingham Trent University, Jefferson, Murray, Faragher and Fleming, eds, Thomas Telford, London (Tables 1, 2 & 6; Figures 1 & 5).
- Chiu, C.F. and Ng, C.W.W. (2003). A state-dependent elasto-plastic model for saturated and unsaturated soils. *Géotechnique* 53, No. 9, 809–829. (Figures 1–16; Tables 3–5).
- Houlsby, G.T. (1997). The work input to an unsaturated granular material. *Géotechnique* 47, No. 1, 193–196 (No tables or figures).
- Ng, C.W.W. and Yung, S.Y. (2007). Determinations of anisotropic shear stiffness of an unsaturated decomposed soil. Provisionally accepted by *Géotechnique* (Figures 2–13; Tables 1–4).
- Ng, C.W.W., Zhan, L.T., Bao, C.G., Fredlund, D.G. and Gong, B.W. (2003). Performance of an unsaturated expansive soil slope subjected to artificial rainfall infiltration. *Géotechnique* 53, No. 2, 143–157. (Figures 1–16; Table 1).

- Ridley, A.M. and Burland, J.B. (1993). A new instrument for the measurement of soil moisture suction.*Géotechnique*, 43(2), 321–324.

National Research Council of Canada from the Canadian Geotechnical Journal as follows:

- Feng, M. and Fredlund, D.G. (2003). Calibration of thermal conductivity sensors with consideration of hysteresis. *Can. Geotech. J.*, 40, 1048–1055. (Figures 1, 12–14)
- Fredlund, D.G. (2000). The 1999 R.M. Hardy Lecture: The implementation of unsaturated soil mechanics into geotechnical engineering. *Can. Geotech. J.*, 37, 963–986. (Figures 16–21; Table 1)
- Fredlund, D.G. and Xing, A. (1994). Equations for the soil–water characteristic curve. *Can. Geotech. J.*, 31(3), 521–532.
- Fredlund, D.G., Xing, A., Fredlund, M.D. and Barbour, S.L. (1996). The relationship of the unsaturated soil shear strength to the soil–water characteristic curve. *Can. Geotech. J.*, 33, 440–445 (Figures 1–15; Table 1)
- Gan, J.K.M. and Fredlund, D.G. (1996). Shear strength characteristics of two saprolitic soils. *Can. Geotech. J.*, 33, 595–609. (No Figures; Tables)
- Khanzode, R.M., Vanapalli, S.K. and Fredlund, D.G. (2002). Measurement of soil–water characteristic curve for fine-grained soils using a small centrifuge. *Can. Geotech. J.*, 39(5), 1209–1217. (Figures 2–4; Tables 1–2)
- Lam, L., Fredlund, D.G. and Barbour, S.L. (1987). Transient seepage model for saturated–unsaturated soil systems: A geotechnical engineering approach. *Can. Geotech. J.*, 24, 565–580 (No Figures; Tables).
- Mancuso, C., Vassallo, R. and d'Onofrio, A. (2002). Small strain behaviour of a silty sand in controlled-suction resonant column-torsional shear tests. *Can. Geotech. J.*, 39, 22–31 (Figures 4, 10; Tables 1, 3).
- Ng, C.W.W. and Pang, Y.W. (2000a). Experimental investigations of the soil–water characteristics of a volcanic soil. *Can. Geotech. J.*, 37, 1252–1264 (Figures 1–12; Tables 1–3).
- Ng, C.W.W., Wang, B. and Tung, Y.K. (2001b). Three-dimensional numerical investigations of groundwater responses in an unsaturated slope subjected to various rainfall patterns. *Can. Geotech. J.*, 38, 1049–1062 (Figures 1–11; Table 1).
- Ng, C.W.W., Zhan, L.T. and Cui, Y.J. (2002a). A new simple system for measuring volume changes in unsaturated soils. *Can. Geotch. J.*, 39, 757–764 (Figures 1–2; No Tables).
- Ng., C.W.W. and Lai, J.C.H. (2004). Effects of state-dependent soil–water characteristic and damming on slope stability. *Proc. 57th Can.*

Geotech. Conf., Géo Québec 2004, Session 5E, 28–35 (Figures 4–9; Table 1).

- Tung, Y.K., Zhang, H., Ng, C.W.W. and Kwok, Y.F. (2004). Transient seepage analysis of rainfall infiltration using a new conjunctive surface-subsurface flow model. *Proc. 57th Can. Geotech. Conf., Géo Québec 2004*, Session 7C, 17–22 (Figures 3–9; Table 1).
- Tripathy, K.S., Subba Rao, K.S. and Fredlund, D.G. (2002). Water content – void ratio swell-shrink paths of compacted expensive soils. *Can. Geotech. J.*, 39, 938–959 (Figures 1–2, 4–6, 10–14, 16–17; Tables 1–5).

McGraw-Hill and Dr. F. Ronald Young from

- Young, F.R. (1989). Cavitation. McGraw-Hill Book Company, London.

Millpress Science Publishers from

- Ng, C.W.W. and Zhou, R.Z.B. (2005). Effects of soil suction on dilatancy of an unsaturated soil. *Proc. 16th Int. Conf. Soil Mech. Geotech. Engrg*, Osaka, Japan, Vol. 2, 559–562. Millpress Science Publishers, Rotterdam.

Japanese Geotechnical Society, Pierre Delage, Yu-Jun Cui and Rui Chen from

- Ng, C.W.W., Cui, Y., Chen, R. and Delage, P. (2007a). The axis-translation and osmotic techniques in shear testing of unsaturated soils: a comparison. *Soils and Foundations*, Vol. 47, No. 4, 657–684, in press.

Texas A&M University Press and Prof. Delwyn G. Fredlund from

- Fredlund, D.G. (1996). *The Emergence of Unsaturated Soil Mechanics*. The Fourth Spencer J. Buchanan Lecture, College Station, Texas, A&M University Press, p. 39.

Soilmoisture Corporation in respect of equipment diagrams and photographs in Figures 2.33, 2.34, 2.35, 3.19, 3.20 and 3.21.

Mills & Boon Ltd, London and TUV NEL from

- Pearsall, I.S. (1972). *Cavitation*. M & B Monograph ME/10. p. 80.

We, the authors, Charles W.W. Ng and Bruce Menzies, do not make any claim whatsoever to any part of this book as being original to us.

Part I

Physical and flow characteristics of unsaturated soils

Part I Frontispiece

(a)

(b)   (c)

Images that represent the content of this Part I: (a) Major landslide in July 2003 along the route of the Three Gorges Dam; (b) Components of modified one-dimensional volumetric pressure plate extractor (Ng and Pang, 2000b); Assembled modified one-dimensional volumetric pressure plate extractor (Ng and Pang, 2000b).

Chapter synopses

Chapter 1: Basic physics, phases and stress state variables

Unsaturated soils, which are the majority of surface or near-surface soils on the earth's land surface, are introduced together with their characteristic of partial saturation giving rise to pore air as well as pore water and hence an air–water interface forming a contractile skin. The importance of stress state variables in defining the engineering behaviour of strength, deformability and transient flow is discussed and the selection of the two independent stress state variables (net normal stress and matric suction) is explained. The associated physics of surface tension and cavitation (and how to avoid it) are described. The case is made that saturated soil is a simplified special case of unsaturated soil and so there are fundamental differences between the two in terms of classification and analysis methods. This important distinction has major implications for practising civil engineers.

Chapter 2: Measurement and control of suction: methods and applications

The measurement and importance of soil suction is highlighted. Principles and limitations of measurement methods are explained including the axis-translation, osmotic and humidity control techniques. A new and simple accurate volume-change measurement system for unsaturated soil is introduced and explained. Examples of using different control techniques are provided. Shear strengths measured by the axis-translation and osmotic techniques are compared. Various laboratory and in situ suction measurement devices are considered. A case study comparing in situ suction measurements using different sensors is described. Now the civil engineer has a wide array of modern sensors that can measure unsaturated soil suctions.

Chapter 3: Flow laws, seepage and state-dependent soil–water characteristics

Flow laws for the seepage of air and water through saturated and unsaturated soil are reviewed. Hydraulic parameters and properties are introduced and explained including the soil–water characteristic curve (SWCC) as well as permeability functions for carrying out transient seepage analysis in unsaturated soil. A new and advanced concept of the state-dependent soil–water characteristic curve (SDSWCC) is introduced to define the capacity of an unsaturated soil to store and release water at different stress states. Experimental techniques and theoretical means of defining and measuring the SDSWCC are introduced. The influence of various hydraulic properties and rainfall conditions on the rainfall infiltration mechanism in both single and two-layer unsaturated soil systems is revealed by an analytical parametric study. This demonstrates that the infiltration process and pore water pressure response are primarily controlled by rainfall infiltration rate, desaturation coefficient and saturated water permeability. Also the influence of antecedent infiltration rate is revealed.

Chapter 1

Basic physics, phases and stress state variables

Sources

This chapter is made up of verbatim extracts from the following sources for which copyright permissions have been obtained as listed in the Acknowledgements.

- Fredlund, D.G. (1996). *The Emergence of Unsaturated Soil Mechanics.* The Fourth Spencer J. Buchanan Lecture, College Station, Texas, A & M University Press, 39 pp.
- Fredlund, D.G. and Rahardjo, H. (1993). *Soil Mechanics for Unsaturated Soils.* John Wiley, New York, 517 pp.
- Houlsby, G.T. (1997). The work input to an unsaturated granular material. *Géotechnique,* 47, 1, 193–196.
- Marinho, F.A.M. and de Sousa Pinto, C. (1997). Soil suction measurement using a tensiometer. *Proc. Conf. on Recent Developments in Soil and Pavement Mechanics,* Almeida, ed., pp. 249–254.
- Marinho, F.A.M. (1995). Cavitation and the direct measurement of soil suction. *Proc. Conf. on Unsaturated Soils,* Paris, E.E. Alonso and P. Delage, eds., Balkema, Rotterdam, pp. 623–630.
- Pearsall, I.S. (1972). *Cavitation.* Mills and Boon, London.
- Young, R.F. (1989). *Cavitation.* McGraw-Hill, London.

Introduction (Fredlund, 1996)

As pointed out by Fredlund and Rahardjo (1993), classical soil mechanics and geotechnical engineering have been often taught with an implicit assumption that soil is either dry (0 per cent saturation) or saturated (100 per cent saturation). Soil behaviour, it is argued, is governed solely by Terzaghi's effective stress principle (Terzaghi, 1936a,b). In fact, dry and saturated states are just two extreme and limiting conditions of a soil. In other words, dry and saturated conditions are just two special cases of an unsaturated (i.e. not-saturated) soil which has a degree of saturation that lies between 0 and 100 per cent. In many engineering problems, however, a soil is often neither saturated nor dry. Relatively, limited research has been conducted on unsaturated soils and only two textbooks have been published on the subject. Clearly, there is an urgent need to improve the understanding of the behaviour and mechanics of an unsaturated soil.

For convenience, the general field of classical soil mechanics is often subdivided into that portion dealing with saturated soils and that portion dealing with unsaturated soils. Although this artificial division between saturated and unsaturated soils can be shown to be unnecessary, it may still be helpful to make use of the knowledge gained from saturated soils as a reference and then to extend it to the broader unsaturated world as shown in Figure 1.1, which provides a visual aid for the generalized world of soil mechanics (Fredlund, 1996). For simplicity, this world of soil mechanics is divided by the water table. Below the water table, soil behaviour is governed by effective stress $(\sigma - u_w)$, whereas the unsaturated soil above the water table is governed by two independent stress variables, net normal stress $(\sigma - u_a)$ and matric suction $(u_a - u_w)$ (Jennings and Burland, 1962; Fredlund and Morgenstern, 1977). Focusing on the soil above the water table, it may be useful to categorize the soil according to its degree of saturation as shown in Figure 1.2. Instead of a two-phase material (i.e. solid and water) when a soil is saturated, an unsaturated soil is recognized by Fredlund and Morgenstern (1977) to have four phases [i.e. solid, water, air and an air–water interface called contractile skin (Paddy, 1969)]. It is obvious that the behaviour of an unsaturated soil is more complex than a saturated soil. It is the intention of this book to simplify the complexities of an unsaturated soil to a digestible level for students and practising engineers.

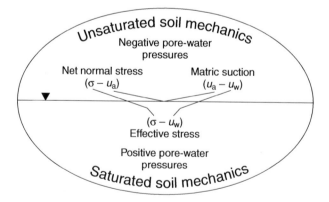

Figure 1.1 A visualization aid for the generalized world of soil mechanics (Fredlund, 1996).

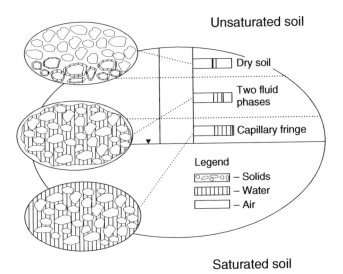

Figure 1.2 Categorization of soil above the water table based on the variation in degree of saturation (Fredlund, 1996).

The vadose zone

According to the representation in Figure 1.1, the geotechnical world is divided by a horizontal line representing the groundwater table. Below the water table, the pore water pressures will be positive and the soil will, in general, be saturated. Above the water table, the pore water pressure will, in general, be negative with respect to the atmospheric pressure (i.e. gauge pressure). The entire soil zone above the water table is called the vadose zone (see Figure 1.3). Immediately above the water table there is a zone called the capillary fringe where the degree of saturation approaches 100 per cent. This zone may range from less than 1 m to approximately 10 m in thickness, depending upon the soil type (Fredlund, 1996). Inside this capillary zone, water phase may be assumed to be continuous whereas air phase is generally discontinuous. Above this capillary zone, a two-phase zone may be identified in which both the water and air phases may be idealized as continuous. Inside this zone, the degree of saturation may vary from about 20 to 90 per cent, depending on soil type and soil state. Above this two-phase zone, the soil becomes dryer and the water phase will be discontinuous whereas the air phase will remain continuous.

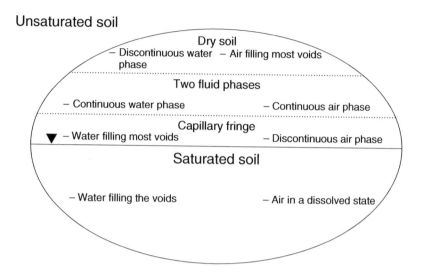

Figure 1.3 A visualization of saturated/unsaturated soil mechanics based on the nature of the fluid phases (Fredlund, 1996).

Climate changes and the vadose zone

The location of the groundwater table is strongly influenced by climatic conditions in a region. If the region is arid or semi-arid, the groundwater table is slowly lowered with time (i.e. may be geologic time scale). If the climate is temperate or humid, the groundwater table may remain quite close to the ground surface. It is the difference between the downward flux (i.e. precipitation) and the upward flux (i.e. evaporation and evapotranspiration) which determines the location of the groundwater table (refer to Figure 1.4).

Regardless of the degree of saturation of the soil, the pore water pressure profile will come to equilibrium at a hydrostatic condition when there is zero net flux from the ground surface. If moisture is extracted from the ground surface (e.g. by evaporation), the pore water pressure profile will be drawn to the left. If moisture enters at the groundwater surface (e.g. by infiltration), the pore water profile will be drawn to the right.

A net upward flux produces a gradual drying, cracking and desiccation of a soil mass, whereas a net downward flux eventually saturates a soil mass. The depth of the water table is influenced, amongst other things, by the net surface flux. A hydrostatic line relative to the groundwater table represents an equilibrium condition where there is no flux at ground surface. During dry periods, the pore water pressures become more negative than those

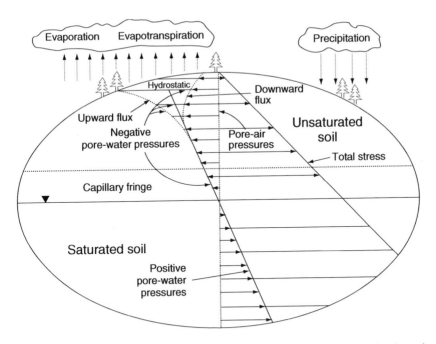

Figure 1.4 A visualization of soil mechanics showing the role of the surface flux boundary condition (Fredlund, 1996).

represented by the hydrostatic line. The opposite condition occurs during wet periods.

Grasses, trees and other plants growing on the ground surface dry the soil by applying a tension to the pore water through evapotranspiration (Dorsey, 1940). Most plants are capable of applying 1–2 MPa (10–20 atm) of tension to the pore water prior to reaching their wilting point (Taylor and Ashcroft, 1972). The tension applied to the pore water acts in all directions, and can readily exceed the lateral confining pressure in the soil. When this happens, a secondary mode of desaturation commences (i.e. cracking). Evapotranspiration also results in the consolidation and desaturation of the soil mass.

Over the years, a soil deposit is subjected to varying and changing environmental conditions. These produce changes in the pore water pressure distribution, which in turn result in shrinking and swelling of the soil deposit. The pore water pressure distribution with depth as shown in Figure 1.4 can take on a wide variety of shapes as a result of environmental changes.

There are many complexities associated with the vadose zone because of its fissured and fractured nature. Traditionally, the tendency in geotechnical engineering has been to avoid or grossly simplify the analysis of this zone, if possible. However, in many cases it is an understanding of this zone which holds the key to the performance of an engineered structure. Historically, classical seepage problems involved a saturated soil where the boundary conditions consist of either a designated head or zero flux. However, the real world for the engineer often involves a ground surface, where there may be a positive or negative flux condition. Geoenvironmental problems have done much to force the engineer to give consideration of saturated/unsaturated transient seepage analyses with flux boundary conditions. The improved computing capability available to the engineer has assisted in accommodating these changes (Fredlund, 1996).

Most of the man-made structures are located on the surface of the earth and as such will have an environmental flux boundary condition. This is the case for a highway where the soil of the embankment and subgrade have an initial set of conditions or stress states. These conditions will change with time primarily because of environmental and climatic (or surface moisture flux) changes. The foundations for light structures are likewise generally placed well above the groundwater table where the pore water pressures are negative. In fact, most of the light engineered structures of the world are placed within the vadose zone which is affected strongly by climate changes.

One of the characteristics of the upper portion of the vadose zone is its ability to slowly release water vapour to the atmosphere at a rate depending on the permeability of the intact portions of soil. At the same time, downward flow of water can occur through fissures under a gradient of unity. There appears to be no impedance to the inflow of water until the soil swells and the mass becomes intact, or until the fissures and cracks filled with water.

A common misconception is that water can always enter the soil at the ground surface. However, if the soil is intact, the maximum flux of water at the ground surface is equal to the saturated coefficient of permeability of the soil. This value may be extremely low. If the ground surface is sloping and the top soil layers are fissured or cracked, the surface layer can become saturated more easily and it will have a higher coefficient of permeability than the underlying soil. As a result, water runs down the top layers of soil on the slope, and may not enter the underlying intact soil.

Some common unsaturated soils

A large portion of the world population is found in the arid regions of the world where the groundwater table is deep because the annual evaporation from the ground surface in these regions exceeds the annual precipitation. There appears to be a strong correlation between the arid regions and the population density (Figure 1.5). The 10°–40° window of the world is defined by +10° and +40° latitude north and 10° and 40° longitude. This window contains approximately 3.1 billion people or 60 per cent of the population

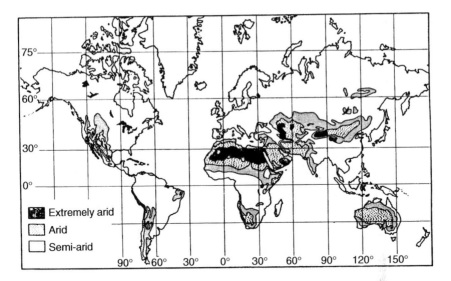

Figure 1.5 Map showing extremely arid, arid and semi-arid areas of the world (Meigs, 1953; Dregne, 1976; after Fredlund, 1996).

of the world and also contains 60 per cent of the countries of the world (Dregne, 1976; Fredlund, 1996).

Among many types of unsaturated soils, some are notorious and problematic for engineers. Some examples are listed as follows:

- Medium to high plastic clays containing a substantial amount of expansive minerals such as Montmorillonite subjected to a changing environment have produced the category of materials known as swelling soils. The shrinkage of the soils may pose an equally severe situation. Expansive plastic clays are commonly found in Colorado, Texas and Wyoming of the USA (Chen, 1988), in Hubei, Guangxi and Shandong of China (Shi *et al.*, 2002; Ng *et al.*, 2003), in Alberta and Saskatchewan of Canada (Chen, 1988; Fredlund and Rahardjo, 1993), in Madrid of Spain and in Gezara, Blue Nile and Kasalla of Sudan (Chen, 1988). A summary of some problematic unsaturated soils is given in Table 1.1.
- Loess soils often undergo collapse when subjected to wetting, and possibly a loading environment. They are commonly found in Missouri, Nebraska and Wisconsin of the USA (Dudley, 1970; Handy, 1995), in Gansu, Ningxia and Shanxi of China (Liu, 1988), and in Kent, Sussex, and Hampshire in the United Kingdom (Jefferson *et al.*, 2001).
- Residual and saprolitic soils located above groundwater table, particularly at many hillsides in Brazil, Portugal and in the Far East such as Hong Kong and Malaysia.

Table 1.1 Distribution of some expansive soils and loess soils worldwide

Expansive soils	Loess soils
China: Hubei, Guangxi, Shandong	China: Gansu, Ningxia, Shanxi
United States: Colorado, Texas, Wyoming	United Kingdom: Kent, Sussex, Hampshire
Canada: Alberta, Saskatchewan, Manitoba	United States: Missouri, Nebraska, Wisconsin
Spain: Andalucia, Madrid	Romania: Galati, Constanta, Tulcea
Sudan: Gezara, Blue Nile, Kasalla	

Apart from natural and geological processes, man-made activities such as excavation, remoulding and recompacting may also lead to the desaturation of saturated soils and hence the formation of unsaturated soils. These natural and man-made materials are difficult to be considered and understood, particularly where volume changes are concerned, within the framework of classical saturated soil mechanics.

The drier climatic regions have become increasingly aware of the uniqueness of their regional soil mechanics problems. In recent years there has also been a shift in emphasis in the developed regions from the behaviour of engineered structures to the impacts of developments on the natural world. This shift in emphasis has resulted in greater need to deal with the vadose zone. There has been an ongoing desire to expand the science dealing with soil mechanics such that it will also embrace the behaviour of unsaturated soils. An all-encompassing saturated and unsaturated soil mechanics has emerged in a number of countries worldwide (Fredlund, 1996).

Definitions of suctions

Soil suction is commonly referred to as the free energy state of soil–water (Edlefsen and Anderson, 1943), which can be measured in terms of its partial vapour pressure. From a thermodynamic standpoint, total suction can be quantitatively described by Kelvin's equation (Sposito, 1981) as follows:

$$\psi = -\frac{RT}{v_{w0}\omega_v} \ln\left(\frac{u_v}{u_{v0}}\right) \tag{1.1}$$

where ψ is total suction (kPa); R is the universal gas constant [J/(mol·K)]; T is the absolute temperature (K); v_{w0} is the specific volume of water or the inverse of the density of water (m³/kg); ω_v is the molecular mass of water vapour (g/mol); u_v is the partial pressure of pore-water vapour (kPa); u_{v0} is the saturation pressure of water vapour over a flat surface of pure water at the same temperature (kPa). The term (u_v/u_{v0}) is called relative humidity, RH (per cent).

Total suction has two components: matric suction $(u_a - u_w)$ and osmotic suction (π), i.e.

$$\psi = (u_a - u_w) + \pi \qquad (1.2)$$

A change of total suction is generally caused by a change of RH in the soil. RH can be reduced due to the presence of a curved water surface produced by capillary phenomenon, i.e. contractile skin (Fredlund and Rahardjo, 1993). The radius of curvature of the curved water surface is inversely proportional to the difference between air pressure (u_a) and water pressure (u_w) across the surface is called the matric suction.

Osmotic suction is a function of the amount of dissolved salts in the pore fluid and is expressed in terms of pressure. Alternatively speaking, a reduction in relative humidity in a pore, due to the presence of dissolved salts in pore water, is referred to as the osmotic suction. In order to keep the subject simple in this book, it is not the intention to provide full details of the definitions and proofs.

Stress state variables

What are stress state variables? According to the *International Dictionary of Physics and Electronics* (Michels, 1961), state variables are defined as follows:

> A limited set of dynamical variables of the system, such as pressure, temperature, volume, etc., which are sufficient to describe or specify the state of the system completely for the considerations at hand.

Fung (1965) describes the state of a system as that 'information required for a complete characterization of the system for the purpose at hand'. Typical state variables for an elastic body are given as those variables describing the strain field, the stress field and its geometry. The state variables must be independent of the physical properties of the material involved.

The stress state variable for saturated soils

An understanding of the meaning of effective stress proves to be valuable when considering the stress state description for unsaturated soils (Fredlund, 1987). Terzaghi's (1936a) statement regarding effective stress defined the stress state variables necessary to describe the behaviour of saturated soils. He stated:

> The stress in any point of a section through a mass of soil can be computed from the total principal stresses, σ_1, σ_2, σ_3, which act at

this point. If the voids of the soil are filled with water under a stress, u, the total principal stresses consist of two parts. One part, u, acts in the water and in the solid in every direction with equal intensity... The balance $\sigma_1' = \sigma_1 - u$, $\sigma_2' = \sigma_2 - u$, $\sigma_3' = \sigma_3 - u$, represents an excess over the neutral stress, u, and it has its seat exclusively in the soil phase of the soil. All the measurable effects of a change in stress, such as compression, distortion and a change in shearing resistance are exclusively due to changes in the effective stress σ_1', σ_2' and σ_3'.

Two aspects of Terzaghi's statement are of particular interest (Fredlund, 1987). The first assertion of interest is that the pore water pressure 'acts in the water and in the solid in every direction'. This becomes meaningful when the concept of continuous stress fields in continuum mechanics is applied to a multiphase system. The second assertion of interest is that 'all the measurable effects... are exclusively due to changes in the effective stress'.

Possible sets of stress state variables for unsaturated soils

Numerous attempts have been made to extend stress concepts useful for saturated soils, into the unsaturated soil range. Table 1.2 contains a summary of some common forms of single-valued equations proposed for unsaturated soils. Soil properties can also be identified in all equations. It could be argued that these equations are constitutive relations and as such 'fall short' of meeting the conditions of a state variable. The difficulties are primarily conceptual in nature, and their adoption gives rise to a deviation from classical continuum mechanics (Fredlund, 1987). Practical difficulties have also been encountered in the use of these effective stress equations in practice.

According to Fredlund (1987), the calling into question of a single-valued effective stress equation for unsaturated soils in the 1960s can be summarized as follows: Coleman (1962) suggested the use of 'reduced' stress variables $(\sigma_1 - u_a)$, $(\sigma_3 - u_a)$ and $(u_a - u_w)$ to represent the axial, confining and pore water pressure, respectively, in triaxial tests. Constitutive relations for volume change were then formulated using the 'reduced' stress variables. In 1963, Bishop and Blight re-evaluated the use of the single-valued effective stress equation and stated that a change in matric suction did not always result in the same change in effective stress. It was also suggested that volume change laboratory data be plotted in terms of the independent stress variables, $(\sigma_1 - u_a)$ and $(u_a - u_w)$. This appears to have initiated a transition towards utilizing the stress variables in an independent manner. This approach was further reinforced by Blight (1965) and Burland (1964, 1965).

Table 1.2 Proposed effective stress equations for unsaturated soils (Fredlund, 1987)

$\sigma' = \sigma - u_a + \chi(u_a - u_w)$	χ = parameter related to degree of saturation u_a = the pressure in gas and vapour phase	Bishop (1959)
$\sigma' = \sigma - \beta' u_w$	β' = holding or bonding factor which is measure of number of bonds under tension effective in contributing to soil strength	Croney, Coleman and Black (1958)
$\sigma = \bar{\sigma} a_m + u_a a_a + u_w a_w$ $+ R - A$	a_a = fraction of total area that is air–air contact $\bar{\sigma}$ = mineral interparticle stress a_m = mineral particle contact area a_w = water phase contact area R = repulsive pore fluid stress due to chemistry A = attractive pore fluid stress due to chemistry	Lambe (1960)
$\sigma' = \sigma + \psi p''$	ψ = parameter with values ranging from zero to one p'' = pore-water pressure deficiency	Aitchison (1961)
$\sigma' = \sigma + \beta p''$	β = statistical factor of same type as contact area; should be measured experimentally in each case	Jennings (1960)
$\sigma' = \sigma - u_a + \chi_m(h_m + u_a)$ $+ \chi_s(h_s + u_a)$	χ_m = effective stress parameter for matric suction h_m = matric suction h_s = solute suction χ_s = effective stress parameter for solute suction	Richards (1966)

Morgenstern (1979) amply summarized the difficulties associated with the use of the first equation listed in Table 1.2, by saying that the Bishop's effective stress equation

... proved to have little impact or practice. The parameter χ when determined for volume change behaviour was found to differ when determined for shear strength. While originally thought to be a function of the degree of saturation and hence bounded by 0 and 1, experiments were conducted in which χ was found to go beyond these bounds. As a result, the fundamental logic of seeking a unique expression for the effective stress independent of degree of saturation has been called into question (Fredlund 1973). The effective stress is a stress variable and hence related to equilibrium considerations alone.

And that it

> ... contains a parameter, χ, that bears on constitutive behaviour. This parameter is found by assuming that the behaviour of a soil can be expressed uniquely in terms of a single effective stress variable and by matching unsaturated behaviour with saturated behaviour in order to calculate χ. Normally, we link equilibrium considerations to deformations through constitutive behaviour and do not introduce constitutive behaviour directly into the stress variable.

Of course, his above comments would also apply to other single-valued equations listed in Table 1.2.

Many subsequent authors have found that while it is relatively easy to relate the shear strength of unsaturated soil to a single stress parameter involving σ, u_a and u_w, the volumetric behaviour is not controlled by the same stress parameter or by any other single stress variable. In particular, it has proved impossible to represent the complex pattern of wetting-induced swelling and collapse in terms of a single effective stress (Wheeler and Karube, 1995) as shown in Figure 1.6. With each of these proposed

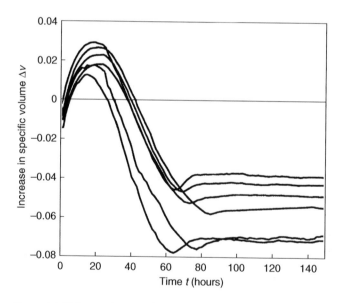

Figure 1.6 Volume change observed during wetting tests on compacted kaolin (Wheeler and Sivakumar, 1995).

definitions of effective stress so far, it is however impossible to represent some of the most fundamental features of unsaturated soil behaviour such as the correct pattern of swelling and collapse on wetting. Recently some researchers such as Molenkamp and Nazemi (2003) and Li (2003) have shown that the inter-particle forces caused by meniscus water bridges would be in static equilibrium with a set of externally applied stresses. However, application of these external stresses would never produce the same set of inter-particle forces (statically indeterminate system) (Wheeler, 2006).

It seems unlikely that it will ever be possible to devise a satisfactory definition of a single effective stress for unsaturated soil. The reason for this is that suction within the pore water and external stress applied to the boundary of a soil element act in qualitatively different ways on the soil skeleton, as noted by Jennings and Burland (1962), and hence these two stresses cannot be combined in a single effective stress parameter. This qualitative difference in the mode of action of suction and external stress is not taken into account in the method of mixtures, which has been used by some authors in attempts to derive a single effective stress expression (Wheeler and Karabe, 1995).

As an example of the different modes of action of suction and external stress, consider the idealized case shown in Figure 1.7 of an unsaturated soil made up of spherical soil particles, with the pore air at atmospheric pressure and the pore water at negative pressure within menisci at the particle contacts (Wheeler and Karube, 1995). External total stress, σ, applied to the boundary of a soil element containing many particles will produce both normal and tangential forces at particle contacts, even if the external stress state is isotropic. Of course, the effectiveness of σ will be influenced by the presence of bulk water inside soil pores. If the external stress is increased sufficiently, the tangential forces at particle contacts can cause inter-particle slippage and plastic strains (this is why soils, unlike most metals, undergo plastic volumetric strains if loaded beyond a pre-consolidation pressure). In contrast, the capillary effect due to meniscus water arising from suction within the menisci produces only an increase in the normal forces at particle contacts.

For any stress variable(s) to capture essential features of unsaturated soil behaviour, two different influences of suction on the mechanical behaviour (refer to Figure 1.8) must be fully recognized:

- Suction modifies the skeleton stresses both normal and tangential of an unsaturated soil through changing the average bulk pore fluid pressure inside its pores;
- Suction provides an additional normal bonding force (stabilising effects) at the particle contacts, attributed to capillary phenomena occurring in the water menisci or contractile skin.

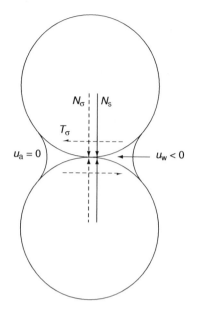

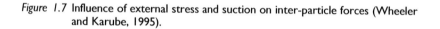

N_σ = normal component of inter-granular force due to external stress
T_σ = tangential component of inter-granular force due to external stress
N_s = inter-granular force due to suction

Figure 1.7 Influence of external stress and suction on inter-particle forces (Wheeler and Karube, 1995).

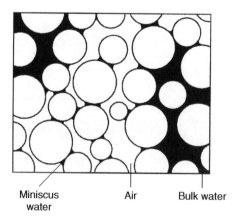

Miniscus water   Air   Bulk water

Figure 1.8 Two forms of liquid water in an unsaturated soil (Wheeler and Karube, 1995; Wheeler, 2006).

It is important to realize that, for these two different mechanisms, two independent stress variables are therefore required. It is known that the effects of suction are influenced by the degree of saturation of the soil. The relative area over which the water and air pressures act depends directly on the degree of saturation (the percentage of pore voids occupied by water), but the same parameter also affects the number and intensity of capillary-induced inter-particle forces (Gallipoli et al., 2003).

In 1977, Fredlund and Morgenstern suggested the use of any two of three possible stress variables, $(\sigma - u_a)$, $(\sigma - u_w)$ and $(u_a - u_w)$, to describe mechanical behaviour of unsaturated soils. The possible combinations are:

(1) $(\sigma - u_a)$ and $(u_a - u_w)$,
(2) $(\sigma - u_w)$ and $(u_a - u_w)$,
(3) $(\sigma - u_a)$ and $(\sigma - u_w)$.

The most common choice is to use net stress $(\sigma - u_a)$ and matric suction $(u_a - u_w)$ as the two independent stress state variables. This approach, which was first used by Coleman (1962), has formed the main basis for development in constitutive modelling of unsaturated soils during the last 30 years (Wheeler and Karube, 1995). However, Houlsby (1997) reported perhaps the most convincing theoretical derivations and justifications of the need of two stress state variables for describing the behaviour of unsaturated soils. The theoretical derivations and justifications are based on the consideration that the rate of input work to the soil is equal to the sum of the products of the stresses with their corresponding strain rates. It should be noted that if the finite compressibilities of the soil grains and pore fluid are to be included, more than two stress state variables may be needed to model the behaviour of unsaturated soils fully.

It is worth considering the merits of selecting $(\sigma - u_a)$ and $(u_a - u_w)$ as stress state variables, rather than $(\sigma - u_w)$ and $(u_a - u_w)$. The former combination has the advantage that the pore air pressure u_a is zero in many practical situations, so that net stress and matric suction simplify to total stress and negative pore water pressure respectively. In addition, the pore water pressure, which is commonly negative, is often very difficult to measure. This leads to uncertainty in the value of only one stress state variable if $(\sigma - u_a)$ and $(u_a - u_w)$ are selected, but uncertainty in the values of both stress state variables if $(\sigma - u_w)$ and $(u_a - u_w)$ are chosen. A counter argument in favour of the combination of $(\sigma - u_w)$ and $(u_a - u_w)$ is however that this choice leads to a slightly easier transition to fully saturated conditions (although it does not solve all the problems associated with this transition (Wheeler and Karube, 1995). Thus, $(\sigma - u_a)$ and $(u_a - u_w)$ are chosen to be the most satisfactory combination from a practical analysis standpoint (Fredlund, 1987).

Several new combinations of two stress state variables have recently been proposed by Wheeler *et al.* (2003) and Gallipoli *et al.* (2003), as modifications to net stress $(\sigma - u_a)$ and matric suction $(u_a - u_w)$. They are mainly in an attempt to account for the influence of degree of saturation on soil behaviour directly. This is vital because of the occurrence of hydraulic hysteresis in the state-dependent soil–water characteristic curve (SDSWCC) (Ng and Pang, 2000a,b) or the state-dependent water retention curve during drying and wetting processes. Merits and shortcomings of these new complex stress state variables are discussed by Wheeler *et al.* (2003) and Gallipoli *et al.* (2003). Moreover, a critical state model for unsaturated soils has been recently published based on two independent stress state variables that appears to overcome some of the limitations of some previous models (Sheng *et al.*, 2007).

Summary of the position on stress state variables

In spite of the many ingenious attempts to find a general expression for effective stresses, it is time to conclude that the description of the full range of behaviour of unsaturated soils requires the simultaneous use of at least two stress state variables to describe the effects of meniscus water and bulk water separately (Figure 1.8). However, in order to avoid too much complexity for engineering applications, the use of the two simple independent stress state variables, $(\sigma - u_a)$ and $(u_a - u_w)$, holds perhaps more promise for geotechnical engineers. Although these two simple independent stress state variables have some limitations, there are convincing experimental verifications and theoretical derivations in support of them in general (Houlsby, 1997). Formulations for some engineering analyses using these two independent stress state variables have been published (Alonso *et al.*, 1990; Wheeler and Sivakumar, 1995; Chiu and Ng, 2003).

Four-phase materials (Fredlund and Rahardjo, 1993)

Introduction

Fredlund and Rahardjo (1993) point out that an unsaturated soil is commonly referred to as a three-phase system. These phases are:

- air
- water
- solid

However, recent research results have realized the important role of the air–water interface (i.e. the contractile skin) which should be warranted as an additional phase when considering certain physical mechanisms. This is because when the air phase is continuous, the contractile skin interacts with the soil particles and provides an influence of the mechanical behaviour of soil. An element of unsaturated soil with a continuous air phase is idealized in Figure 1.9.

The mass and volume of each phase can be schematically represented by a phase diagram as shown in Figure 1.10.

The thickness of the contractile skin is in the order of only a few molecular layers. Therefore the physical subdivision of the contractile skin is considered as part of the water phase without any significant error. A simplified three-phase diagram is used when referring to the summation of masses and volumes of all soil particles.

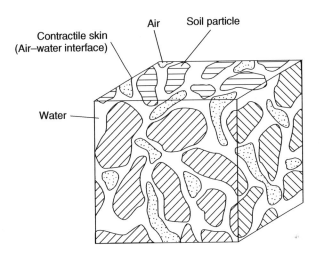

Figure 1.9 An element of unsaturated soil with a continuous air phase (after Fredlund and Rahardjo, 1993).

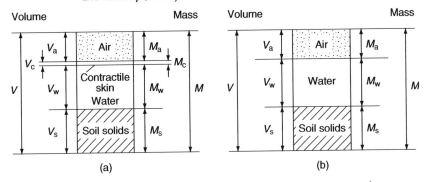

Figure 1.10 Rigorous and simplified phase diagrams for an unsaturated soil. (a) Rigorous four-phase unsaturated soil system; (b) simplified three phase diagram (after Fredlund and Rahardjo, 1993).

Surface tension (Fredlund and Rahardjo, 1993)

The air–water interface (i.e. contractile skin) possesses a property called surface tension. Surface tension results from the inter-molecular forces which are different from those that act on molecules in the interior of the water, as shown in Figure 1.11.

A molecule in the interior of the water experiences equal forces in all directions, which means there is no unbalanced force. A water molecule within the contractile skin experiences an unbalanced force towards the interior of the water. In order for the contractile skin to be in equilibrium, a tensile pull is generated along the contractile skin. The property of the contractile skin that allows it to exert a tensile pull is called its surface tension, T_s. Surface tension is measured as the tensile force per unit length of the contractile skin (i.e. units of N/m). Surface tension is tangential to the contractile skin surface. Its magnitude decreases as temperature increases. Table 1.3 gives surface tension values for the contractile skin at different temperatures (Kaye and Laby, 1973).

The surface tension causes the contractile skin to behave like an elastic membrane (Figure 1.12). This behaviour is similar to an inflated balloon which has a greater pressure inside the balloon than outside. If a flexible two-dimensional membrane is subjected to different pressures on each side, the membrane must assume a concave curvature towards the larger pressure and exert a tension in the membrane in order to be in equilibrium. The pressure difference across the curved surface can be related to the surface tension and the radius of curvature of the surface by considering equilibrium across the membrane.

The pressures acting on the membrane are u and $(u + \Delta u)$. The membrane has a radius of curvature of R_s and a surface tension, T_s. The horizontal

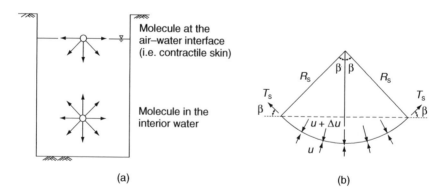

(a)   (b)

Figure 1.11 Surface tension at the air–water interface. (a) Inter-molecular forces on contractile skin and water; (b) pressures and surface tension acting on a curved two-dimensional surface (after Fredlund and Rahardjo, 1993).

Table 1.3 Surface tension of contractile skin i.e. air–water interface (from Kaye and Laby, 1973; after Fredlund and Rahardjo, 1993)

Temperature, t^0 (°C)	Surface tension, T_s (mN/m)
0	75.7
10	74.2
15	73.5
20	72.75
25	72.0
30	71.2
40	69.6
50	67.9
60	66.2
70	64.4
80	62.6
100	58.8

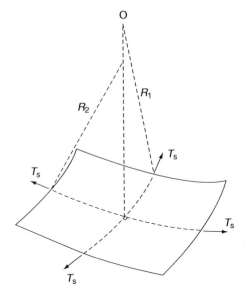

Figure 1.12 Surface tension on a warped membrane (after Fredlund and Rahardjo, 1993).

forces along the membrane balance each other. Force equilibrium in the vertical direction requires that

$$2T_s \sin \beta = 2\Delta u R_s \sin \beta \qquad (1.3)$$

where $2R_s \sin \beta$ is the length of the membrane projected onto the horizontal plane. Rearranging the above equation gives

$$\Delta u = \frac{T_s}{R_s} \tag{1.4}$$

The above equation gives the pressure difference across a two-dimensional curved surface with a radius, R_s, and a surface tension, T_s. For a warped or saddle-shaped surface (i.e. three-dimensional membrane),

$$\Delta u = T_s \left(\frac{1}{R_1} + \frac{1}{R_2} \right) \tag{1.5}$$

where R_1 and R_2 are radii of curvature of a warped membrane in two orthogonal principal planes.

If the radius of curvature is the same in all directions (i.e. R_1 and R_2 are equal to R_s), The immediate above equation becomes

$$\Delta u = \frac{2T_s}{R_s} \tag{1.6}$$

In an unsaturated soil, the contractile skin would be subjected to an air pressure, u_a, which is greater than the water pressure, u_w. The pressure difference $(u_a - u_w)$, is referred to as matric suction. The pressure difference causes the contractile skin to curve in accordance with the immediate equation:

$$(u_a - u_w) = \frac{2T_s}{R_s} \tag{1.7}$$

where $(u_a - u_w)$ is matric suction or the difference between pore air and pore water pressures acting on the contractile skin. This equation is referred to as Kelvin's capillary model equation. As the matric suction of a soil increases, the radius of curvature of the contractile skin decreases. The curved contractile skin is often called a meniscus. When the pressure difference between the pore air and pore water goes to zero, the radius of curvature, R_s, goes to infinity. Therefore, a flat air–water interface exists when the matric suction goes to zero.

Cavitation (Young, 1989; Marinho, 1995; Marinho and de Sousa Pinto, 1997)

Overview

Young (1989) points out that the pore water in soils can sustain very high negative pressures (suctions), and it is possible to estimate this pressure using many different indirect methods (Marinho and Chandler, 1995). However, attempts to measure directly suction greater than one atmosphere are often unsuccessful until recently, due to cavitation in the measuring system. It is recognized that water has a high tensile strength, and this characteristic is in conflict with the phenomenon of cavitation observed in measuring systems. The lack of inter-communication between the various sciences involved in measuring negative gauge pressure in liquids has delayed the development of direct suction measurement system for soils and the understanding of the behaviour of unsaturated soils, which generally have negative gauge pore water pressures.

Definitions

Young (1989) reports that cavitation is the formation and activity of bubbles (or cavities) in a liquid. Here the word 'formation' refers, in a general sense, both to the creation of a new cavity or to the expansion of a pre-existing one to a size where macroscopic effects can be observed. These bubbles may be suspended in the liquid or may be trapped in tiny cracks either in the liquid's boundary surface or in solid particles suspended in the liquid.

The expansion of the minute bubbles may be affected by reducing the ambient pressure by static or dynamic means. The bubbles then become large enough to be visible to the unaided eye. The bubbles may contain gas or vapour or a mixture of both gas and vapour. If the bubbles contain gas, then the expansion may be by diffusion of dissolved gases from the liquid into the bubble, or by pressure reduction, or by temperature rise. If, however, the bubbles contain mainly vapour, reducing the ambient pressure sufficiently at essentially constant temperature causes an 'explosive' vaporization into the cavities which is the phenomenon that is called *cavitation*, whereas raising the temperature sufficiently causes the mainly vapour bubbles to grow continuously producing the effect known as *boiling*. This means that 'explosive' vaporization or boiling do not occur until a threshold is reached.

There are thus four ways of inducing bubble growth (Young, 1989):

- For a gas-filled bubble, by pressure reduction or increase in temperature. This is called *gaseous cavitation*.
- For a vapour-filled bubble, by pressure reduction. This is called *vaporous cavitation*.
- For a gas-filled bubble, by diffusion. This is called *degassing* as gas comes out of the liquid.
- For a vapour-filled bubble, by sufficient temperature rise. This is called *boiling*.

The situation is complicated because the bubble usually contains a mixture of gas and vapour.

Cavitation inception and nuclei

Pearsall (1972) observes that in theory, a liquid will vaporize when the pressure is reduced to its vapour pressure. In practice, the pressure at which cavitation starts is greatly dependent on the liquid's physical state. If the liquid contains much dissolved air, then as the pressure is reduced the air comes out of solution and forms cavities in which the pressure will be greater than the vapour pressure of the liquid. Even if there are no visible air bubbles, the presence of submicroscopic gas bubbles may provide nuclei which cause cavitation at pressures above the vapour pressure. Each cavitation bubble grows from a nucleus to a finite size and collapses again, the entire cycle taking place perhaps within a few milliseconds. Bubbles may follow each other so rapidly that they appear to the eye to form a single continuous bubble.

In the absence of nuclei, the liquid may withstand negative pressures or tensions without undergoing cavitation. In theory, a liquid should be able to withstand tensions equivalent to thousands of atmospheres. It is estimated that water, for example, will withstand a tension ranging from 500 to 10,000 atmospheres (Harvey et al., 1947). In practice, even when water has been subjected to rigorous filtration and pre-pressurized to several hundred atmospheres, it has ruptured at tensions of 300 atmospheres. According to Plesset (1969), water without nuclei theoretically will withstand tensions of 15,000 atmospheres, but the probability of this happening is low unless the bubbles are of molecular dimensions. However, when solid non-wetted nuclei of size 10^{-8} cm are present, it is likely that water will rupture at tensions of the order of tens of atmospheres. It follows from this that it should be possible to raise water under vacuum to a height greater than that corresponding to atmospheric pressure. This has actually been achieved by pre-pressurizing the liquid (Hayward, 1970).

Physics of pore water under tension in unsaturated soils

Marinho and Chandler (1995) observe that measurement of matric suction, using a tensiometer, requires the water inside the measuring system to be at the same value of tensile stress as the soil pore water, and this may induce cavitation inside the system. In soil suction measurement, cavitation is the result of pressure reduction and has the effect that, after cavitation has occurred, the pressure measured will be approximately equal to the pressure of gas in the system. This is probably why the belief that it is impossible to measure suction greater than 1 atmosphere.

The conditions under which a liquid with a flat liquid–gas interface (free water) will vaporize are given by the vapour pressure–temperature curve for that particular liquid. However, if the liquid–gas interface is not flat, as in the case of a vapour cavity (bubbles) in a liquid, it is possible to reach a

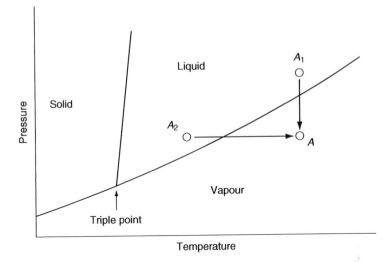

Figure 1.13 Phase diagram for a simple substance (after Marinho and Chandler, 1995).

state normally associated with the vapour phase without the vapour phase developing (Apfel, 1970).

A phase diagram for a simple substance, as shown in Figure 1.13 indicates the regions where under a given pressure and temperature combination, the substance is in the solid, liquid or vapour phase. The substance at point A would be a vapour. However, under certain circumstances it is possible, starting from A_1, to reduce the pressure and reach A, at a pressure lower than the vapour pressure, and for the substance to remain in liquid. In the same way, it is possible to go from A_2 to A remaining a liquid. When the phase boundary is crossed without the substance changing phase, the phase boundary is said to have been transgressed. If transgression has occurred, the system is said to be in a metastable state (Apfel, 1970). Water under tension is in a metastable condition and this metastability can be destroyed if nucleation occurs. Nucleation is the formation of vapour cavities within the liquid itself or at their boundaries (Trevena, 1987).

Cavitation in a metastable liquid can result from two types of nucleation: nucleation in the pure liquid, and nucleation caused by impurities in the liquid. Impurities can be other pure substances, solid impurities or even radiation acting on the liquid. Nucleation within a pure liquid is called 'homogeneous' nucleation and nucleation due to impurities is called 'heterogeneous' nucleation. Heterogeneous nucleation is more common, and it is responsible for most cavitation which occurs in suction measurement systems.

How to avoid cavitation

Marinho and Sousa Pinto (1997) observe that although water could sustain tension, attempts to measure soil suction higher than 1 atm, using tensiometers, have failed. The reason for that is associated with the entrapment of air inside microcrevices in the system. There are many theories that try to explain the entrapment of air between liquid and a solid container. Harvey *et al.* (1944) presented the most accepted model to justify the presence of air and how the air nuclei can be stabilized.

Figure 1.14 presents a schematic representation of air trapped during a usual saturation procedure of a tensiometer. The air trapped is called cavitation nuclei. In order to dissolve this air, high positive water pressure should be applied. However, due to some particular aspects of geometry of the crevice, the air may not be dissolved. In this case the application of positive pressure may 'stabilize' the cavitation nuclei, as shown in Figure 1.14b.

The stabilization of the cavitation nuclei increases the level of the suction that can be applied. It is not clear how the process occurs, and thus cannot be precisely controlled (Marinho and Sousa Pinto, 1997).

Usually the pressure required to stabilize the cavitation nuclei is higher than 5 MPa (Ridley and Burland, 1993). Marinho and Sousa Pinto (1997) suggested a chemical procedure to be used with the usual technique for stabilizing the cavitation nuclei inside a suction probe (see Figure 1.15). They reported that the chemical procedure used reduced the level of positive pressure to be applied. The system used by Marinho and Sousa Pinto (1997) was pressurized to a maximum pressure of 3.5 MPa by a hand pump as shown in Figure 1.16. The pressure was kept for 24 h. The system was then cycled 10 times from 3.5 MPa to zero pressure. After that it was able to sustain suction up to 650 kPa. The porous stone used has a nominal air-entry value of 500 kPa. Marinho and Sousa Pinto (1997) believed that chemical action can help to eliminate cavitation nuclei.

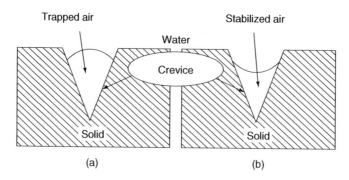

Figure 1.14 Cavitation nuclei (after Marinho and de Sousa Pinto, 1997).

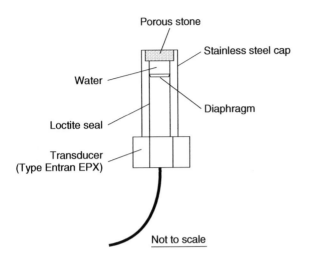

Figure 1.15 Schematic representation of the suction probe (after Marinho and de Sousa Pinto, 1997).

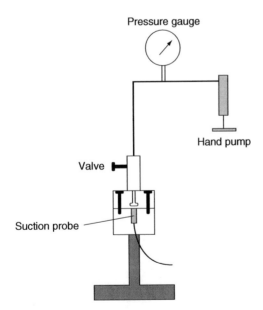

Figure 1.16 System for saturating the tensiometer using high positive pressure (after Marinho and de Sousa Pinto, 1997).

Therefore, to reduce the likelihood of trapping permanent undissolved air in the system, the following measures are recommended by Marinho and Chandler (1995):

- The use of de-aired water is important to avoid air saturation (boiling the water is an appropriate method).
- The water and all surfaces within the measurement system must be extremely pure and clean (Henderson and Speedy, 1980).
- The surfaces in contact with the water must be as smooth as possible to avoid or reduce the number and size of crevices. The smaller the surface area, the easier it is to avoid cavitation.
- The system should be evacuated by vacuum application in order to remove the maximum amount of air entrapped in the crevices, though it is unlikely that all the air will be removed (Jones et al., 1981).
- The system should be cycled from positive to zero (or negative) pressure. This may help in dissolving persistent bubbles (Chapman et al., 1975; Richards and Trevena, 1976).
- Pre-pressurization of the system to high pressure is required in order to dissolve all the free air (Harvey et al., 1944).

It is important to point out that cavitation can be delayed if the nuclei are stabilized against tension. The procedures above probably only stabilize the nuclei, which should allow the measurement of a certain level of suction before cavitation occurs. It should be noted that cavitation will not occur if no cavitation nuclei are present. Examples of the success and failure of measuring soil suction higher than 1 atm using tensiometers are given and discussed in detail by Take and Bolton (2003) and Zhou et al. (2006).

Chapter 2

Measurement and control of suction: methods and applications

Sources

This chapter is made up of verbatim extracts from the following sources for which copyright permissions have been obtained as listed in the Acknowledgements.

- Feng, M. and Fredlund, D.G. (2003). Calibration of thermal conductivity sensors with consideration of hysteresis. *Can. Geotech. J*, 40, 1048–1055.
- Fredlund, D.G. and Rahardjo, H. (1993). *Soil mechanics for unsaturated soils*. John Wiley & Sons, Inc., New York, p. 517.
- Harrison, B.A. and Blight, G.E. (2000). A comparison of in situ soil suction measurements. *Proc. Conf. on Unsaturated Soils for Asia*, Singapore, Rahardjo, Toll and Leong eds, Balkema, Rotterdam, pp. 281–284.
- Ng, C.W.W. and Chen, R. (2005). Advanced suction control techniques for testing unsaturated soils (in Chinese). Keynote lecture, *2nd Nat. Conf. on Unsat. Soils*, Hangzhou, China, Zhejiang University Press, 144–167.
- Ng, C.W.W. and Chen, R. (2006). Advanced suction control techniques for testing unsaturated soils (in Chinese). *Chinese J. of Geotech. Eng.*, Vol. 28, No. 2, 123–128.
- Ng, C.W.W. and Pang, Y.W. (2000a). Experimental investigation of soil–water characteristics of a volcanic soil. *Can. Geotech. J.*, 37(6), 1252–1264.
- Ng, C.W.W., Cui, Y.J., Chen, R. and Delage, P. (2007). The axis-translation and osmotic techniques in shear testing of unsaturated soils: a comparison. *Soils and Foundations*. Vol. 47, No. 4, 675–684.
- Ridley, A.M. and Wray, W.K. (1995). Suction measurement: a review of current theory and practices. *Proc. Conf. on Unsaturated Soils*, Paris, Alonso and Delage eds, Balkema, Rotterdam, pp. 1293–1322.
- Ridley, A.M. and Burland, J.B. (1993). A new instrument for the measurement of soil moisture suction. *Géotechnique*, 43(2), 321–324.

Theory of soil suction (Fredlund and Rahardjo, 1993)

Soil consists of solids and voids. Inside the voids of unsaturated soil, there are generally two fluids, i.e. air and water. As discussed and given by Equation (1.1) in Chapter 1, total suction ψ can be quantitatively described by Kelvin's equation (Sposito, 1981) as follows:

$$\psi = -\frac{RT}{v_{w0}\omega_v}\ln\left(\frac{\bar{u}_v}{\bar{u}_{v0}}\right) \tag{1.1}$$

At 20°C, Equation (1.1) can be rewritten to give a fixed relationship between total suction in kilopascals and relative vapour pressure as follows:

$$\psi = -135,022\ln\left(\frac{\bar{u}_v}{\bar{u}_{v0}}\right) \tag{2.1}$$

The capillary model

Equilibrium at the soil–water interface (Ridley and Wray, 1995)

The commonly depicted picture of the soil–water system is that of granular particles separated by water. At an air–water interface, a meniscus will form between adjacent soil particles in a similar manner to water in a capillary tube (Ridley and Wray, 1995). It is therefore understandable that this model became known as the capillary model (Buckingham, 1907).

For equilibrium at the air–water interface of the capillary tube shown in Figure 2.1, the downward force exerted by the air must be equal to the upward force exerted by the water. The curved shape of the interface is the result of the upward force which exists at the boundary due to the wetting

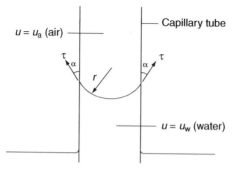

Figure 2.1 Capillary suction (after Ridley and Wray, 1995).

of the surface. It is this upward force that holds the column of water up above the flat water surface outside the tube. Therefore

$$u_a \pi r^2 = u_w \pi r^2 + 2\pi r T_s \sin \alpha \qquad (2.2)$$

where T_s is the surface tension at the boundary, and α is the angle of contact between the water and the boundary. For a perfectly spherical meniscus, the latter will be 90° and the equilibrium will reduce to

$$u_a - u_w = \frac{2T_s}{r} \qquad (2.3)$$

In the majority of cases, the air pressure will be atmospheric (or zero gauge), and the pressure which exists on a water molecule in the meniscus is a direct result of the surface tension and the radius of the capillary.

Now consider the equilibrium that exists at the air–water interface between the liquid water molecules and the water vapour molecules. To escape from the surface of the liquid, a water molecule must have an energy equal to or greater than the latent heat of evaporation for water. If the space above the air–water interface is a closed system, equilibrium will be reached when it becomes saturated with water vapour molecules. If the water is pure and its surface is flat, then the partial pressure of the vapour at equilibrium is equal to the saturated vapour pressure of the liquid at the temperature of the system. However, soil–water is held in a meniscus and so the additional force caused by the surface tension of the curved surface reduces the vapour pressure reached in the enclosed system at equilibrium. The stress holding a water molecule in the meniscus (i.e. the soil suction) is then directly related to the relative humidity (RH) in the space surrounding the soil by Equation (1.1).

Equilibrium of the water column and capillary height (Fredlund and Rahardjo, 1993)

For the vertical force equilibrium of the capillary water in the tube shown in Figure 2.1, the vertical resultant of the surface tension (i.e. $2\pi r T_s \cos \alpha$) is responsible for holding the weight of the water column, which has a height of h_c (i.e. $\pi r^2 h_c \rho_w g$):

$$2\pi r T_s \cos \alpha = \pi r^2 h_c \rho_w g \qquad (2.4)$$

where
h_c = capillary height
g = gravitational acceleration.

The equation can be rearranged to give the maximum height of water in the capillary tube, h_c:

$$h_c = \frac{2T_s \cos \alpha}{\rho_w g r} \tag{2.5}$$

The contact angle between the contractile skin for pure water and tube is zero (i.e. $\alpha = 0$). When the α angle is zero, the capillary height of pure water in a clean tube is

$$h_c = \frac{2T_s}{\rho_w g r} \tag{2.6}$$

The radius of the tube is analogous to the pore radius in soils. Equation (2.6) shows that the smaller the pore radius in the soil, the higher will be the capillary height, as illustrated in Figure 2.2.

Assuming that the contact angle is zero, the capillary height can be plotted against the pore radius as shown in Figure 2.3 (Fredlund and

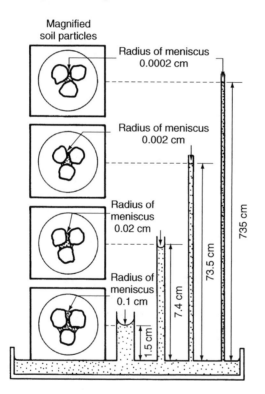

Figure 2.2 Capillary tubes showing the air–water interfaces at different radii of curvature of menisci (from Janssen and Dempsey, 1980; after Fredlund and Rahardjo, 1993).

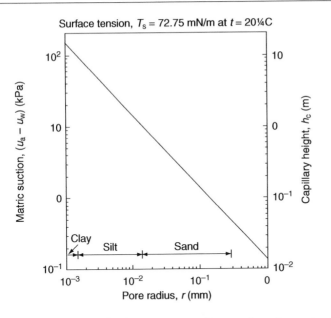

Figure 2.3 Relationship between pore radius, matric suction and capillary height (after Fredlund and Rahardjo, 1993).

Rahardjo, 1993). The above explanation has demonstrated the ability of the surface tension to support a column of water, h_c, in a capillary tube. The surface tension associated with the contractile skin results in a reaction force on the wall of the capillary tube as shown in Figure 2.4 (Fredlund and

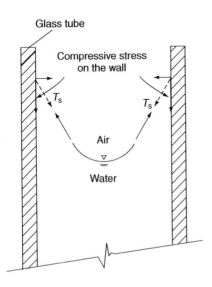

Figure 2.4 Forces acting on a capillary tube (after Fredlund and Rahardjo, 1993).

Rahardjo, 1993). The vertical component of this reaction force produces compressive stresses on the wall of the tube. In other words, the weight of the water column is transferred to the tube through the contractile skin. In the case of a soil having a capillary zone, the contractile skin results in an increased compression of the soil structure. As a result, the presence of matric suction in an unsaturated soil increases the shear strength of the soil, provided there is sufficient contractile skin present in the soil (i.e. the soil is not too dry).

Suction control and measurement methods (Ng and Chen, 2005, 2006; Ng et al., 2007a)

Axis-translation technique

Overview

In laboratory studies on unsaturated soils, an important issue is how to control or measure suction in an unsaturated soil specimen. Generally, total suction can be controlled by using the humidity control technique (Esteban and Saez, 1988). Matric suction can be controlled by using the axis-translation technique (Hilf, 1956) and the osmotic technique (Zur, 1966). Osmotic suction can be controlled by using different solutions as pore fluids or changing solute concentrations of pore fluid in the soil. In most geotechnical engineering applications, chemistry of pore fluids in the soil is not changed and soil–water content varies within a range in which concentrations of pore fluids are not altered significantly, and so osmotic suction appears not to be sensitive to changes in soil–water content. Therefore, it is expected to control total suction and matric suction in most geotechnical testing for unsaturated soils. The most commonly used technique is axis-translation, followed by osmotic and humidity control.

In this section, the working principle, the development and applications of the three suction control techniques in the laboratory are introduced and reviewed. Experimental data using the axis-translation and osmotic techniques are compared and discussed. No matter which technique is used, suction equalization is a vital stage in testing unsaturated soils. To illustrate the influence of suction equalization on subsequent shearing behaviour, two direct shear tests were performed on a compacted expansive soil applying different durations of suction equalization under the same applied vertical stress and matric suction.

Working principle

Matric suction may be considered as an important variable in defining the state of stress in an unsaturated soil. Therefore, it is necessary to control or measure matric suction in laboratory studies on unsaturated soils. However, difficulties associated with the measurement and control of negative pore water pressure present an important practical limitation. Water is normally thought to have little tensile strength and may start to cavitate when the magnitude of gauge pressure approaches -1 atm. Under some suitable conditioning (see Chapter 1), water can stand tensions of the order of 40–300 atm (Temperley and Chambers, 1946; Young, 1989). As cavitation occurs, water phase becomes discontinuous, making the measurements unreliable or impossible. Because it is required to control the matric suction variable over a range far greater than 1 atm for many soil types and their applications, alternatives to measurement or control of negative water pressure are desirable.

Hilf (1956) introduced the axis-translation technique of elevating pore air pressure u_a to increase pore water pressure to be positive, preventing cavitation in the water drainage system. Total stress σ is increased together with air pressure at the same amount so the net stress $(\sigma - u_a)$ remains unchanged. As shown in Figure 2.5, stresses on an unsaturated soil in the field are total stress σ^1, pore air pressure u_a^1 (generally equal to atmospheric pressure) and pore water pressure u_w^1 (generally negative gauge pressure). When applying the axis-translation technique, total stress is increased from σ^1 to σ^2, pore air pressure is increased from u_a^1 to u_a^2, and pore water pressure is increased from u_w^1 to u_w^2 (generally positive gauge pressure). The net stress $(\sigma - u_a)$ and matric suction $(u_a - u_w)$ remain unchanged. This process is referred as 'axis-translation'. Based on the axis-translation principle, the matric suction variable $(u_a - u_w)$ can be controlled over a range far greater than the cavitation limit for water under negative pressure.

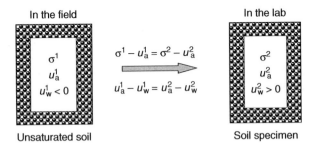

Figure 2.5 Schematic diagram illustrating the axis-translation principle (Ng and Chen, 2005).

Axis-translation is accomplished by separating the air and water phases of the soil through porous material with a high air-entry value. When saturated, these materials allow water passage but prevent flow of free air when the applied matric suction does not exceed the air-entry value of the porous material, which can be as high as 1,500 kPa for sintered ceramics or 15 MPa for special cellulose membranes (Zur, 1966).

Applications

MEASUREMENTS OF SWCC AND SDSWCC

The axis-translation technique has been successfully applied by numerous researchers to study the soil–water characteristic properties of unsaturated soils (Fredlund and Rahardjo, 1993; Ng and Pang, 2000a,b), as well as the volume change and shear strength properties of unsaturated soils (Fredlund and Rahardjo, 1993; Gan *et al.*, 1988; Ng and Chiu, 2001; Ng and Chiu, 2003a,b; Chiu, 2001; Zhan, 2003; Ng and Zhou, 2005).

The soil–water characteristic curve (SWCC) is the relationship between suction and water content or degree of saturation for an unsaturated soil. It is now generally accepted that unsaturated soil behaviour is governed by two independent stress state variables, i.e. net stress and matric suction (Fredlund and Morgenstern, 1977). Therefore, it is necessary to consider the influence of net stress on the SWCC. However, the SWCC of a soil is conventionally measured by means of a pressure plate extractor in which no external stress is applied, and volume change of the soil specimen is assumed to be zero. To investigate the influence of net stress on SWCC, Ng and Pang (2000a,b) developed a total stress controllable one-dimensional volumetric pressure plate extractor based on the axis-translation principle at the Hong Kong University of Science and Technology (HKUST) (see Figure 2.6). This apparatus can be applied to measure the SWCCs at various vertical stresses under K_0 condition. An oedometer ring equipped with a high air-entry ceramic plate at its base is located inside an airtight chamber. Vertical stress is applied through a loading frame to a soil specimen inside the oedometer ring. To eliminate the error due to side friction of the loading piston, a load cell is attached near the end of the piston inside the airtight chamber for determining the actual vertical load applied to the soil specimen. Because the radial deformation is zero for the K_0 condition, the total volume change of the specimen is measured from the vertical displacement of the soil specimen using a dial gauge. Using this apparatus, state-dependent soil–water characteristics curves (SDSWCCs) can be measured, and the assumption of zero volume change is no longer required. Similar to the conventional volumetric pressure plate extractor, pore air pressure u_a is controlled through a coarse porous stone together with a coarse geotextile located at the top of the specimen. Pore water pressure u_w is controlled at the atmospheric

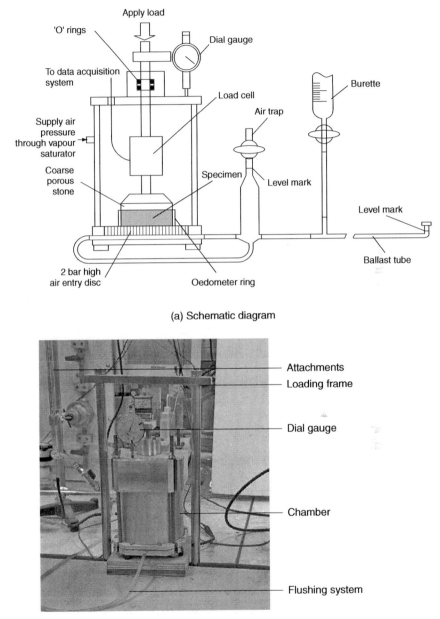

(a) Schematic diagram

(b) Photograph

Figure 2.6 A new total stress controllable one-dimensional volumetric pressure-plate extractor at HKUST (after Ng and Pang, 2000b; Ng and Chen, 2005).

pressure through the high air-entry ceramic plate mounted at the base of the specimen. In addition, some attachments are used for the purpose of studying the hysteresis of the SWCCs associated with the drying and wetting of the soil. These are a vapour saturator, air trap, ballast tube and burette. The vapour saturator is used to saturate the in-flow air to the airtight chamber to prevent the soil from drying by evapouration. The air trap is attached to collect air that may diffuse through the high air-entry disc. The ballast tube serves as a horizontal storage for water flowing in or out of the soil specimen. The burette is used to store or supply water and to measure the water volume change in the soil specimen.

Figure 2.7 shows the influence of stress state on the soil–water characteristics of natural completely decomposed volcanic (CDV) specimens (Ng and Pang, 2000a). The size of the hysteresis loops does not seem to be governed by the applied stress level. The specimens subjected to higher applied stresses possess a slight higher air-entry value and lower rates of desorption and adsorption as a result of smaller pore-size distribution. These experimental results demonstrate that the stress state has a substantial influence on the soil–water characteristics of unsaturated soils. Ng and Pang (2000b) adopted the measured wetting SDSWCCs to perform numerical analyses on slope stability.

They found that during a prolonged low intensity rainfall, the FOS (factor of safety) predicted by using the SDSWCC is substantially lower than predicted by using the conventional drying SWCC. Under highly intensive but short duration rainfalls, however, the two predicted FOS values are close. Therefore, in studying soil–water characteristics of unsaturated soils

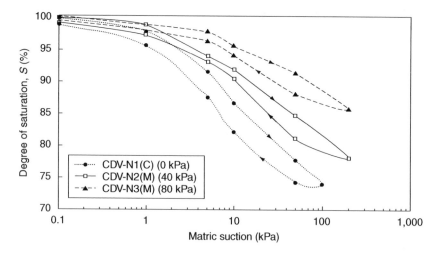

Figure 2.7 Influence of stress state on soil–water characteristics of natural CDV specimens (after Ng and Pang, 2000a; Ng and Chen, 2005).

and their applications, the effect of two independent stress state variables, i.e. net stress and suction, should be considered simultaneously.

STRESS PATH TESTING IN THE TRIAXIAL APPARATUS

The triaxial test and the direct shear test (shear box) are two commonly used shear strength tests. Figure 2.8 illustrates a triaxial system applying the axis-translation technique for testing unsaturated soils at HKUST (Zhan, 2003). It is composed of a triaxial cell, four GDS pressure controllers, six transducers, a digital transducer interface (DTI) and a computer, a new total volume change measuring system and a diffused air volume indicator (DAVI).

As shown in Figure 2.8, the triaxial cell is a Bishop and Wesley stress-path cell for testing up to 100 mm diameter specimens. Of the four GDS pressure controllers, two are automatic pneumatic controllers for controlling cell pressure and pore air pressure, and the other two are digital hydraulic

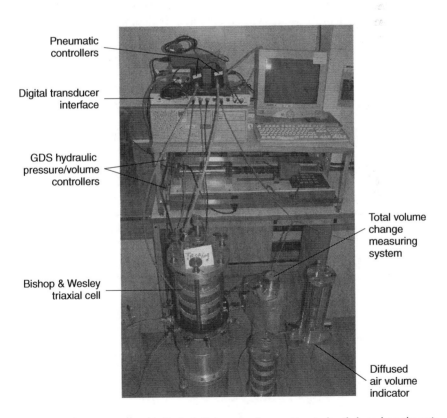

Figure 2.8 A computer-controlled triaxial system for unsaturated soils based on the axis-translation principle at HKUST (after Ng and Chen, 2005, 2006).

pressure/volume controllers for controlling back pressure and axial stress. With these four GDS controllers, cell pressure, axial stress, pore air pressure and pore water pressure can be controlled independently. The six transducers consist of an internal load cell (measurement of axial force), a linear variable difference transformer (LVDT) (measurement of axial displacement), a differential pressure transducer (DPT) (a component of the total volume change measuring system, Ng *et al.*, 2002a) and three pressure transducers (monitoring cell pressure, pore water pressure and pore air pressure). All the six transducers are connected to the DTI for data acquisition. The DTI as well as the two pneumatic controllers are connected to the computer through a multiplexer. The two digital hydraulic pressure/volume controllers are connected to the computer by an IEEE interface card for computer control. All of these form a closed-loop controlling and feedback system, which is capable of performing strain-controlled and stress-path tests in triaxial stress space.

As shown in Figure 2.9a, a high air-entry disk was sealed onto the pedestal bottom of the triaxial cell using an epoxy. A spiral groove was produced at the bottom of the pedestal to serve as a water compartment as well as a channel for flushing air bubbles that may be trapped or accumulated as a result of air diffusion (Chiu, 2001; Zhan, 2003). The DAVI is used to measure the amount of diffused air (Fredlund, 1975; Zhan, 2003). Matric suction is applied to the test specimen through one water pressure controller and one air pressure controller using the axis-translation principle. Pore water pressure is applied or measured at the base of the specimen through the ceramic disc and the compartment. Pore air pressure is applied at the top of the specimen through a sintered copper filter.

Figure 2.9 shows the setup of the total volume measuring system (Ng *et al.*, 2002a). The basic principle of the double-cell total volume measuring system is that the overall volume change in an unsaturated–saturated specimen is measured by recording the differential pressures between the water inside an open-ended, bottle-shaped inner cell and the water inside a reference tube using a high-accuracy DPT. The inner cell is sealed onto the pedestal of the outer cell in the triaxial apparatus. The high-accuracy DPT is connected to the inner cell and to a reference tube, in order to record changes in differential pressures between the water pressure change inside the inner cell due to a volume change in the specimen and the constant water pressure in the reference tube. Detailed calibrations have been carried out to account for apparent volume changes due to changes in cell pressure, fluctuation in the ambient temperatures, creep in the inner cell wall and relative movement between the loading ram and the inner cell (Ng *et al.*, 2002a). The estimated accuracy of the volume change measuring system is in the order of $31.4\,mm^3$ if the system is properly calibrated. For a 100 mm diameter by 200 mm high test specimen, this corresponds to a volumetric strain of 0.002 per cent. In this system, apparent volume

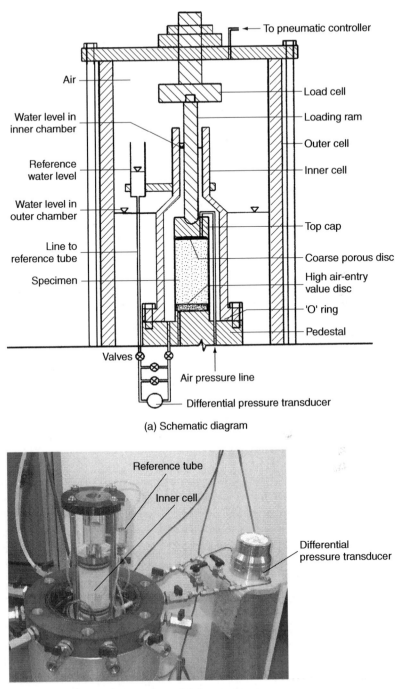

(a) Schematic diagram

(b) Photograph

Figure 2.9 A new double-cell total volume change measuring system for unsaturated soils at HKUST (after Ng et al., 2002a).

changes due to changes in cell pressure, fluctuations due to variations of ambient temperature and creeping are all smaller than other existing double-cell volume change measuring systems. More detailed comparisons are made by Ng *et al.* (2002a).

Based on the axis-translation principle, Ng and Chiu (2001) conducted triaxial stress path tests on a loosely compacted unsaturated CDV specimen. Figure 2.10 shows the results of field stress path (wetting) tests at constant deviator stress (i.e. field stress path tests which simulate rainfall infiltration on a slope element). As shown in Figure 2.10a, when the suction decreases from its initial value (i.e. 150 kPa), there exists a threshold suction above which only small axial strain is mobilized. As the suction drops below this threshold value, the rate of increase in axial strain accelerates towards the end of the test. The threshold suction increases with applied net mean stress. Figure 2.10b shows the variation of volumetric strain with suction. Similar to the variation of axial strain with suction, there exists also a threshold suction above which only small volumetric strain is mobilized. As the suction drops below such threshold suction, contractive behaviour is observed for the two specimens (ua1 and ua2) compressed to net mean stress smaller than 100 kPa. In contrast, the other two specimens (ua3 and ua4) show dilative behaviour.

SHEARING TEST IN THE DIRECT SHEAR BOX

Compared to the triaxial test, the direct shear test is simpler to perform and requires shorter test durations due to the smaller drainage paths. Figure 2.11 shows a direct shear apparatus for unsaturated soils based on the axis-translation principle at HKUST. It is modified from a conventional direct shear box for testing saturated soils (Gan *et al.*, 1988; Zhan, 2003). The cylindrical pressure chamber is built of stainless steel and was designed for pressure of up to 1,000 kPa. Three holes are drilled through to provide the necessary housing for a vertical loading ram and two horizontal pistons. Teflon ring seals are installed to ensure the airtightness around the loading ram and pistons. A high air-entry value ceramic disk is installed in the lower portion of the shear box. The water chamber beneath the ceramic disk is designed to serve as a water compartment as well as a channel for flushing air, similar to the pedestal of triaxial apparatus for testing unsaturated soils (refer to Figure 2.9). The desired matric suction is applied to a soil specimen by maintaining a constant air pressure in the air pressure chamber and a constant water pressure in the water chamber below the ceramic disk. The pore air pressure and pore water pressure in the soil are then allowed to come into equilibrium with these applied pressures. Matric suction in the soil is equal to the difference between the applied air and water pressures. The modified direct shear apparatus has five measuring devices. There are two LVDTs for horizontal and vertical

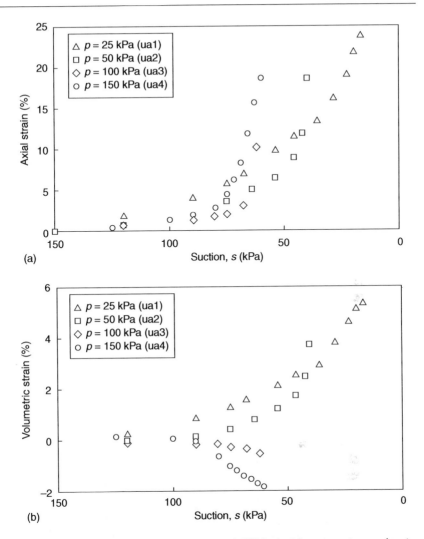

Figure 2.10 Stress path tests on unsaturated CDV triaxial test specimens showing the relationships between (a) axial strain and suction; (b) volumetric strain and suction (after Ng and Chiu, 2001).

displacements, a load cell for shear force, a pressure transducer for water pressure and a water volume indicator for water volume changes in the soil specimen. All these transducers are connected to a data logger for data acquisition.

Figure 2.12 shows direct shear results for a completely decomposed granite (CDG) test specimen (Ng and Zhou, 2005). In Figure 2.12, stress ratio

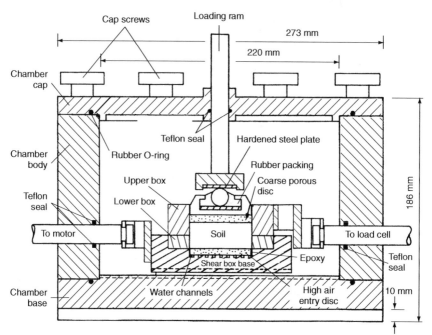

(a) Schematic diagram

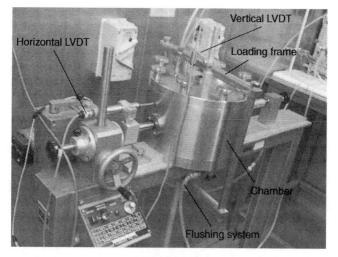

(b) Photograph

Figure 2.11 A direct shear apparatus for unsaturated soil test specimens: (a) schematic diagram (after Gan *et al.*, 1988); (b) photograph at HKUST (after Zhan, 2003). Interpreting the results uses the axis-translation principle (Ng and Chen, 2005, 2006).

is defined as the ratio of shear stress (τ) to net vertical stress ($\sigma_v - u_a$). Dilatancy is defined as the ratio ($\delta y/\delta x$) of incremental vertical displacement (δy) to incremental horizontal displacement (δx). Negative sign (or negative dilatancy) means dilative behaviour. As shown in Figure 2.12a, at zero suction and suctions of 10 and 50 kPa, the stress ratio–displacement curve displays strain hardening behaviour. As suction increases, strain softening behaviour is observed at suctions of 200 and 400 kPa. Generally, measured peak and ultimate stress ratios increase with suction, except the ultimate stress ratio measured at suction of 200 kPa. Figure 2.12b shows the effects of suction on dilatancy of the CDG in the direct shear tests. Under the saturated conditions, the soil specimen shows contractive behaviour (i.e. positive dilatancy). On the other hand, under unsaturated conditions, all soil specimens exhibit contractive behaviour initially but then dilative behaviour as horizontal displacement continues to increase. The measured maximum negative dilatancy is increased by an increase in suction. The measured peak stress ratio in each test does not correspond with its maximum negative dilatancy.

Advantages and limitations of the axis-translation method

In the axis-translation technique, both pore water pressure and pore air pressure are controlled and measured independently. This enables controlled

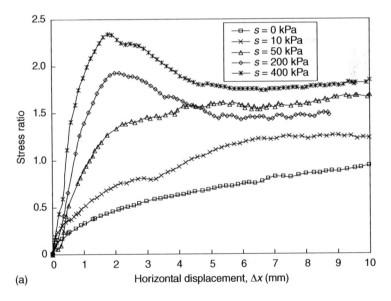

(a)

Figure 2.12 Evolution of (a) stress ratio and (b) dilatancy of CDG subjected to direct shear under different controlled suctions (after Ng and Zhou, 2005).

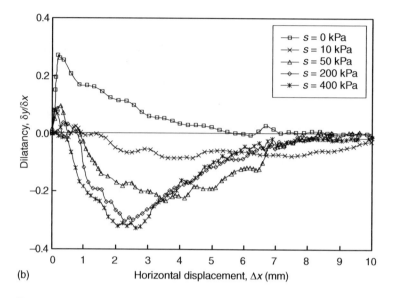

Figure 2.12 (Continued).

variation of suction. When a feedback system is used, suction can be controlled automatically. The majority of experimental results for unsaturated soils have been obtained by the application of axis-translation technique because of the easy measurement and control of suction.

One limitation of the axis-translation technique pertains to the maximum value of suction that can be applied. It is limited by the maximum value of cell pressure and the air-entry value of porous material. Hence, this technique generally is used for controlling suction in the order of several hundred kilopascals.

Another disadvantage of the axis-translation technique is that by elevating the pore water pressure from a negative to a positive value, the possibility of cavitation is prevented not only in measuring system but also within soil pores. This implies that any influence of cavitation on the pore water by altering the desaturation mechanism of a soil under in situ conditions is not accounted for in the laboratory tests using the axis-translation technique (Dineen and Burland, 1995). It is then fundamental to understand whether experimental results obtained by using the axis-translation technique can be extrapolated to unsaturated soils under atmospheric conditions in the field. Some researchers (Zur, 1966; Williams and Shaykewich, 1969; Ng et al., 2007) compared experimental data by using the axis-translation technique and the osmotic technique where the pore air pressure is at the atmospheric pressure. It is found that there are some differences in the experimental results by using the two techniques. The comparisons will be discussed later.

Osmotic technique

Working principle

Due to limitations of the axis-translation technique, an alternative method, the osmotic technique, has been used in testing unsaturated soils to control matric suction. Delage *et al.* (1998) reported that this technique was initially developed by biologists (Lagerwerff *et al.*, 1961) and then adopted by soil scientists (Williams and Shaykewich, 1969) and geotechnical researchers (Kassif and Ben Shalom, 1971; Komornik *et al.*, 1980; Cui and Delage, 1996; Delage *et al.*, 1998; Ng *et al.*, 2007).

Figure 2.13 illustrates the principle of the osmotic technique for controlling suction in a soil specimen. The specimen and an osmotic solution are separated by a semi-permeable membrane. The membrane is permeable to water and ions in the soil but impermeable to large solute molecules and soil particles (Zur, 1966). Therefore, at equilibrium, the component of osmotic suction related to soil salts is the same on both sides of the membrane (i.e. $\pi_{salt}^1 = \pi_{salt}^2$), and the component of osmotic suction related to the solute is zero in the soil. Then the difference of osmotic suction on both sides of the membrane is equal to the component of osmotic suction related to the solute in the solution (i.e. π_{solute}^1). The pore air pressure u_a in the soil is at atmospheric pressure, similar to the natural condition in the field. Zur (1966) presented the principal of the osmotic technique in a review of energy analysis on soil–water. When water exchange through

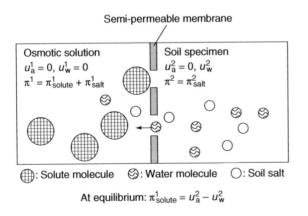

At equilibrium: $\pi_{solute}^1 = u_a^2 - u_w^2$

Figure 2.13 Schematic diagram illustrating the osmotic technique (Ng and Chen, 2005).

the membrane is in equilibrium, the energy potential in soil–water is equal to that in solution water, i.e. total suction in the soil is equal to that in the solution. Therefore, the difference of osmotic suction is equal to the difference of matric suction on both sides of membrane. Since the matric suction in the solution is zero, the matric suction $(u_a^2 - u_w^2)$ in the soil is equal to the difference of osmotic suction on both sides of the membrane, i.e. equal to the component of osmotic suction related to the solute π_{solute}[1]. Polyethylene glycol (PEG) is the most commonly used solute in biological, agricultural and geotechnical testing for its safety and simplicity. The osmotic suction (or osmotic pressure) in the PEG solution is determined by the concentration of PEG. The maximum value of osmotic pressure in PEG solution was reported to be above 10 MPa (Delage et al., 1998).

Applications

In geotechnical testing, the osmotic technique was successfully adapted in an oedometer (Dineen and Burland, 1995; Kassif and Ben Shalom, 1971), a hollow cylinder triaxial apparatus (Komornik et al., 1980) and a modified triaxial apparatus (Cui and Delage, 1996; Ng et al., 2007).

A collaborative research was developed between HKUST and the Ecole Nationale des Ponts et Chaussees (ENPC) to investigate the reliability of osmotic technique and compare the axis-translation and osmotic technique (Ng et al., 2007). Figure 2.14a illustrates a modified triaxial apparatus applying the osmotic technique at ENPC. A soil specimen is put in contact with semi-permeable membranes on both bottom and top surfaces. Concentric grooves are machined in the base pedestal and top cap for a circulation of PEG solution. A thin sieve mesh is placed over the grooves and covered by the membranes, which are glued with an epoxy resin. The top cap is connected to the base of the cell by two flexible tubes. PEG solution is contained in a closed circuit comprising the serial connection of the base of the cell, the top cap, a reservoir and a pump. The reservoir is big enough to ensure a relatively constant concentration in spite of water exchanges occurring through the membranes between the specimen and the solution. The reservoir is closed with a rubber cap, pierced by three glass tubes. Two of these tubes are for inflow and outflow of the solution to the triaxial cell; the third is connected to a graduated capillary tube for monitoring water exchanges. The stabilization of the level of the solution in the tube indicates that an equilibrium suction has been reached in the whole specimen. An air vent was machined on the base of the cell in order to ensure within the specimen a constant pore air pressure equal to the atmospheric pressure. At HKUST, the triaxial apparatus is similar to that at ENPC, except that a double-cell volume change measuring system (see Figure 2.9) and an electronic balance for measuring water volume change are used. Figure 2.14b shows a photograph of the triaxial apparatus at HKUST.

Figure 2.15 shows the stress point failure envelops obtained from the traixial shear results using the osmotic technique at HKUST and ENPC on an expansive soil (Ng *et al.*, 2007). In the figure, $s = (\sigma_1 + \sigma_3)/2 - u_a$ and $t = (\sigma_1 - \sigma_3)/2$. Under the same net confining pressure and suction, tests OH4 and OE3 exhibit similar shear strength. The comparison between tests OH5 and OE6 shows the same conclusion. Consistency of the results from HKUST and ENPC confirms the reliability of osmotic technique. If the tests at HKUST and ENPC are regarded as one test series, four stress point failure envelops at different suctions can be presented. Parallel stress envelops suggest a constant internal friction angle ϕ' within a suction range from 0 to 165 kPa. The increasing intercepts on the t axis indicate that apparent cohesion increases with suction. There also appears a small intercept of 19.4 kPa on the t axis when the failure envelop at zero suction is extended linearly. This apparent intercept is likely due to experimental error or a curved fail-

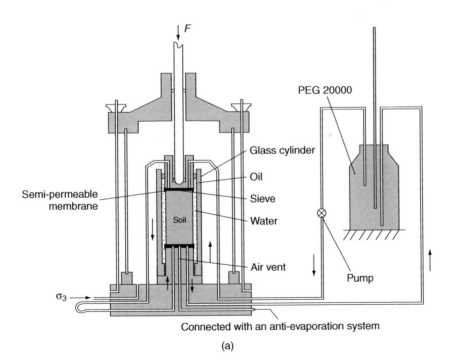

(a)

Figure 2.14 Triaxial test systems using the osmotic technique: (a) schematic for the ENPC system (after Cui and Delage, 1996); (b) photograph of the HKUST system (Ng and Chen, 2005, 2006).

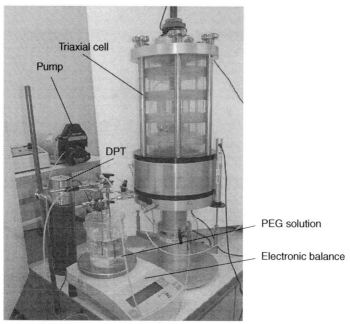

(b)

Figure 2.14 (Continued).

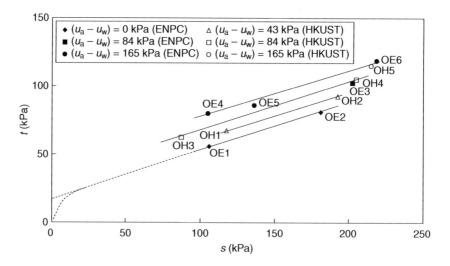

Figure 2.15 Triaxial test results using the osmotic technique at ENPC and HKUST (after Ng et al., 2007).

ure envelope within the range of small stress. The comparison of test results between the axis-translation and osmotic techniques will be discussed later.

Advantages and limitations

As compared with the axis-translation technique, the main advantage of the osmotic technique is that pore water pressure within the soil is maintained at its negative value. A second advantage is that high values of suction can be applied without the use of very high cell pressure. The maximum applied suction in the osmotic technique is determined by the maximum osmotic pressure of PEG solution, which was reported to be above 10 MPa (Delage et al., 1998). However, an evaluation on the performance of three different semi-permeable membranes by Tarantino and Mongiovi (2000a) indicated that all membranes experienced a chemical breakdown as the osmotic pressure of PEG solution exceeded a threshold value, which was found to depend on the type of membrane. Beyond this value, solute molecules were no longer retained by the semi-permeable membrane and passed into the soil specimen, resulting in a reduction of concentration gradient and a decay of soil suction. Accordingly, the maximum applied suction in the osmotic technique is also limited by the performance of semi-permeable membranes.

A further limitation is that the osmotic technique in its current form cannot be used to control suction in a continuous manner because in existing technology, suction changes are applied in steps by exchanging PEG solution with different concentrations manually.

Moreover, calibration of the osmotic pressure against concentration of PEG solution is essential in the osmotic technique. Calibration for PEG solution can be carried out with a psychrometer (Williams and Shaykewich, 1969), an osmotic tensiometer (Peck and Rabbidge, 1969), a high-suction probe (Dineen and Burland, 1995) or an osmotic pressure cell (Slatter et al., 2000; Ng et al., 2007). All these methods involve a semi-permeable membrane except the method using the psychrometer, which requires strict temperature control. It has been found that the relationship between osmotic pressure and the concentration of PEG solution is affected significantly by the calibration method (Dineen and Burland, 1995; Slatter et al., 2000; Ng et al., 2007). Dineen and Burland (1995) suggested a need for some means for direct measuring negative pore water pressure in soil specimen when using the osmotic technique. Therefore, the crucial issues in the osmotic technique are not only the calibration of osmotic pressure against PEG concentration but also the direct measurement of applied suction in the soil.

Humidity control techniques

Working principle

Based on the thermodynamic definition in Equation (1.1), total suction can be imposed on an unsaturated soil specimen by controlling RH of the atmosphere surrounding the soil. Humidity can be controlled by using aqueous solutions (Esteban and Saez, 1988; Oteo-Mazo *et al.*, 1995; Delage *et al.*, 1998; Al-Mukhtar *et al.*, 1999) or by mixing vapour-saturated gas with dry gas via a feedback system (Kunhel and van der Gaast, 1993).

Delage *et al.* (1998) reported that humidity control with solutions was initially developed by soil scientists and the first application to geotechnical testing was by Esteban and Saez (1988). This technique was generally used to study the combined effect of high suction and stress on mechanical properties of various expansive soils (Oteo-Mazo *et al.*, 1995; Al-Mukhtar *et al.*, 1999; Lloret *et al.*, 2003).

Figure 2.16 shows a schematic diagram illustrating the work principle of humidity control with solutions. A soil specimen is placed in a closed thermodynamic environment containing an aqueous solution of a given chemical compound. Depending on the physico-chemical properties of the compound, a given RH is imposed within the sealed environment. Water exchanges occur by vapour transfer between the solution and the specimen, and the given suction is applied to the specimen when vapour equilibrium is achieved. The solution can be the same product at various concentrations (i.e. unsaturated solutions, sulphuric acid or sodium chloride for instance – Oteo-Mazo *et al.*, 1995), or various saturated saline solutions (Delage *et al.*, 1998). Saturated saline solutions have the practical advantage over unsaturated solutions of being able to liberate or

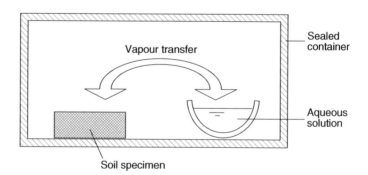

Figure 2.16 Schematic diagram illustrating the principle of humidity control with solutions (Ng and Chen, 2005).

adsorb relative large quantities of water without significantly affecting the equilibrium RH.

Kunhel and van der Gaast (1993) reported a system for XRD (X-ray diffraction) analyses under controlled humidities by mixing vapour-saturated gas with dry gas. Figure 2.17 shows a feedback system for controlling humidity in a soil specimen. Humidity is controlled by proportioned mixing of vapour-saturated nitrogen gas and desiccated nitrogen gas in a closed environmental chamber. The vapour-saturated and desiccated gas streams are reintroduced in a three-neck flask where the resulting gas stream has a RH that is a direct function of the wet to dry gas flow ratio. The humid gas stream is routed into an acrylic environmental chamber containing a soil sample. An effluent gas vent on the top cap of the chamber allows the influent humid gas to escape after flowing around the soil. RH and temperature in the chamber are continuously monitored with a polymer capacitance probe. Signals from the probe form a feedback loop with a control computer for automated regulation of the wet to dry gas flow ratio, enabling control of RH. Humidity variation is controlled to approximately 0.6 per cent RH in this system.

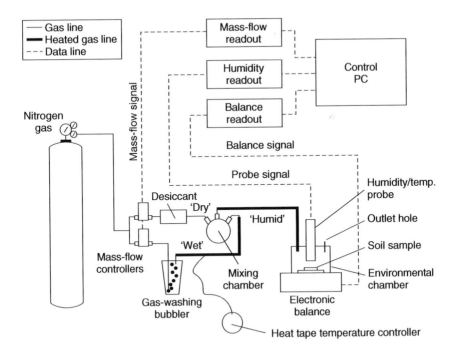

Figure 2.17 General layout diagram for an automatic humidity control system (after Likos and Lu, 2003).

Applications

In the humidity control technique with solutions, vapour exchanges are quite slow due to the very low kinetics of vapour transfer. Therefore, this technique is extremely time consuming. Oedometers are often adopted to achieve a small drainage path of soil specimens when applying this technique. Figure 2.18 shows oedometer results for a highly compacted smectite (Na-laponite) under constant RHs (Al-Mukhtar *et al.*, 1999). The maximum applied total suction was 298 MPa and the maximum applied vertical stress was 10 MPa. The combined high suction and stress on volume change properties of the expansive soil is illustrated in Figure 2.18. The results show that

- void ratio increases as suction decreases (RH increases);
- increasing axial stress reduces the difference in the void ratio of the tested samples;
- the slope of compression curve decreases as suction increases, indicating that suction stiffens the soil.

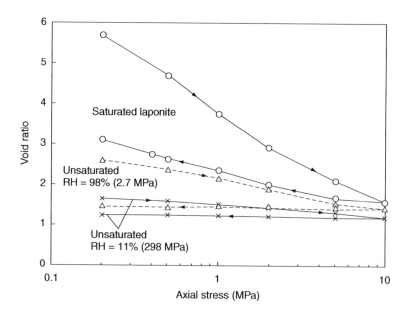

Figure 2.18 Oedometer results for Na-laponite tested under controlled humidity by using solutions (after Al-Mukhtar *et al.*, 1999).

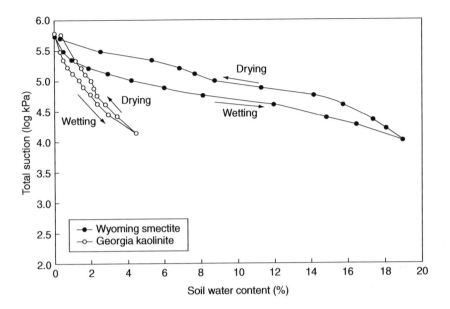

Figure 2.19 Measured total suction characteristic curves for Wyoming smectite and Geogia kaolinite under controlled humidity by using a feedback system (after Likos and Lu, 2003).

Applications of a feedback system to control humidity in an unsaturated specimen are scarce. Likos and Lu (2003) used the system described by Al-Mukhtar *et al.* (1999) to determine total suction characteristic curves in high suction range for four types of clay, ranging from highly expansive smectite to non-expansive kaolinite. Figure 2.19 shows the measured total suction characteristic curves for the expansive smectite and non-expansive kaolinite. Both drying and wetting paths can be performed using the testing system. The values of applied total suction are within a relatively large range. The expansive smectite contains more water than the kaolinite under the same applied total suction, exhibiting larger capability of water adsorption. The expansive smectite shows larger values of the rate of change in water content due to suction and larger hysteresis than the kaolinite.

Advantages and limitations

Similar to the osmotic technique, humidity control techniques can maintain the pore water pressure within a soil specimen at its negative value while high values of suction can be applied without use of a very high cell pressure. When using solutions to control humidity, very high values of suction (up to 1000 MPa) can be applied. However, Delage *et al.* (1998) point out that uncertainties in this technique limit applications under 10 MPa suction. When using a feedback system to control humidity, the suction control range is determined by the measuring range and accuracy of the humidity probe. For the results given in Figure 2.17, the suction control range was from 7 to 700 MPa (Likos and Lu, 2003).

When using solutions to control humidity, only fixed values of suction can be applied and cannot be used for varying suction in a continuous manner. If the humidity is controlled by proportioning vapour-saturated gas with dry gas via a feedback system, suction can be controlled automatically. The accuracy of the method depends on both the accuracy of the humidity probe and the resolution of the control loop for automated regulation of RH.

When humidity is controlled by using solutions, as activity of the solutions is very sensitive to thermal fluctuation, temperature must be strictly controlled during testing. The control of suction by this technique is much slower than the techniques involving liquid transfer (axis-translation and osmotic techniques) due to the very low kinetics of vapour transfer. When humidity is controlled by proportioning vapour-saturated gas with dry gas, due to active gas circulation, test durations can be reduced. Likos and Lu (2003) reported that equilibrium water contents in several clays were reached within 12 h when changes of humidity in the soils were controlled by the system discussed in Figure 2.12. The short suction equalization duration (within 12 h) is very questionable, since it has been reported that testing durations of months were required when humidity was controlled by using solutions (Oteo-Mazo *et al.*, 1995).

Characteristics of the three suction control techniques (axis-translation, osmotic and humidity control) are summarized in Table 2.1.

Experimental comparisons between the axis-translation and osmotic techniques (Ng et al., 2007)

Overview

Unsaturated soil behaviour is governed by two independent stress variables, net stress $(\sigma - u_a)$ and matric suction $(u_a - u_w)$, and not by component stresses. Since both axis-translation and osmotic techniques control matric suction, test results obtained by these two techniques should be the same provided the applied stress is the same. However, few direct experimental comparisons between these two techniques are available. Zur (1966) found that for a sandy loam, gravimetric SWCCs obtained by using these two techniques showed good agreement. However for a clay, equilibrium water contents obtained by using the axis-translation technique were higher than those obtained by using the osmotic technique under the same applied

Table 2.1 Characteristics of three suction control techniques (Ng and Chen, 2005, 2006)

	Axis-translation	Osmotic	Humidity control	
			Using solutions	Using feedback system
Controlled suction	Matric	Matric	Total	Total
Automatisation	Automatic	Manual	Manual	Automatic
Suction range	Zero to hundreds of kPa	Zero to 10 MPa (maximum is limited by performance of semi-permeable membrane)	10 MPa to 1,000 MPa	Determined by measuring range and accuracy of humidity probe
Similarity with field condition	Positive pore water pressure	Negative pore water pressure (similar to field condition)	Negative pore water pressure (similar to field condition)	Negative pore water pressure (similar to field condition)
Verification	Valid (continuous air phase); controversial (occluded air phase)	NA	NA	NA
Requirement	Continuous air and water phases	Calibration of PEG solution or direct measurement of negative pore water pressure	Strict temperature control; time-consuming	Accurate humidity probe and feedback loop

nominal suction. Zur (1966) ascribed this difference to test duration, suggesting that 2 days were not sufficient for suction equilibrium in the axis-translation technique. When the test duration for suction equilibrium was extended to 3 days, better agreement was achieved. However, Williams and Shaykewich (1969) found that even when the suction equilibrium duration was extended to more than 10 days in the axis-translation technique, the obtained equilibrium gravimetric water content was up to 10 per cent higher than that obtained by using the osmotic technique for a clay. To validate the axis-translation and osmotic techniques for shearing unsaturated soils, a collaborative research was developed at HKUST and ENPC (Ng *et al.*,

2007). In the research, triaxial shearing tests were performed on a recompacted expansive clay at high initial degree of saturation (i.e. 0.83), applying both axis-translation and osmotic techniques. Based on the extended Mohr–Coulomb shear strength formulation (Fredlund et al., 1978), the results by using the two techniques were compared. In determining internal friction angle ϕ', there is no difference in the results by using the two techniques. However, in terms of the average value of ϕ^b at different suctions (an angle with respect to changes in $(u_a - u_w)$ when $(\sigma - u_a)$ is held constant), the value obtained by using the axis-translation technique is 13° within a suction range from 0 to 100 kPa while that obtained by using the osmotic technique is 8° within a suction range from 0 to 165 kPa. Perhaps the difference is due to smaller values of void ratio at failure in the tests by using the axis-translation technique (Ng et al., 2007).

Influence of suction equalization (Ng and Zhou, 2005)

No matter which method is used to control suction, a vital issue in testing unsaturated soils is suction equalization. To investigate the influence of suction equalization on subsequent shearing behaviour, two direct shear tests were performed on an expansive soil subjected to different durations of suction equalization.

The expansive soil was taken from Zaoyang, Hubei province of China. It has 3 per cent sand, 58 per cent silt and 39 per cent clay, which is classified as a silty clay (BSI, 1990). The liquid limit and plasticity index of the soil are 50.5 and 31 per cent, respectively. Soil specimens for direct shear tests were statically compacted to a dry density of $1.56 \, \text{g/cm}^3$ with an initial degree of saturation of 69 per cent. The initial matric suction of the specimens was about 540 kPa which was measured by a high-suction probe (Zhan, 2003).

The direct shear apparatus based on the axis-translation principle as shown in Figure 2.11b was used to perform direct shear tests. Zero suction was applied to two specimens and their suction equalization durations were 2 h for Test A and 1 day for Test B. Then the two specimens were sheared at a rate of 0.0019 mm/min under the same applied suction (i.e. 0 kPa) and vertical stress (i.e. 50 kPa).

Figure 2.20 shows the results of direct shear tests subjected to different suction equalization durations. Figure 2.20a shows the relationships between shear stress and horizontal displacement. The curves of the two tests show distinct behaviour. The curve of Test A exhibits strain softening behaviour. The curve of Test B shows strain hardening behaviour. The influence of suction equalization duration on the shear stress–displacement behaviour is similar to that of suction as shown in Figure 2.12a. Each curve reaches a steady value at a displacement in excess of 2.5 mm. The steady value decreases with increasing suction equalization duration. The difference in the steady values of shear stress is about 12 per cent. The difference in the maximum values of shear stress is about 18 per cent. The differences may be due to non-equilibrium of suction at the end of suction equalization duration, which will be discussed later.

Figure 2.20b shows the relationships between vertical displacement (positive value denotes soil contraction) and horizontal displacement. The curves of the two tests show contrary behaviour in terms of volume change. When the suction equalization duration increases from 2 h to 1 day, the volume change behaviour changes from dilative to contractive. Again, the influence

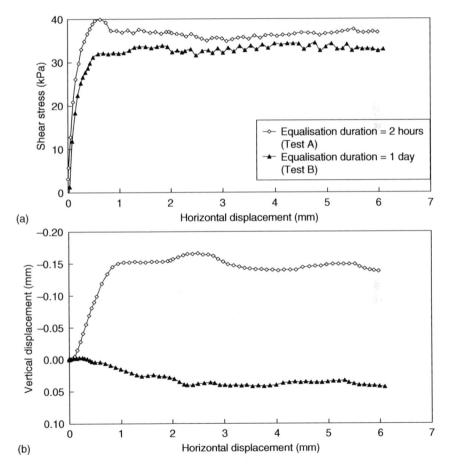

(a)

(b)

Figure 2.20 Influence of suction equalization duration on shearing tests: (a) stress–displacement curves; (b) vertical displacement versus horizontal displacement; (c) water content change versus displacement. Test results from the HKUST direct shear apparatus using the axis-translation principle (Ng and Chen, 2005, 2006).

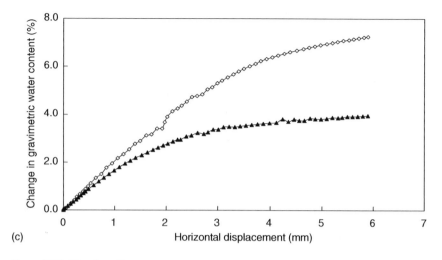

(c)

Figure 2.20 (Continued).

of suction equalization duration on volume change is similar to that of suction as shown in Figure 2.12b. The volume change of the two tests tends to a steady volume at a horizontal displacement in excess of 2.5 mm. Figure 2.20c shows the change in gravimetric water content with horizontal displacement. Both tests showed a continuous absorption of water during shearing. The magnitude of water absorption decreases as the applied suction equalization duration increases.

Figure 2.21 shows the degree of saturation of Tests A and B just before shearing. Wetting and drying SWCCs of this expansive soil are also shown in this figure. Since the value of matric suction after specimen preparation was relatively large (i.e. 540 kPa), the specimens in Tests A and B were subjected to a wetting path at the applied suction of 0 kPa. The degree of saturation of the two specimens increases with the applied suction equalization duration and are smaller than the value obtained from the wetting SWCC at the suction of 0 kPa. The degree of saturation is considered to indicate actual suction in the specimens. The actual suction decreases with the applied suction equalization duration, however, and is larger than the applied suction (i.e. 0 kPa) for the same degree of saturation. This is similar to the influence of suction as shown in Figure 2.12. The higher value of actual suction in Test A resulted in larger strength, more dilative behaviour and stronger water adsorption during the shearing stage as compared with Test B. These tests illustrate that it is vital to obtain suction equalization before shearing.

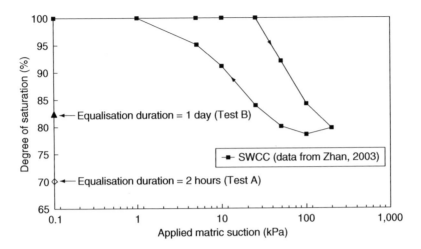

Figure 2.21 Degree of saturation before shearing in direct shear tests (Zhan, 2003; Ng and Chen, 2005, 2006).

Suction measuring devices (Fredlund and Rahardjo, 1993; Ridley and Wray, 1995; Feng and Fredlund, 2003)

Overview

There are two distinct levels at which suction can be measured. The first involves a measurement of the energy required to move a water molecule within the soil matrix – it is termed the *matric suction*. The second is a measure of the energy required to remove a soil–water molecule from the soil matrix into the vapour phase – it is termed the *total suction*.

In a salt-free granular material the total suction and the matric suction are equal (refer to Equation (1.2)). If the pore water contains ions (as is generally the case for clay soils), the vapour pressure (u_v) is reduced and the energy required to remove a water molecule from the water phase of the soil (i.e. the total suction) is increased. The additional stress caused by the dissolved salts is termed the *osmotic suction*. Ideally, if the vapour pressure of a sample of soil–water (extracted from the soil) is measured, the stress implied from Equation (1.1) will be equal to the difference between the total suction and the matric suction of the soil. However, for this to be so the composition of the soil, water must not change during the extraction process. At the present time, the only satisfactory method of estimating the osmotic suction is by subtracting the matric suction from the total suction.

As suggested by Ridley and Wray (1995), in the absence of a truly semi-permeable membrane, the presence of dissolved salts within the pore water should not result in an appreciable movement of the pore water. Therefore, dissolved salts do not cause a change in the energy required to move a water molecule (while maintaining its state) within the soil matrix, and the matric suction is independent of the osmotic suction.

In deciding whether an instrument is measuring total or matric suction, it is necessary to determine if the instrument is making direct contact with the pore water in the soil. If no contact is established between the measuring instrument and the soil, the dissolved salts cannot move from the pore fluid to the measuring instrument. When the instrument being used to measure suction makes contact with the pore water, the dissolved salts will be free to move between the two. Therefore, unless it is certain that the concentration of dissolved salts is everywhere the same, their effect on the suction measured in this way is unquantifiable. However, the effect is generally believed to be negligible and if this is so then it is the matric suction that will be measured in this way. Moreover, at high values of suction, the meniscus may recede completely inside the soil and an instrument that is making contact with the soil will not necessarily be making close contact with the soil–water. Therefore, contact between soil and instrument does not guarantee that matric suction will be measured (Ridley and Wray, 1995).

Methods for measuring soil suction (Ridley and Wray, 1995)

Overview

Generally suction measurement devices fall into two categories depending on whether the measurement is direct or indirect. A direct measurement is one that measures the relevant quantity under scrutiny (e.g. the pore water energy). With indirect measurements another parameter (e.g. RH, resistivity, conductivity or moisture content) is measured and related to the suction through a calibration against known values of suction. These devices will now be considered.

Psychrometers

A psychrometer is an instrument that measures humidity and can therefore be used to measure total suction. In its simplest form, it consists of a thermometer that has a wetted bulb from which evaporation into the adjacent air reduces the temperature of the bulb to a value lower than the ambient temperature. When evaporation ceases and equilibrium with the ambient vapour pressure is reached, the measured temperature is compared with the temperature of a dry-bulb thermometer placed in the same environment.

The difference between the temperatures of the dry-bulb and the wet-bulb is related to the RH and can be calibrated using salt solutions of known concentrations.

Electrical devices now exist that measure the heat (which is a function of the electrical current) generated in an electrical junction when water evaporates from or condenses onto the junction. They may be subdivided into two groups:

- thermistor/transistor psychrometers (see Figure 2.22) and
- thermocouple psychrometers (see Figure 2.23).

THERMISTOR/TRANSISTOR PSYCHROMETERS

A thermistor is a temperature sensitive resistor. In the thermistor psychrometer, two identical (matched) thermistors are employed. Onto the first is placed a drop of water (the wet-bulb) and the other is left dry (the dry-bulb). If both thermistors are then exposed to an enclosed environment, evapouration from or condensation onto the wet thermistor will result in an e.m.f. being generated in the thermistor that can be related to the RH of

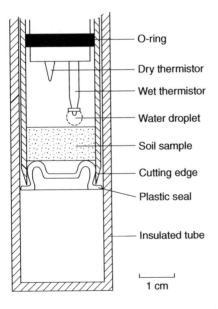

Figure 2.22 A thermistor psychrometer (after Ridley and Wray, 1995).

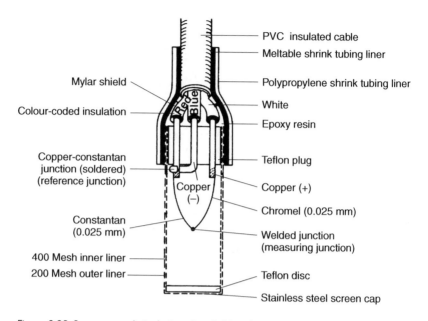

Figure 2.23 Screen-caged single-junction Peltier thermocouple psychrometer (from Brown and Collins, 1980; after Fredlund and Rahardjo, 1993).

the environment. Richards (1965) gave a detailed description of the theory, construction and calibration of the thermistor psychrometers in Australia.

THERMOCOUPLE PSYCHROMETERS

The introduction of the thermocouple psychrometer is attributed to Spanner (1951). Working in the field of plant physiology, he recognized the potential for using the Peltier effect and the Seebeck effect to measure the suction pressure in plant leaves.

In 1834, Peltier discovered that on passing an electrical current across the junction between two different metal wires, there is a change of temperature, the sign of which is dependent on the direction of flow of the current. If such a junction is placed in an atmosphere of humid air and a current is passed in the direction which causes cooling of the junction sufficient to cause condensation of water onto it, then it essentially becomes a 'wet-bulb' thermometer. When the circuit is broken, the water will evaporate, and the change in temperature of the wet junction will generate an e.m.f. (the Seebeck effect) between it and a reference junction which is at the ambient temperature (i.e. a 'dry-bulb'). This e.m.f. can be measured with a sensitive galvanometer. The characteristic output curve for this type of psychrometer has a plateau voltage at the 'wet-bulb' temperature corresponding to the vapour pressure of the ambient air. Unfortunately, the evaporation continues and the plateau only lasts for a period of time determined by the RH (and therefore by the suction) being measured. At suctions less than 500 kPa, the plateau is difficult to detect. Details of the operating principles and procedures for this type of psychrometer are given by Fredlund and Rahardjo (1993).

Thermocouple psychrometers are seriously affected by temperature fluctuations (Rawlins and Dalton, 1967; Fredlund and Rahardjo, 1993). When the measuring junction is cooled, heat is generated in the reference junction. It is important for the correct measurement of RH that the reference junction is held at the ambient temperature. Temperature gradients between the cooling junction and the reference junction at the start of the test can result in a large error in the measured humidity. Measuring the initial potential difference between the two junctions will give an indication of any temperature gradient. It is recommended that a zero offset greater than $1\,\mu V$ represents an excessive temperature gradient (Ridley and Wray, 1995). In order to obtain satisfactory performance of a psychrometer, a constant temperature water bath which can be regulated to $\pm 0.001°C$ may be required to measure total suction as low as 100 kPa (Fredlund and Rahardjo, 1993). The useful range over which the total suction can be made lies between about 100 and 8,000 kPa.

The efficiency of the thermocouple psychrometer is also temperature dependent, and the calibration needs adjustment relative to the temperature at which the probe was calibrated. The effect of the absolute temperature on the calibration of thermocouple psychrometers was investigated by Brown and Bartos (1982). It was found that the calibration was more sensitive at higher temperatures.

The filter paper method (Fredlund and Rahardjo, 1993; Ridley and Wray, 1995)

According to Fredlund and Rahardjo (1993), the filter paper method for measuring soil suction was developed in the soil science discipline, and has since been used primarily in soil science and agronomy (Gardner, 1937; Fawcett and Collis-George, 1967; McQueen and Miller, 1968; Al-Khafaf and Hanks, 1974). Attempts have also been made to use the filter paper method in geotechnical engineering (Ching and Fredlund, 1984; Gallen, 1985; McKeen, 1985; Chandler and Gutierrez, 1986; Crilly et al., 1991; Gourley and Schreiner, 1995).

Ridley and Wray (1995) reported that McQueen and Miller (1968) proposed to use the filter paper method to measure either the total or the matric suction of a soil. The filter paper is used as a sensor. The filter paper method is classified as an 'indirect method' for measuring soil suction. The filter paper method is based on the assumption that a filter paper will come into

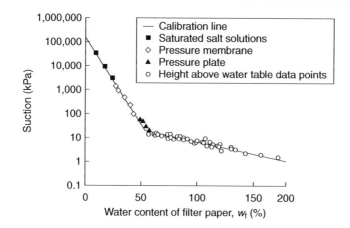

Figure 2.24 Contact and non-contact filter paper methods for measuring matric suction and total suction (from Al-Khafaf and Hanks, 1974; after Fredlund and Rahardjo, 1993).

equilibrium (with respect to moisture flow) with a soil having a specific suction. Equilibrium can be reached by either liquid or vapour moisture exchange between the soil and the filter paper. When a dry filter paper is placed in direct contact with a soil specimen, it is assumed that water flows from the soil to the paper until equilibrium is achieved (Figure 2.24). When a dry filter paper is suspended above a soil specimen (i.e. no direct contact with the soil), vapour flow of water will occur from the soil to the filter paper until equilibrium is achieved. Having established equilibrium conditions, the water content of the filter paper is measured. Once the water content is known, soil suction can be obtained from a calibration curve. The two commonly adopted types of filter paper are *Whatman* No. 42 and *Schleicher and Schuell* No. 589. The method was first formally reported by Gardner (1937) and has recently been accepted as a standard method of measuring soil suction (ASTM, 1993).

The calibration of a filter paper is usually achieved by either (a) equilibrating it on a pressure plate brought to a known suction, or (b) enclosing it in a sealed container with a salt solution of known vapour pressure. In practice, the former approach (that involves an in-contact measurement) has been used for suctions up to 1,500 kPa and the latter approach (that involves an out-of-contact measurement) has been used above 1,500 kPa (Ridley and Wray, 1995). The widely accepted calibrations arrived at using approaches (a) and (b) are shown (combined) in Figure 2.25 for *Whatman* No. 42 and *Schleicher* and *Schuell* No. 589. Each, initially air dry filter paper was left to equilibrate for 7 days before measuring its water content. The resulting relationship between suction and filter paper water content is bi-linear with the change in sensitivity occurring at a water content of about

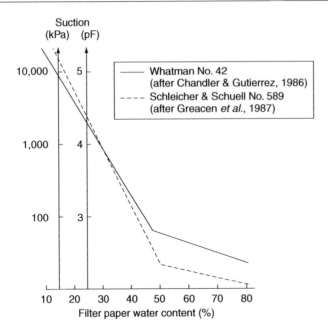

Figure 2.25 Filter paper calibrations for papers placed in contact with soil (after Ridley and Wray, 1995).

47 per cent for *Whatman* No. 42 paper (Chandler and Gutierrez, 1986) and about 54 per cent for *Schleicher* and *Schuell* No. 589 paper (Greacen *et al.*, 1987). These relationships have remained remarkably consistent (Swarbrick, 1995) and are now universally accepted as the representative calibrations for in-contact measurements.

Sibley and Williams (1990) compared the calibration curves for five materials; *Whatman* No. 42 filter paper, unwashed dialysis tubing, washed dialysis tubing and two new materials; *Milipore* MF 0.025 and MF 0.05 filtration membranes. The results show that each material has an increased sensitivity over different ranges of suction. The materials were categorized as most accurate over the ranges shown in Table 2.2. However, Sibley

Table 2.2 Appropiate filter materials (after Sibley and Williams; Ridley and Wray, 1995)

Applicable range of measurement	Appropriate type of paper
Up to 30 kPa	Whatman No. 42
30 kPa to 300 kPa	Millipore 0.025
30 kPa to 1,000 kPa	Millipore 0.05
1,000 kPa to 10 MPa	Washed dialysis tubing
10 MPa to 100 MPa	Unwashed dialysis tubing

and Williams stated that 'if only one material is to be used over the entire suction range, then *Whatman's* No. 42 is the most appropriate'. Details of calibration and measurement procedures using the filter paper method in the laboratory and in the field are given by Fredlund and Rahardjo (1993) and Ridley and Wray (1995).

Until fairly recently, it has been generally assumed that the calibrations for both methods of absorption (i.e. capillary flow and vapour flow) are identical for all levels of suction. This assumption, however, was shown to be not reliable by a number of researchers (El-Ehwany and Houston, 1990; Lee and Wray, 1992; Houston *et al.*, 1994; Harrison and Blight, 1995). It is recommended that separate calibration curves should be used.

It should be emphasized that the filter paper technique is highly user-dependent, and great care must be taken when measuring the water content of the filter paper (Fredlund and Rahardjo, 1993). The balance to determine the water content of the filter paper must be able to weigh to the nearest 0.0001 g. Each dry filter paper has a mass of about 0.52 g, and at a water content of 30 per cent, the mass of water in the filter paper is about 0.16 g.

Porous blocks (e.g. Gypsum block)

The electrical resistance of an absorbent material changes with its moisture content. This is the principle of the porous block method of measuring soil matric suction. The block (which is placed inside the soil sample) consists of two concentric electrodes buried inside a porous material (Figure 2.26). If the suction in the soil is higher than the suction in the block, the latter will lose

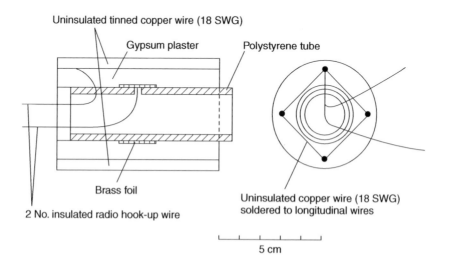

Figure 2.26 A typical gypsum block (after Ridley and Wray, 1995).

moisture to the soil until the soil and the block has the same matric suction. Alternatively, if the block is initially dry, it will absorb moisture from the soil until an equilibrium condition is reached. The electrical resistance of the block can then be related to the suction. The calibration is performed by burying each block inside a soil sample that is subsequently subjected to known suctions inside a pressure plate apparatus.

Gypsum (in the form of plaster of Paris) was found to be the most suitable medium for the measurement of electrical resistance. It took the shortest time to saturate, was the quickest to respond when placed in the ground and had the most stable electrical properties. However, it had the distinct disadvantage of softening when saturated (Ridley and Wray 1995).

The gypsum block is a low-cost device that is easy to handle and install. Measurements are simple and correction factors, for the effects of temperature and salt content, can be applied where necessary. However, the blocks are susceptible to hysteresis, and their response to a change in suction can be slow (2–3 weeks) which precludes the use of them as accurate indicators of absolute suction values in a rapidly changing moisture environment. Therefore, it is recommended that only a small portion of the calibration curve should be used (i.e. the instruments should not be used in situations where large fluctuations in soil suction are likely).

The ideal conditions for the use of gypsum blocks are in a soil that is relatively non-saline, having a suction in the range 50–3,000 kPa, and that is subjected to small and uni-directional moisture changes. Conversely, the most adverse conditions for the use of the blocks are in relatively saline soils subjected to rapid changes of water content. These rather narrow constraints have led to a diminished use of porous blocks for measuring soil suction.

Thermal conductivity sensors (Feng and Fredlund, 2003)

A thermal conductivity soil suction sensor consists of a cylindrical porous tip containing a miniature heater and a temperature-sensing element (Phene et al., 1971). Figure 2.27 shows the structure of a thermal conductivity sensor developed at the University of Saskatchewan (Shuai et al., 1998). The porous tip is a specially designed and manufactured ceramic with an appropriate pore-size distribution corresponding to the range of soil suctions to be measured. The heater at the centre of the ceramic tip converts electrical energy to thermal energy. A portion of the thermal energy will be dissipated within the ceramic tip. The undissipated thermal energy results in a temperature rise at the centre of the ceramic tip. The temperature sensor (i.e. the integrated circuit IC in Figure 2.27) measures the temperature rise with respect to time in terms of output voltage. Since water has a much higher thermal conductivity than air, the rate of dissipation of the thermal energy within the ceramic tip increases the water content of the ceramic.

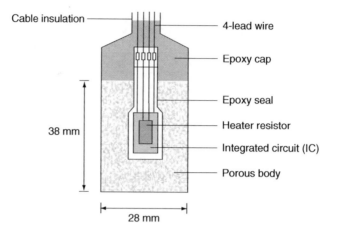

Figure 2.27 A cross-sectional diagram of the newly developed thermal conductivity sensor (from Shuai et al., 1998; after Feng and Fredlund, 2003).

Higher water contents result in a lower temperature rise at the centre of the ceramic and, consequently, a lower output voltage from the temperature sensor. Since the suction in the sensor is equal to the suction in the surrounding soil, the voltage output of the temperature sensor (i.e. the output of the thermal conductivity suction sensor) can be calibrated against matric suction in the surrounding soil.

Calibration is the first step leading towards the use of the thermal conductivity sensor for field measurements of soil suction. Presumably, the calibration should reproduce the actual field conditions as closely as possible. The sensors are conventionally calibrated following a drying process. The calibration curve represents the relationship between the sensor output and the applied matric suction for the specific drying process. It is recognized, however, that the water content (and consequently the output voltage) versus matric suction relationships for a porous material exhibit hysteresis between the wetting and drying processes (Fredlund et al., 1994).

Several hypothetical hysteresis models have been proposed in the literature (Feng, 1999). An examination of some of the hysteresis models using the measured hysteresis data from the ceramic sensors showed that the models either require too much calibration data or cannot provide a reasonable prediction of hysteresis (Feng, 1999).

The experimental results showed that the hysteresis curves were consistent from one suction sensor to another. If a prediction model can be developed that fits the measured calibration curves of the sensors with known hysteresis characteristics, the model can be used to predict hysteresis curves of other sensors of the same type. The following equation is proposed to fit the main drying and main wetting curves:

$$V(\psi) = \frac{ab + c\psi^d}{b + \psi^d} \qquad (2.7)$$

where a is the sensor reading at suction equal to zero on the main hysteresis loop, c is the sensor reading when the ceramic tip is in a dry condition. Parameters a and c are easy to measure and remain the same for the main wetting and main drying curves, respectively. With one branch of the main hysteresis loop measured, only two parameters, b and d, remain unknown for the other branch. If two points on the unknown branch are measured, this branch can be estimated using Equation (2.7).

With one branch of the main hysteresis loop measured and the other branch estimated, the following equations are used to fit to the scanning curves:

$$V_d(\psi, \psi_1) = V_d + \left(\frac{\psi_1}{\psi}\right)^\alpha (V_w - V_d) \qquad (2.8)$$

$$V_w(\psi, \psi_1) = V_w + \left(\frac{\psi_1}{\psi}\right)^\alpha (V_w - V_d) \qquad (2.9)$$

where $V_d(\psi, \psi_1)$ is the output voltage at suction ψ on the drying scanning curve that starts at a suction value ψ_1; ψ_1 is the soil suction at which the scanning curve starts; V_w and V_d are the output voltages at suction ψ on the main wetting and drying curves, respectively, and α is an empirical parameter that controls the degree of curvature of the scanning curves and is the only unknown parameter in Equations (2.8) and (2.9).

The measured hysteresis curves of the six sensors were fitted using Equations (2.7)–(2.9). Similar prediction results were obtained for the six sensors. The predicted curves of the main hysteresis loop and primary scanning curves for sensors 1–3 are shown in Figures 2.28–2.30, respectively.

A best-fit value of α equal to 1.8 was used for both the primary drying scanning curves and the primary wetting scanning curves of all six sensors. Figures 2.28–2.30 show that the predicted curves are close to the measured curves. The errors between the predicted and measured values are under 5 per cent. The α value of 1.8 appears to be reasonable for predicting the primary scanning curves for the newly developed suction sensors. It should be noted that the α value could be different for sensors other than the newly developed suction sensor. It is necessary to investigate the hysteresis properties using typical sensors to estimate the α value when calibrating other types of thermal conductivity sensors.

Feng and Fredlund (2003) recommend the following procedure for calibrating the thermal conductivity suction sensors:

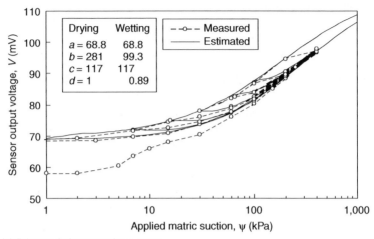

(a) Primary drying scanning curves

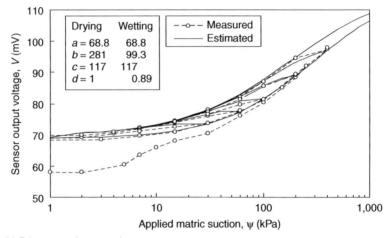

(b) Primary wetting scanning curves

Figure 2.28 Measured and predicted primary scanning curves for sensor 1 (after Feng and Fredlund, 2003).

- saturate the ceramic sensor tip by submerging it in water for 2 days or more;
- place the sensor ceramic tip in the pressure plate cell and apply a suction of 50–100 kPa;
- when equilibrium has been reached, reduce the applied suction to zero;

- after equilibrium at zero suction, increase the applied suction in increments following the conventional calibration procedure to measure the *main drying curve*; and
- rewet the ceramic sensor to obtain two points on the main wetting curve.

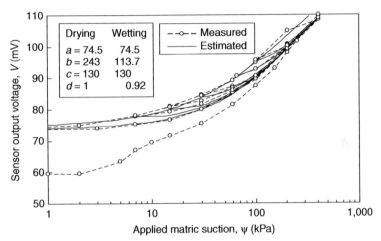

(a) Primary drying scanning curves

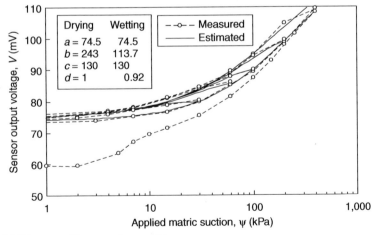

(b) Primary wetting scanning curves

Figure 2.29 Measured and predicted primary scanning curves for sensor 2 (after Feng and Fredlund, 2003).

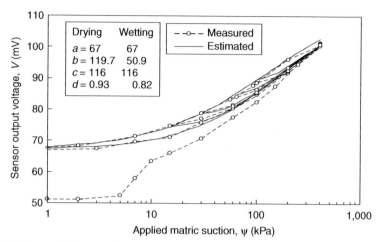

(a) Primary drying scanning curves

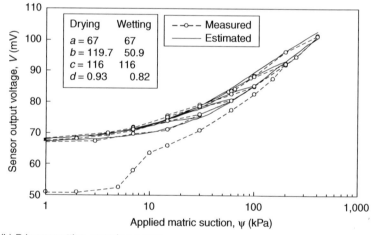

(b) Primary wetting scanning curves

Figure 2.30 Measured and predicted primary scanning curves for sensor 3 (after Feng and Fredlund, 2003).

The main wetting curve is estimated using Equation (2.7). The primary scanning curves are estimated using Equations (2.8) and (2.9), assuming an α value of 1.8.

A similar procedure can be used to calibrate other types of thermal conductivity suction sensors. However, a study of the hysteretic properties of the ceramic sensor output voltage versus matric suction should be carried out to establish the value of the α parameter for Equations (2.8) and (2.9).

Suction plates and pressure plates (Ridley and Wray, 1995)

In its simplest form, the suction plate consists of a saturated flat porous ceramic filter disc that separates a soil specimen from a reservoir of water and a mercury manometer. The soil which, by virtue of its suction, is deficient of water will imbibe water from the porous disc, causing a drop in the water pressure in the reservoir which is measured using the manometer. When the pore water pressure in the soil and the tension in the reservoir water are in equilibrium, there will be no further flow of water, and the soil suction can be interpreted from the manometer reading. As a result of the water exchange, the measured suction will most probably be lower than the actual suction in the soil prior to the test.

The main limitation of the suction plate is its range of usefulness; it fails to prevent the formation of air in the reservoir at suctions greater than 1 atm. When air is present in the reservoir, the tension measured by the manometer will remain less than or equal to 1 atm, and any further moisture exchange will only result in a change in the volume of air present. Schofield (1935) recognized this limitation and proposed extending the useful range by enclosing the specimen in a chamber and applying an air pressure to the inside of the chamber.

The pressure plate apparatus (Figure 2.31) consists of a base plate with a porous ceramic filter set into it. The air-entry pressure of the ceramic should be higher than the maximum suction to be measured (usually 500 kPa, but 1,500 kPa ceramic filters are available). Beneath the filter is a water reservoir

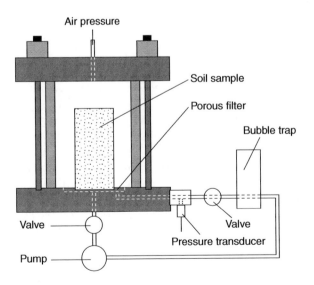

Figure 2.31 A typical pressure plate apparatus (after Ridley and Wray, 1995).

Table 2.3 Methods for measuring soil suction (after Ridley and Wray, 1995)

Device	Measurement mode	Range (kPa)	Approximate equilibrium time
Thermocouple psychrometer	Total	100 to 7,500	Minutes
Thermisor/Transistor psychrometer	Total	100 to 71,000	Minutes
Filter paper (in-contact)	Matric	30 to 30,000	7 days
Filter paper (no-contact)	Total	400 to 30,000	7–14 days
Porous block	Matric	30 to 3,000	Weeks
Thermal conductivity probe	Matric	0 to 300	Weeks
Pressure plate	Matric	0 to 1,500	Hours
Standard tensiometer	Matric	0 to −100	Minutes
Tensiometer suction probe	Matric	0 to −1,800	Minutes

connected to a pressure measurement device and a drainage system. Onto the base plate is placed an airtight chamber into which compressed air can be piped.

When a soil specimen is placed on the ceramic filter and the air pressure in the chamber is raised, the water pressure in the soil specimen is raised by an amount roughly equal to the difference between the air pressure in the chamber and atmospheric pressure. This technique is called axis-translation (Hilf, 1956). Therefore, if the difference between the air pressure in the chamber and the original atmospheric pressure is greater than the suction in the soil, the final water pressure in the soil specimen will be positive and can be measured using standard pore pressure measuring equipment. Table 2.3 summarizes and compares the characteristics of various measuring devices.

Tensiometers

A conventional tensiometer measures the absolute negative pore water pressure in a similar manner to the suction plate. It is principally used in the field but has found some applications in the laboratory (Tadepalli and Fredlund, 1991; Chiu et al., 1998). It works by allowing water to be extracted from a reservoir in the tensiometer, through a porous ceramic filter and into the soil, until the stress holding the water in the tensiometer is equal to the stress holding the water in the soil (i.e. the soil suction). At equilibrium, no further flow of water will occur between the soil and the tensiometer. The suction will then manifest itself in the reservoir as a tensile stress in the water and can be measured using any stress measuring instrument. Stannard (1992) has reviewed the theory, construction and usage of common tensiometers.

All of the commercially available tensiometer devices, designed for use in the field, consist of a porous ceramic cup, a fluid reservoir and a pressure

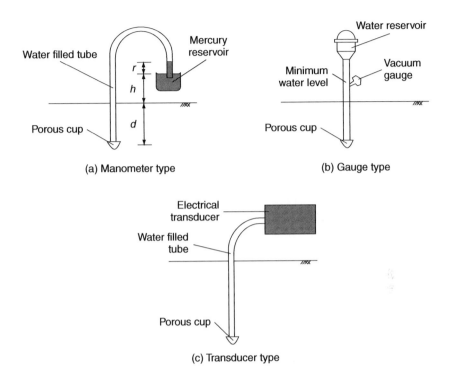

(a) Manometer type

(b) Gauge type

(c) Transducer type

Figure 2.32 Typical tensiometer arrangements (after Ridley and Wray, 1995).

measuring device (Figure 2.32). It is with the last constituent that the main differences lie. Three types of measuring instrument are commonly used:

- a mercury manometer,
- a vacuum gauge and
- an electronic pressure transducer.

MERCURY MANOMETER

The manometer type uses a water filled tube to connect the porous cup to a mercury reservoir (Figure 2.32a). The suction imparted at the porous cup causes the mercury to rise in the tube above the level of the free surface. Using the free mercury surface as a datum, the suction is related to the rise of the mercury and the depth of the porous cup below the free surface as shown in Equation (2.10). Referring to Figure 2.31a, this suction may be expressed as:

$$\text{suction} = \left(\rho_{Hg} - \rho_{H_2O}\right) r - \rho_{H_2O} \left(h + d\right) \tag{2.10}$$

where ρ_{Hg} and ρ_{H_2O} are the density of mercury and of water, respectively.

VACUUM GAUGE

A vacuum gauge is the most commonly used device in commercially available tensiometers. In the vacuum gauge tensiometer (Figure 2.32b), the system is sealed and measurements are made on a standard vacuum gauge. The porous cup is connected to the vacuum gauge via a rigid pipe that extends above the level of the gauge to help in refilling the system when air penetrates into the reservoir. The advantage of this approach is that, provided water continuity is maintained between the vacuum gauge and the porous cup, the gauge will read directly the soil suction in the porous cup. However, the range of suction over which the tensiometer can be used is limited when long lengths of tube are needed (e.g. the correction required is 10 kPa/m between the tip and the gauge). The vacuum gauge will normally have an adjustable zero that is used to compensate for the difference in height between the gauge and the porous cup. The adjustment is made with the cup immersed, in water, to its mid-height. Without adjustment, the gauge can be used to measure small positive pressures (should they occur), but this practice will also reduce the range over which suctions can be recorded.

PRESSURE TRANSDUCER

Tensiometers employing electronic pressure transducers (Figure 2.32c) are similar to the vacuum gauge type except that pressure measurements are converted to a voltage measurable by a digital voltmeter. This means that automatic data logging by computer or chart recorder can be used. The sensitivity of the transducer can make it susceptible to the thermal expansion/contraction of the water in the system, and this can result in errors when continuous monitoring is necessary. The sensor used in electronic transducers is normally a semi-conductive resistor located inside or on the surface of a diaphragm. The resistance of these devices is (a) strain dependent, (b) a non-linear function of pressure and (c) highly sensitive to changes in temperature. Connecting the circuit to a Wheatstone bridge arrangement smooths the non-linearity and reduces the temperature sensitivity. However the zero and occasionally the sensitivity may change with time (a condition known as drift) and therefore frequent re-calibration may be required. Using an electronic transducer means that the sensor can be located as close to the porous cup as is practical, negating the need for a depth correction.

DE-AIRING (FREDLUND AND RAHARDJO, 1993)

To measure the suction accurately and quickly, it is necessary to de-air the tensiometer. To do this, the porous filter should be de-aired by immersing it in water, under a vacuum before assembling the equipment. Once the tube is full of water, a hand-held vacuum pump (see Figure 2.33) can be used to remove any large air bubbles which become trapped during the filling process. If the tube is carefully filled with water that has been previously de-aired, the operation is made considerably easier. If the sensor is buried (as in the case of electronic sensors), a purging system will be required to flush out any air that is trapped in the reservoir.

The response time of tensiometers is affected by (a) the sensitivity of the measuring gauge, (b) the volume of water in the system, (c) the permeability of the porous filter and (d) the amount of undissolved air in the system.

In the tensiometers manufactured by Soil Moisture Equipment, Inc. USA, the gauge is either a standard Bourdon type or an electronic transducer, the volume of water is variable depending on the tube diameter and length, and the porous ceramic filter has an air-entry value of 100 kPa. Response times in these tensiometers are about a few minutes, but they are limited to measuring suctions less than 100 kPa, limited by the air-entry value of porous cup and cavitation.

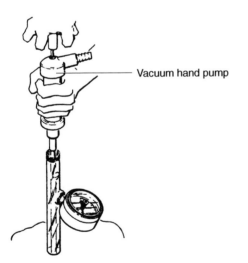

Vacuum hand pump

Figure 2.33 De-airing the tensiometer using a hand-held vacuum pump (courtesy Soilmoisture Equipment Corp.; after Fredlund and Rahardjo, 1993).

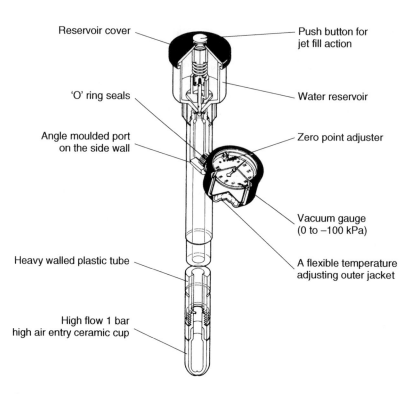

Reservoir cover

Push button for
jet fill action

'O' ring seals

Water reservoir

Angle moulded port
on the side wall

Zero point adjuster

Vacuum gauge
(0 to −100 kPa)

Heavy walled plastic tube

A flexible temperature
adjusting outer jacket

High flow 1 bar
high air entry ceramic cup

Figure 2.34 Jet-fill tensiometer (courtesy Soilmoisture Equipment Corp.; after
Fredlund and Rahardjo, 1993).

JET FILL TENSIOMETER (FREDLUND AND RAHARDJO, 1993)

Figure 2.34 shows a jet fill-type tensiometer. The jet fill type is an improved
model of the regular tensiometer. A water reservoir is provided at the top of
the tensiometer tube for the purpose of removing the air bubbles. The jet fill
mechanism is similar to the action of a vacuum pump. The accumulated air
bubbles are removed by pressing the button at the top to activate the jet fill
action. The jet fill action causes water to be injected from the water reservoir
to the tube of the tensiometer, and air bubbles move upward to the reservoir.

SMALL TIP TENSIOMETER (FREDLUND AND RAHARDJO, 1993)

A small tip tensiometer with a flexible coaxial tubing is shown in Figure 2.35.
The tensiometer is prepared for installation using a similar procedure to
that described for the regular tensiometer tube. A vacuum pump can be
initially used to remove air bubbles from the top of the tensiometer tube.
Subsequent removal of air bubbles can be performed by flushing through
the coaxial tube.

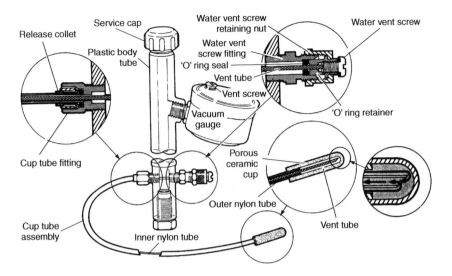

Figure 2.35 Small tip tensiometer with flexible coaxial tubing (courtesy Soilmoisture Equipment Corp.; after Fredlund and Rahardjo, 1993).

HIGH-SUCTION TENSIOMETER PROBE (IMPERIAL COLLEGE, RIDLEY AND BURLAND, 1993)

Ridley and Burland (1993) introduced a new high-suction tensiometer (see Figure 2.36) to measure suctions greater than 100 kPa directly in the laboratory. By pre-pressurizing the water in the reservoir to inhibit the formation of air within the tensiometer and taking advantage of the high tensile strength of water (Harvey *et al.*, 1944; Marinho and Chandler, 1995; Guan and Fredlund, 1997), matric suction of up to 1,800 kPa can be measured. Subsequently, Ridley and Burland (1996) also extended the use of this type of high-suction tensiometer probe in the field, together with a technique for excavating a borehole with a flat, horizontal bottom in compacted London clay. Ng *et al.* (2002b), Take and Bolton (2003) and Zhou et al. (2006) have developed miniature tensiometer suction probes to use in centrifuge model tests successfully.

Squeezing technique (Fredlund and Rahardjo, 1993)

The *osmotic suction* of a soil can be indirectly estimated by measuring the electrical conductivity of the pore water from the soil (Fredlund and Rahardjo, 1993). Pure water has a low electrical conductivity in comparison to pore water which contains dissolved salts. The electrical conductivity of the pore water from the soil can be used to indicate the total

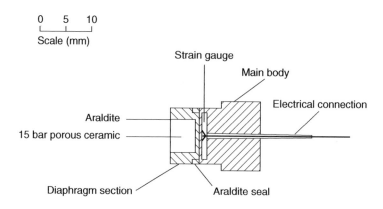

Figure 2.36 The Imperial College tensiometer (after Ridley and Burland, 1995; Ridley and Wray, 1995).

concentration of dissolved salts which is related to the osmotic suction of the soil.

The pore water in the soil can be extracted using a pore fluid squeezer which consists of a heavy-walled cylinder and piston squeezer (Figure 2.37). The electrical resistivity (or electrical conductivity) of the pore water is then measured. A calibration curve (Figure 2.38) can be used to relate the electrical conductivity to the osmotic pressure of the soil. The results of squeezing technique measurements appear to be affected by the magnitude of the extraction pressure applied. Krahn and Fredlund (1972) used an extraction pressure of 34.5 MPa in the osmotic suction measurements on the glacial till and Regina clay.

Concluding comments (Ridley and Wray, 1995)

Ridley and Wray (1995) point out that before choosing a particular instrument, it is important that the engineer decides whether to measure the total suction, the matric suction or both. The direct measurement of osmotic suction can be misleading, and it is probably sufficient to derive a value from the difference between the measured total and matric suctions. An appropriate device can then be chosen based on its range of measurement, simplicity (primarily for installation) and expense. Table 2.3 summarizes the principal methods of suction measurement in terms of the type of suction they measure, the range over which they are useable and the response time. In general, those instruments that make contact with the soil will usually measure matric suction, and those which do not make contact will usually measure total suction.

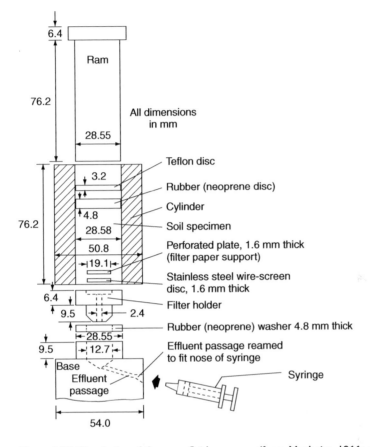

Figure 2.37 The design of the pore fluid squeezer (from Manheim, 1966; after Fredlund and Rahardjo, 1993).

The measurement of total suction requires special attention to the calibration procedure and equipment design. The psychrometer and the out-of-contact filter paper method are the only techniques available for measuring total suction, but both present difficulties if used in the low suction range.

It is in the interests of the engineer to measure the soil suction directly. Difficulties associated with doing this at atmospheric pressure coupled with uncertainties regarding the effect of raising the ambient pressure have resulted in the adoption by many (particularly for in situ measurements), of indirect methods (e.g. filter paper method, psychrometer and thermal conductivity sensor). However, the accuracy of the indirect methods is not always good and a large number of tests are often necessary to produce a reasonable degree of confidence in the estimate of suction. In addition,

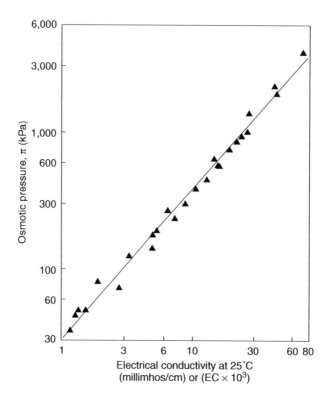

Figure 2.38 Osmotic pressure versus electrical conductivity relationship for pore water containing mixtures of dissolved salts (from *USDA Agricultural Handbook No. 60*, 1950; after Fredlund and Rahardjo, 1993).

indirect techniques such as the filter paper method and psychrometer rely on another indirect method (i.e. salt solutions) for calibration.

Tensiometers are the only devices that can be said to truly measure the matric suction in a direct manner, but theoretically, the maximum suction which standard tensiometers can measure is 100 kPa. Recent advances in the understanding of the direct measurement of soil suction have extended the range of such measurements to about 1.8 MPa in the laboratory environment.

Soil suction is quantitatively defined as the difference between the ambient air pressure and the soil pore water pressure. Although the pressure plate apparatus is regarded as a direct measurement of matric suction because it measures the absolute difference between ambient air pressure and soil pore water pressure, it does not measure negative water pressures directly. Having said that the axis-translation technique is the most common method of matric suction measurement/control in the laboratory, but in the field, it is not very practical to introduce a stable raised air pressure down a borehole.

Case study: comparisons of in situ suction measurements (Harrison and Blight, 2000)

Introduction

A number of soil suction measuring devices were employed to measure total and matric suction in situ by Harrison and Blight (2000). Two sites were selected, each underlain by different subsoil and groundwater conditions. Measurements of suction were taken over a period of 2 years, from which it was established that none of the instruments yielded comparable values of absolute suction or suction change. The instruments did, however, record suctions that follow seasonal wet and dry cycles.

Descriptions of the sites

Climate

Two sites, referred to as Site A and Site B, were selected to carry out the soil suction measurements in situ. Both sites are in South Africa and enjoy warm temperate conditions with summer rain. The area is in a water deficient region where annual evaporation exceeds precipitation. Some climatic statistics are as follows:

> Annual average rainfall: 713 mm
> Annual Apan evaporation: 2,306 mm
> Annual average temperature: 16°C
> Mean of warmest month: 22°C
> Mean of coldest month: 10°C

Subsoils

The subsoil underlying Site A comprises a deep residual andesite in the form of a slightly clayey silt, which classifies as MH according to the Unified Soil Classification (USC) system. The regional water table is at some 15–20 m below the ground surface in this area. Site B is underlain by transported colluvium comprising clayey sand, which classifies as SC according to the USC system. A ferruginised (ferricrete) layer is present beneath the colluvium at a depth of approximately 1 m below the ground surface. The regional water table in the area is fairly deep. However, the ferricrete horizon forms a relatively impermeable layer on top of which a perched water table develops in the wet summer seasons.

In situ suction measurements

The instruments used to measure the soil suction in situ at the two sites included thermocouple psychrometer probes, gypsum resistance blocks and fibreglass moisture cells, all of which employ indirect methods to measure soil suction. In addition to these instruments, direct suction measurements were carried out at Site B employing a jet fill tensiometer and Imperial College (IC) tensiometer.

Each psychrometer, gypsum block and fibre cell was inserted into a 100 mm diameter hand augered hole drilled to a depth of 600 mm below the ground surface. The soil removed from the hole was lightly tamped around each of the sensors and the wire electrical leads were sealed in a jar buried just below the surface.

The jet fill tensiometer was inserted into a pre-drilled hole of the same diameter as the tensiometer probe. The hole was prepared just prior to inserting the instrument. The IC tensiometer was bedded into the base of a hole augered near to the sensors, which was also drilled just prior to taking a reading.

Periodic readings from the buried sensors were taken over a period of 2 years and, based on the calibration data, converted to suction pressures. These in situ suction readings are presented in Figure 2.39a and b for Sites A and B, respectively. Neither the jet fill nor the IC tensiometers were able to function at Site A; due to the high suctions encountered in the area, however, some readings were possible with the jet fill tensiometer at Site B when a perched water table developed in the wet summer seasons. Cavitation occurred in the IC tensiometer and no meaningful measurements were possible with this instrument for the soil encountered at this site.

Suctions recorded with the jet fill tensiometer are illustrated in Figure 2.40. Readings were only possible when a perched water table developed on the underlying ferricrete layer, and the instrument was unable to record soil suctions above 75 kPa.

Superimposed on the Figures 2.39 and 2.40 is the mean monthly rainfall for the region. It is evident from the two figures that none of the devices measure comparable soil suctions. Whilst it can be argued that since the psychrometer measures total suction and the others matrix, the psychrometer should record higher suctions than the other instruments, which in most instances it does. However, total suctions should be consistently greater than matric suctions, which is not evident from the data. Matric suction measurements are similarly not in agreement for each of the two indirect and direct instruments and different calibration techniques. Possible reasons for this include hysteresis and moisture continuity.

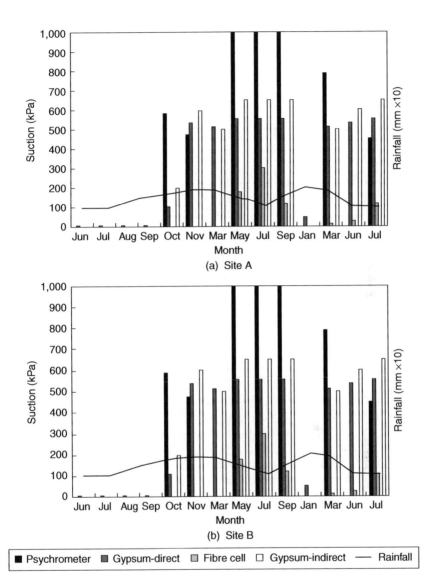

Figure 2.39 In situ suction readings at Sites A and B (after Harrison and Blight, 2000).

Effects of hysteresis and moisture continuity

Hysteresis

Unsaturated soils undergo substantial hysteresis in suction during wetting and drying, as do the porous sensors of the measuring instruments.

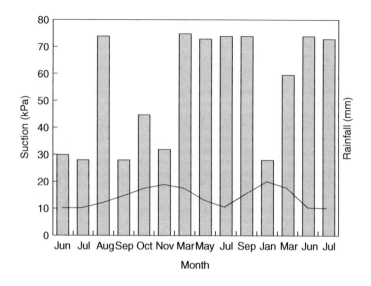

Figure 2.40 Suction readings using the jet fill tensiometer at Site B (after Harrison and Blight, 2000).

Thus, depending on whether the soil is wetting or drying, its suction for the same water content will be different. Similarly, the readout from the porous measuring sensor will also record a different value when it is wetting to that when it is drying. Due to the large time (monthly) intervals over which the suction readings were taken, it was not possible to establish whether the soil was in the process of wetting up or drying out at the time of taking the readings. An attempt was nevertheless made to account for the hysteresis effects of the instruments by assuming that they wet up in the summer rainfall months and dry out as the drier winter approaches.

Moisture continuity

The devices employed to indirectly measure matric suction utilize a porous sensor, its purpose being to communicate, via the pores, with the moisture in the soil. For this to occur, moisture must be transferred to the soil from the sensor or from the sensor to the soil. If moisture continuity is not maintained between the measuring device and the soil, then the instrument will not record the matric suction of the soil.

As far as possible, the suction–water content curve of a porous sensor should be similar to that of the soil, and with similar hysteresis character-

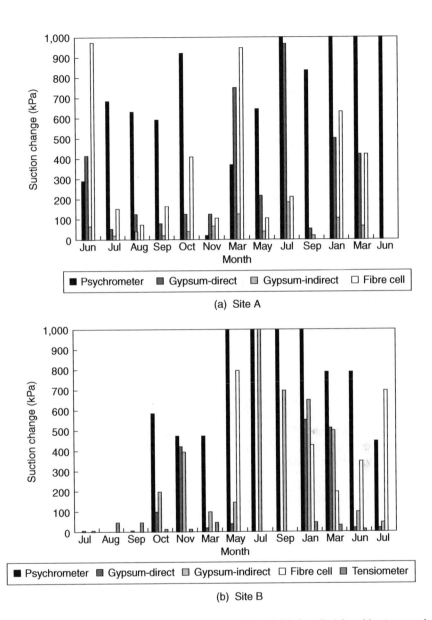

Figure 2.41 Monthly suction changes at (a) Site A and (b) Site B (after Harrison and Blight, 2000).

istics. However, the limited range of sensors commercially available makes this difficult to achieve.

It may be argued that in many instances absolute suction is less relevant to the prediction of unsaturated soil behaviour than the moduli of suction change. Month-to-month suction changes within the soil at each of the two sites have been plotted in Figure 2.41a and b. Again it is evident from the figure that none of the devices recorded similar changes in suction over the 2-year period.

What is perhaps noteworthy, and the very least to be expected, is that suction measurements do change in sympathy with the rainfall patterns, with relatively high and low suctions being recorded in the dry and wet months respectively.

Summary of the case study

Three indirect and one direct in situ soil suction measuring devices were employed to record, over a period of 2 years, the soil suction at two sites underlain by two soil types with different groundwater conditions. None of the instruments recorded comparable suctions nor were suction changes inferred from the various instruments comparable. Suction was, however, seen to change in response to the wet and dry seasons. There are many reasons for the lack of agreement between the various devices, some of which have been discussed earlier, while others have yet to be identified.

The study shows that after close to a century of efforts to measure soil suction in situ, it seems that there is still no consistent, simple and reliable method of doing this. The various devices that are available indicate the direction of suction change, but they still lack a method for reliably measuring suction magnitude.

Chapter 3

Flow laws, seepage and state-dependent soil–water characteristics

Sources

This chapter is made up of verbatim extracts from the following sources for which copyright permissions have been obtained as listed in the Acknowledgements.

- Fredlund, D.G. (2000). The 1999 R.M. Hardy Lecture: The implementation of unsaturated soil mechanics into geotechnical engineering. *Can. Geotech. J.*, 37, 963–986.
- Fredlund, D.G. (2002). Use of soil–water characteristic curves in the implementation of unsaturated soil mechanics. *Proc. 3rd Int. Conf. on Unsat. Soils, Recife, Brazil*, Vol. 3, Balkema, Rotterdam, pp. 887–902.
- Fredlund, D.G. and Rahardjo, H. (1993). *Soil Mechanics for Unsaturated Soils*. John Wiley & Sons, Inc., New York, p. 517.
- Fredlund, D.G., Fredlund, M.D. and Zakerzadeh, N. (2001b). Predicting the permeability function for unsaturated soils. *Proc. Int. Symp. on Suction, Swelling, Permeability and Structured Clays*, IS-Shizuoka, Shizuoka, Japan, pp. 215–222.
- Fredlund, D.G., Rahardjo, H., Leong, E.C. and Ng, C.W.W. (2001a). Suggestions and recommendations for the interpretation of soil–water characteristic curves. *Proc. 14th Southeast Asian Geotech. Conf., Hong Kong*, Vol. 1, Balkema, Rotterdam, pp. 503–508.
- Hillel, D. (1998). *Introduction to Environmental Soil Physics*. Academic Press, San Diego, CA, USA, p. 364.
- Khanzode, R.M., Vanapalli, S.K. and Fredlund, D.G. (2002). Measurement of soil–water characteristic curve for fine-grained soils using a small centrifuge. *Can. Geotech. J.*, 39(5), 1209–1217.
- Lam, L., Fredlund, D.G. and Barbour, S.L. (1987). Transient seepage model for saturated–unsaturated soil systems: a geotechnical engineering approach. *Can. Geotech. J.*, 24, 565–580.
- Ng, C.W.W. and Pang, Y.W. (2000a). Experimental investigations of the soil–water characteristics of a volcanic soil. *Can. Geotech. J.*, 37, 1252–1264.

- Ng, C.W.W. and Shi, Q. (1998). A numerical investigation of the stability of unsaturated slopes subject to transient seepage. *Computers and Geotechnics*, Vol. 22, 1–28.
- Ng, C.W.W., Wang, B. and Gong, B.W. (2001a). A new triaxial apparatus for studying stress effects on soil–water characteristics of unsaturated soils. *Proc. 15th Conf. of Int. Soc. Soil Mech. and Geotech. Engrg.*, Istanbul, Vol. 1, Balkema, Rotterdam 611–614.
- Zhan, L.T. and Ng, C.W.W. (2004). Analytical analysis of rainfall infiltration mechanism in unsaturated soils. *Int. J. Geomech., ASCE.* Vol. 4, No. 4, 273–284.

Flow laws for water and air (Fredlund and Rahardjo, 1993)

Introduction

In the pores of an unsaturated soil, there are two fluid phases: water and air. The analysis of fluid flow requires a law to relate the flow rate to driving potential using appropriate coefficients. The air in an unsaturated soil may be in an occluded form when the degree of saturation is relatively high. At lower degrees of saturation, the air phase is predominantly continuous. The form of the flow laws may vary for each of these cases. In addition, there may be the movement of air through the water phase, which is referred to as air diffusion through the pore water in Figure 3.1.

A knowledge of the driving potentials that cause air and water to flow or to diffuse is necessary for understanding the flow mechanisms. The driving

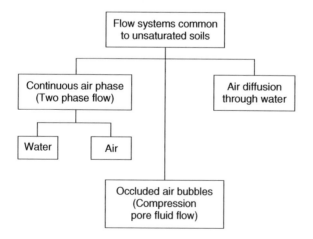

Figure 3.1 Flow systems common to unsaturated soils (after Fredlund and Rahardjo, 1993).

potentials of the water phase are given in terms of 'heads' in this chapter. Water flow is caused by a hydraulic head gradient, where the hydraulic head consists of an elevation head plus a pressure head. A diffusion process is usually considered to occur under the influence of a chemical concentration or a thermal gradient. Water can also flow in response to an electrical gradient (Casagrande, 1952).

The concept of hydraulic head and the flow of air and water through unsaturated soils are presented here. A brief discussion on the diffusion process is also presented, together with its associated driving potential. Flows due to chemical, thermal and electrical gradients are beyond the scope of this book.

Flow of water in soils (Fredlund and Rahardjo, 1993)

Several concepts have been used to explain the flow of water through an unsaturated soil. These are

- a water content gradient
- a matric suction gradient
- a hydraulic head gradient.

They have all been considered as driving potentials. However, it is important to use the form of the flow law that most fundamentally governs the movement of water.

Water content gradient

A gradient in water content has sometimes been used to describe the flow of water through unsaturated soils. It is assumed that water flows from a point of high water content to a point of lower water content. This type of flow law, however, does not have a fundamental basis since water can also flow from a region of low water content to a region of high water content when there are variations in the soil types involved, hysteretic effects, or stress history variations are encountered. Therefore, a water content gradient should not be used as fundamental driving potential for the flow of water (Fredlund, 1981).

Matric suction gradient

In an unsaturated soil, a matric suction gradient has sometimes been considered to be the driving potential for water flow. However, the flow of water does not fundamentally and exclusively depend upon the matric suction gradient. Three hypothetical cases where the air and water pressure gradients

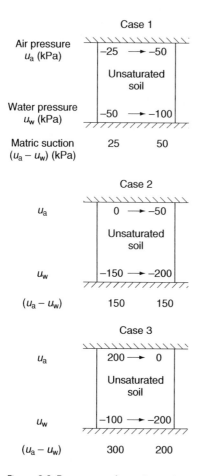

Figure 3.2 Pressure and matric suction gradients across an unsaturated soil element (after Fredlund and Rahardjo, 1993).

are controlled across an unsaturated soil element at a constant elevation are illustrated in Figure 3.2.

In all cases, the air and water pressures on the left-hand side are greater than the pressures on the right-hand side.

The matric suction on the left-hand side may be smaller than on the right-hand side (Case 1), equal to the right-hand side (Case 2) or larger than on the right-hand side (Case 3). However, air and water will flow from left to right in response to the pressure gradient in the individual phases, regardless of the matric suction gradient. Even in Case 2, where the matric suction gradient is zero, air and water will still flow.

Hydraulic head gradient

Flow can be defined more appropriately in terms of a hydraulic head gradient (i.e. a pressure head gradient in this case) for *each* of the phases. Therefore, the matric suction gradient is not the fundamental driving potential for the flow of water in an unsaturated soil. In the special case where the air pressure gradient is zero, the matric suction gradient is numerically equal to the pressure gradient in the water. This is the common situation in nature and is probably the reason for the proposal of the matric suction form for water flow. However, the elevation head component has then been omitted.

The flow of water through a soil is not only governed by the pressure gradient but also by the gradient due to elevation differences. The pressure and elevation gradients are combined to give a hydraulic head gradient as the fundamental driving potential. The hydraulic head gradient in a specific fluid phase is the driving potential for flow in that phase. This is equally true for saturated and unsaturated soils.

Driving potential for water phase

The driving potential for the flow of water defines the energy or capacity to do work. The energy at a point is computed relative to a datum. The datum is chosen arbitrarily because only the gradient in the energy between two points is of importance in describing flow.

A point in the water phase had three primary component of energy, namely, gravitational, pressure and velocity. Figure 3.3 shows point A in the water phase which is located at an elevation, *y*, above an arbitrary datum.

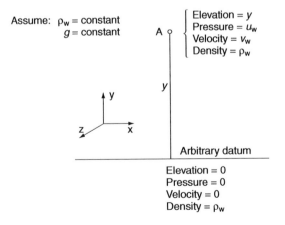

Figure 3.3 Energy at point A in the *y*-direction relative to an arbitrary datum (after Fredlund and Rahardjo, 1993).

The total energy [i.e. gravitational (potential) energy, pressure energy and velocity energy] at point A can be expressed as energy per unit weight, which is called a potential or a hydraulic head. The hydraulic head, h_w, at point A is obtained by dividing the energy equation by the weight of water at point A:

$$h_w = y + \frac{u_w}{\rho_w g} + \frac{v_w^2}{2g} \tag{3.1}$$

where
h_w = hydraulic head or total head
g = gravitational acceleration
u_w = pore water pressure at point A
v_w = flow rate of water at point A (i.e. in the y direction)
y = elevation of point A above the datum
ρ_w = density of water at point A.

The hydraulic head consists of three components, namely, the gravitational head, y, the pressure head $(u_w/\rho_w g)$ and the velocity head $(v_w^2/2g)$. The velocity head in a soil is negligible in comparison with the gravitational and the pressure heads. The above equation can therefore be simplified to yield an expression for the hydraulic head at any point in the soil mass:

$$h_w = y + \frac{u_w}{\rho_w g} \tag{3.2}$$

The heads expressed in this equation have the dimension of length. Hydraulic head is a measurable quantity, the gradient of which causes flow in saturated and unsaturated soils. To illustrate how water flow through a soil mass, Figure 3.4 considers two arbitrary points A and B at which a tensiometer and a piezometer are used to measure its in situ pore water pressure, respectively. The tensiometer at A is used to measure the pore water pressure when the pressure is negative, whereas the piezometer at B is used to measure the pore water pressure when the pore water pressure is positive.

The distance between the elevation of the point under consideration and the datum indicates the elevation head (i.e. y_A and y_B).

The water level in the measuring device will rise or drop, depending upon the pore water pressure at the point under consideration. For example, the water level in the piezometer rises above the elevation of point B at a distance equal to the positive pore water pressure head at point B. Alternately, the water level in the tensiometer drops below the elevation of point A to a distance equal to the negative pore water pressure head at point A. The distance between the water level in the measuring device and the datum is the sum of the gravitational and pressure heads (i.e. the hydraulic head).

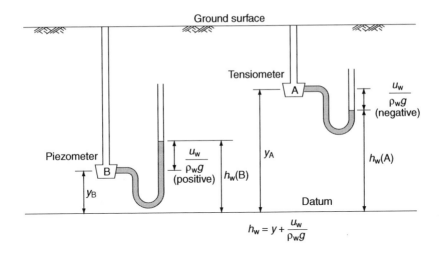

Figure 3.4 Concept of potential and head for saturated soils (after Fredlund and Rahardjo, 1993).

In Figure 3.4, point A has a higher total head than point B [i.e. $h_w(A) > h_w(B)$]. Water will flow from point A to point B due to the total head gradient between these two points. The driving potential causing flow in the water phase has the same form for both saturated (i.e. point B) and unsaturated (i.e. point A) soils (Freeze and Cherry, 1979). Water will flow from a point of high total head to a point of low total head, regardless of whether the pore water pressures are positive or negative.

Osmotic suction has sometimes been included as a component in the total head equation for flow. However, it is better to visualize the osmotic suction gradient as the driving potential for the osmotic diffusion process (Corey and Kemper, 1961). Osmotic diffusion is a process where ionic or molecular constituents move as a result of their kinetic activity. For example, an osmotic gradient across a semi-permeable membrane causes the movement of water through the membrane. On the other hand, the bulk flow of solutions (i.e. pure water and dissolved salts) in the absence of a semi-permeable membrane is governed by the hydraulic head gradient. Therefore, it would appear superior to analyse the bulk flow of water separately from the osmotic diffusion process since two independent mechanisms are involved (Corey, 1977).

Darcy's law for unsaturated soils

The flow of water in a saturated soil is commonly described using Darcy's law (1856). He postulated that the rate of water flow through a soil mass was proportional to the hydraulic head gradient:

$$v_w = -k_w \frac{\partial h_w}{\partial y} \tag{3.3}$$

where
v_w = flow rate of water
k_w = coefficient of permeability with respect to the water phase
$\partial h_w / \partial y$ = hydraulic head gradient in the y-direction, which can be designated as i_{wy}.

The coefficient of proportionality between the flow rate of water and the hydraulic head gradient is called the coefficient of permeability, k_w. The coefficient of permeability is relatively constant for a specific saturated soil. The above equation can also be written for the x- and z-directions. The negative sign in the equation indicates that water flows in the direction of a decreasing hydraulic head.

Darcy's law also applies for the flow of water through an unsaturated soil (Buckingham, 1907; Richards, 1931; Childs and Collis-George, 1950). In a *saturated* soil, the coefficient of permeability is a function of the void ratio (Lambe and Whitman, 1979). However, the coefficient of permeability of a saturated soil is generally assumed to be a constant when analysing problems such as transient flow. In an *unsaturated* soil, the coefficient of permeability is significantly affected by combined changes in the void ratio and the degree of saturation (or water content) of the soil. Water flows through the pore space filled with water; therefore, the percentage of the voids filled with water is an important factor. As a soil becomes unsaturated, air first replaces some of the water in the large pores, and this causes the water to flow through the smaller pores with an increased tortuosity to the flow path. A further increase in the matric suction of the soil leads to a further decrease in the pore volume occupied by water. In other words, the air–water interface is drawn closer and closer to the soil particles as shown in Figure 3.5. As a result, the coefficient of permeability with respect to the water phase decreases rapidly as the space available for water flow reduces.

Coefficient of permeability with respect to the water phase

The coefficient of permeability with respect to the water phase, k_w, is a measure of the space available for water to flow through the soil. The coefficient of permeability depends upon the properties of the fluid and the properties of the porous medium. Different types of fluid (e.g. water and oil) or different types of soil (e.g. sand and clay) produce different values for the coefficient of permeability, k_w.

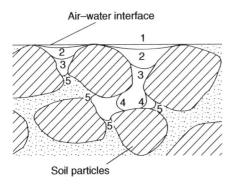

Figure 3.5 Development of an unsaturated soil by the withdrawal of the air–water interface at different stages of matric suction or degree of saturation (i.e. stages 1–5) (from Childs, 1969; after Fredlund and Rahardjo, 1993).

FLUID AND POROUS MEDIUM COMPONENTS

The coefficient of permeability with respect to the water phase, k_w, can be expressed in terms of the intrinsic permeability, K:

$$k_w = \frac{\rho_w g}{\mu_w} K \tag{3.4}$$

where
μ_w = absolute (dynamic) viscosity of water
K = intrinsic permeability of the soil.

The above equation shows the influence of the fluid density, ρ_w, and the fluid viscosity, μ_w, on the coefficient of permeability, k_w. The intrinsic permeability of a soil, K, represents the characteristics of the porous medium and is independent of the fluid properties.

   The fluid properties are commonly assumed to be constant during the flow process. The characteristics of the porous medium are a function of the volume–mass properties of the soil. The intrinsic permeability is used in numerous disciplines. However, in geotechnical engineering, the coefficient of permeability, k_w, is the most commonly used term.

RELATIONSHIP BETWEEN PERMEABILITY AND VOLUME–MASS
PROPERTIES

The coefficient of permeability, k_w, is a function, f, of any two of three possible volume–mass properties (Lloret and Alonso, 1980; Fredlund, 1981):

$$k_w = f_1(S, e) \quad \text{or} \quad k_w = f_2(e, w) \quad \text{or} \quad k_w = f_3(w, S)$$

where
S = degree of saturation
e = void ratio
w = water content
f = a mathematical function.

EFFECT OF VARIATIONS IN DEGREE OF SATURATION ON PERMEABILITY

The coefficient of permeability of an unsaturated soil can vary considerably during a transient process as a result of changes in the volume–mass properties. If the change in void ratio in an unsaturated soil may be assumed to be small, its effect on the coefficient of permeability may be secondary. However, the effect of a change in degree of saturation may be highly significant. As a result, the coefficient of permeability is often described as a singular function of the degree of saturation, S, or the volumetric water content, θ_w.

A change in matric suction can produce a more significant change in the degree of saturation or water content than can be produced by a change in net normal stress. The degree of saturation has been commonly described as a function of matric suction. The relationship is called the matric suction versus degree of saturation curve, as shown in Figure 3.6.

Numerous semi-empirical equations for the coefficient of permeability have been derived using either the matric suction versus degree of saturation curve or the soil–water characteristic curve (SWCC). In either case, the soil pore size distribution forms the basis for predicting the coefficient of permeability. The pore size distribution concept is somewhat new to geotechnical engineering. The pore size distribution has been used in other disciplines to give reasonable estimates of the permeability characteristics of a soil (Fredlund and Rahardjo, 1993).

The prediction of the coefficient of permeability from the matric suction versus degree of saturation curve is discussed first, followed by the coefficient of permeability prediction using the SWCC.

RELATIONSHIP BETWEEN COEFFICIENT OF PERMEABILITY AND DEGREE OF SATURATION

Coefficient of permeability functions obtained from the matric suction versus degree of saturation curve have been proposed by Burdine (1952) and Brooks and Corey (1964). The matric suction versus degree of saturation curve exhibits hysteresis. Only the drainage curve is used in their derivations. In addition, the soil structure is assumed to be incompressible.

There are three soil parameters that can be identified from the matric suction versus degree of saturation curve. These are the air entry value of the soil, $(u_a - u_w)_b$, the residual degree of saturation, S_r, and the pore size distribution index, λ. These parameters can readily be visualized if

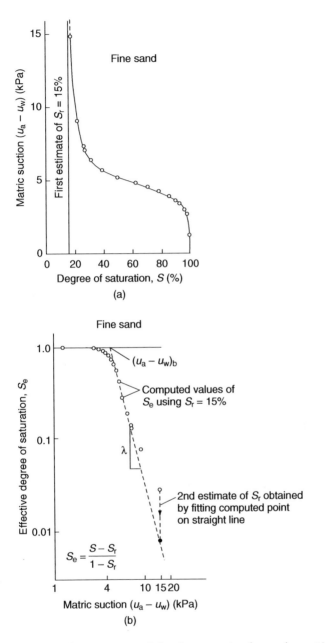

Figure 3.6 Determination of the air-entry value $(u_a - u_w)_b$, residual degree of saturation S_r and pore air size distribution index λ. (a) Matric suction versus degree of saturation curve; (b) effective degree of saturation versus matric suction curve (from Brooks and Corey, 1964; after Fredlund and Rahardjo, 1993).

the saturation condition is expressed in terms of an effective degree of saturation, S_e, (Corey, 1954) (see Figure 3.6b):

$$S_e = \frac{S - S_r}{1 - S_r} \tag{3.5}$$

where
S_e = effective degree of saturation
S_r = residual degree of saturation.

The residual degree of saturation, S_r, is defined as the degree of saturation at which an increase in matric suction does not produce a significant change in the degree of saturation. The values for all degree of saturation variables used in the above equation are in decimal form.

The effective degree of saturation can be computed by first estimating the residual degree of saturation (see Figure 3.6b). The effective degree of saturation is then plotted against the matric suction, as illustrated in Figure 3.6b. A horizontal and a sloping line can be drawn through the points. However, points at high matric suction values may not lie on the straight line used for the first estimate of the residual degree of saturation. Therefore, the point with the highest matric suction must be forced to lie on the straight line by estimating a new value of S_r (see Figure 3.6b). A second estimate of the residual degree of saturation is then used to recompute the values for the effective degree of saturation. A new plot of matric suction versus effective degree of saturation curve can then be obtained. The above procedure is repeated until all of the points on the sloping line constitute a straight line. This usually occurs by the second estimate of the residual degree of saturation.

The air entry value of the soil, $(u_a - u_w)$, is the matric suction value that must be exceeded before air recedes into the soil pores. The air entry value is also referred to as the 'displacement pressure' in petroleum engineering or the 'bubbling pressure' in ceramics engineering (Corey, 1977). It is a measure of the maximum pore size in a soil. The intersection point between the straight sloping line and the saturation ordinate (i.e. $S_e = 1.0$) in Figure 3.6b defines the air entry value of the soil. The sloping line for the points having matric suctions greater than the air entry value can be described by the following equation:

$$S_e = \left\{ \frac{(u_a - u_w)_b}{(u_a - u_w)} \right\}^\lambda \quad \text{for } (u_a - u_w) \geq (u_a - u_w)_b \tag{3.6}$$

where
λ = pore size distribution index, which is defined as the negative slope of the effective degree of saturation, S_e, versus matric suction, $(u_a - u_w)$, curve.

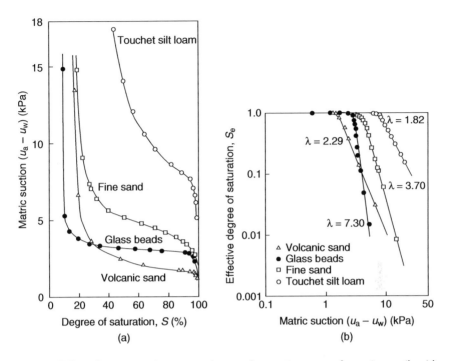

Figure 3.7 Typical matric suction versus degree of saturation curves for various soils with their corresponding λ values. (a) Matric suction versus degree of saturation curves; (b) effective degree of saturation versus matric suction (from Brooks and Corey, 1964; after Fredlund and Rahardjo, 1993).

Soils with a wide range of pore sizes have a small value for λ. The more uniform the distribution of the pore sizes in a soil, the larger is the value for λ. Some typical λ values for various soils which have been obtained from matric suction versus degree of saturation curves are shown in Figure 3.7.

The coefficient of permeability with respect to the water phase, k_w, can be predicted from the matric suction versus degree of saturation curves as follows (Brooks and Corey, 1964):

$$k_w = k_s \quad \text{for } (u_a - u_w) \leq (u_a - u_w)_b \tag{3.7}$$

$$k_w = k_s S_e^\delta \quad \text{for } (u_a - u_w) \geq (u_a - u_w)_b \tag{3.8}$$

where
k_s = coefficient of permeability with respect to the water phase for the soil at saturation (i.e. $S = 100$ per cent)
δ = an empirical constant.

Table 3.1 Suggested values of the constant δ and the pore size distribution index λ for various soils (after Fredlund and Rahardjo, 1993)

Soils	δ values	λ values	Source
Uniform sand	3.0	∞	Irmay (1954)
Soil and porous rocks	4.0	2.0	Corey (1954)
Natural sand deposits	3.5	4.0	Averjanov (1950)

The empirical constant, δ, is related to the pore size distribution index:

$$\delta = \frac{2 + 3\lambda}{\lambda} \tag{3.9}$$

Table 3.1 presents several δ values and their corresponding pore size distribution indices, λ, for various soil types.

RELATIONSHIP BETWEEN WATER COEFFICIENT OF PERMEABILITY AND MATRIC SUCTION

The coefficient of permeability with respect to the water phase, k_w, can also be expressed as a function of the matric suction by substituting the effective degree of saturation, S_e, into the permeability function (Brooks and Corey, 1964). Several other relationships between the coefficient of permeability and matric suction have also been proposed (Gardner, 1958a; Arbhabhirama and Kridakorn, 1968) and these are summarized in Table 3.2.

Table 3.2 Relationships between water coefficient of permeability and matric suction (after Fredlund and Rahardjo, 1993)

Equations	Source	Symbols
$k_w = k_s$ for $(u_a - u_w) \leq (u_a - u_w)_b$ $$k_w = k_s \left\{ \frac{(u_a - u_w)_b}{(u_a - u_w)} \right\}^{\eta}$$ for $(u_a - u_w) > (u_a - u_w)_b$	Brooks and Corey (1964)	η = empirical constant $\eta = 2 + 3\lambda$
$$k_w = \frac{k_s}{1 + a\left\{ \frac{(u_a - u_w)}{\rho_w g} \right\}^n}$$	Gardner (1958a)	a, n = constant
$$k_w = \frac{k_s}{\left\{ \frac{(u_a - u_w)}{(u_a - u_w)_b} \right\}^{n'} + 1}$$	Arbhabhirama and Kridakom (1968)	n' = constant

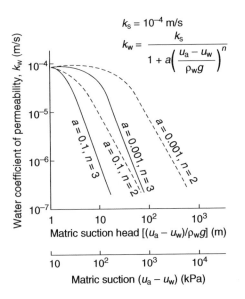

$$k_s = 10^{-4} \text{ m/s}$$

$$k_w = \frac{k_s}{1 + a\left(\dfrac{u_a - u_w}{\rho_w g}\right)^n}$$

Figure 3.8 Gardner's equation for the water coefficient of permeability as a function of the matric suction (after Fredlund and Rahardjo, 1993).

The relationship between the coefficient of permeability and matric suction proposed by Gardner (1958a) is presented in Figure 3.8. The equation provides a flexible permeability function which is defined by two constants, 'a' and 'n.' The constant 'n' defines the slope of the function, and the 'a' constant is related to the breaking point of the function. Four typical functions with differing values of 'a' and 'n' are illustrated in the above figure. The permeability functions are written in terms of matric suction; however, these equations could also be written in terms of total suction.

RELATIONSHIP BETWEEN WATER COEFFICIENT OF PERMEABILITY AND VOLUMETRIC WATER CONTENT

The water phase coefficient of permeability, k_w, can also be related to the volumetric water content, θ_w (Buckingham, 1907; Richards, 1931; Moore, 1939). A coefficient of permeability function, $k_w(\theta_w)$, has been proposed using the configurations of the pore space filled with water (Childs and Collis-George, 1950). The soil is assumed to have a random distribution of pores of various sizes and an incompressible soil structure. The permeability function, $k_w(\theta_w)$, is written as the sum of a series of terms obtained from the statistical probability of interconnections between water-filled pores of varying sizes.

The volumetric water content, θ_w, can be plotted as function of matric suction, $(u_a - u_w)$, and the plot is called the SWCC. Therefore, the permeability function, $k_w(\theta_w)$, can also be expressed in terms of matric suction (Marshall, 1958; Millington and Quirk, 1959, 1961). In other words, the soil–water characteristic curve can be visualized as an indication of the configuration of water-filled pores. A theoretical relationship between the water coefficient of permeability and the volumetric water content can be obtained using the SWCC (Kunze et al., 1968; Fredlund and Rahardjo, 1993). The comparisons between the computed and measured water permeabilities for a sand specimen are illustrated in Figure 3.9.

HYSTERESIS OF THE PERMEABILITY FUNCTION

The coefficient of permeability, k_w, is generally assumed to be uniquely related to the degree of saturation, S, or the volumetric water content, θ_w.

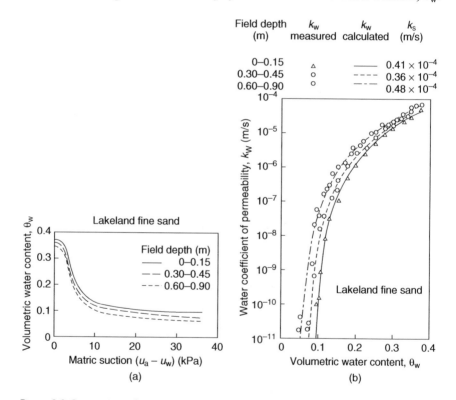

Figure 3.9 Comparisons between calculated and measured unsaturated permeabilities for Lakeland fine sand. (a) Soil–water characteristic curves; (b) water coefficient of permeability as a function of volumetric water content (from Elzeftawy and Cartwright, 1981; after Fredlund and Rahardjo, 1993).

This assumption may be reasonable since the volume of water flow is a direct function of the volume of water in the soil. The relationships between the degree of saturation (or volumetric water content) and the coefficient of permeability appear to exhibit little hysteresis (Nielsen and Biggar, 1961; Topp and Miller, 1966; Corey, 1977; Hillel, 1982). Nielsen *et al.* (1972) stated, 'The function $k_w(\theta_w)$ is well behaved, inasmuch as for coarse-textured soils, it is approximately the same for both wetting and drying.' However, this is not the case for the relationship between the water coefficient of permeability, k_w, and matric suction, $(u_a - u_w)$. Since there is hysteresis in the relationship between the volume of water in a soil and the stress state [i.e. namely, $(u_a - u_w)$], there will also be hysteresis in the relationship between the coefficient of permeability and matric suction.

The degree of saturation or volumetric water content shows significant hysteresis when plotted versus matric suction, as shown in Figure 3.10.

As a result, the coefficient of permeability, which is directly related to the volumetric water content or degree of saturation, will also show significant hysteresis when plotted versus matric suction. The above figures demonstrate a similar hysteresis form for both the volumetric water content, θ_w, and the coefficient of permeability, k_w, when plotted against matric suction. However, if the coefficient of permeability is cross-plotted against volumetric water content, the resulting plot shows essentially no hysteresis, as demonstrated in Figure 3.11.

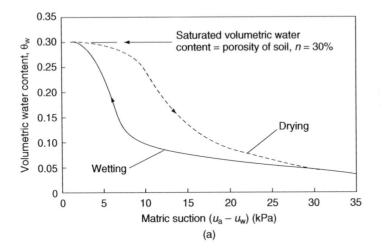

Figure 3.10 Similar hysteresis forms in the volumetric water content and water coefficient of permeability when plotted as a function of matric suction for a naturally deposited sand. (a) Volumetric water content versus matric suction; (b) water coefficient of permeability versus matric suction (from Liakopoulos, 1965; after Fredlund and Rahardjo, 1993).

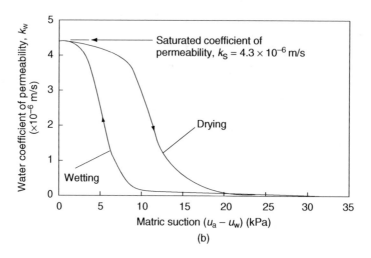

Figure 3.10 (Continued).

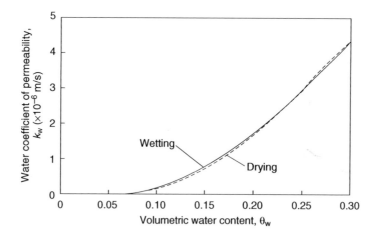

Figure 3.11 Essentially no hysteresis is shown in the relationship between water coefficient of permeability versus volumetric water content (after Fredlund and Rahardjo, 1993).

Flow of air in soils (Fredlund and Rahardjo, 1993)

Introduction

Fredlund and Rahardjo (1993) point out that the air phase of an unsaturated soil can be found in two forms:

- the continuous air phase form
- the occluded air bubble form.

The air phase generally becomes continuous as the degree of saturation reduces to around 85 per cent or lower (Corey, 1957). The flow of air through an unsaturated soil commences at this point.

Under naturally occurring conditions, the flow of air through a soil may be caused by factors such as variations in barometric pressure, water infiltration by rain that compresses the air in the soil pores and temperature changes. The flow of air in compacted fills may be due to applied loads.

When the degree of saturation is above about 90 per cent, the air phase becomes occluded, and air flow is reduced to a diffusion process through the pore water (Matyas, 1967).

Driving potential for air phase

The flow of air in the continuous air phase form is governed by a concentration or pressure gradient. The elevation gradient has a negligible effect. The pressure gradient is most commonly considered as the only driving potential for the air phase. Both Fick's and Darcy's laws have been used to describe the flow of air through a porous media.

Fick's law for air phase

Fick's law (1855) is often used to describe the diffusion of gases through liquids. A modified form of Fick's law can be applied to the air flow process. Fick's first law states that the rate of mass transfer of the diffusing substance across a unit area is proportional to the concentration gradient of the diffusing substance.

In the case of air flow through an unsaturated soil, the porous medium (i.e. soil) can be used as the reference in order to be consistent with the permeability concept for the water phase. This means that the mass rate of flow and the concentration gradient in the air are computed with respect to a unit area and a unit volume of the soil:

$$J_a = -D_a \frac{\partial C}{\partial y} \tag{3.10}$$

where
J_a = mass rate of air flowing across a unit area of the soil
D_a = transmission constant for air flow through a soil
C = concentration of the air expressed in terms of the mass of air per unit volume of soil
$\partial C/\partial y$ = concentration gradient in the y-direction.

The negative sign in Equation (3.10) indicates that air flows in the direction of a decreasing concentration gradient. Equation (3.10) can similarly be written for the x- and z-directions.

The concentration of air with respect to a unit volume of the soil can be written as

$$C = \frac{M_a}{V_a/(1-S)n} \tag{3.11}$$

where
M_a = mass of air in the soil
V_a = volume of air in the soil
S = degree of saturation
n = porosity of the soil.

Substituting the density of air, ρ_a, for (M_a/V_a) in Equation (3.11) gives

$$C = \rho_a(1-S)\,n \tag{3.12}$$

Air density is related to the absolute air pressure in accordance with the gas law $\rho_a = \omega_a \bar{u}_a/RT$. Therefore, the concentration gradient can also be expressed with respect to a pressure gradient in the air. The gauge air pressure is used in reformulating the original Fick's law, since only the gradient is of importance.

$$J_a = -D_a \frac{\partial C}{\partial u_a} \frac{\partial u_a}{\partial y} \tag{3.13}$$

where
u_a = pore air pressure
$\partial u_a/\partial y$ = pore air pressure gradient in the y-direction (or similarly in the x- and z-directions).

A modified form of Fick's law is obtained from Equation (3.13) by introducing a coefficient of transmission for air flow through soils, D_a^*:

$$D_a^* = D_a \frac{\partial C}{\partial u_a} \tag{3.14}$$

or

$$D_a^* = D_a \frac{\partial[\rho_a(1-S)n]}{\partial u_a}$$

(3.15)

The coefficient of transmission, D_a^*, is a function of the volume–mass properties of the soil (i.e. S and n) and the air density. Substituting D_a^* into Equation (3.15) results in the following form:

$$J_a = -D_a^* \frac{\partial u_a}{\partial y}$$

(3.16)

This modified form of Fick's law has been used in geotechnical engineering to describe air flow through soils (Blight, 1971).

The coefficient of transmission, D_a^*, can be related to the air coefficient of permeability which is given the symbol, k_a. The air coefficient of permeability, k_a, is the value measured in the laboratory.

A steady-state air flow can be established through an unsaturated soil specimen with respect to an average matric suction or an average degree of saturation. The soil specimen is treated as an element of soil having one value for its air coefficient of permeability that corresponds to the average matric suction or degree of saturation. This means that the air coefficient of permeability is assumed to be constant throughout the soil specimen. Steady-state air flow is produced by applying an air pressure gradient across the two ends of the soil specimen. The amount of air flowing through the soil specimen is measured at the exit point as a flow under constant pressure conditions (i.e. usually at 101.3 kPa absolute or zero gauge pressure) (Matyas, 1967). In other words, the mass rate of the air flow is measured at a constant air density, ρ_{ma}. Equation (3.16) can be rewritten for this particular case as follows:

$$\rho_{ma} \frac{\partial V_a}{\partial t} = -D_a^* \frac{\partial u_a}{\partial y}$$

(3.17)

or

$$v_a = -D_a^* \frac{1}{\rho_{ma}} \frac{\partial u_a}{\partial y}$$

(3.18)

where
ρ_{ma} = constant air density corresponding to the pressure used in the measurement of the mass rate (i.e. at the exit point of flow)
$\partial V_a/\partial t$ = volume rate of the air flow across a unit area of the soil at the exit point of flow; designated as flow rate, v_a.

The pore air pressure, u_a, in Equation (3.18) can also be expressed in terms of the pore air pressure head, h_a, using a constant air density, ρ_{ma}:

$$v_a = -D_a^* g \frac{\partial h_a}{\partial y} \tag{3.19}$$

where
h_a = pore air pressure head (i.e. $u_a / \rho_{ma} g$)
$\partial h_a / \partial y$ = pore air pressure head gradient in the y-direction; designated as i_{ay}.

This equation has the same form as Darcy's equation for the air phase:

$$v_a = -k_a \frac{\partial h_a}{\partial y} \tag{3.20}$$

where the relationship between the air coefficient of transmission, D_a^*, and the air coefficient of permeability, k_a, is defined as follows:

$$k_a = D_a^* g \tag{3.21}$$

The hydraulic head gradient, $\partial h_a / \partial y$, consists of the pore air pressure head gradient as the driving potential. Equation (3.20) has been used in geotechnical engineering to compute the air coefficient of permeability, k_a (Barden, 1965; Matyas, 1967; Langfelder et al., 1968; Barden et al., 1969; Barden and Pavlakis, 1971).

Air permeability measurements can be performed at various matric suctions or different degrees of saturation in order to establish the functional relationship, $k_a(u_a - u_w)$, or $k_a(S)$. This relationship also applies to the air coefficient of transmission, D_a^*, since the two coefficients are related by a constant, 'g'.

Experimental verifications have been carried out for Fick's and Darcy's laws by Blight (1971).

Coefficient of permeability with respect to air phase

Several relationships have been proposed between the air coefficient of permeability and the volume–mass properties of a soil. The coefficient of transmission, D_a^*, can either be computed in accordance with the Equation (3.21) or measured directly in experiments. The coefficient of permeability for the air phase, k_a, is a function of the fluid (i.e. air) and soil volume–mass properties, as described by Equation (3.4). The fluid properties are generally considered to be constant during the flow process. Therefore, the air coefficient of permeability can be expressed as a function of the volume–mass properties of the soil. In this case, the volumetric percentage of air in the pores is an important factor since air flows through the pore space filled with air. As the matric suction increases or the degree of saturation decreases, the air coefficient of permeability increases.

RELATIONSHIP BETWEEN AIR COEFFICIENT OF PERMEABILITY AND DEGREE OF SATURATION

The prediction of the air coefficient of permeability based on the pore size distribution and the matric suction versus degree of saturation curve has also been proposed for the air phase. The air coefficient permeability function, k_a, is essentially the inverse of the water coefficient of permeability function, k_w. The following equation has been used by Brooks and Corey (1964) to describe the $k_a(S_e)$ function:

$$k_a = 0 \quad \text{for } (u_a - u_w) \leq (u_a - u_w)_b \tag{3.22}$$

$$k_w = k_d(1 - S_e)^2(1 - S_e^{(2+\lambda)/\lambda}) \quad \text{for } (u_a - u_w) > (u_a - u_w)_b \tag{3.23}$$

where
k_d = coefficient of permeability with respect to the air phase for a soil at a degree of saturation of zero.

RELATIONSHIP BETWEEN AIR COEFFICIENT OF PERMEABILITY AND MATRIC SUCTION

Another form of Equation (3.23) is obtained when the effective degree of saturation, S_e, is expressed in terms of matric suction (Brooks and Corey, 1964),

$$k_a = 0 \quad \text{for } (u_a - u_w) \leq (u_a - u_w)_b \tag{3.24}$$

$$k_w = k_d\left[1 - \left\{\frac{(u_a - u_w)_b}{(u_a - u_w)}\right\}^{\lambda}\right]^2\left[1 - \left\{\frac{(u_a - u_w)_b}{(u_a - u_w)}\right\}^{2+\lambda}\right] \tag{3.25}$$

$$\text{for } (u_a - u_w) > (u_a - u_w)_b$$

The following figure illustrates the agreement between measured data and the theoretical air coefficient of permeability function described using the immediate above equation. Several studies have been conducted on the air permeability of compacted soils. The coefficient of permeability with respect to air, k_a, decreases as the soil water content or degree of saturation increases (Ladd, 1960; Olson, 1963; Langfelder et al., 1968; Barden and Pavlakis, 1971). The comparisons between the air and water coefficients of permeability for a soil compacted at different water contents or matric suction values are illustrated Figure 3.12.

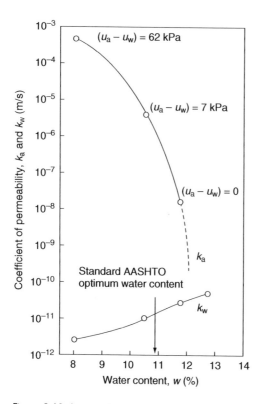

Figure 3.12 Air coefficients of permeability k_a and water coefficients of permeability k_w as a function of the gravimetric water content for Westwater soil (from Barden and Pavlakis, 1971; after Fredlund and Rahardjo, 1993).

The air and water coefficients of permeability, k_a and k_w, were measured on the same soil specimen during steady-state flow conditions induced by small pressure gradients. The air coefficient of permeability, k_a, decreases rapidly as the optimum water content is approached. At this point, the air phase becomes occluded, and the flow of air takes place as a diffusion of air through water. The occluded stage for soils with a high clay content usually occurs at water contents higher than the optimum water content (Matyas, 1967; Barden and Pavlakis, 1971).

Although the air coefficient of permeability decreases and the water coefficient of permeability increases with increasing water content, the air permeability values remain significantly greater than the water permeability values for all water contents. The difference in air and water viscosities is one of the reasons for the air coefficient of permeability being greater than the water coefficient of permeability.

The coefficient of permeability is inversely proportional to the absolute (i.e. dynamic) viscosity of the fluid, μ, as shown in the equation, $k_w = (\rho g\, K/\mu_w)$. The absolute viscosity of water, μ_w, is approximately 56 times the absolute viscosity of air, μ_a, at an absolute pressure of 101.3 kPa and a temperature of 20°C. Assuming that the volume–mass properties of a soil do not differ for completely saturated and completely dry conditions, the saturated water coefficient of permeability would be expected to be 56 times smaller than the air coefficient of permeability at the dry condition (Koorevar et al., 1983). It should be noted that this may not be the case for many soils.

Another factor affecting the measured air coefficient of permeability is the method of compaction. A dynamically compacted soil usually has a higher air coefficient of permeability than a statically compacted soil at the same density.

The air coefficient of transmission, D_a^*, can be obtained by dividing the air coefficient of permeability, k_a, by the gravitational acceleration, g. If the gravitational acceleration is assumed to be constant, D_a^* functions are similar to the above air permeability, k_a, functions.

Diffusion (Fredlund and Rahardjo, 1993)

Introduction

The diffusion process occurs in response to a concentration gradient (Fredlund and Rahardjo, 1993). Ionic or molecular movement will take place from regions of higher concentration to regions of lower concentration. The air and water phase in a soil (i.e. soil voids) are the conducting media for diffusion processes. On the other hand, the soil structure determines the path length and cross-sectional area available for diffusion. The transport of gases (e.g. O_2 and CO_2), water vapour and chemicals are examples of diffusion processes in soils.

There are two diffusion mechanisms common to unsaturated soil behaviour.

- The first type of diffusion involves the flow of air through the pore water in a saturated or unsaturated soil (Barden and Sides, 1967; Matyas, 1967). Another example of air diffusion involves the passage of air through the water in a high air entry ceramic disk. This type of diffusion involves gases dissolving into water and subsequently coming out of water.
- The second type of diffusion involves the movement of constituents through the water phase due to a chemical concentration gradient or an osmotic suction gradient.

Air diffusion through water

Fick's law can be used to describe the diffusion process. The concentration gradient which provides the driving potential for the diffusion process is expressed with respect to the soil voids (i.e. air and water phases). In other words, the mass rate of diffusion and the concentration gradient are expressed with respect to a unit area and a unit volume of the soil voids, respectively.

The formulation of Fick's law for diffusion in the y-direction is as follows:

$$\frac{\partial M}{\partial t} = -D \frac{\partial C}{\partial y} \tag{3.26}$$

where

$\partial M / \partial t$ = mass rate of the air diffusion across a unit area of the soil voids

D = coefficient of diffusion

C = concentration of the diffusion air expressed in terms of mass per unit volume of the soil voids

$\partial C / \partial y$ = concentration gradient in the y-direction (or similarly in the x- or z- direction).

The diffusion equation can appear in several forms, similar to the forms presented for the flow of air through a porous medium. The concentration gradient for gas or water vapour (i.e. $\partial C / \partial y$) can be expressed in terms of their partial pressures. Consider a constituent diffusing through the pore water in a soil. Equation (3.26) can be rewritten with respect to the partial pressure of the diffusing constituent:

$$\frac{\partial M}{\partial t} = -D \frac{\partial C}{\partial \bar{u}_i} \frac{\partial \bar{u}_i}{\partial y} \tag{3.27}$$

where

$\bar{u}_i$ = partial pressure of the diffusing constituent

$\partial C / \partial \bar{u}_i$ = change in concentration with respect to a change in partial pressure

$\partial \bar{u}_i / \partial y$ = partial pressure gradient in the y-direction (or similarly in the x- or z-direction).

The mass rate of the constituent diffusing across a unit area of the soil voids (i.e. $\partial M / \partial t$) can also be determined by measuring the volume of the diffused constituent under constant pressure conditions. The ideal gas law is applied to the diffusing constituent in order to obtain the mass flow rate:

$$\frac{\partial M}{\partial t} = \bar{u}_{fi} \frac{\omega_i}{RT} \frac{\partial V_{fi}}{\partial t} \tag{3.28}$$

where
$\bar{u}_{fi}$ = absolute constant pressure used in the volume measurement of the diffusing constituent
ω_i = molecular mass of the diffusing constituent
R = universal (molar) gas constant
T = absolute temperature
$\partial V_{fi}/\partial t$ = flow rate of the diffusing constituent across a unit area of the soil voids
V_{fi} = volume of the diffusing constituent across a unit area of the soil voids.

The change in concentration of the diffusing constituent relative to a change in partial pressure (i.e. $\partial C/\partial \bar{u}_i$) is obtained by considering the change in density of the dissolved constituent in the pore water. The density of the dissolved constituent in the pore water is the ratio of the mass of dissolved constituent to the volume of water:

$$\frac{\partial C}{\partial \bar{u}_i} = \frac{\partial(M_{di}/V_w)}{\partial \bar{u}_i} \tag{3.29}$$

where
M_{di} = mass of the dissolved constituent in the pore water
V_w = volume of water.

Applying the ideal gas law, Equation (3.29) yields the following equation:

$$\frac{\partial C}{\partial \bar{u}_i} = \frac{\partial\left(\dfrac{V_{di}}{V_w}\bar{u}_i\dfrac{\omega_i}{RT}\right)}{\partial \bar{u}_i} \tag{3.30}$$

where
V_{di} = volume of the dissolved constituent in the pore water.

The ratio of the volume of dissolved constituent to the volume of water (i.e. V_{di}/V_w) is referred to as the volumetric coefficient of solubility, h. Under isothermal conditions, h is essentially a constant.

$$\frac{\partial C}{\partial \bar{u}_i} = h\frac{\omega_i}{RT} \tag{3.31}$$

where
h = volumetric coefficient of solubility for the constituent in water.

By substitutions and rearranging the above equations, van Amerongen (1946) obtained the following diffusion equation:

$$v_{fi} = -\frac{Dh}{\bar{u}_{fi}} \frac{\partial \bar{u}_i}{\partial y} \tag{3.32}$$

where
v_{fi} = flow rate of the diffusing constituent across a unit area of the soil voids, $(\partial V_{fi}/\partial t)$.

Equation (3.32) can be applied to air or gas diffusion through the pore water in a soil or through free water or some other material such as a rubber membrane (Poulos, 1964). The partial pressure in the above equation can be expressed in terms of the partial pressure head, $h_{fi} = \bar{u}_{fi}/\rho_{fi}g$ with respect to the constituent density, ρ_{fi}. The density, ρ_{fi}, corresponds to the absolute constant pressure, $\bar{u}_{fi}$, used in the measurement of the diffusing constituent volume. The absolute constant pressure, $\bar{u}_{fi}$, is usually chosen to correspond to atmospheric conditions (i.e. 101.3 kPa), and ρ_{fi} is the constituent density at the corresponding pressure.

$$v_{fi} = -Dh\frac{\rho_{fi}g}{\bar{u}_{fi}} \frac{\partial h_{fi}}{\partial y} \tag{3.33}$$

where
ρ_{fi} = constituent density at the constant pressure, $\bar{u}_{fi}$, used in the volume measurement of the diffusing constituent
h_{fi} = partial pressure head $(\bar{u}_{fi}/\rho_{fi}g)$.

Equation (3.33) has a similar form to Darcy's law. Therefore, Equation (3.33) can be considered as a modified form of Darcy's law for air flow through an unsaturated soil with occluded air bubbles where the air flow is of the diffusion form.

$$v_{fi} = -k_{fi}\frac{\partial h_{fi}}{\partial y} \tag{3.34}$$

where k_{fi} = diffusion coefficient of permeability for air through an unsaturated soil with occluded air bubbles.

The diffusion coefficient of permeability can then be written as follows:

$$k_{fi} = Dh\frac{\rho_{fi}g}{\bar{u}_{fi}} \tag{3.35}$$

Substituting the ideal gas law into Equation (3.35) results in another form for the diffusion coefficient of permeability:

$$k_{fi} = D\frac{h\omega_i g}{RT} \tag{3.36}$$

This equation indicates that under isothermal conditions, the coefficient of permeability (i.e. diffusion type) is directly proportional to the coefficient of diffusion since the term $(h\omega_i g/RT)$ is a constant.

The diffusion coefficients, D, for air through water were computed in accordance with the equation, $v_{fi} = -k_{fi}\partial h_{fi}/\partial y$ (Barden and Sides, 1967). The diffusion values for porous media (e.g. soils) appear to be smaller than the diffusion values for free water. This has been attributed to factors such as the tortuosity within the soil and the higher viscosity of the adsorbed water close to the clay surface. As a result, the diffusion values decrease as the soil water content decreases.

Chemical diffusion through water

The flow of water induced by an osmotic suction gradient (or a chemical concentration gradient) across a semi-permeable membrane can be expressed as follows:

$$\frac{\partial M}{\partial t} = -D_o \frac{\partial \pi}{\partial y} \tag{3.37}$$

where
$\partial M/\partial t$ = mass rate of pure water diffusion across a unit area of a semi-permeable membrane
D_o = coefficient of diffusion with respect to osmotic suction (i.e. $D_o = \partial C/\partial \pi$)
C = concentration of the chemical
π = osmotic suction
$\partial \pi/\partial y$ = osmotic suction gradient in the y-direction (or similarly in the x- or z-direction).

A semi-permeable membrane restricts the passage of the dissolved salts, but allows the passage of solvent molecules (e.g. water molecules). Clay soils may be considered as 'leaky' semi-permeable membranes because of the negative charges on the clay surfaces (Barbour, 1987). Dissolved salts are not free to diffuse through clay particles because of the adsorption of the cations to the clay surfaces and the repulsion of the anions. This, however, may not completely restrict the passage of dissolved salts, as would be the case for a perfectly semi-permeable membrane. Therefore, pure water diffusion through a perfect, semi-permeable membrane may not fully describe the flow mechanism related to the osmotic suction gradient in soils.

Summary of flow laws (Fredlund and Rahardjo, 1993)

Several flow laws related to the fluid phases of an unsaturated soil have been described in the preceding sections. A summary of the flow laws is given in Table 3.3 (after Fredlund and Rahardjo, 1993).

Table 3.3 Summary of flow laws (after Fredlund and Rahardjo, 1993)

Type of flow	Phase	Driving potential	Flow law (in y-direction)	Comments
Bulk flow	Water	$h_w = y + \dfrac{u_w}{\rho_w g}$	$v_w = -k_w \dfrac{\partial h_w}{\partial y}$	Darcy's law
		u_a	$\dfrac{\partial M}{\partial t} = -D_a^* \dfrac{\partial u_a}{\partial y}$	Fick's law
	Air	$h_a = \dfrac{u_a}{\rho_a g}$	$v_a = -k_a \dfrac{\partial h_a}{\partial y}$	Darcy's law $k_a = D_a^* g$
Gas constituent diffusion	Gases, including water vapor and air through the pore–water in a soil	C	$\dfrac{\partial M}{\partial t} = -D \dfrac{\partial C}{\partial y}$	Fick's law $k_{fi} = D \dfrac{h\omega_i g}{RT}$
		u_i	$v_{fi} = -k_{fi} \dfrac{\partial h_{fi}}{\partial y}$	Darcy's law (obtained from Fick's, Henry's, and Darcy's laws)
Chemical diffusion	Pure water in osmotic diffusion	π	$\dfrac{\partial M}{\partial t} = -D_o^* \dfrac{\partial \pi}{\partial y}$	Fick's law

Soil–water characteristic curves and water permeability functions (Fredlund et al., 2001a, 2002)

Introduction

What is a SWCC? It is a relationship between water content and suction of a soil. Hydraulically and physically, it means how much equilibrium water a soil can take at a given suction originally. Although SWCCs are taking on an increasingly important role in the application of unsaturated soil mechanics to geotechnical and geoenvironmental engineering, there is a lack of consistency in the terminology used to describe the relationship between the amount of water in a soil and soil suction (Fredlund et al., 2001a). The lack of a consistent terminology is primarily the result of having numerous disciplines involved with measuring and using the data. The desire, however, is to encourage greater consistency on terminology for the disciplines of geotechnical and geoenvironmental engineering.

Some of the terms used when referring to the relationship between the amount of water in the soil and soil suction are as follows (Fredlund et al., 2001a):

- soil–water characteristic curve
- suction–water content relationship
- moisture retention curves
- soil moisture retention curves and
- water retention curves.

The recommendation is that the term 'soil–water characteristic curve' be used in civil engineering related disciplines. The preference is to use 'soil–water' because it is the water content of the soil that is measured. The preference is to use the word 'characteristic' simply because it appears to have historically been the most common term used in engineering. In addition, the 'characteristic' implies that the curve describes the character or the behaviour of the soil. The word 'retention' is more closely related to retaining water for plant growth in agriculture.

Conventionally, a SWCC is determined from a drying path without the consideration of the effects of volume changes and stress state. Ng and Pang (2000a,b) have demonstrated experimentally and numerically the relevance and importance of considering these two effects and introduced the concept of stress-dependent soil–water characteristic curve (SDSWCC). In this book, SWCC is reserved for a drying curve without including the effects of volume changes and stress state as usual, unless stated otherwise. On the other hand, the term 'SDSWCC' is exclusively used for a state-dependent soil–water characteristic curve which incorporates the effects of two independent stress state variables (i.e. suction and net mean stress) and volume changes.

Components of soil suction

There are numerous factors that have been suggested as contributing factors to soil suction. At the same time, there appears to be two primary components of soil suction: matric suction and osmotic suction. Krahn and Fredlund (1972) tested two soils and independently measured the matric, osmotic and total suctions of numerous specimens (Figure 3.13). The test results showed that the matric plus the osmotic components of suction, essentially add up to total suction. While there may be other components of soil suction, it would appear that the matric and osmotic components are the dominant components making up total suction.

From thermodynamic considerations, it has been hypothesized that there are two primary mechanisms that can result in a vapour pressure reduction at the air–water interphase (i.e. contractile skin). The mechanisms are a

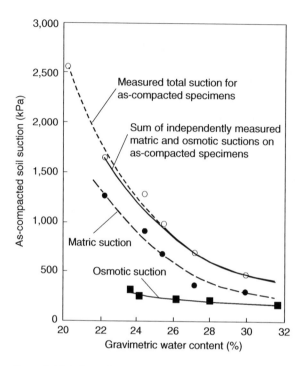

Figure 3.13 Independent measurements of matric, osmotic and total suctions for a clayey soil compacted at various water contents (after Fredlund *et al.*, 2001a).

curved meniscus giving rise to matric suction and salts in the pore fluid giving rise to osmotic suction. Geotechnical engineers are most commonly solving problems related to changes in the negative pore water pressure of the soil. Consequently, it is the matric suction component that is of most relevance in unsaturated soil mechanics, just as positive pore water pressures are of greatest importance in saturated soil mechanics.

There are two types of plots found in the literature where soil suction is plotted versus water content. Figure 3.13 shows the initial components of soil suction on as-compacted soil specimens. This is NOT a SWCC since it is possible to take each of the data points and continue to measure the SWCC (Fredlund *et al.*, 2001a)

Components of soil suction on the soil–water characteristic curve

Considerable confusion has arisen in the application of SWCC information in geotechnical engineering. The confusion arises primarily over the mixed

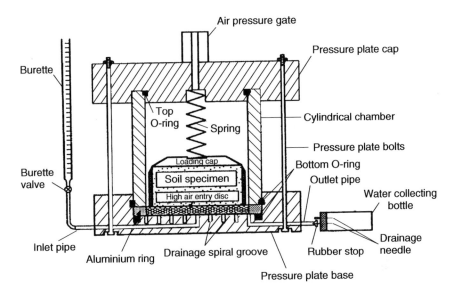

Figure 3.14 A single specimen, 500-kPa pressure plate cell developed at the University of Saskatchewan, Canada (after Fredlund, 2002).

components of suction commonly plotted to make up the SWCC. Pressure plate apparatuses (see Figure 3.14) are generally used to apply suctions up to 1,500 kPa. The water contents at equilibrium are measured. Vacuum desiccators (see Figure 3.15) are generally used to equilibrate small soil specimens for suctions greater than 1,500 kPa. The equilibrium water content corresponds to total suctions. The matric and total suction data obtained in the laboratory are routinely plotted on the same graph. In other words, the lower values of soil suction are matric suction while the higher values of suction are total suction.

The total suction values are generally in the residual suction range for the soil. Figure 3.16 shows how the two components of soil suction interrelate to total suction for a silty clay soil. It is not suggested that there be any change in the manner in which SWCC data is routinely plotted. There simply needs to be a recognition that mixed components of suction are being plotted and that the mechanisms related to classic soil mechanics (i.e. seepage, shear strength and volume change) are quite different in the low and high soil suction ranges. For example, there can be hydraulic flow of liquid water in a soil up to residual suction conditions, but it appears that vapour flow moisture is of most significance in the residual suction range.

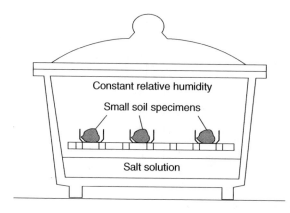

Figure 3.15 Vacuum desiccator for equilibrating small soil specimens in a constant relative humidity environment above a controlled salt solution (after Fredlund, 2002).

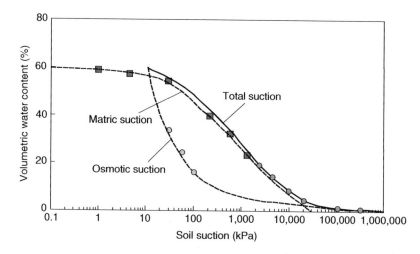

Figure 3.16 The components and total suction for a typical silty clay soil (after Fredlund et al., 2001a).

Units of soil suction

One of the first units used as a measure of soil suction was pF. The pF unit is equal to the logarithm (to the base 10) of the absolute value of a column of water in centimetres that equilibrates with the suction in the soil. The pF scale was widely used in agronomy and soil physics for many years. More recently, there appears to be a substantial move towards an acceptable SI system of units.

There are numerous disadvantages to using the pF system of stress measurements for engineering. For example, the pF scale does not translate across the water table from positive to negative pore water pressures. This gives the perception that there is a division between positive pore water pressure and negative pore water pressure conditions (i.e. between saturated soil mechanics and unsaturated soil mechanics). Also, the logarithm is being taken of a negative water head. But most importantly, pF is not an acceptable SI unit of measurement.

The most commonly used unit for soil suction appears to be kilopascals, kPa. When referring to high suctions, it might be preferable to use megapascals (i.e. $1,000 \, kPa = 1 \, MPa$). There are many other units of stress used in the literature for soil suction (e.g. psi, psf, kgf/cm^2, tsf, etc.). There may be need to use different units for soil suction in some cases, but there are definite benefits related to ease and accuracy of interpretation related to using a consistent set of units such as kilopascals.

Range of units of soil suction

There is an upper and lower limit to the range for soil suction. The lower limit is zero. This corresponds to the case of zero pore water pressure and (theoretically) no salts in the pore fluid when considering total suction. The upper limit of soil suction corresponds to a dry soil (i.e. zero water content). The upper limit of soil suction is generally in the order of $1,000,000 \, kPa$. Soil suction approaches approximately $1,000,000 \, kPa$ when a soil has lost all its water and the relative humidity approaches zero.

SWCC curve data are usually plotted with soil suction on a logarithmic scale since the data vary over several orders of magnitude. Therefore, the maximum soil suction has a logarithm of 6 on the SI scale. Unfortunately, there is no lower limit on the logarithm scale as soil suction approaches zero. The logarithm scale is particularly valuable for determining the air-entry value and residual conditions for the soil.

It is recommended that SWCC data always be plotted on a logarithmic scale for interpretation purposes. This applies regardless of the soil suction range over which measurement have been made. It is recommended that the lower limit of soil suction required for most cases is about $0.1 \, kPa$. This is essentially the lowest suction that can be applied to a soil through a high air-entry disk. Figure 3.17 shows an example of a – SWCC data set (desorption or drying curve) with soil suction plotted from 0.1 to $1,000,000 \, kPa$. Even if there may not be any data in the high suction range, it is still possible to get a perspective or 'feel' for the entire SWCC.

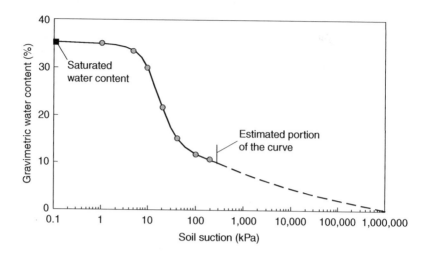

Figure 3.17 Typical soil–water characteristic curve data set (desorption or drying branches) for a sandy soil (after Fredlund *et al.*, 2001a).

Descriptions of water content

There have been numerous designations used as a measure for the amount of water in the soil. The three most basic measures are the following:

- volumetric water content, θ_w,
- gravimetric water content, w, and
- degree of saturation, S.

All three of the above variables will convey similar information to the engineer provided the structure of the soil is essentially incompressible. Clayey soils may undergo volume change as a result of a soil suction increase. In this case, it is the degree of saturation variable that provides the indication of the air-entry value of the soil.

Volumetric water content has most commonly been used in agriculture-related disciplines when plotting SWCC data. Volumetric water content is defined as an instantaneous water content in the sense that the volume of water in the soil is referenced to the present total volume. However, it is generally not the instantaneous volumetric water content that is being plotted since the volume of water is usually referenced back to the original volume of soil, and volume changes are not considered generally. Consequently, both the gravimetric water content representation and the volumetric water content representation portray similar information.

It is recommended that gravimetric water content be used when plotting the SWCC when there are no continuous volume measurements. At the same time, it is recognized that volumetric water content appears in the formulation of transient flow processes when using a referential element. Care must be exercised when using volumetric water content to ensure that continuous volume measurements are made and the correct reference volume is used in both the mathematical formulations and the laboratory measurements. If so, θ_w, or degree of saturation, S, can be used in plotting the SWCC. If the volume of water in the soil is always referenced back to the original total volume, then it is better that this variable simply be referred to as, V_w/V_t where V_w is the present volume of water and V_t is the initial total volume of the soil specimen.

If the SWCC data is to be used to estimate unsaturated soil property functions, it is of benefit to normalize or dimensionalize the water content scale. The dimensionless water content, Θ_d, can be defined as follows:

$$\Theta_d = \frac{w}{w_s} \tag{3.38}$$

where $w = $ any gravimetric water content and $w_s = $ gravimetric water content at saturation. The total range of water contents from the saturated state under zero confining pressure to zero water content are reduced to a scale of 1.0–0.0, respectively. The normalized water content Θ_n, can be defined as follows:

$$\Theta_n = \frac{w - w_r}{w - w_s} \tag{3.39}$$

where $w_r = $ gravimetric water content at the commencement of residual conditions.

Terminology used for a soil–water characteristic curve

Variables that should be defined from the SWCC, if possible, are the:

- saturated water content, w_s
- air-entry value, ψ_{aev}
- inflection point suction, ψ_{inf}
- inflection point water content, w_{inf}
- slope at the inflection point, n_{inf} (rate of desorption or drying),
- residual suction, ψ_r and
- residual water content, w_r.

Figure 3.18 shows an ideal data set along with the definition of each variable for a drying SWCC. Straight lines on a semi-log plot are necessary for the constructions. The first line is horizontal through saturation water content. The second line passes tangent to the steepest portion of the SWCC data.

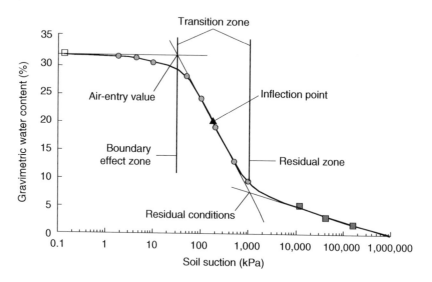

Figure 3.18 An ideal data set along with the definition of each variable for a drying SWCC (after Fredlund *et al.*, 2001a).

The third line passes through 1,000,000 kPa and extends back through data points at high suctions (provided the data are available).

Conventional methods for measurement of the soil–water characteristic curve

Overview

The drying conventional SWCC curve with continuous volume measurements and applying stresses can be measured in the laboratory with relative ease (Fredlund and Rahardjo, 1993; Barbour, 1998). The test equipment is commonly found in soil science laboratories and has also found its way into some soil mechanics laboratories. The cost of performing the tests is somewhat less than the costs associated with performing a one-dimensional consolidation test. Of course, some advanced measuring devices, which are capable of applying one-dimensional, isotropic and deviatoric stresses with continuous volume measurements (Ng and Pang, 2000a, b; Ng *et al.*, 2001a) do provide more realistic and appropriate results for engineering applications but with increased complexity.

   The experimental measurement of a drying SWCC can be divided into two parts, namely the regions where the suctions are less than approximately 1,500 kPa and the regions where the suctions are greater than 1,500 kPa. Suctions greater than 1,500 kPa are generally established using an osmotic desiccator (see Figure 3.15). The water content corresponding to high suction values (i.e. suctions greater than 1,500 kPa) is determined by allowing small soil specimens to come to equilibrium in an osmotic desiccator containing a salt solution.

Water contents corresponding to low suction values are usually determined using an acrylic pressure plate device (often referred to as a Tempe Cell, see Figure 3.19) with a 1 bar (100 kPa) high-air-entry disk. A second commercially available pressure plate device is the Volumetric Pressure Plate (see Figure 3.20), which has a 2 bar (200 kPa) high-air-entry disk. A third pressure plate device is available with a 15 bar (1,500 kPa) ceramic or a pressure membrane (See Figure 3.21). Each of the above pieces of equipment was designed for use in areas other than geotechnical engineering. As such, each apparatus has certain limitations, and several attempts have been made to develop an apparatus more suitable for geotechnical engineering. Two such devices are the one-dimensional stress controllable volumetric pressure plate (see Figure 3.22) and the triaxial-stress controllable apparatus (see Figure 3.23). Details of these advanced devices are described later.

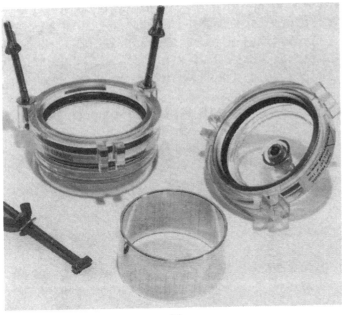

(a)

Figure 3.19 Assemblage of the Tempe pressure cell. (a) Disassembled components of a Tempe cell; (b) assembled Tempe cell (courtesy Soilmoisture Equipment Corp.; after Fredlund and Rahardjo, 1993).

(b)

Figure 3.19 (Continued).

Figure 3.20 Disassembled components of a volumetric pressure plate extractor (courtesy Soilmoisture Equipment Corp.; after Fredlund and
Rahardjo, 1993).

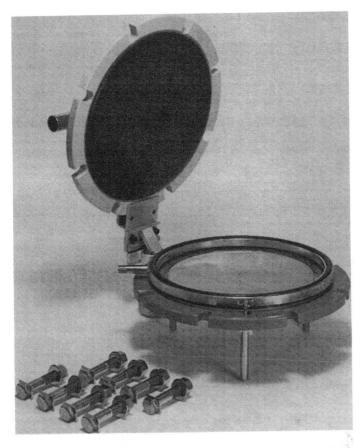

Figure 3.21 Pressure membrane extractor (courtesy Soilmoisture Equipment Corp.; after Fredlund and Rahardjo, 1993).

Test procedures for measuring the soil–water characteristic curve

The test procedure for measuring the SWCC was originally developed in soil science and agronomy, but the test procedures adopted within geotechnical engineering have remained similar until late 1990s (Fredlund, 2000). The attention given to the initial preparation of soil specimens is quite different between geotechnical engineering and soil science. The agriculture-related disciplines have generally not paid much attention to the initial state (or structure) of the soil specimens. On the other hand, many of the soil mechanics theories used in geotechnical engineering are predicated on the assumption that it is possible to obtain undisturbed soil samples that can be tested to measure in situ physical soil properties. SWCC data collected

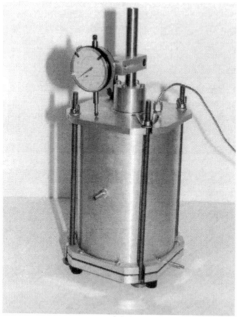

(a)

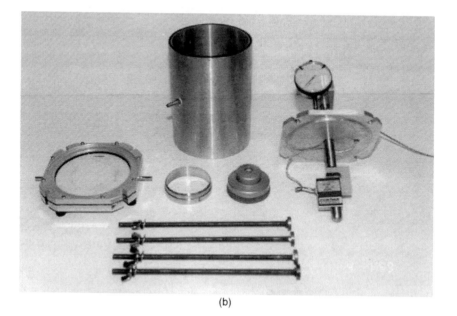

(b)

Figure 3.22 (a) Assembled volumetric pressure plate extractor; (b) Components of modified volumetric pressure-plate extractor (Ng and pang, 2000b).

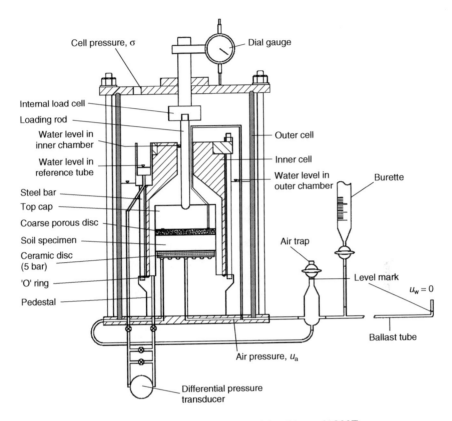

Figure 3.23 Triaxial pressure plate testing system (after Ng *et al.*, 2007).

from a variety of sources will generally have used several different specimen preparation procedures.

Initial preparation states for a soil specimen can be categorized as follows (Fredlund, 2000):

- natural (or so-called undisturbed) samples that retain the in situ soil structure,
- completely remoulded specimens where the soil is mixed with water to form a semi-liquid paste and
- remoulded and compacted samples where the initial water content is near to the plastic limit.

Regardless of which of the above conventional procedures is used, the soil specimens are placed into the pressure plate apparatus, covered with water and allowed to saturate. Therefore, the soil suction is reduced to zero prior to commencing the formal test.

The difference between the various specimen preparation procedures may not be of great concern for a sandy soil but the opposite is true for a clayey soil. The soil structure and secondary microstructure become of increasing concern for clay-riched soils. Different testing procedures should be carefully considered and adopted. Figure 3.24 shows the effect that the above initial states can have on the drying curves for a clayey soil (Fredlund, 2000). It is well recognized that there is no single or unique SWCC for a particular soil. More details of sample preparations and testing procedures are given by Fredlund and Rahardjo (1993).

Mathematical forms of soil–water characteristic curve (Fredlund and Xing, 1994)

Fredlund (2000) reviews several mathematical equations that have been proposed to describe the SWCC. The equation of Gardner (1958) was originally proposed for defining the unsaturated coefficient of permeability function, and its application to the SWCC is inferred (see Table 3.4). The

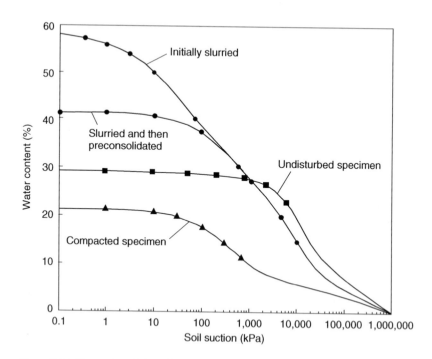

Figure 3.24 Illustration of the influence of initial state on the soil–water characteristic curve (after Fredlund, 2000).

Table 3.4 Summary of some of the mathematical equations proposed for the soil–water characteristic curve (after Fredlund, 2000)

Author(s)	Equation	Soil parameter
Gardner (1958)	$w = \dfrac{w_s}{1 + \left(\dfrac{\psi}{a_g}\right)^{n_g}}$	a_g, n_g
van Genuchten (1980)	$w = \dfrac{w_s}{\left[1 + \left(\dfrac{\psi}{a_{vg}}\right)^{n_{vg}}\right]^{m_{vg}}}$	a_{vg}, n_{vg}, m_{vg}
Mualem (1976)	$w = \dfrac{w_s}{\left[1 + \left(\dfrac{\psi}{a_m}\right)^{n_m}\right]^{m_m}}$	$a_m, n_m, m_m = 1/(1 - n_m)$
Burdine (1952)	$w = \dfrac{w_s}{\left[1 + \left(\dfrac{\psi}{a_b}\right)^{n_b}\right]^{m_b}}$	$a_b, n_b, m_b = 2/(1 - n_b)$
Fredlund and Xing (1994)	$w = C(\psi)\dfrac{w_s}{\left[\ln\left(e + \left(\dfrac{\psi}{a_f}\right)^{n_f}\right)\right]^{m_f}}$	$a_f, n_f, m_f, C(\psi)$

mathematical equations proposed by Burdine (1952) and Maulem (1976) are two-parameter equations that become special cases of the more general three-parameter equation proposed by van Genuchten (1980). These equations are asymptotic to horizontal lines in the low soil suction range and a suction beyond residual conditions. As such, these equations are not forced through zero water content at 1,000,000 kPa of suction. A correction factor, C_r, has been applied to the mathematical equation proposed by Fredlund and Xing (1994). The correction factor forces the SWCC function through a suction of 1,000,000 kPa at a water content of zero. All of the proposed equations in Table 3.4 provide a reasonable fit of SWCC data in the low- and intermediate-suction ranges (Leong and Rahardjo, 1997a). In all cases, the a parameter bears a relationship to the air-entry value of the soil and usually refers to the inflection point along the curve. The n parameter corresponds to the slope of the straight-line portion of the main desorption (i.e. drying) or adsorption portion of the SWCC.

The Fredlund and Xing (1994) mathematical function applies over the entire range of soil suctions from 0 to 1,000,000 kPa. The relationship is essentially empirical and, similar to earlier models, is based on the assumption that the soil consists of a set of interconnected pores that are randomly distributed. Subsequent discussions on the SWCC are restricted to the Fredlund and Xing's equation.

The Fredlund and Xing (1994) equation, written in terms of gravimetric water content, w, is as follows:

$$w = C\left(\psi\right) \frac{w_s}{\left[\ln\left(e + \left(\frac{\psi}{a}\right)^n\right)\right]^m} \tag{3.40}$$

where w_s is the saturated gravimetric water content;
a is a suction value corresponding to the inflection point on the curve and is somewhat greater than the air-entry value;
n is a soil parameter related to the slope of the SWCC at the inflection point;
ψ is the soil suction (i.e. matric suction at low suctions and total suction at high suctions);
m is a fitting parameter related to the results near to residual water content;
e is the natural number, 2.71828...; and
$C(\psi)$ is the correction function that causes the SWCC to pass through a suction of 1,000,000 kPa at zero water content.

The correction function $C(\psi)$ is defined as

$$C\left(\psi\right) = \left[1 - \frac{\ln\left(1 + \dfrac{\psi}{\psi_r}\right)}{\ln\left(1 + \dfrac{1\,000\,000}{\psi_r}\right)}\right] \tag{3.41}$$

where ψ_r is the suction value corresponding to residual water content, w_r. Residual suction can be estimated as 1,500 kPa for most soils, unless the actual value is known, Equation (3.41) can be written in a dimensionless form by dividing both sides of the equation by the saturated gravimetric water content (i.e. $\Theta = w/w_s$, where Θ is the dimensionless water content):

$$\Theta = C\left(\psi\right) \frac{1}{\left[\ln\left(e + \left(\frac{\psi}{a}\right)^n\right)\right]^m} \tag{3.42}$$

Equation (3.42) can be used to best fit the desorption or adsorption branches of the SWCC data over the entire range of suctions. The fitting parameters (i.e. a, n and m values) can be determined using a nonlinear regression procedure such as the one proposed by Fredlund and Xing (1994).

The character of the Fredlund and Xing (1994) Equation (3.42) can be observed by varying each of the curve-fitting parameters (i.e. a, n and m). Figure 3.25 illustrates the lateral translation of the SWCCs as a result of

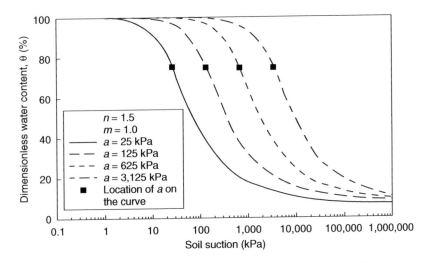

Figure 3.25 Plot of the Fredlund and Xing (1994) equation with *m* and *a* constant and *n* varying (after Fredlund, 2000).

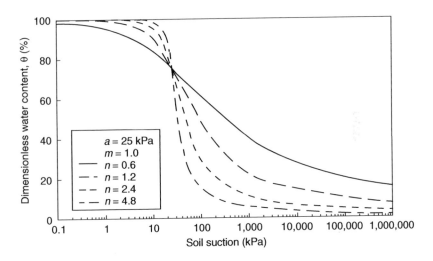

Figure 3.26 Plot of the Fredlund and Xing (1994) equation with *m* and *n* constant and *a* varying (after Fredlund, 2000).

varying the *a* parameter with the *n* parameter fixed at 1.5. Figure 3.26 illustrates the change in slope of the SWCC as a result of changing the *n* parameter with the *a* parameter fixed at 25 kPa. Figure 3.27 illustrates the rise in the SWCC as the *m* parameter is varied. The correction factor, $C(\psi)$,

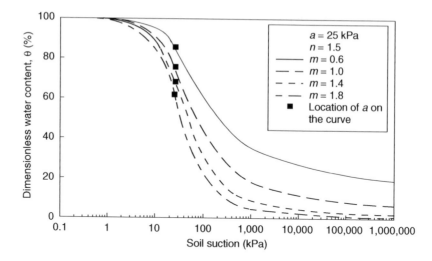

Figure 3.27 Plot of the Fredlund and Xing (1994) equation with *a* and *n* constant and *m* varying (after Fredlund, 2000).

was set at 1.0 for each of the three illustrative examples. If C(ψ) is computed using Equation (3.41), each of the SWCCs would pass through a soil suction of 1,000,000 kPa at zero water content.

Hysteresis of soil–water characteristic curve

Of course, the real soil behaviour is far more complex. The hysteresis associated with the drying and wetting of a soil illustrates that there is no unique SWCC (Haines, 1930; Hillel, 1998; Fredlund, 2000, 2002). As discussed by Hillel (1998), the relation between matric potential and soil wetness can be obtained in two ways: (1) in drying *(desorption)*, by starting with a saturated sample and applying increasing suction, in a step-wise manner, to gradually dry the soil while taking successive measurements of wetness versus suction; and (2) in wetting *(sorption)*, by gradually wetting an initially dry soil sample while reducing the suction incrementally. Each of these methods yields a continuous curve, but the two curves will generally not be identical. The equilibrium soil wetness at a given suction is greater in drying than in sorption wetting. This dependence of the equilibrium content and state of soil–water upon the direction of the process leading up to it is called *hysteresis* (Haines, 1930).

A description of hysteresis in physical terms was offered by Poulovassilis (1962): If a physical property Y depends on an independent variable X, then it may occur that the relationship between Y and X is unique, and, in particular, independent of whether X is increasing or decreasing. Such

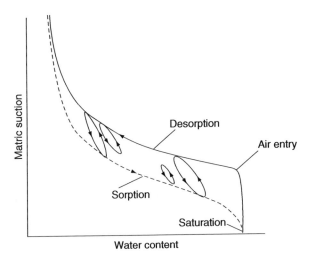

Figure 3.28 Suction versus water content curves in sorption and desorption. The intermediate loops are scanning curves representing complete or partial transitions between the main branches (after Hillel, 1998).

a relationship is *reversible*. Many physical properties are, however, *irreversible*. Even when the changes of X are made very slowly, the curve for increasing X does not coincide with that for decreasing X. The phenomenon of hysteresis is commonly observed in magnetism, for instance.

Figure 3.28 shows a typical soil–moisture characteristic curve and illustrates the hysteresis effect in the soil wetness-matric suction equilibrium relationship. The hysteresis effect may be attributed to several causes (Hillel, 1998):

- The geometric non-uniformity of the individual pores (which are generally irregularly shaped voids interconnected by smaller passages), resulting in the 'ink bottle' effect, illustrated in Figure 3.29.
- The contact–angle effect (Hillel, 1998), by which the contact angle and the radius of curvature are greater in the case of an advancing meniscus than in the case of a receding one. A given water content will tend therefore to exhibit greater suction in desorption than in sorption. (Contact–angle hysteresis can arise because of surface roughness, the presence of adsorbed impurities on the solid surface, and the mechanism by which liquid molecules adsorb or desorb when the interface is displaced.)
- The encapsulation of air in 'blind' or 'dead-end' pores, which further reduces the water content of newly wetted soil. Failure to attain true equilibrium (though not, strictly speaking, true hysteresis) can accentuate the hysteresis effect.

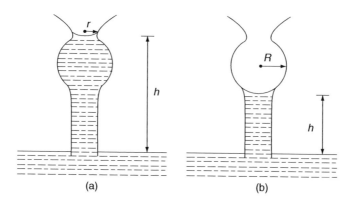

Figure 3.29 The 'ink bottle' effect determines the equilibrium height of water in a variable-width pore: (a) in capillary drainage (desorption) and (b) in capillary rise (sorption) (after Hillel, 1998).

- Swelling, shrinking, or aging phenomena, which result in differential changes of soil structure, depending on the wetting and drying history of the sample (Hillel and Mottes, 1966). The gradual solution of air, or the release of dissolved air from soil–water, can also have a differential effect on the suction–wetness relationship in wetting and drying systems.

Of particular interest is the ink bottle effect. Consider the hypothetical pore shown in Figure 3.29. This pore consists of a relatively wide void of radius R, bounded by narrow channels of radius r. If initially saturated, this pore drains abruptly the moment the suction exceeds $(u_a - u_w)_r$, where $(u_a - u_w)_r = 2T_s/r$. For this pore to rewet, however, the suction must decrease to below $(u_a - u_w)_R = 2T_s/R$, whereupon the pore abruptly fills. Since $R > r$, it follows that $(u_a - u_w)_r > (u_a - u_w)_R$. Drying depends on the narrow radii of the connecting channels, whereas sorption depends on the maximum diameter of the large pores. These discontinuous spurts of water, called *Haines jumps* (after W.B. Haines, who first noted the phenomenon in 1930), can be observed readily in coarse sands. The hysteresis effect is in general more pronounced in coarse-textured soils in the low-suction range, where pores may empty at an appreciably larger suction than that at which they fill.

The two complete characteristic curves, from saturation to dryness and vice versa, are the *main branches* of the hysteretic soil moisture characteristic. When a partially wetted soil commences to drain, or when a partially desorbed soil is rewetted, the relation of suction to moisture content follows some intermediate curve as it moves from one main branch to the other. Such intermediate spurs are called *scanning curves*.

Cyclic changes often entail wetting and drying scanning curves, which may form loops between the main branches (Figure 3.29). The matric suction–volumetric water content relationship can thus become very complicated. Because of its complexity, the hysteresis phenomenon is too often ignored.

Among the alternative concepts of hysteresis are the theory of independent domains (Poulovassilis, 1962) and the theory of dependent domains (Mualem and Miller, 1979; Mualem, 1984). The former is based on the assumption that all pores are free to drain independently. The latter approach recognizes that only pores with access to the atmosphere can drain, and that access depends on whether the surrounding (or interceding) pores are water filled or air filled. To account for that dependence, a domain-dependence factor is applied.

In the past, hysteresis was generally disregarded in the practice of soil physics. This may be justifiable in the treatment of processes involving monotonic wetting (e.g. infiltration) or monotonic drying (e.g. evaporation). But hysteresis may be important in cases of composite processes where wetting and drying occur simultaneously or sequentially in various parts of the soil profile (e.g. redistribution). Two soil layers of identical texture and structure may be at equilibrium with each other (i.e. at identical energy states) and yet they may differ in wetness if their wetting–drying histories have been different. Furthermore, hysteresis can affect the dynamic, as well as the static, properties of the soil (i.e. hydraulic conductivity and flow phenomena).

As revealed by many researchers, the wetting and drying branches form the extreme bounds for the SWCC, and there are an infinite number of inter-mediate (drying and wetting) scanning curves (Fredlund, 2000). Figure 3.30 shows idealized drying and wetting scanning curves become asymptotic to the bounding curves. A natural soil (undisturbed) sample from the field may have a soil suction that lies somewhere between the bounding curves. When a soil is sampled, the engineer does not know whether the soil is on a drying path or a wetting path. The difference between the drying and wetting branches of the SWCC may be as much as one to two orders of magnitude (Fredlund, 2002). For example, a particular water content may correspond to a soil suction ranging from 10 to 100 to 1,000 kPa. Slight changes in the nature of the soil can also produce a lateral shift in the SWCC. Consequently, it is extremely difficult to use the SWCC for the estimation of in situ soil suctions. There has been very limited success in the estimation of in situ soil suctions.

Correlations between soil–water characteristic curve and water permeability functions (Fredlund et al., 2001b)

Fredlund et al. (2001b) observe that although some field and laboratory methods are available to measure the relationship between water

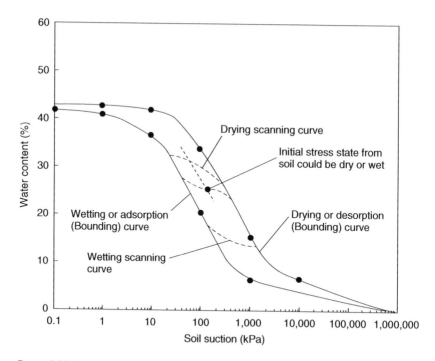

Figure 3.30 Idealized drying and wetting scanning curves become asymptotic to the bounding curves (Fredlund, 2000).

permeability and suction of unsaturated soils (Fredlund and Rahardjo, 1993), it is extremely difficult and time-consuming to the measure the relationship accurately due to the low water permeability of the soils. Very often, semi-empirical correlations are used to determine a permeability function from a SWCC. The shape of the water permeability function bears a relationship shape of the SWCC. Figure 3.31 compares the SWCCs and water permeability functions for sand and clayey silt. The water permeability for both soils remains relatively constant from zero suction up to the air-entry value of the soil. The change in water permeability of a soil occurs at approximately the air-entry value of the soil. The water permeability decreases rapidly beyond the air-entry value, for both soils. The decrease in the water permeability is due to the reduction in the cross-sectional area of flow. The initial water permeability, or saturated water permeability, k_s, of the sand can be two or more orders of magnitude greater than that of the clayey silt. As the suction increases, it is possible that the water permeability of sand will decrease by more than two orders of magnitude. Under certain conditions, it is possible for the clayey silt to be more permeable than the sand.

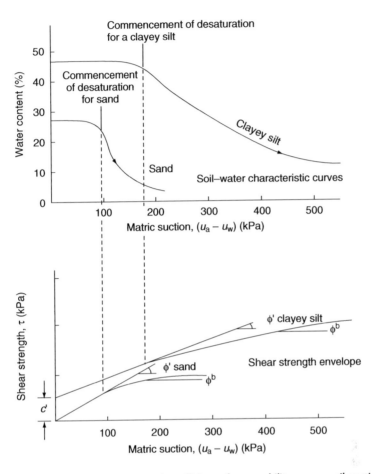

Figure 3.31 Water content and coefficient of permeability versus soil suction (after Fredlund *et al.*, 2001b).

The water permeability function of an unsaturated soil can be predicted with sufficient accuracy for many engineering applications with a knowledge of the saturated coefficient of water permeability and the SWCC. The SWCC equation developed by Fredlund *et al.*, (1994) along with the saturated coefficient of permeability can be used to compute the relationship between water permeability and soil suction. Previous studies (Leong and Rahardjo, 1997; Benson and Gribb, 1997) have shown that proposed integration procedures involving the use of the Fredlund and Xing's (1994) SWCC provide a good estimate of the permeability function.

The calculation of the permeability function is performed by dividing the water content versus suction relationship into several water content increments. This is equivalent to integrating along the water content axis. The numerical integration procedure can be used to compute data points for a permeability function for the unsaturated soil.

$$k_r(w) = \frac{\int_{w_r}^{w} \frac{w-x}{\psi^2 x}\,dx}{\int_{w_r}^{w_s} \frac{w_s-x}{\psi^2 x}\,dx} \tag{3.43}$$

where k_r = relative coefficient of permeability and x = a variable of integration representing water content.

The accuracy of the prediction of the permeability function was shown to improve when the complete SWCC was used. Although the permeability function can be computed down to zero water content, it should be noted that the function may be more indicative of vapour flow in the region beyond the residual stress state. As a result, it may be more reasonable to leave the water permeability as a constant, k_{res}, beyond the residual state. When calculating the permeability function, it is convenient to perform the integration along the soil suction axis as shown in Equation (3.44).

$$k_r(\psi) = \frac{\int_{\psi_r}^{\psi} \frac{w(y)-w(\psi)}{y^2} w'(y)\,dy}{\int_{\psi_{aev}}^{\psi_r} \frac{w(y)-w_s}{y^2} w'(y)\,dy} \tag{3.44}$$

where ψ_{aev} = air-entry value of the soil under consideration; ψ_r = suction corresponding to the residual water content w_r; ψ = a variable of integration representing suction; y = a variable of integration representing the logarithm of suction; and w' = the derivative of the SWCC. Leong and Rahardjo (1997) reported that the 'computed coefficient of permeability for this statistical model (i.e. from Equations 3.43 and 3.44 above) showed good agreement with the measured coefficient of permeability'.

To avoid the numerical difficulties associated with performing the integration over the entire soil suction range, it is more convenient to perform the integration on a logarithm scale. The proposed models have been found to be most satisfactory for sandy soils whereas agreement with experimental data may prove to be less satisfactory for clayey soils. Equation (3.44) can be multiplied by the dimensionless water content raised to a power (i.e. Θ^p) in order to provide greater flexibility in computing the permeability function. The additional parameter, p, is assumed to account for tortuosity in the soil pores (Maulem, 1986); it is a parameter whose magnitude must be assumed.

$$k_r(\psi) = \Theta^p(\psi) \frac{\int\limits_{\ln(\psi)}^{b} \frac{w(e^y) - w(\psi)}{e^y} w'(e^y)\, dy}{\int\limits_{\ln(\psi_{aev})}^{b} \frac{w(e^y) - w_s}{e^y} w'(e^y)\, dy} \tag{3.45}$$

Based on research work by Kunze *et al.* (1968), the value of the power, *p*, can be assumed to be one, unless there is reason to assume otherwise.

Leong and Rahardjo (1997) suggested that rather than performing the above integration, the dimensionless equation for the SWCC simply be raised to a power, *q*. Therefore, the permeability function can be written in the following form.

$$k_r(\psi) = k[\Theta(\psi)]^q \tag{3.46}$$

where $\Theta(\psi) =$ dimensionless form of the SWCC and $q =$ a new soil parameter.

This form for the permeability function is obviously attractive due to its simplicity. The equation is simple to use and clearly illustrated the relationship between the SWCC and the permeability function. In order to use Equation (3.46), it is necessary to know what value to use for the *q* soil fitting parameter. In the original study undertaken by Leong and Rahardjo (1997), the data from six soils were used to assess the magnitude of the *q* soil fitting parameter. The results of their study are shown in Table 3.5. More details of how to determine water permeability can be found from Fredlund *et al.* (1994) and Leong and Rahardjo (1997).

The state-dependent soil–water characteristic curve: a laboratory investigation (Ng and Pang, 2000a,b)

Overview

As discussed previously, only the drying SWCCs of soil specimens are often measured under zero net normal stress conditions. However, in the actual

Table 3.5 Summary of typical parameters from the study by Leong and Rahardjo (1997)

Soil type	a (kPa)	n	m	q
Beit Netofa clay	389	0.69	1.176	52.12
Rehovot sand	2.25	4.32	1.235	6.04
Touchet silt loam	7.64	7.05	0.506	4.55
Columbia sandy loam	5.81	10.59	0.381	5.79
Superstition sand	2.66	6.86	0.525	6.21
Yolo light clay	2.93	2.11	0.379	9.57

field situations, the influence of stress state may not be negligible. Even though the total net normal stress on the soil elements in an unsaturated slope is seldom altered, the stress state at each element at different depths is different, which may affect the soil–water characteristic of these elements. To more realistically predict pore water pressure distributions and hence the slope stability of an unsaturated slope, it is thus essential to investigate the influence of stress state on soil–water characteristics (Ng and Pang, 2000a,b; Ng et al., 2001). Although it is theoretically recognized that the stress state of a soil has some influence on soil–water characteristics (Fredlund and Rahardjo, 1993), few experimental results can be found in the literature, with the exception of those by Vanapalli et al. (1998, 1999), who conducted a series of experiments to investigate the influence of soil structure and stress history on the soil–water characteristics of a recompacted clayey till. Recompacted fine-grained soil specimens were first loaded in an oedometer to different void ratios before their soil–water characteristics were measured in a conventional pressure plate extractor. The results of Vanapalli et al. suggest that the rate of change of water content with respect to matric suction of the specimens recompacted at dry of optimum water content is higher than that of wet of optimum specimens due to their different soil structures. Moreover, the stress history appears to have a significant effect on the soil–water characteristics of the specimens recompacted at dry of optimum, but not at wet of optimum. In their experiments, Vanapalli et al. essentially considered the over-consolidation effects but not, strictly speaking, the actual stress effects on SWCCs.

Here, the influences of initial dry density and initial water content, history of drying and wetting, soil structure and the stress state upon the drying (desorption) and wetting (adsorption) soil–water characteristics of a completely decomposed volcanic soil in Hong Kong are examined and discussed. The experimental results presented are obtained by using a conventional volumetric pressure plate extractor and a newly modified one-dimensional stress-controllable pressure plate extractor with deformation measurements (Ng and Pang, 2000a).

Descriptions of soil specimens

The volcanic soil tested was a completely decomposed ash tuff. The soil can be classified as a firm, moist, orangish brown, slightly sandy silt/clay with high plasticity, as per Geoguide 3 (GCO, 1988). The soil specimens were obtained from an 'undisturbed' $200 \times 200 \times 200 \, mm^3$ block sample excavated from a hillside slope in Shatin, Hong Kong for failure investigations after a major landslide took place in July 1997 (Sun and Campbell, 1998). The grain size distribution, which is determined by sieve analysis (BS, 1990) and hydrometer analysis (BS, 1990), is shown in Figure 3.32. In addition, some index properties of the soil are summarized in Table 3.6. The gravel, sand, silt and clay contents determined in the tuff are 4.9,

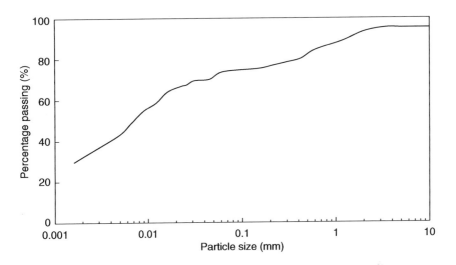

Figure 3.32 Particle size distribution of a completely decomposed volcanic (CDV) soil (after Ng and Pang, 2000a,b).

Table 3.6 Index properties of a completely decomposed volcanic (CDV) soil (sandy silt–clay – after Ng and Pang, 2000a,b)

Specific gravity (Mg/m^3)	2.62
Maximum dry density (kg/m^3)	1,603
Optimum moisture content (%)	22
Initial moisture content (%)	30
Gravel content ($\leq$62 mm, %)	4.9
Sand content ($\leq$2 mm, %)	20.1
Silt content ($\leq$63 μm, %)	36.6
Clay content ($\leq$2 μm, %)	37.1
Liquid limit (%)	55.4
Plastic limit (%)	33.4
Plasticity index (%)	22

20.1, 36.6 and 37.1 per cent, respectively, and it can be classified as a well-graded soil.

Experimental program

An experimental program was undertaken to investigate the soil–water characteristics of a volcanic soil. In addition to three natural (or undisturbed) soil specimens, seven recompacted soil specimens with different initial conditions

Table 3.7 Summary of the testing program for determining soil–water characteristics (after Ng and Pang, 2000a,b)

Specimen identity[a]	Initial dry density (Mg/m^3)	Initial void ratio	Relative compac- tion (%)	Initial water content (%)	Applied stress (kPa)	No. of wetting– drying cycles	Max. suction applied (kPa)	Measured final void ratio
CDV-R1(C)	1.47	0.782	92	30.3	0	3	400	0.758
CDV-R2(C)	1.50	0.747	94	30.3	0	1	200	0.736
CDV-R3(C)	1.53	0.712	95	30.3	0	1	200	0.703
CDV-R4(C)	1.47	0.782	92	22.0	0	1	200	0.767
CDV-R5(C)	1.47	0.782	92	15.0	0	1	200	0.761
CDV-R2(M)	1.47	0.782	92	30.3	40	1	200	0.724
CDV-R3(M)	1.47	0.782	92	30.3	80	1	200	0.701
CDV-N1(C)	1.47	0.782	92	30.3	0	1	100	0.771
CDV-N2(M)	1.47	0.782	92	30.3	40	1	200	0.743
CDV-N3(M)	1.47	0.782	92	30.3	80	1	200	0.695

Note

a N and R denote natural and recompacted specimens, respectively; C and M in parentheses denote conventional and modified one-dimensional volumetric pressure-plate extractor, respectively.

were used for testing. Details of the testing program are summarized in Table 3.7.

Equipment

Three different types of extractor were used for measuring SWCC in this study. A conventional volumetric pressure plate extractor was used to measure the soil–water characteristics of unsaturated soils for suction ranges of 0–200 kPa. In the context of this chapter, the phase – 'conventional volumetric pressure plate extractor' – refers to a commercially available pressure extractor in which the principle of axis translation is applied to control matric suction in a soil specimen and no external load is imposed on the specimen (Fredlund and Rahardjo, 1993). This extractor was also adopted to study the hysteresis of the soil–water characteristics associated with the drying and wetting of the soil. For measuring the soil–water characteristic for suction ranges greater than 200 kPa (only one test), a 500-kPa pressure plate extractor was adopted.

The conventional volumetric pressure plate extractor was not equipped for applying any vertical or confining stress to the specimen. Hence, an extractor for measuring the stress-dependent soil–water characteristic curve (SDSWCC) of unsaturated soils under one-dimensional (K_o) stress conditions was designed (Ng and Pang, 2000a,b) to investigate the effects of stress on the soil–water characteristic of the volcanic soil. A schematic diagram of the modified pressure plate extractor is shown in Figure 3.33. An oedometer

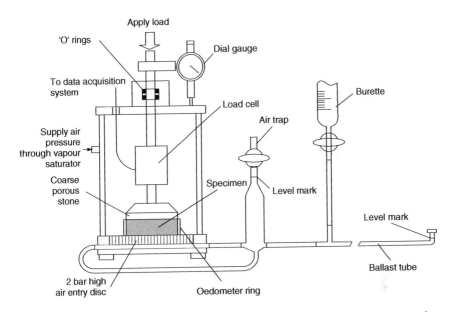

Figure 3.33 Schematic diagram of the modified one-dimensional volumetric pressure plate extractor (after Ng and Pang, 2000a).

ring equipped with a high air-entry ceramic plate at its base is located inside an airtight chamber. Vertical load via a loading piston is applied to a soil specimen inside the oedometer ring through a loading frame. Possible pulling away of soil specimen from the sides of the oedometer ring was minimized by applying a vertical load and by testing the samples under small suction ranges. The airtightness of the chamber is maintained using rubber o-rings at the openings. Along the piston, a load cell is attached inside the airtight chamber for determining the actual vertical net normal stress applied to the soil specimen. A dial gauge is attached for measuring the vertical displacement of the soil specimen. The size of specimen that can be accommodated in this extractor is 70 mm in diameter and 20 mm in height.

As in the conventional volumetric pressure plate extractor, the pore air pressure, u_a, is controlled through a coarse porous stone together with a thin woven geotextile located at the top of the specimen. The compression of the geotextile was accounted for by calibration using a steel dummy specimen in the extractor. The air applied is saturated using an air-saturator. The pore water pressure, u_w, is controlled at the atmospheric pressure through the high air-entry ceramic plate mounted at the base of the specimen. The high air-entry ceramic plate will remain saturated if the matric suction does not exceed the air-entry value of the plate (200 kPa). The matric suction, $(u_a - u_w)$, imposed on the soil specimen will be the difference between the applied air and pore water pressures. The net normal stress can be controlled one-dimensionally and axial deformation can be measured by using this extractor. Other details of this modified extractor are available in Ng and Pang (2000a).

Preparation of soil specimens

The soil was first dried in an oven at 45 °C for 48 h before preparing a 'recompacted' soil specimen. The soil was then pulverized using a rubber hammer. The desired water content (see Table 3.7) was added, and the soil was thoroughly mixed. After keeping the mixed soil in a plastic bag for moisture equalization for about 24 h in a temperature and moisture controlled room, the soil was then statically recompacted to the desired density in an oedometer ring of 70 mm in diameter and 20 mm in height. A 'natural' soil specimen, on the other hand, was directly cut from the block sample into an oedometer ring with a cutting edge. Both natural and recompacted soil specimens were then submerged in de-aired water inside a desiccator subjected to a small vacuum for about 24 h to ensure almost full saturation. The 24-h duration was sufficient to ensure almost full saturation as the soil has a high saturated hydraulic conductivity of 2.88×10^{-6} m/s, measured by Ng and Pang (1999).

Testing procedures for using a conventional volumetric pressure-plate extractor

A complete drying and wetting cycle was imposed on each specimen during the tests. Initially the specimen was subjected to an increasing value of matric suction in a series of steps to measure the soil–water characteristic along the drying path. As the matric suction increased, water was expelled from the soil specimen into the ballast tube (see Figure 3.33). The movement of water in the ballast tube was continuously monitored by using a ruler fixed along the ballast tube until the equilibrium condition was reached. Typically, 2–7 days were required to achieve equilibrium at a given suction. Any air trapped was removed via the air-trap removal attachment. After reaching equilibrium, the water levels in both the air-trap and the ballast tube were adjusted to their respective level marks. Change of volume of water in the specimen between two successive suctions was then precisely measured using the marked burette, which has a resolution of 0.1 ml. In fact, it could be estimated to 0.05 ml with confidence. For the specimens tested, the estimated accuracy of the measured volumetric water content was better than 0.2 per cent. After reaching the required maximum suction, a wetting process was then started by decreasing the value of matric suction in a series of steps. Water in the ballast tube was absorbed by the soil specimen. The change of the volume of water during wetting was determined as previously described. At the end of the test, each soil specimen was oven-dried at

45 °C for 24 h to determine its water content. The degrees of saturation at various matric suctions were then determined from the change of volume of water and the final water content (i.e. backward calculations). It should be noted that volume change of the soil specimen was assumed to be zero for determining the volumetric water content and the degree of saturation using the conventional volumetric plate extractor.

Similar procedures cannot be used for the CDV-R1(C) soil specimen. This is due to the reason that the soil–water characteristic for suction ranges greater than 200 kPa cannot be obtained by using the volumetric pressure plate extractor. Hence, after the suction reached 200 kPa, the soil specimen was removed from the volumetric pressure plate extractor and weighed immediately. It was then placed in a 500 kPa pressure plate extractor and subjected to a suction value of 400 kPa. Upon reaching the equilibrium condition, the soil specimen was weighed and put back to the conventional volumetric pressure plate extractor to continue the wetting path. In total, three repeated cycles of drying and wetting were performed on this specimen.

Testing procedures for using the modified volumetric pressure plate extractor

The saturated soil specimens (i.e. CDV-R2(M), CDV-R3(M), CDV-N2(M) and CDV-N3(M)) were loaded to the required net normal stresses (i.e. 40 and 80 kPa) in an oedometer for measuring the stress-dependent soil–water characteristics. These specimens were allowed to have free drainage at the top and bottom for 24 h and hence they were pre-consolidated under their required corresponding net normal stresses. The purpose of this pre-consolidation was to eliminate the effects of consolidation on volume changes during the subsequent measurements of soil–water characteristics at various suctions. After consolidation, the soil specimens were then removed from the oedometer and subjected to the same applied net normal stresses in the modified volumetric pressure plate extractor. Consolidation carried out in the oedometer, which is equipped with a low air-entry value porous disc, was to reduce the testing time.

Vertical deformation of the specimen was continuously monitored in the oedometer to ensure the end of primary consolidation was reached by studying the relationship between deformation and root time. It was found that 24 h was sufficient for the completion of primary consolidation. After placing the consolidated specimen in the modified volumetric pressure plate extractor, good contacts between the specimen and the ceramic plate were ensured by the applied net normal stress.

For each drying and wetting path, the required net normal stress was maintained constant throughout the test, and the vertical load and axial displacement of the soil specimen were measured (see Figure 3.33). To account for deformation of the loading piston subjected to an applied load, calibration was carried out. Thus, the actual vertical deformation and hence the actual volume change of the soil specimen was measured.

Soil–water characteristics of recompacted soils

Influence of initial dry density on soil–water characteristics

The soil–water characteristics of specimens recompacted at the same initial water content but different dry densities (i.e. CDV-R1(C), CDV-R2(C) and CDV-R3(C)) are compared in Figure 3.34. The structures of these three specimens are not identical as different compaction efforts were used during the preparation of specimens (Lambe, 1958). However, this series of tests is appropriate for investigating the soil–water characteristics of a fill slope or embankment formed by various degrees of compaction at different depths.

The air-entry values for all three specimens fall within a small range of matric suction values (see Figure 3.34). The air-entry value for each specimen can be estimated by using a 'judging-by-eye' method to extend a line from the constant slope portion of the first drying SWCC to intersect the suction axis at 100 per cent saturation. The estimated values from all three first drying curves vary between 4 and 5 kPa (refer to Table 3.8). Only a very small increase in the air-entry value is observed between CDV-R1 (dry density $= 1.47\,\text{Mg/m}^3$) and CDV-R3 (dry density $= 1.53\,\text{Mg/m}^3$). This is likely due to the small increase in dry density (i.e. only 4 per cent increase in dry density between the two specimens). By studying the experimental

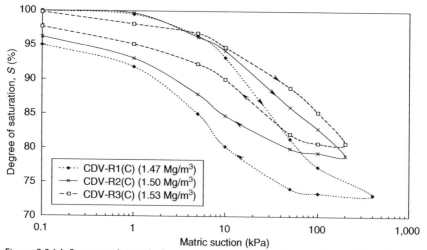

Figure 3.34 Influence of initial density on soil–water characteristics (after Ng and Pang, 2000a).

results presented in Table 4 of Tinjum *et al.* (1997), two clay specimens compacted at 19.7 per cent water content (dry unit weight $= 17.1 \, kN/m^3$) and at 20.0 per cent water content (dry unit weight $= 19.8 \, kN/m^3$) of soil M are suitable for comparisons. It is found that an increase in dry unit weight of 15.7 per cent results in an increase in the air-entry value from 2.5 to 7.6 kPa. Croney and Coleman (1954) also reported that an initial high compacted density silty sand specimen has a higher air-entry value than that of an low-density one. Regarding the rate of desorption, the higher the initial dry density, the slower the rate is. This is consistent with the current test results.

As shown in Figure 3.34, there is a marked hysteresis between the drying and wetting curves for all soil specimens. This marked hysteresis is likely attributed to the geometric non-uniformity of the individual pores, resulting in the so-called 'ink bottle' effect (Hillel, 1982, 1998). Also the difference in the contact angles at the receding soil–water interface during drying and at the advancing soil–water interface during wetting may contribute to the observed hysteresis. Moreover, any trapped air in the 'blind' or 'dead-end' pores inside soil specimen or in the testing system may account for the observed hysteresis. There is a general trend that the size of the hysteresis loop decreases as the initial void ratio reduces. Similar test results were also reported by Croney and Coleman (1954). The 'ink bottle' effect may be more pronounced in soils with a large than that with a small pore size distribution. In addition, an average large pore size distribution in a loose specimen may lead to a larger difference in the receding and advancing contact angles than that in a dense one.

At the end of the wetting path (i.e. 0.1 kPa suction), none of the wetting curves reaches full saturation. The looser the sample, the lower the degree of saturation is. The 'non-return' of the wetting paths may be attributed to air trapped in the specimens. For the observed difference in the degree of saturation achieved between a loose and a dense specimen, air trapped in large pores is more difficult to be displaced by capillary force than that in the small pores, and this may account for the observed difference. Full saturation is very difficult to achieve in loose specimen through the capillary action alone. Moreover, the 'ink bottle' effect is likely to be more pronounced in a loose than in a dense specimen.

Influence of initial water content on soil–water characteristics

Soil specimens must be recompacted to the same initial water content and dry density in order to be qualified as 'identical' (Lambe, 1958). Specimens recompacted to the same dry density at different initial water contents or using different compaction efforts are different soils due to different inherent structures. Thus, samples A, B and C shown in Figure 3.35 are not identical

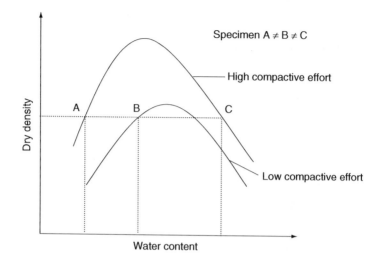

Figure 3.35 Schematic compaction curves at various dry densities and water contents (after Ng and Pang, 2000a,b).

even though they have the same dry density. The influence of the soil structure, which is defined as the arrangement of soil particles, fabric and bonding, on soil–water characteristics may be investigated by comparing the soil–water characteristics of the specimens that have the same dry density but were recompacted at different initial water contents or with different compaction efforts.

Figure 3.36 shows the variation in the degree of saturation with respect to matric suction for specimens recompacted to the same dry density at optimum, dry of optimum and wet of optimum initial water content. According to Benson and Daniel (1990) and Vanapalli *et al.* (1999), the CDV-R5(C) specimen, which is recompacted at dry of optimum initial water content, should have relatively large pores between the clods of soil. On the contrary, the CDV-R1(C) specimen, which is recompacted at wet of optimum initial water content, should have no visible, relatively large interclod pores and thus this soil would be relatively homogenous. As all three specimens have the same initial dry density, it is perhaps reasonable to assume that the total amount of voids would also be the same. However, their pore size distributions are likely to be different. A capillary tube analogy (assuming that $r_1 \geq r_2 \geq r_3$) shown in Figure 3.37a and b may be used to illustrate the difference in pore size distributions of CDV-R5(C) and CDV-R1(C), respectively. It is likely that the pore size distribution of CDV- R4(C) lies somewhere in between.

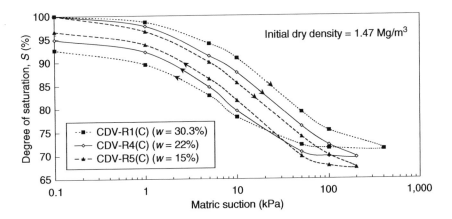

Figure 3.36 Influence of initial water content on soil–water characteristics (after Ng and Pang, 2000a).

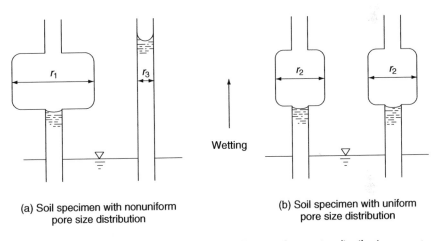

Figure 3.37 Capillary analogy to illustrate the influence of pore size distribution on saturation characteristics (after Ng and Pang, 2000a).

It can be seen from Table 3.8 that the air-entry value of CDV-R5(C) is lower than that of CDV-R1(C). A decreasing air-entry value is consistent with decreasing compaction water content. This is because a smaller air pressure is required to enter the relatively large interclod pores (i.e. r_1 in Figure 3.37a) in the CDV-R5(C) specimen. The current experimental data are consistent with results published by Tinjum *et al.* (1997) and Vanapalli *et al.* (1999). Regarding the rate of desorption, there is no significant difference between the three specimens, however.

Table 3.8 Estimated air-entry values for a completely decomposed volcanic (CDV) soil (after Ng and Pang, 2000a)

Specimen identity	Path	Estimated air-entry value (kPa)
CDV-R1(C)	1st drying	4
CDV-R1(C)	2nd drying	2
CDV-R1(C)	3rd drying	2
CDV-R2(C)	1st drying	4
CDV-R3(C)	1st drying	5
CDV-R4(C)	1st drying	2
CDV-R5(C)	1st drying	1.5
CDV-R2(M)	1st drying	4
CDV-R3(M)	1st drying	5
CDV-N1(C)	1st drying	1.5
CDV-N2(M)	1st drying	3
CDV-N3(M)	1st drying	5

During the wetting process, the degree of saturation increases as the matric suction decreases. The end points of the wetting paths at low suction (0.1 kPa) are different from those at the drying paths, due to air trapped in the soil. This difference is higher for specimen recompacted with larger initial water content. The overall 'ink bottle' effect would be more pronounced in the uniform pore size distribution than that in the non-uniform pore size distribution (Figure 3.37a), assuming that the two samples have the same amount of total voids. This leads to a large hysteresis loop for the specimen recompacted at wet of optimum than that recompacted at dry of optimum.

Influence of the drying and wetting history on soil–water characteristics

The soil–water characteristics of the soil specimen CDV-R1(C) subjected to three repeated drying and wetting cycles are shown in Figure 3.38. In all the three cycles, marked hysteresis loops between the drying and wetting paths can be seen. The size of the hysteresis loop is the largest in the first cycle but it seems to become independent of the drying and wetting history after the completion of the first cycle. As the matric suction decreases during wetting, none of the three wetting curves returns to its original degree of saturation. As discussed earlier, it is difficult to attain full saturation conditions mainly because of trapped air. The degree of saturation at low suction (i.e. 0.1 kPa) decreases as the number of drying and wetting cycles increases, but at a reduced rate.

The desorption characteristic with respect to matric suction is dependent on the drying and wetting history. It can be seen that the rate of desorption is relatively high during the first drying cycle compared to that at the second and third drying cycles. This may be due to the presence of relatively large

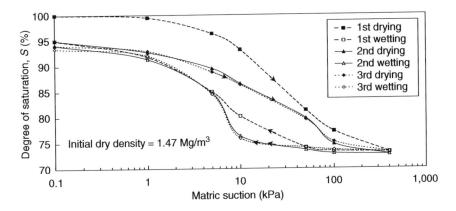

Figure 3.38 Influence of drying and wetting cycles on soil–water characteristics of specimen CDV-RI(C) (after Ng and Pang, 2000a; Ng *et al.*, 2000a).

voids initially. During the first wetting, significant volume change is likely to take place, and this results in smaller voids present in the sample due to collapse of soil structures of a virgin soil (Chiu *et al.*, 1998; Ng *et al.*, 1998). Thus, a smaller rate of desorption for the second and third drying and wetting cycles is expected. Moreover, the estimated air-entry value reduces during the second drying, but it remains almost unchanged in the subsequent drying cycles (see Table 3.8).

The adsorption characteristic of the first wetting process also seems to be different from that of the subsequent wetting processes. The value of matric suction at which the soil starts to absorb water significantly is relatively higher during the first wetting cycle (i.e. about 50 kPa) than (i.e. about 10 kPa) during the subsequent cycles. The rates of adsorption are substantially different for the first and for the subsequent wetting cycles at suctions ranging from 50 to 10 kPa. This might be caused by some soil structure changes after the first drying cycle. According to Bell (1992), drying initiates cementation by aggregation formation, leading to some relatively large inter-pores formed between the aggregated soil lumps. These large inter-pores reduce the specimen's rate of absorption along a certain range of the wetting path.

Influence of stress state on recompacted and natural soils

Verification of constant volume assumption

Soil–water characteristics are conventionally determined using a pressure plate extractor with the assumption that no volume change takes place throughout the test. This assumption is studied and verified using the newly modified extractor.

The test results from the recompacted (i.e. CDV-R2(M), CDV-N3(M)) and natural (i.e. CDV-N2(M), CDV-N3(M)) soil specimens are shown in Figures 3.39 and 3.40 respectively. The curves labelled 'constant volume assumed' are obtained by ignoring volume change taken place throughout the tests. From the figures, it can be seen that there is no significant difference

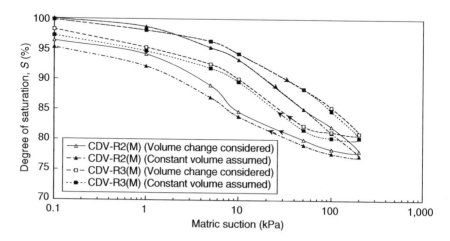

Figure 3.39 Comparison of measured soil–water characteristics with and without volume change considered for the recompacted soil specimens (after Ng and Pang, 2000a).

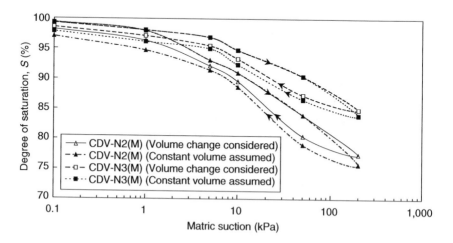

Figure 3.40 Comparison of measured soil–water characteristics with and without volume change considered for the natural soil specimens (after Ng and Pang, 2000a).

between the drying paths with and without volume change corrections until the matric suction reaches 100 kPa, at which measurable shrinkage of the soil starts to occur. The traditional method of interpretation by neglecting any volume reduction clearly underpredicts the volumetric water content and degree of saturation present in the soil specimen. Also it leads to a higher rate of desorption than that from a test considering volume changes.

During the wetting process, the difference between the two wetting paths does not stay constant, indicating that different degrees of swelling take place throughout. The influence of volume change for the natural soil specimens seems to be consistent with that for the recompacted soil specimens. In general, the higher the normal stress applied to a soil specimen, the smaller the average pore size of the specimen, the stiffer the specimen. In turn, it would have a higher resistance to volume change due to drying. From now on, volume change corrections are applied for the test results obtained from the modified pressure volumetric plate extractor in this chapter.

Influence of net normal stress

By using both the conventional and the modified volumetric pressure extractor, soil–water characteristic and stress-dependent soil–water characteristic curves (SDSWCC) are measured and compared. The soil–water characteristics of the recompacted soil specimens vertically loaded with 0-, 40-, 80-kPa net normal stresses under K_0 conditions (CDV-R1(C), CDV-R2(M) and CDV-R3(M)) are shown in Figure 3.41. It can be seen that there is a general tendency for the soil specimen subjected to higher stress to possess a slightly larger air-entry value (Table 3.8). As the matric suction keeps increasing, all

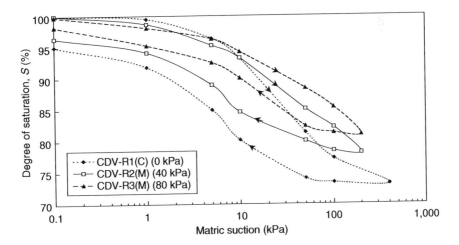

Figure 3.41 Influence of stress state on soil–water characteristics of recompacted CDV specimens (after Ng and Pang, 2000a).

specimens start to desaturate but at different rates. The higher the applied load on the specimen, the lower the rate of desorption. This is likely to be caused by the presence of a smaller average pore size distribution in the soil specimen under higher applied load.

Upon the completion of the drying phase, the tests proceed to the wetting phase. There is a marked hysteresis between the drying and wetting curves for all soil specimens, as expected. The size of the hysteresis loops seems to have reduced with the increase in applied stress, for the range of the net normal stresses considered. This is consistent with the results shown in Figure 3.34 that the denser the specimen, the smaller the size of the hysteresis loop. At the end of the tests, the end points of all three wetting curves are lower than their corresponding starting points. This is likely to be caused by trapped air. In addition, the denser the specimen or the higher the applied load, the closer are the end points to the starting points.

Figure 3.42 shows the influence of stress state on the soil–water characteristics for the natural soil specimens (CDV-N1(C), CDV-N2(M) and CDV-N3(M)). The size of the hysteresis loops does not seem to be governed by the applied stress level. Similar to the recompacted soil specimens, a natural soil specimen, which is subjected to higher applied stresses, possesses a slight larger air-entry value (see Table 3.8) and lower rates of desorption and adsorption, as a result of smaller pore size distribution. However, the influence of stress state on the soil–water characteristics of the natural specimens appears to be more significant than that on the soil–water characteristics of the recompacted specimens. This may be attributed to the fact that the recompacted specimens, which were recompacted at the wet side of the optimum (30.3 per cent), would have a relatively uniform pore

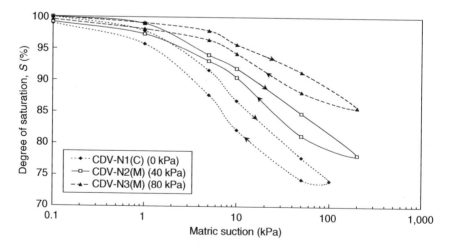

Figure 3.42 Influence of stress state on soil–water characteristics of natural CDV specimens (after Ng and Pang, 2000a,b).

size distribution, whereas the structure of the natural residual soil generally involves a wide range of pore sizes (Bell, 1992). The natural specimens, therefore, can be reasonably postulated to have relatively non-uniform pore size distributions. As a load is applied, relatively large pores in the natural specimens may be significantly reduced, and the specimen would probably become more homogenous. On the contrary, the pore size distributions of the recompacted specimens may not be affected substantially as compared with those in the natural specimens. Hence, the influence of stress state on soil–water characteristics appears to be more significant in natural specimens than in recompacted soil specimens.

Comparison between natural and recompacted soils

The soil–water characteristics of a recompacted (CDV-R2(M)) and a natural (CDV-N2(M)) specimen are compared in Figure 3.43. The CDV-R2(M) specimen was recompacted to the same density at the same initial moisture contents as the natural specimen. The size of the hysteresis loop in the recompacted soil is considerably larger than that in the natural specimen. Soil recompacted wet of optimum (i.e. CDV-R2(M)) is generally believed to be more homogenous, whereas the natural soil specimen (CDV-N2(M)) has relatively non-uniform pore size distributions due to various geological processes such as leaching in the field. As the two specimens have the same density, it is reasonable to postulate that CDV-N2(M) would

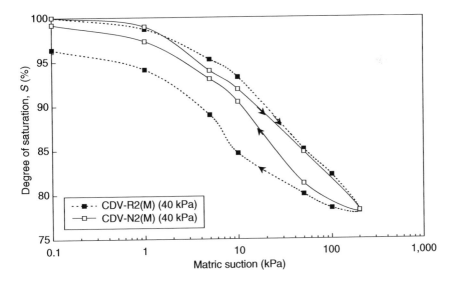

Figure 3.43 Influence of soil structure on soil–water characteristics (after Ng and Pang, 2000a).

have larger pores than those in the CDV-R2(M), statistically. Thus, CDV-N2(M) has a slightly lower air-entry value (see Table 3.8) and a higher rate of desorption than CDV-R2(M) does for suctions up to 50 kPa. The rates of desorption of the two soil specimens appear to be the same for high suctions.

On the other hand, the rates of adsorption for the two specimens are considerably different. The rate of wetting curve obtained from the natural soil specimen is substantially higher than that of the recompacted specimen. This observed behaviour may be explained by the capillary tube analogy discussed previously (Figure 3.37). The difference in pore size distributions between CDV-N2(M) and CDV-R2(M) is illustrated in Figure 3.37a and b. As the natural soil specimen has a non-uniform pore size distribution in which some relatively large pores (r_1) together with some relatively small pores (r_3), the presence of these small pores would facilitate the ingress of water to the specimen as the soil suction reduces. On the contrary, the lack of small pores and the presence of the relatively uniform medium-sized pores (r_2) in the recompacted specimen would slow down the speed of water entering the soil specimen. The rates of adsorption of the two soil specimens appear to be the same for small suctions (less than 5 kPa).

Summary

In order to gain a fundamental understanding of the desorption and adsorption characteristics of a volcanic soil in Hong Kong, an experimental program was carried out to test both recompacted and natural soil specimens. Various influencing factors on soil–water characteristics were considered in the experimental program, including initial dry density, initial water content, soil structure, history of drying and wetting and stress state. Based on the experimental investigations undertaken, the following conclusions can be drawn:

1. There is a marked hysteresis between the drying and wetting curves for all soil specimens recompacted to three different densities at the same water content. The size of the hysteresis loops decreases as the initial density of the specimen increases. This is probably attributed to the difference in the interconnection voids among the soil specimens. Although no apparent difference in the air-entry value of the specimens is observed (due to the small increase in density), the rate of desorption appears to be governed by the soil density. The soil specimens with a lower density exhibits a higher rate of desorption.

2. For soil specimens recompacted to the same density at optimum, dry and wet of optimum water contents, the soil specimen recompacted at wet of optimum, which has a relatively uniform pore size distribution resulting in more air entrapment, has the largest hysteresis loop among the three samples. The soil specimen recompacted at dry of optimum,

which has a relatively non-uniform pore size distribution, possesses the smallest hysteresis loop, whereas the soil specimen recompacted at optimum lies somewhere between the two loops. Although the air-entry value is somewhat a function of the initial moisture contents, the rates of desorption and adsorption do not seem to be significantly affected by the initial water contents for the ranges of suction and water content considered.

3. Soil–water characteristics of the recompacted soil highly depend on the history of drying and wetting. There is a distinct difference in both desorption and adsorption characteristics between the first and the second drying and wetting cycles. The rates of desorption and adsorption are substantially higher at the first than at the second drying and wetting cycle. This may be caused by changes in the soil structure during the first drying and wetting cycle. However, the soil–water characteristics of the subsequent drying and wetting cycles appear to remain approximately the same.

4. The soil–water characteristics of natural specimen are very different from these of the same soil recompacted to the same density at the same water content. The size of the hysteresis loop for the natural specimen is considerably smaller than that for the recompacted soil. Also the natural specimen has a slightly lower air-entry value and a higher rate of desorption than that of recompacted one for suction up 50 kPa. The rates of desorption of the two soil specimens appear to be the same for high suctions. On the other hand, the wetting curve of the natural soil specimen is substantially higher than that of the recompacted specimen. This observed behavior may be explained by the capillary tube analogy.

5. For recompacted specimens subjected to different stress states under various applied loads, the higher the applied load on the specimen, the lower the rate of desorption. This is likely caused by the presence of an average smaller pore size distribution in the soil specimen under the higher applied load. As expected, there is a marked hysteresis between the drying and wetting curves for all soil specimens under various applied loads. The size of the hysteresis loops seems to have reduced with increase in the applied load.

6. The size of the hysteresis loops does not seem to be affected by the net normal stresses applied to natural specimens. Similar to the recompacted soil specimens, the natural soil specimen subjected to a higher applied load, possesses a slight larger air-entry value and lower rates of desorption and adsorption, due to the presence of an average smaller pore size distribution.

7. Traditionally, SWCC has been interpreted by neglecting any volume change throughout the test. By using the newly modified volumetric pressure plate extractor, it is found that the traditional method slightly

underpredicts the volumetric water content or degree of saturation in soil specimens for suctions higher than 100 kPa.

Generalised triaxial apparatus for determination of the state-dependent soil–water characteristic curve (Ng et al., 2001a)

Overview

Ng et al. (2001) observed that many traditional methods for measuring SWCCs were originated from soil science (Fredlund and Rahardjo, 1993). It is understandable that stress effects have not been taken into account and have rarely been studied. Recent work by Vanapalli et al. (1999) has found that the stress history and soil fabric have significant influence on the measured SWCCs of a compacted till. The SWCCs were determined on over-consolidated soil specimens under zero net normal stress, simulating the effects of over-consolidation. Constant volume of soil samples was assumed in the interpretation of test results. More recently, the effects of one-dimensional (ID) stresses on SWCCs have been investigated and found to have a significant influence on the SWCCs of a sandy silt/clay and hence on transient seepage and slope stability (Ng and Pang, 2000a).

As far as the authors are aware, no apparatus is available commercially for measuring drying and wetting soil–water characteristics of unsaturated soils under both isotropic and deviatoric stress states with volume change measurements. Following on the development of a recently modified one-dimensional stress controllable volumetric pressure plate extractor (Ng and Pang, 2000a), it was decided to build a new triaxial apparatus for measuring drying and wetting soil–water characteristics of soils under isotropic and deviatoric stress conditions.

Figure 3.44 shows a schematic diagram of the apparatus. Matric suction $(u_a - u_w)$ is controlled using the axis translation technique (Hilf, 1956), where u_a and u_w are air and water pressures, respectively. The sample size is designed to be 70 mm in diameter and 20 mm in height in order to minimize testing time. With a 5-bar air-entry value ceramic disk mounted inside the lower pedestal, matric suction up to 500 kPa can be adjusted by varying the air pressure via the top cap whereas water pressure is kept equal to the atmospheric pressure.

Various isotropic stresses can be applied by controlling the confining cell pressure (σ_3) acting on the soil. De-aerated water is used as cell fluid. To maintain a constant net normal stress $(\sigma_3 - u_a)$, if required, cell pressure and air pressure can be adjusted simultaneously. For controlling any deviatoric stress, an axial load can be applied via a stainless steel rod by using a loading frame. An internal load cell is used to verify the applied load.

Before and during any test, it is important to remove air bubbles trapped in the system. Beneath the ceramic disk, the surface of the base is engraved

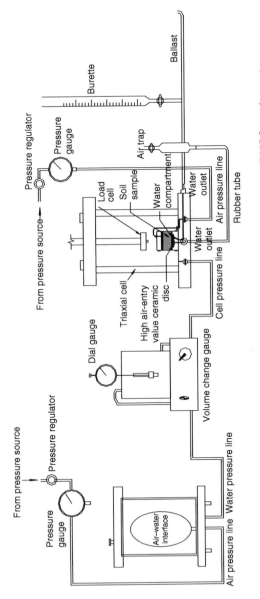

Figure 3.44 Schematic diagram of the newly modified triaxial apparatus for measuring SWCCs under various stresses (after Ng et al., 2001a).

with a spiral groove. Any water stored in the water compartment can flow out through two small apertures connected to the two water outlets. The groove is to drive out accumulated air bubbles in the water compartment (see Figure 3.44). Before taking volume measurements, air bubbles should be driven out by rolling the plastic tube between the two outlets and the air-trap. The volume change of the sample is measured using a volume gage by recording inflow/outflow of water. The deformation of the cell due to an applied isotropic stress is corrected by calibration. The resolution of the dial gauge for the volume gage is 0.01 mm, which is equivalent to 0.041 cm^3 of volume change. Other attachments such as the burette, air-trap and ballast are similar to a commercially available 2-bar volumetric pressure plate extractor (Soil Moisture Equipment Corp. Ltd, 1994). During the tests, temperature variations are controlled within ±1°C.

Measurements of the state-dependent soil–water characteristic curve in a centrifuge (Khanzode et al., 2002)

Background

Briggs and McLane (1907) appear to be the first investigators to use the centrifuge technique for measuring the relationship between soil suction and the water content retained by a soil (Khanzode et al., 2002). Gardner (1937) measured the capillary tension in a soil over a wide range of water contents by determining the equilibrium water content of calibrated filter papers that were in contact with the moist soil. The filter papers were calibrated by determining their water content when brought to equilibrium with a free water surface in a centrifugal field.

Russell and Richards (1938) improved the technique introduced by Briggs and McLane (1907) for measuring the amount of water retained in a soil at different values of applied suction. Hassler and Brunner (1945) used a centrifuge method to obtain the relationship between capillary pressure and saturation for small, consolidated core specimens. Croney et al. (1952) studies showed that the use of a solid ceramic cylinder in a centrifuge, in comparison to a hollow cylinder, considerably reduced the time required to attain equilibrium conditions.

Typically, in a conventional water-retention centrifuge technique, SWCCs are measured by draining a saturated soil specimen. Different values of equilibrium water content conditions (i.e. lower than the saturated water content condition) can be rapidly achieved by varying the distance of the soil specimen from the centre of rotation of the centrifuge and the speed of rotation of the centrifuge. An increase in the applied soil suction results in a decrease in the water content of the soil specimens. In other words, data required for the SWCC (i.e. water content versus suction relationship) can be obtained using the centrifuge technique. The Gardner (1937) equation can be used to estimate the suction in the soil specimen, and the water content can be determined using conventional procedures. In the described centrifuge testing procedure, there is a water content variation along the length of the soil specimen. However, the relative changes in water content and suction values over the thickness of the soil specimen are relatively small if thin specimens are used (i.e. 10–15 mm).

Principle of the centrifuge technique

A high gravity field is applied to an initially saturated soil specimen in the centrifuge. The soil specimen is supported on a saturated, porous ceramic column. The base of the ceramic stone rests in a water reservoir that is at atmospheric pressure conditions. The water content profile in the soil specimen after attaining equilibrium is similar to water draining under in situ conditions to a groundwater table where gravity is increased several times.

Figure 3.45 demonstrates the principle used in the centrifuge method for measuring soil suction. The suction in the soil specimen in a centrifuge can be calculated using Equation (3.47) as proposed by Gardner (1937).

$$\psi = \frac{\rho\omega^2}{2g}\left(r_1^2 - r_2^2\right) \tag{3.47}$$

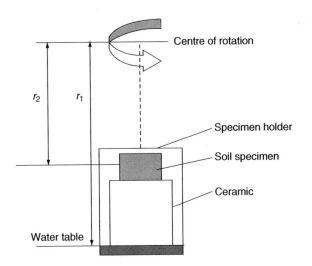

Figure 3.45 Soil suction measurement principle of the centrifuge (after Khanzode et al., 2002).

where
ψ is the suction in the soil specimen
r_1 is the radial distance to the mid-point of the soil specimen
r_2 is the radial distance to the free water surface
ω is the angular velocity
ρ is the density of the pore fluid and
g is the acceleration due to gravity.

Equation (3.47) defines a non-linear relationship between soil suction and centrifugal radius. The soil suction, ψ, becomes a function of the difference of the squares of the centrifugal radii, r_1 and r_2, while the density, ρ, and angular velocity, ω, are constant. The distance from the centre of the rotation to the free water surface, r_2, is a constant.

Different values of suction can be induced in a soil specimen by varying the radial distance to the mid-point of the soil specimen, r_1. In other words, ceramic cylinders of different heights can be used to achieve different suction applied to the soil specimen at a single speed of rotation. Higher values of soil suction can be subsequently induced into soil specimens by increasing the test speed (i.e. angular velocity, ω). It must be pointed out that the specimen located at r_1 from the centre of rotation is also subjected to an elevated centrifugal acceleration of $r_1\omega^2$, i.e. radial stress. Therefore, the SWCC determined is, in fact, a drying SDSWCC. Improvement can be made to measure volume change of the specimen in-flight automatically and continuously.

Description of the apparatus

Figure 3.46 shows a commercially available Beckman J6-HC small-scale medical centrifuge with an operable radius of 254 mm. The JS-4.2 rotor assembly of the centrifuge consists of six swinging type buckets capable of carrying six test specimens in one test run. The buckets in the centrifuge can be subjected to angular velocities varying from 50 to 4200 rpm. The maximum suction that can be applied to the specimen at 4200 rpm is 2,800 kPa.

The swinging buckets of the centrifuge assume horizontal positions when the centrifuge is spinning. All of the six buckets can be used simultaneously with six specimen holders available for testing. Six data points of water content versus soil suction can be obtained from a single test run of the centrifuge at a constant angular velocity, ω. The mass in all of the specimen holders, however, should be essentially the same to avoid rotary imbalance. Identical soil specimens must be placed at different heights in the six specimen holders to obtain six data points of water contents for different suction values. This also means that these six specimens are subjected to different radial centrifugal accelerations (or stresses), according to their radii from the centre of rotation. Interpretations of test data should be made with caution.

Figure 3.46 Small scale medical centrifuge rotor assembly with six swinging type buckets (after Khanzode *et al.*, 2002).

Specially designed soil specimen holders were constructed to accommodate the soil specimens in the two centrifuge buckets (see two soil specimen holders in two opposite buckets in Figure 3.46). The water content in the specimen can be measured after attaining equilibrium conditions, and the soil suction in the specimen is computed using Equation (3.27). Higher values of soil suction and radial stress can be induced in the same soil specimen by increasing the speed of rotation and centrifuging the specimens until new equilibrium conditions are attained.

Soil specimen holders

Khanzode *et al.* (2002) reported two specially designed aluminium soil specimen holders for the centrifuge to hold 12 mm thick soil specimens at different heights. Figure 3.47 shows the typical aluminium soil specimen holders used. The soil specimen holder consists of outer rings and a drainage plate with a free water surface reservoir to accommodate a ceramic cylinder. The outer rings have an inner diameter of 75 mm and are 15 mm thick. A reservoir cup serves as a collection area for water extracted from the soil specimens at the base of the holder.

A porous ceramic cylinder was designed to act as a filter while allowing the movement of water from the specimen to the drainage plate. This plate facilitates drainage into the reservoir cup through eight evenly spaced drainage

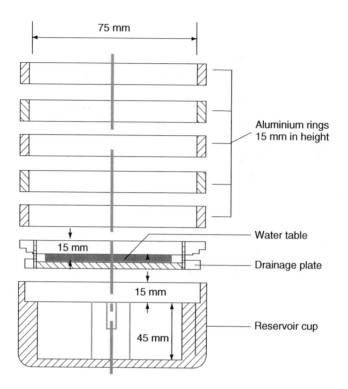

Figure 3.47 Details of aluminium soil specimen holder (after Khanzode *et al.*, 2002).

ports drilled horizontally through the sides of the plate. The horizontal overflow ports are connected to vertically drilled drainage holes to allow for the removal of water. The water then flows down from the drainage plate into the reservoir cup.

Ceramic cylinders

The ceramic cylinders used in the drainage plate were made of 60 per cent kaolinite and 40 per cent aluminium oxide. The porosity of the ceramic cylinders was approximately 45 per cent. Four ceramic cylinders with heights of 15, 30, 45, and 60 mm were made to keep the soil specimens at four different distances from the centre of rotation to induce four different suction values in the specimens at one speed of rotation. Table 3.9 summarizes the soil suctions associated with varying test speeds using different ceramic cylinders. Equation (3.47) was used to calculate the soils suction values.

Two ceramic cylinders of different heights were used in one test run to position the soil specimens at two different distances from the centre of rotation in the two opposite buckets of the centrifuge (see Figure 3.46).

Table 3.9 Soil suctions associated with different test speeds and different ceramic cylinders (after Khanzode et al., 2002)

Test speed (rpm)	Suction in the soil specimen (kPa)			
	15 mm cylinder	30 mm cylinder	45 mm cylinder	60 mm cylinder
300	3.28	5.91	8.33	10.53
500	9.08	16.43	23.15	29.24
1,000	36.34	65.74	92.52	116.99
1,500	81.78	147.94	208.43	263.27
2,000	145.35	262.9	370.43	467.89
2,500	227.0	410.74	578.68	730.94

The soil specimens were subjected to two different centrifugal forces and different values of soil suction were induced in two identical soil specimens placed in the soil specimen holders, subjected to the same angular velocity, ω.

Equation (3.27), used for calculating the equilibrium suction values in the soil specimen, does not take into account the shift in the centre of gravity of the soil specimen due to radial effects. The centre of gravity of a solid cylinder (or test specimens used for measuring the SWCCs) with parallel bases lies along centre line $(a - a)$ connecting the centres of the top and bottom circular bases of the cylinder (Figure 3.48). In other words, the centre of gravity of a soil specimen will be along the plane $(b - b)$, which lies at mid-height of the soil specimen. In spite of centrifugation, the centre of gravity will be along the mid-height plane and may shift towards the boundary of the soil specimen (i.e. away from the axis $a - a$). In such situations, r_1' should be used in Equation (3.47) instead of r_1 (Figure 3.48). Table 3.10 summarizes the suction values in soil specimens using r_1 and r_1' for a 15-mm height ceramic cylinder. The errors associated due to radial effects are less than 6 per cent for a 50-mm-diameter specimen for a test speed range of 0–2,500 rpm. The errors associated with the use of 30, 45, and 60 mm height ceramic cylinders are 5.2, 3.7 and 2.9 per cent, respectively. These errors from a practical perspective are not significant. Hence, the suctions calculated using Equation (3.47) at the mid-height of the specimen approximately represent the suction at the centre of gravity of the test specimen. It must be pointed out that the test specimen is also subjected to an elevated centrifugal acceleration of $r_1\omega^2$, i.e. radial stress. Therefore, the centrifuge provides data for the SDSWCC but not for the conventional SWCC. Other details of the apparatus, test procedures and experimental results are given by Khanzode et al. (2002).

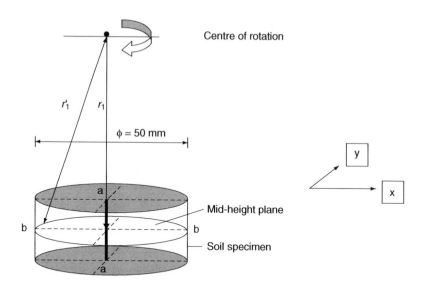

Figure 3.48 Schematic diagram to demonstrate the effects of centrifugation on the shift of the centre of gravity of solid cylindrical specimens (after Khanzode et al., 2002).

Table 3.10 Calculation of suction values in specimens using r_1 and r_1' for the 15 mm height ceramic cylinder (after Khanzode et al., 2002)

Test speed (rpm)	Suction at the midpoint of the specimen using r_1 for 15 mm cylinder (kPa)	Suction at the boundary of the specimen using r_1' for 15 mm cylinder (kPa)	Maximum possible error in %
300	3.28	3.08	6.0
500	9.08	8.57	5.6
1,000	36.34	34.30	5.61
1,500	81.78	77.18	5.62
2,000	145.35	137.17	5.60
2,500	227.0	214.29	5.60

Steady-state and transient flows (Fredlund and Rahardjo, 1993)

Introduction

Engineers are often interested in knowing the direction and quantity of flow through porous media. The pore pressure variations resulting from the flow process are vital for determining the stability and safety of soil structures.

This is because the information is required in predicting the volume change and shear strength change associated with the flow of water or air.

Seepage problems are usually categorized as steady-state or unsteady-state flow analyses.

- Steady-state analyses, the hydraulic head and the coefficient of permeability at any point in the soil mass remain constant with respect to time.
- Unsteady-state flow analyses, the hydraulic head (and possibly the coefficient of permeability) change with respect to time. Changes are usually in response to a change in the boundary conditions with respect to time.

The quantity of flow of an incompressible fluid such as water is expressed in terms of a flux, q. Flux is equal to a flow rate, v, multiplied by a cross-sectional area, A. On the other hand, the quantity of flow of a compressible fluid such as air is usually expressed in terms of a mass rate. The governing partial differential seepage equations are derived in a manner consistent with the conservation of mass. The conservation of mass for steady-state seepage of an incompressible fluid dictates that the flux into an element must equal the flux out of an element. In other words, the net flux must be zero at any point in the soil mass. For a compressible fluid, the net mass rate through an element must be zero in order to satisfy the conservation of mass for steady-state seepage conditions.

Steady-state water flow (Fredlund and Rahardjo, 1993)

The slow movement of water through soil is commonly referred to as seepage or percolation. Seepage analyses may form an important part of studies related to slope stability, groundwater contamination control and earth dam design. Seepage analyses involve the computation of the rate and direction of water flow and the pore water pressure distributions within the flow regime.

Water flow through unsaturated soils is governed by the same law as flow through saturated soils (i.e. Darcy's law). The main difference is that the water coefficient of permeability is assumed to be a constant for saturated soils, while it must be assumed to be a function of suction, water content or some other variable for unsaturated soils.

Also, the pore water pressure generally has a positive gauge value in a saturated soil and a negative gauge value in an unsaturated soil. In spite of these differences, the formulation of the partial differential flow equation is similar in both cases. There is also a smooth transition when going from the unsaturated to the saturated case (Fredlund, 1981).

Variation of coefficient of permeability with space for an unsaturated soil

For steady-state seepage analyses, the coefficient of permeability is a constant with respect to time at each point in a soil. However, the coefficient of permeability usually varies from one point to another in an unsaturated soil. A spatial variation in permeability in a saturated soil can be attributed to a heterogeneous distribution of the *soil solids*. For unsaturated soils, it is more appropriate to consider the heterogeneous volume distribution of the *pore-fluid* (i.e. pore water). This is the main reason for a spatial variation in the coefficient of permeability. Although the soil solid distribution may be homogeneous, the pore-fluid volume distribution can be heterogeneous due to spatial variations in matric suction. A point with a high matric suction (or a low water content) has a lower water coefficient of permeability than a point having a low matric suction.

Several functional relationships between the water coefficient of permeability and matric suction [i.e. $k_w(u_a - u_w)$] or volumetric water content [i.e. $k_w(\theta_w)$] have been described earlier. Coefficients of permeability for different points in a soil are obtained from the permeability function. The magnitude of the coefficient of permeability depends on the matric suction (or water content). In addition, the coefficient of permeability at a point may vary with respect to direction. This condition is referred to as anisotropy. The largest coefficient of permeability is called the major coefficient of permeability. The smallest coefficient of permeability is in a direction perpendicular to the largest permeability and is called the minor coefficient of permeability.

HETEROGENEOUS, ISOTROPIC STEADY-STATE SEEPAGE

Permeability conditions in unsaturated soils can be classified into three groups, as illustrated in Figure 3.49.

This classification is based on the pattern of permeability variation. A soil is called heterogeneous, isotropic if the coefficient of permeability in the x-direction, k_x, is equal to the coefficient of permeability in the y-direction, k_y, at any point in the soil mass (i.e. $k_x = k_y$ at A and $k_x = k_y$ at B). However the magnitude of the coefficient of permeability can vary from point A to point B, depending upon the matric suction in the soil.

HETEROGENEOUS, ANISOTROPIC STEADY-STATE SEEPAGE

The second case is that the ratio of the coefficient of permeability in the x-direction, k_x, to the coefficient of permeability in the y-direction, k_y, is a constant at any point (i.e. (k_x/k_y) at $A = (k_x/k_y)$ at $B =$ a constant not equal

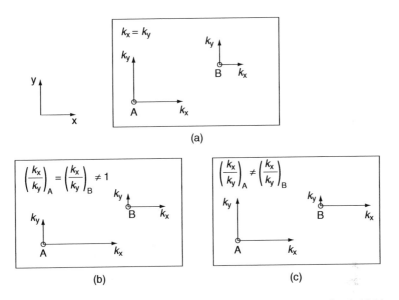

Figure 3.49 Coefficient of permeability variations in an unsaturated soil: (a) Heterogeneous, isotropic conditions; (b) heterogeneous, anisotropic conditions;
(c) continuous variation on permeability with space (after Fredlund and
Rahardjo, 1993).

to unity). The magnitude of the coefficients of permeability, k_x and k_y, can
also vary with matric suction from one location to another, but their ratio
is assumed to remain constant. Anisotropic conditional can also be oriented
in any two perpendicular directions. The general case of any orientation
for the major coefficient of permeability is not considered in the following
formulations.

The third case is where there is a continuous variation in the coefficient
of permeability. The permeability ratio (k_x/k_y) may not be a constant from
one location to another (i.e. (k_x/k_y) at $A \neq (k_x/k_y)$ at B), and different directions may have different permeability functions. The following steady-state
seepage formulations deal with the heterogeneous, isotropic and heterogeneous, anisotropic cases. The case where there is a continuous variation
in permeability with space requires further study and is not presented in
this lecture. All of the steady-state seepage analyses assume that the pore
air pressure has reached a constant equilibrium value. Where the equilibrium pore air pressure is atmospheric, the water coefficient of permeability
function with respect to matric suction, $k_w(u_a - u_w)$, has the same absolute value as the permeability function with respect to pore water pressure,
$k_w(-u_w)$.

One-dimensional flow

There are numerous situations where the water flow is predominantly in one direction. Let us consider a covered ground surface, with the water table located at a specified depth as shown in Figure 3.50 in which the surface cover prevents any vertical flow of water from the ground surface.

The pore water pressures are negative under static equilibrium conditions with respect to the water table. The negative pore water pressure head has a linear distribution with depth (i.e. line 1). Its magnitude is equal to the gravitational head (i.e. elevation head) measured relative to the water table. In other words, the hydraulic head (i.e. the gravitational head plus the pore water pressure head) is zero throughout the soil profile. This means that the change in head, and likewise the hydraulic gradient, is equal to zero. Therefore, there can be no flow of water in the vertical direction (i.e. $q_{wy} = 0$).

If the cover were removed from the ground surface, the soil surface would be exposed to the environment. Environmental changes could produce flow in a vertical direction, and subsequently alter the negative pore water pressure head profile. Steady-state evaporation would cause the pore water pressures to become more negative, as illustrated by line 2 in the figure. The hydraulic head changes to a negative value since the gravitational head remains constant. The hydraulic head has a non-linear

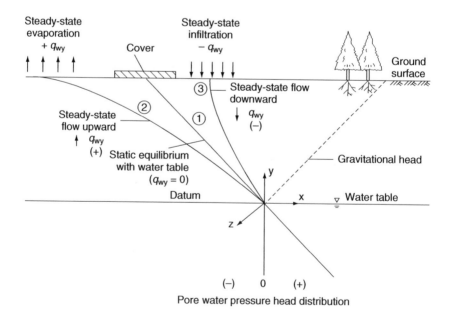

Figure 3.50 Static equilibrium and steady-state flow conditions in the zone of negative pore water pressures (after Fredlund and Rahardjo, 1993).

distribution from a zero value at the water table to a more negative value at ground surface. An assumption is made that the water table remains at a constant elevation. The non-linearity of the hydraulic head profile is caused by the spatial variation in the coefficient of permeability. Water flows in the direction of the decreasing hydraulic head. In other words, water flows from the water table upward to the ground surface. The upward constant flux of water is designated as positive for steady-state evaporation.

Steady-state infiltration causes a downward water flow. The negative pore water pressure increases from the static equilibrium condition. This condition is indicated by line 3 in the figure. The hydraulic head profile starts with a positive value at ground surface and decreases to zero at the water table. Therefore, water flows downward with a constant, negative flux for steady-state infiltration.

The above one-dimensional flow cases involve flux boundary conditions. A steady rate of evaporation or infiltration can be used as the boundary condition at ground surface. The water table acts as the lower boundary condition, giving a fixed zero pore water pressure head.

The steady-state procedure for measuring the water coefficient of permeability in the laboratory is also a one-dimensional flow example. In the laboratory measurement of the coefficient of permeability, however, the hydraulic heads are controlled as boundary conditions at the top and bottom of the soil specimen. Techniques to analyse both head and flux boundary conditions are explained in the following sections.

FORMULATION FOR ONE-DIMENSIONAL FLOW

Consider an unsaturated soil element with one-dimensional water flow in the y-direction, as shown in Figure 3.51.

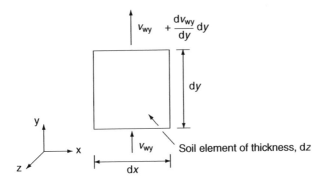

Figure 3.51 One-dimensional water flow through an unsaturated soil element (after Fredlund and Rahardjo, 1993).

The element has infinitesimal dx, dy, and dz dimensions. The flow rate, v_{wy}, is assumed to be positive when water flows upward in the y-direction. Continuity requires that the volume of water flowing in and out of the element must be equal for steady-state conditions:

$$\left(v_{wy} + \frac{dv_{wy}}{dy} dy \right) dx dz - v_{wy} dx\, dz = 0 \tag{3.48}$$

where

v_{wy} = water flow rate across a unit area of the soil in the y-direction
dx, dy, dz = dimensions in the x-, y- and z-directions, respectively.

The net flow can be written as follows:

$$\left(\frac{dv_{wy}}{dy} \right) dx\, dy\, dz = 0 \tag{3.49}$$

Substituting Darcy's law into Equation (3.49) yields

$$\left(\frac{d\left\{ -k_{wy} \dfrac{dh_w}{dy} \right\}}{dy} \right) dx\, dy\, dz = 0 \tag{3.50}$$

where
k_{wy} = water coefficient of permeability as a function of matric suction, $(u_a - u_w)$, which varies with location in the y-direction
dh_w/d_y = hydraulic head gradient in the y-direction
h_w = hydraulic head (i.e. gravitational head plus pore water pressure head).

Equation (3.50) can be used to solve for the hydraulic head distribution in the y-direction through a soil mass. Since matric suction varies from one location to another, the coefficient of permeability also varies. Rewriting Equation (3.50) and considering the non-zero dimensions for dx, dy and dz gives the following non-linear differential equation:

$$k_{wy} \frac{d^2 h_w}{dy^2} + \frac{dk_{wy}}{dy} \frac{dh_w}{dy} = 0 \tag{3.51}$$

where dk_{wy}/dy = change in water coefficient of permeability in the y-direction due to a change in matric suction.

The non-linearity of Equation (3.51) is caused by its second term, which accounts for the variation in permeability with respect to space.

When the soil becomes *saturated*, the water coefficient of permeability, k_{wy}, can be taken as being equal to a single-valued, saturated coefficient of permeability, k_s. If the saturated soil is *heterogeneous* (e.g. layered soil), the coefficient of permeability, k_s, will again vary with respect to location. In a saturated soil, the heterogeneous distribution of the *soil solids* is the primary factor producing a varying coefficient of permeability. As a result, the flow equation can be written as follows:

$$k_s \frac{d^2 h_w}{dy^2} + \frac{dk_s}{dy} \frac{dh_w}{dy} = 0 \qquad (3.52)$$

where k_s = saturated coefficient of permeability.

A comparison of the Equations (3.51) and (3.52) reveals a similar form. In other words, the non-linearity in the unsaturated soil flow equation produces the same form of equation as that required for a heterogeneous, saturated soil. In an *unsaturated* soil, the variation in the coefficient of permeability is caused by the *heterogeneous* distribution of the *pore-fluid* volume occurring as a result of different matric suction values.

If a saturated soil is *homogeneous*, the coefficient of permeability is constant for the soil mass. Substituting a non-zero, constant coefficient of permeability into Equation (3.52) produces a linear differential equation:

$$\frac{d^2 h_w}{dy^2} = 0 \qquad (3.53)$$

Equations similar to Equation (3.53) can also be derived for one-dimensional flow in the x- and z-directions.

Two-dimensional flow

Seepage through an earth dam is a classical example of two-dimensional flow. Water flow is in the cross-sectional plane of the dam, while flow perpendicular to the plane is assumed to be negligible. Until recently, it has been conventional practice to neglect the flow of water in the unsaturated zone of the dam. The analysis presented herein assumes that water flows in both the saturated and unsaturated zones in response to a hydraulic head driving potential.

The following two-dimensional formulation is an expanded form of the previous one-dimensional flow equation. The formulation is called an uncoupled solution since it only satisfies continuity. For a rigorous formulation of two-dimensional flow, continuity should be coupled with the force equilibrium equations.

FORMULATION FOR TWO-DIMENSIONAL FLOW

The following derivation is for the general case of a heterogeneous, anisotropic, unsaturated soil. The coefficients of permeability in the x-direction, k_{wx}, and the y-direction, k_{wy}, are assumed to be related to the matric suction by the same permeability function, $f(u_a - u_w)$. The ratio of the coefficients of permeability in the x- and y-directions, (k_{wx}/k_{wy}), is assumed to be constant at any point within the soil mass.

A soil element with infinitesimal dimensions of dx, dy and dz is considered, but flow is assumed to be two-dimensional, as shown in Figure 3.52.

TWO-DIMENSIONAL WATER FLOW THROUGH AN UNSATURATED SOIL ELEMENT

The flow rate, v_{wx}, is positive when water flows in the positive x-direction. The flow rate, v_{wy}, is positive for flow in the positive y-direction. Continuity for two-dimensional, steady-state flow can be expressed as follows:

$$\left(v_{wx} + \frac{\partial v_{wx}}{\partial x} dx - v_{wx}\right) dy\, dz + \left(v_{wy} + \frac{\partial v_{wy}}{\partial y} dy - v_{wy}\right) dx\, dz = 0 \quad (3.54)$$

where v_{wx} = water flow rate across a unit area of the soil in the x-direction.

Therefore, the net flux in the x- and y-directions is

$$\left(\frac{\partial v_{wx}}{\partial x} + \frac{\partial v_{wy}}{\partial y}\right) dx\, dy\, dz = 0 \quad\quad\quad (3.55)$$

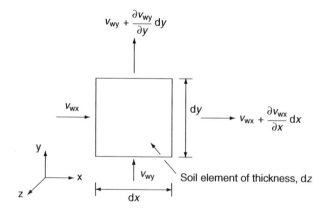

Figure 3.52 Two-dimensional water flow through an unsaturated soil element (after Fredlund and Rahardjo, 1993).

Substituting Darcy's laws into Equation (3.55) results in a non-linear partial differential equation:

$$\frac{\partial}{\partial x}\left\{k_{wx}\frac{\partial h_w}{\partial x}\right\} + \frac{\partial}{\partial y}\left\{k_{wy}\frac{\partial h_w}{\partial y}\right\} = 0 \qquad (3.56)$$

where
k_{wx} = water coefficients of permeability as a function of matric suction, $(u_a - u_w)$; the permeability can vary with location in the x-direction.
$\partial h_w/\partial x$ = hydraulic head gradient in the x-direction.

Equation (3.56) describes the hydraulic head distribution in the x–y plane for steady-state water flow. The non-linearity of the equation becomes more obvious after an expansion of the equation:

$$k_{wx}\frac{\partial^2 h_w}{\partial x^2} + k_{wy}\frac{\partial^2 h_w}{\partial y^2} + \frac{\partial k_{wx}}{\partial x}\left\{\frac{\partial h_w}{\partial x}\right\} + \frac{\partial k_{wy}}{\partial y}\left\{\frac{\partial h_w}{\partial y}\right\} = 0 \qquad (3.57)$$

where $\partial k_{wx}/\partial x$ = change in water coefficient of permeability in the x-direction.

The spatial variation of the coefficient of permeability given in the third and fourth terms in Equation (3.57) produces non-linearity in the governing flow equation.

For the heterogeneous, isotropic case, the coefficients of permeability in the x- and y-directions are equal (i.e. $k_{wx} = k_{wy} = k_w$). Therefore, Equation (3.57) can be written as follows:

$$k_w\left(\frac{\partial^2 h_w}{\partial x^2} + \frac{\partial^2 h_w}{\partial y^2}\right) + \frac{\partial k_w}{\partial x}\left\{\frac{\partial h_w}{\partial x}\right\} + \frac{\partial k_w}{\partial y}\left\{\frac{\partial h_w}{\partial y}\right\} = 0 \qquad (3.58)$$

where k_w = water coefficient of permeability in the x- and y-directions.

Seepage through a dam involves flow through the unsaturated and saturated zones. Flow through the saturated soil can be considered as a special case of flow through an unsaturated soil. For the *saturated* portion, the water coefficient of permeability becomes equal to the saturated coefficient of permeability, k_s. The saturated coefficients of permeability in the x- and y-directions, k_{sx} and k_{sy}, respectively, may not be equal due to anisotropy. The saturated coefficients of permeability may vary with respect to location due to heterogeneity. A summary of steady-state equations for *saturated* soils under different conditions is presented in the Table 3.11.

The equations listed in Table 3.11 are specialized forms that can be derived from the steady-state flow equation for unsaturated soils, i.e. Equation (3.41). Therefore, steady-state seepage through saturated–unsaturated soils can be analysed simultaneously using the same governing equation.

Table 3.11 Two-dimensional steady-state equations for saturated soils (Fredlund and Rahardjo, 1993)

Anisotropic	Isotropic
Heterogeneous	
$k_{sx}\dfrac{\partial^2 h_w}{\partial x^2} + k_{sy}\dfrac{\partial^2 h_w}{\partial y^2} + \dfrac{\partial k_{sx}}{\partial x}\dfrac{\partial h_w}{\partial x}$	$k_s\left(\dfrac{\partial^2 h_w}{\partial x^2} + \dfrac{\partial^2 h_w}{\partial y^2}\right) + \dfrac{\partial k_s}{\partial x}\dfrac{\partial h_w}{\partial x}$
$+\dfrac{\partial k_{sy}}{\partial y}\dfrac{\partial h_w}{\partial y} = 0$	$+\dfrac{\partial k_s}{\partial y}\dfrac{\partial h_w}{\partial y} = 0$
Homogeneous	
$k_{sx}\dfrac{\partial^2 h_w}{\partial x^2} + k_{sy}\dfrac{\partial^2 h_w}{\partial y^2} = 0$	$\dfrac{\partial^2 h_w}{\partial x^2} + \dfrac{\partial^2 h_w}{\partial y^2} = 0$

Flow in an infinite slope (Fredlund and Rahardjo, 1993)

A slope of infinite length is illustrated in Figure 3.53. Let us consider the case where steady-state water flow is established within the slope and the phreatic line is parallel to the ground surface. Water flows through both the saturated and unsaturated zones and is parallel to the phreatic line. The direction of the water flow indicates that there is no flow perpendicular to the phreatic line. In other words, the hydraulic head gradient is equal to zero in a direction perpendicular to the phreatic line. In this case, the lines drawn normal to the phreatic line are equipotential lines.

Isobars are parallel to the phreatic line. This is similar to the condition in the central section of a homogeneous dam, as shown earlier. The coefficient of permeability is essentially independent of the pore water pressure in the saturated zone. Therefore, the saturated zone can be subdivided into several flow channels of equal size. An equal amount of water (i.e. water flux, q_w,) flows through each channel. Lines separating the flow channels are referred to as flow lines.

The water coefficient of permeability depends on the negative pore water pressure or the matric suction in the unsaturated zone. The pore water pressure decreases from zero at the phreatic line to some negative value at ground surface. Similarly, the permeability decreases from the phreatic line to ground surface. As a result, increasingly larger flow channels are required in order to maintain the same quantity of water flow, q_w, as ground surface is approached.

The water flow in each channel is one-dimensional, in a direction parallel to the phreatic line. The coefficient of permeability varies in the direction perpendicular to flow. This condition can be compared to the previous case of water flow through a vertical column. In the case of the vertical

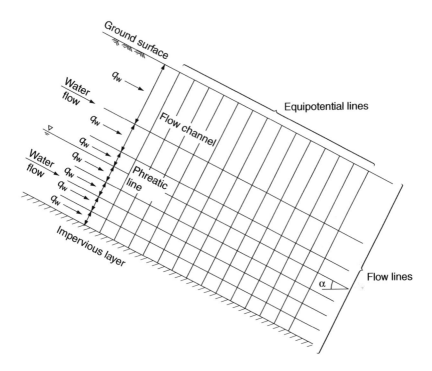

Figure 3.53 Steady-state water flow through an infinite slope (after Fredlund and Rahardjo, 1993).

column, the coefficient of permeability varied in the flow direction, and the equipotential lines were not equally distributed throughout the soil column.

The above examples illustrate that equipotential lines, and flow lines intersect at right angles for unsaturated flow problems, as long as the soil is isotropic. Heterogeneity with respect to the coefficient of permeability results in varying distances between either the flow lines or the equipotential lines; however, these lines cross at 90°.

The pore water pressure distribution in the unsaturated zone can be analysed by considering a horizontal datum through an arbitrary point (e.g. point A in Figure 3.54) on the phreatic line. The pore water pressure distribution in a direction perpendicular to the phreatic line (i.e. in the a-direction) is first examined. The results are then used to analyse the pore water pressure distribution in the y-direction (i.e. vertically). The gravitational head distribution in the a-direction is zero at point A (i.e. datum) and increases linearly to a gravitational head of ($H \cos^2 \alpha$) at ground surface. The pore water pressure head at a point in the a-direction must be negative and equal in magnitude to its gravitational head because the hydraulic heads are zero

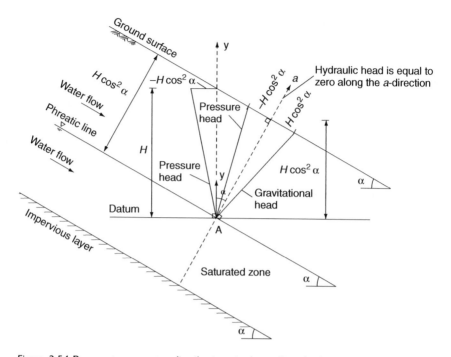

Figure 3.54 Pore water pressure distributions in the undisturbed zone of an infinite slope during steady-state seepage (after Fredlund and Rahardjo, 1993).

in the *a*-direction. Therefore, the pore water pressure head distribution in the *a*-direction must start at zero at the datum (i.e. point *A*) and decrease linearly to $(-H \cos^2 \alpha)$ at ground surface. A pore water pressure head of $(-H \cos^2 \alpha)$ applies to any point along the ground surface since every line parallel to the phreatic line is also an isobar.

The pore water pressure head distribution in a vertical direction also commences with a zero value at point *A*, and decreases linearly to a head of $(-H \cos^2 \alpha)$ at ground surface. However, the pore water pressure head is distributed along a length, $(H \cos \alpha)$, in the *a*-direction, while the head is distributed along a length, *H*, in the vertical direction. The negative pore water pressure head at a point on a vertical plane can therefore be expressed as follows:

$$h_{pi} = -y \cos^2 \alpha \tag{3.59}$$

where

h_{pi} = negative pore water pressure head on a vertical plane (i.e. the *y*-direction) for an infinite slope

$y =$ vertical distance from the point under consideration to the datum
(i.e. point A)
$\alpha =$ inclination angle of the slope and the phreatic line.

When the ground surface and the phreatic line are horizontal (i.e. $\alpha = 0$ or
$\cos \alpha = 1$), the negative pore water pressure head at a point along a vertical
plane, h_{ps}, is equal to $-y$. This is the condition of static equilibrium above
and below a horizontal water table. The ratio between the pore water pressure heads on a vertical plane through an infinite slope (i.e. $h_{pi} = -y \cos^2 \alpha$)
and the pore water pressure heads associated with a horizontal ground
surface (i.e. $h_{ps} = -y$) is plotted in Figure 3.55. This ratio indicates the
reduction in the pore water pressures on a vertical plane as the slope, α,
becomes steeper.

The gravitational head at a point along a vertical plane is equal to its
elevation from the datum, y (Figure 3.55). The hydraulic head is computed
as the sum of the gravitational and pore water pressure heads:

$$h_{wi} = \left(1 - \cos^2 \alpha\right) y \qquad (3.60)$$

Equation (3.60) indicates that there is a decrease in the hydraulic head as
the datum is approached. In other words, there is a vertical downward
component of water flow.

The above analysis also applies to the pore water pressure conditions
below the phreatic line. Using the same horizontal line through point A,
positive pore water pressure heads along a vertical plane can be computed

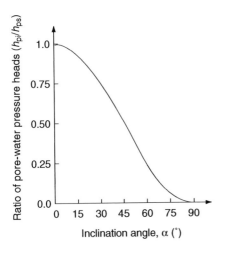

Figure 3.55 Effect of slope inclination on the pore water pressure distribution along
a vertical plane (after Fredlund and Rahardjo, 1993).

in accordance with $h_{pi} = -y\cos^2\alpha$. The hydraulic head $[h_{wi} = (1 - \cos^2\alpha)y]$ is zero at the phreatic line, and decreases linearly with depth along a vertical plane.

Steady-state air flow (Fredlund and Rahardjo, 1993)

Introduction

The bulk flow of air can occur through an unsaturated soil when the air phase is continuous. In many practical situations, the flow of air may not be of concern. However, it is of value to understand the formulations for compressible flow through porous media.

The air coefficient of transmission, D_a^*, or the air coefficient of permeability, k_a (i.e. $D_a^* g$), is a function of the volume–mass properties or the stress state of the soil. The relationships between the air coefficient of permeability, k_a, and matric suction [i.e. $k_a(u_a - u_w)$] or degree of saturation [i.e. $k_a(S_e)$] are described above. The value of k_a or D_a^* may vary with location, depending upon the distribution of the pore air volume in the soil. Possible variations in the air coefficient of permeability in an unsaturated soil are described in the beginning of this lecture. The air coefficient of permeability at a point can be assumed to be constant with respect to time during steady-state air flow.

The section presents the steady-state formulations for one- and two-dimensional air flow using Fick's law. Heterogeneous, isotropic and anisotropic situations are presented. Steady-state air flow is analysed by assuming that the pore water pressure has reached equilibrium. The following air flow equations can be solved using numerical methods such as the finite difference or the finite element methods. The manner of solving the equations is similar to that described in the previous sections.

One-dimensional flow of air

Consider an unsaturated (i.e. heterogeneous) soil element with one-dimensional air flow in the y-direction, as shown in Figure 3.56.

The air flow has a mass rate of flow, J_{ay}, under steady-state conditions. The mass rate is assumed to be positive for an upward air flow. The principle of continuity states that the mass of air flowing into the soil element must be equal to the mass of air flowing out of the element:

$$\left(J_{ay} + \frac{dJ_{ay}}{dy}\,dy\right)dx\,dz - J_{ay}dx\,dz = 0 \tag{3.61}$$

where J_{ay} = mass rate of air flow across a unit area of the soil in the y-direction.

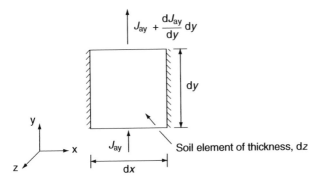

Figure 3.56 One-dimensional steady-state air flow through an unsaturated soil element (after Fredlund and Rahardjo, 1993).

Rearranging Equation (3.61) gives the net mass rate of air flow

$$\left(\frac{dJ_{ay}}{dy}\right) dx\, dy\, dz = 0 \tag{3.62}$$

Substituting Fick's law for the mass rate of flow into Equation (3.62) yields a non-linear differential equation:

$$\left(d\left\{-D^*_{ay}\frac{du_a}{dy}\right\}/dy\right) dx\, dy\, dz = 0 \tag{3.63}$$

where
D^*_{ay} = air coefficient of transmission as a function of matric suction, $(u_a - u_w)$
du_a/dy = pore air pressure gradient in the y-direction
u_a = pore air pressure.

The spatial variation of D^*_{ay} causes non-linearity in Equation (3.63):

$$D^*_{ay}\frac{d^2 u_a}{dy^2} + \frac{dD^*_{ay}}{dy}\frac{du_a}{dy} = 0 \tag{3.64}$$

where dD^*_{ay}/dy = change in the air coefficient of transmission in the y-direction.

The above two equations describe the pore air pressure distribution in the soil mass in the y-direction. The second term in the last equation accounts for the spatial variation in the coefficient of transmission. The coefficient of transmission is obtained by dividing the air coefficient of permeability,

k_{ay}, by the gravitational acceleration (i.e. $D^*_{ay} = k_{ay}/g$). In other words, the coefficients D^*_{ay} and k_{ay} have similar functional relationships to matric suction.

The measurement of the air coefficient of permeability using a triaxial permeameter cell is an application involving one-dimensional, steady-state air flow. In this case, however, the air coefficient of permeability is assumed to be constant throughout the soil specimen. Neglecting the change in the air coefficient of permeability with respect to location, Equation (3.64) is reduced to a linear differential equation:

$$\frac{d^2 u_a}{dy^2} = 0 \tag{3.65}$$

The pore air pressure distribution in the y-direction is obtained by integrating Equation (3.65) twice:

$$u_a = C_1 y + C_2 \tag{3.66}$$

where
C_1, C_2 = constants of integration related to the boundary conditions
y = distance in the y-direction.

The pore air pressure distribution within a soil specimen during an air permeability test is illustrated in Figure 3.57.

The air pressures at both ends of the specimen (i.e. $u_a = u_{ab}$ at $y = 0.0$ and $u_a = u_{at} = 0.0$ at $y = h_s$) are the boundary conditions. Substituting the boundary conditions into the last equation results in a linear equation for the pore air pressure along the soil specimen (i.e. $u_a = (1 - y/h_s)\, u_{ab}$).

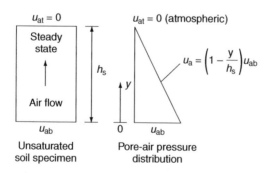

Figure 3.57 Pore air pressure distribution during the measurement of the air coefficient permeability, k_a (after Fredlund and Rahardjo, 1993).

Two-dimensional flow of air

Two-dimensional, steady-state air flow is first formulated for the hetero-geneous, anisotropic condition. The air coefficients of transmission in the x- and y-directions, D_{ax}^* and D_{ay}^*, are related to matric suction using the same transmission function, $D_a^* = f(u_a - u_w)$. The (D_{ax}^*/D_{ay}^*) ratio will be assumed to be constant at any point within the soil mass.

An element of soil subjected to two-dimensional air flow is shown in the Figure 3.58.

Satisfying continuity for steady-state flow yields the following equation:

$$\left(J_{ax} + \frac{\partial J_{ax}}{\partial x} dx - J_{ax}\right) dy\, dz + \left(J_{ay} + \frac{\partial J_{ay}}{\partial y} dy - J_{ay}\right) dx\, dz = 0 \qquad (3.67)$$

where J_{ax} = mass rate of air flowing across a unit area of the soil in the x-direction.

Rearranging Equation (3.67) results in the following equation:

$$\left(\frac{\partial J_{ax}}{\partial x} + \frac{\partial J_{ay}}{\partial y}\right) dx\, dy\, dz = 0 \qquad (3.68)$$

Substituting Fick's law for the mass rates, J_{ax} and J_{ay}, into Equation (3.68) gives the following non-linear partial differential equation:

$$\frac{\partial}{\partial x}\left\{D_x^* \frac{\partial u_a}{\partial x}\right\} + \frac{\partial}{\partial y}\left\{D_y^* \frac{\partial u_a}{\partial y}\right\} = 0 \qquad (3.69)$$

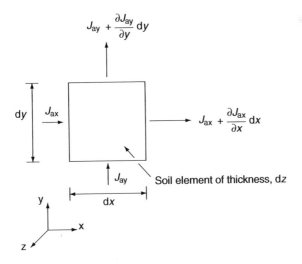

Figure 3.58 An element subjected to two-dimensional air flow (after Fredlund and Rahardjo, 1993).

where
$D_{ax}^* =$ air coefficient of transmission as a function of matric suction, $(u_a - u_w)$ and $\partial u_a / \partial x =$ pore air pressure gradient in the x-direction.

The coefficient of transmission, D_{ax}^*, is related to the air coefficient of permeability, k_{ax}^*, by the gravitational acceleration (i.e., $D_{ax}^* = k_{ax}^*/g$). Expanding Equation (3.69) results in the following flow equation:

$$\left\{ D_x^* \frac{\partial^2 u_a}{\partial x^2} \right\} + \left\{ D_y^* \frac{\partial^2 u_a}{\partial y^2} \right\} + \frac{\partial D_{ax}^*}{\partial x} \left\{ \frac{\partial u_a}{\partial x} \right\} + \frac{\partial D_{ay}^*}{\partial y} \left\{ \frac{\partial u_a}{\partial y} \right\} = 0 \qquad (3.70)$$

where $\partial D_{ax}^* / \partial x =$ change in air coefficient of transmission in the x-direction.

Spatial variations in the coefficients of transmission are accounted for by the third and fourth terms in the last equation. These two terms produce the non-linearity in the flow equation. When solved, the last equation describes the pore air pressure distribution in the x–y plane of the soil mass during two-dimensional, steady-state air flow.

For the heterogeneous, isotropic case, the coefficients of transmission in the x- and y-directions are equal (i.e., $D_{ax}^* = D_{ay}^* = D_a^*$), and Equation (3.70) becomes

$$D_a^* \left\{ \frac{\partial^2 u_a}{\partial x^2} + \frac{\partial^2 u_a}{\partial y^2} \right\} + \frac{\partial D_a^*}{\partial x} \left\{ \frac{\partial u_a}{\partial x} \right\} + \frac{\partial D_a^*}{\partial y} \left\{ \frac{\partial u_a}{\partial y} \right\} = 0 \qquad (3.71)$$

where $D_a^* =$ air coefficient of transmission in the x- and y-directions.

These partial differential equations for air flow are similar in form to those previously presented for water flow.

Two-dimensional transient flow of water (Fredlund and Rahardjo, 1993)

Introduction

Unlike the steady-state water flow, transient flow alters the moisture content or volumetric water content in an soil element as shown Figure 3.59.

The flow rate, v_{wx}, is positive when water flows in the positive x-direction. The flow rate, v_{wy}, is positive for flow in the positive y-direction. For unsteady-state flow (transient):

$$\left(v_{wx} + \frac{\partial v_{wx}}{\partial x} dx - v_{wx} \right) dy\, dz + \left(v_{wy} + \frac{\partial v_{wy}}{\partial y} dy - v_{wy} \right) dx\, dz$$

$$+ q(t)dx\, dy\, dz + \frac{\partial \theta_w}{\partial t} dx\, dy\, dz = 0 \qquad (3.72)$$

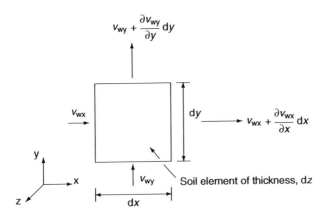

Figure 3.59 Two-dimensional transient water flow through an unsaturated soil element. (after Fredlund and Rahardjo, 1993)

where
θ_w = volumetric water content, which is a function of matric suction, $(u_a - u_w)$
v_{wx}, v_{wy} = water flow rate across a unit area of the soil in the x- and y-direction, respectively
$q(t)$ = applied boundary flux.

Therefore, the net flux in the x- and y-directions is

$$\left(\frac{\partial v_{wx}}{\partial x} + \frac{\partial v_{wy}}{\partial y} + \frac{\partial \theta_w}{\partial t} + q(t) \right) dx\, dy\, dz = 0 \tag{3.73}$$

Substituting Darcy's laws into Equation (3.73) results in a non-linear partial differential equation for transient water flow:

$$\frac{\partial}{\partial x} \left\{ k_{wx} \frac{\partial h_w}{\partial x} \right\} + \frac{\partial}{\partial y} \left\{ k_{wy} \frac{\partial h_w}{\partial y} \right\} + q(t) + \frac{\partial \theta_w}{\partial t} = 0 \tag{3.74}$$

If it is assumed that no external loads are added to the soil mass during the transient process and that the air phase is continuous in the unsaturated zone,

$$d\theta_w = m_1 d(u_a - u_w) \tag{3.75}$$

where m_1 is the slope of the $(u_a - u_w)$ versus θ_w plot.

For no external loads, $u_a = 0$, therefore, the governing equation becomes

$$\frac{\partial}{\partial x}\left\{k_{wx}\frac{\partial h_w}{\partial x}\right\} + \frac{\partial}{\partial y}\left\{k_{wy}\frac{\partial h_w}{\partial y}\right\} + q(t) = m_1\frac{\partial u_w}{\partial t} \qquad (3.76)$$

or

$$\frac{\partial}{\partial x}\left\{k_{wx}\frac{\partial h_w}{\partial x}\right\} + \frac{\partial}{\partial y}\left\{k_{wy}\frac{\partial h_w}{\partial y}\right\} + q(t) = m_1\rho_w g\frac{\partial h_w}{\partial t} \qquad (3.77)$$

Analytical analysis of rainfall infiltration mechanism in unsaturated soils (Zhan and Ng, 2004)

Introduction

The distribution and variation of negative pore water pressure in unsaturated soils are governed by the water flow, which in turn is affected by many intrinsic and external factors. The intrinsic factors are mainly the hydraulic properties of the soils, including water retention characteristics and water coefficient of permeability. The external factors mainly refer to climatic conditions, such as rainfall intensity and duration, rainfall pattern and evapotranspiration rate. The effects of these factors on rainfall infiltration and hence on the pore water pressure responses have been partially investigated by several researchers as presented in the following paragraphs.

The simple infiltration model proposed by Green and Ampt (1911) indicates that the depth of the wetting front during rainfall infiltration is proportional to the saturated permeability (k_s) but inversely proportional to the specific yield in the ground. The semi-empirical model developed by Leach and Herbert (1982) suggests that the pore water pressure response to rainfall in a given geometry is a function of the ratio k_s/S_s (where, k_s is saturated permeability of a soil and S_s is specific storage of the soil). The higher the value of the ratio (k_s/S_s), the faster the hydraulic heads will rise as a result of rainfall infiltration, and the more rapid will be the pore water pressure response to storm events. In this model, the dependence of water coefficient of permeability on soil suction (or water content) was not taken into account.

Kasim et al. (1998) performed a numerical simulation to investigate the influence of the coefficient of permeability function on the steady-state pore water pressure distributions in horizontal and inclined unsaturated soil layer. The coefficient of permeability function is characterized by a saturated permeability (k_s), an approximate air-entry value (a) and a desaturation rate (n). The study shows that the ratio of rainfall infiltration rate to saturated permeability (i.e. q_f/k_s) and the air-entry value of the soil are the dominant factors affecting the steady-state pore water pressure distributions. For a given q_f/k_s ratio, the computed steady-state suction values increase with an air-entry value. For simplicity, the influence of air-entry value is not considered again in this chapter.

Ng and Shi (1998) and Ng et al. (1999) accounted for more parameters in their numerical studies on the effects of rain infiltration on pore water pressure in unsaturated slopes and hence on slope stability. The parameters considered in their study included intensity and duration of rainfall, saturated permeability and its anisotropy, the presence of an impeding layer and conditions of surface cover. They found that the response of negative pore water pressure and the groundwater table is mainly governed by the ratio of q/k_s and k_s/m_w (where, q is the infiltration flux, k_s is saturated permeability and m_w is the slope of the SWCC) as well as the initial and boundary conditions. They also found that for a given slope, there existed a critical saturated permeability and a critical rainfall duration with regards to the effect of rainfall infiltration on slope stability. However, SWCC (i.e. the relationship between water content and matric suction) was kept the same in their studies, i.e. the effects of different SWCCs not investigated. In addition, the relative sensitivity and combined effects of some relevant parameters were not investigated.

Effects of hysteresis of a SWCC on unsaturated flows were studied by Ng and Pang (2000b) and Rahardjo and Leong (1997). Ng and Pang (2000b) showed that during a prolonged rainfall, the analysis using a wetting SWCC would predict adverse pore water pressure distributions with depth than those from an analysis using a drying SWCC. Rahardjo and Leong (1999) demonstrated that the use of a drying SWCC with a higher permeability function would result in a deeper infiltration depth. Both articles suggested that it would be more appropriate to use a wetting SWCC for infiltration problems.

In spite of all the valuable work reviewed above, effects of hydraulic parameters and rainfall conditions on the infiltration into an unsaturated ground, and hence on the pore water pressure responses, are still not fully understood. The reasons for this may be (1) high non-linearity and variability of hydraulic parameters involved in the transient unsaturated flow system; (2) spatial and temporal variations in the initial and boundary conditions of the system.

Parametric studies of rainfall infiltration into an unsaturated ground can be carried out either by using an analytical solution or by using numerical simulations. Analytical solutions generally have advantages of explicitness and simplicity over numerical simulations. Of course, because of the high non-linearity of hydraulic parameters involved in the governing equation of unsaturated flow, analytical solutions for the infiltration problem can only be obtained by making some assumptions and under some given initial and boundary conditions (Philip, 1957; Broadbridge and White, 1988; Pullan, 1990; Sander, 1991; Srivastava and Yeh, 1991; Boger, 1998). In here, an analytical solution for one-dimensional rainfall infiltration in homogeneous and two-layer soils derived by Srivastava and Yeh (1991) was selected and applied for parametric study. The reasons for this selection are (1) almost all the key parameters involved in the infiltration system (i.e. saturated permeability, desaturation rate, water storage capacity, antecedent and subsequent rainfall infiltration rates) were taken into account in this analytical solution; (2) the pore water pressure responses associated with rainfall infiltration can be directly calculated by this solution.

Here, hydraulic properties of unsaturated soils are first discussed to clarify the key hydraulic parameters controlling an unsaturated flow (i.e. saturated permeability, desaturation rate and water storage capacity). Then, the analytical solution for one-dimensional rainfall infiltration in homogeneous and two-layer soils derived by Srivastava and Yeh (1991) is reviewed. Thereafter, parametric studies are carried out to investigate the influences of the key parameters (i.e. saturated permeability, desaturation rate, water storage capacity, antecedent and subsequent rainfall infiltration rates) on the transient pore water responses in an infiltration system. The investigation includes the individual effect of each parameter, the possible combined effects of some relevant parameters and a comparison of the effects among the relevant parameters. Insights into the mechanism of the infiltration process and the associated pore water pressure response in unsaturated soils are discussed.

Hydraulic properties of unsaturated soils

Similar to the flow through a saturated soil, water flow through an unsaturated soil is generally governed by Darcy's law (Fredlund and Rahardjo, 1993). However, comparing the water flow in an unsaturated soil with the saturated flow, two major differences stand out (1) there exits a storage term which represents the variation of water content with matric suction; (2) the water coefficient of permeability depends strongly on matric suction. It should be noted that no volume change in soil is considered during the infiltration process. The storage term in unsaturated flow is not a constant but dependent on the suction (or water content) in an unsaturated soil, and it can be characterized by the SWCC. Therefore, SWCC and water coefficient of permeability are the most important hydraulic properties for unsaturated soils.

Figure 3.60 shows an idealized SWCC with two characteristic points, A^* and B^*. Point A^* corresponds to the air-entry value $((u_a - u_w)_b)$, and B^* corresponds to the residual water content (w_r). As shown in Figure 3.60, prior to A^*, the soil is saturated or nearly saturated, so it can be treated as a saturated soil with a compressible fluid due to the existence of occluded air bubbles. Beyond B^*, there is little water in the soil, so the effects of water content or negative pore water pressure on soil behaviour may be negligible.

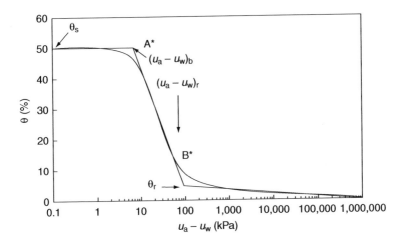

Figure 3.60 Idealized soil–water characteristic curve (Zhan and Ng, 2004).

Therefore, the soil at these two unsaturated stages is not the key focus of unsaturated soil behaviour (Bao *et al.*, 1998). What is of great concern in unsaturated soils is the stage between A^* and B^* in which both air and water phases are continuous or partially continuous, and hence the soil properties are strongly related to its water content or negative pore water pressure. It can be seen from Figure 3.60 that the SWCC between A^* and B^* is nearly a straight line on a semi-logarithmic scale. The linear part of the SWCC can be approximately represented with the air-entry value $((u_a - u_w)_b)$, the saturated and residual volumetric water content and the desaturation rate of the SWCC (i.e. the slope of the linear part).

For an unsaturated soil, the water coefficient of permeability depends on the degree of saturation or negative pore water pressure of the soil. Water flows only through the pore space filled with water (air-filled pores is not conductive to water), so the percentage of the voids filled with water (i.e. degree of saturation) is an important factor. The relationship of the degree of saturation to negative pore water pressure can be represented by a SWCC. Therefore the water coefficient of permeability for unsaturated soils with respect to negative pore water pressure bears a relationship to the SWCC, and it can be estimated from the saturated permeability and the SWCC (Marshall, 1958; Brooks and Corey, 1964; Van Genuchten, 1980; Fredlund *et al.*, 1994). Hence, a water permeability function of unsaturated soils can be approximated in terms of saturated permeability, air-entry value, desaturation rate, saturated and residual volumetric water contents.

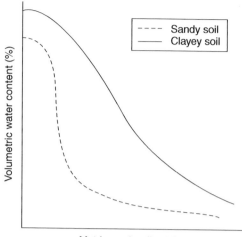

Figure 3.61 Soil–water characteristic curves for different soils (based on Hillel, 1998; Zhan and Ng, 2004).

Figure 3.61 illustrates the general trend of SWCCs for two different types of soils (Hillel, 1998). It can be seen that the shape and magnitude of the SWCC depends strongly on soil type. The greater the clay content, in general, the larger the air-entry value, the greater the water retention ability at a given matric suction and the more gradual the slope of the curve (i.e. the lower the desaturation rate). In a sandy soil, most of the pores are relatively large, and once these larger pores are emptied at a given negative pore water pressure, only a small amount of water remains. In a clayey soil, the pore size is relatively small and the distribution of pores is more uniform, with more water being adsorbed, so that the increase in matric suction causes a more gradual decrease in water content.

Figure 3.62 shows the general trend of the water permeability curves for different soils (Hillel, 1998). It is seen that, although the saturated permeability of the sandy soil (k_{s1}) is typically greater than that of the clayey soil (k_{s2}), the unsaturated permeability of the sandy soil decreases more sharply with the increasing matric suction (or negative pore water pressure) and eventually becomes lower than that of the clayey soil. This is attributed to the larger desaturation rate of the sandy soil than the clayey soil (Figure 3.61). Since air-filled pores are non-conductive channels to the flow of water, a sharp reduction in water content in sandy soil will result in a sharp decrease of water coefficient of permeability with the increasing negative pore water pressure.

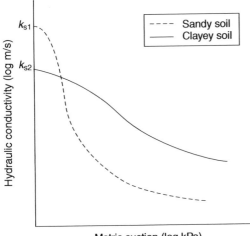

Matric suction (log kPa)

Figure 3.62 Relationships between water permeability and matric suction for different soils (Hillel, 1998; from Zhan and Ng, 2004).

Analytical analysis for one-dimensional infiltration in unsaturated soils

Review of analytical solution

The governing equation for one-dimensional vertical infiltration in unsaturated soils is given by the following equation (Richards, 1931):

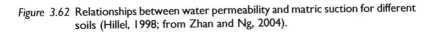

$$\frac{\partial}{\partial z}\left[k\left(\psi\right)\frac{\partial\left(\psi+z\right)}{\partial z}\right]=\frac{\partial\theta}{\partial t} \tag{3.78}$$

where, ψ is pore water pressure head (negative for unsaturated flow);
θ is volumetric water content of a soil;
t is time;
z is elevation head.

The water coefficient of permeability (k) and the volumetric water content (θ) in an unsaturated soil are non-linear functions of negative pressure head (ψ), which results in the non-linearity of this equation. To obtain a closed form analytical solution, some simplifying assumptions are required. Over the past decades, many researchers have made great efforts to obtain analytical solutions for the above equation (Philip, 1957; Broadbridge and White, 1988; Pullan, 1990; Sander, 1991). Most of them assumed an exponential relationship between water coefficient of permeability and the negative pressure head (ψ). Srivastava and Yeh (1991) assumed that the relationship of the water coefficient of permeability (k) and that of volumetric water content (θ) to ψ are both exponential,

$$\theta = \theta_r + (\theta_s - \theta_r)\, e^{\alpha\psi} \tag{3.79a}$$

$$k = k_s e^{\alpha\psi} \tag{3.79b}$$

where
θ_s and θ_r are saturated and residual volumetric water content of the soil, respectively;
k_s is saturated permeability;
α is a coefficient representing the desaturation rate of the SWCC, defined here as the desaturation coefficient.

For many soils, these two functions can be used to approximate their hydraulic characteristics with the four parameters (i.e. α, θ_s, θ_r and k_s) over a wide range of negative pore water pressures (Philip, 1969). However, the two functions do not account for the air-entry value of a SWCC probably because of the ease of obtaining an analytical solution of Equation (3.61). Since only wetting process is considered in here, wetting path of the SWCC is investigated and hence hysteresis of the SWCC is not studied.

With these two functions, the Richards's equation can be transformed to the following linear equation:

$$\frac{\partial^2 k}{\partial z^2} + \alpha \frac{\partial k}{\partial z} = \frac{\alpha\,(\theta_s - \theta_r)}{k_s} \frac{\partial k}{\partial t} \tag{3.80}$$

A one-dimensional transient infiltration problem generally involves one initial and two boundary conditions. In the study by Srivastava and Yeh (1991), the one-dimensional infiltration problem in homogeneous soil is defined in Figure 3.63. The lower boundary is located at the stationary

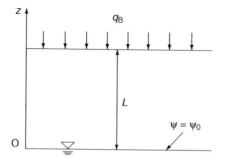

Figure 3.63 Homogeneous soil profile for infiltration calculation (Zhan and Ng, 2004).

groundwater table, where negative pore water pressure is equal to 0 (i.e. $\psi_0 = 0$). Its boundary condition can be written as:

$$k|_{z=0} = k_s \tag{3.81}$$

The upper boundary at the ground surface is subjected to a rainfall infiltration (q), which is kept constant throughout the duration of rainfall considered. Its boundary condition can be expressed as:

$$\left(\frac{1}{\alpha}\frac{\partial k}{\partial z} + k\right)_{z=L} = q \tag{3.82}$$

Conventionally, the initial pore water pressure distribution (i.e. initial condition) is obtained by performing a steady state analysis, in which the upper boundary is subjected to a small infiltration rate. The small infiltration rate is called antecedent infiltration rate (q_A) in here. The thickness of the unsaturated soil layer is L. For a two-layered soil system, the lower and upper layers are L_l and L_u in thickness, respectively (see Figure 3.64). Its boundary conditions and initial conditions remain the same as the case of homogeneous soil layer. The analytical solution for these two kinds of soil systems can be obtained through Laplace's transformation (Srivastava and Yeh, 1991). The solution for a homogeneous soil system is written as:

$$k^* = q_B^* - \left[q_B^* - e^{\alpha\psi_0}\right] \cdot e^{-z^*} - 4(q_B^* - q_A^*) \cdot e^{(L^*-z^*)/2} \cdot e^{-t^*/4}$$

$$\sum_{n=1}^{\infty} \frac{\sin(\lambda_n z^*) \sin g(\lambda_n L^*) \cdot e^{-\lambda_n^2 t^*}}{1 + (L^*/2) + 2\lambda_n^2 L^*} \tag{3.83}$$

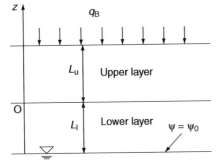

Figure 3.64 Two-layer soil profile for infiltration calculation (Zhan and Ng, 2004).

in which

$$\begin{cases} z^* = \alpha z, L^* = \alpha L \\ k^* = k/k_s \\ q_A{}^* = q_A/k_s, q_B{}^* = q_B/k_s \\ t^* = \dfrac{\alpha k_s t}{\theta_s - \theta_r} \end{cases}$$

λ_n is the nth root of the following characteristic equation

$$\tan(\lambda_n L) + 2\lambda_n = 0 \tag{3.84}$$

The solution for a two-layered soil system is given as follows:

$$\begin{cases} k_1^* = q_{Bl}^* - [q_{Bl}^* - e^{\alpha\psi_0}] \cdot e^{-(L_1 + z^*)} - 4(q_{Bl}^* - q_{Al}^*) \cdot \\ \quad e^{(L_u^* - z^*)/2} \cdot [RA_1 + (RB_1 \, or \, RC_1)] \\ k_u^* = q_{Bu}^* - [q_{Bu}^* - q_{Bl}^* + (q_{Bl}^* - e^{\alpha\psi_0})e^{-L_1^*}] \cdot e^{-z^*} \\ \quad -4(q_{Bl}^* - q_{Al}^*) \cdot e^{(L_u^* - z^*)/2} \cdot [RA_u + (RB_u \, or \, RC_u)] \end{cases} \tag{3.85}$$

where

$$\begin{cases} z^* = \alpha_1 z \quad \text{for } -L_1^* \le z^* \le 0 \text{ so that } L_1^* = \alpha_1 L_1 \\ z^* = \alpha_u z \quad \text{for } 0 \le z^* \le L_u^* \text{ so that } L_u^* = \alpha_u L_u \\ k_1^* = k_1/k_{sl} \quad q_{Al}^* = q_{Al}/k_{sl} \quad q_{Bl}^* = q_{Bl}/k_{sl} \\ k_u^* = k_u/k_{su} \quad q_{Au}^* = q_{Au}/k_{su} \quad q_{Bu}^* = q_{Bu}/k_{su} \\ t^* = \dfrac{\alpha_1 k_{sl} t}{\theta_{sl} - \theta_{rl}} \end{cases} \tag{3.86}$$

RA_1, RB_1, RA_u and RB_u are functions of z^* and t^*. The details of these functions can be found in Srivastava and Yeh (1991).

Then, pore water pressure heads (ψ) in the ground can be calculated from Equation (3.79b) as follows

$$\psi = \frac{\ln k^*}{\alpha} \tag{3.87}$$

Volumetric water contents (θ) in the ground can be calculated from Equation (3.79a) as follows

$$\theta = \theta_r + (\theta_s - \theta_r) \, e^{\alpha\psi} \tag{3.88}$$

Introductory analysis for parametric studies

An introductory analysis which is helpful for understanding the results from a parametric study is discussed here.

The governing equation for one-dimensional vertical transient infiltration in unsaturated soils [Equation (3.78)] can be rewritten in terms of ψ (Hillel, 1998),

$$\frac{\partial}{\partial z}\left[k\left(\psi\right)\frac{\partial\left(\psi+z\right)}{\partial z}\right]=\frac{\partial\theta}{\partial\psi}\frac{\partial\psi}{\partial t} \tag{3.89}$$

where $\partial\theta/\partial\psi$ is a storage term, representing the variation of volumetric water content with respect to negative pore water pressure. It is also called the specific moisture capacity in soil science (Hillel, 1998).

For steady-state infiltration, both the two variables (ψ and θ) in Equations (3.87) and (3.88) are independent of time, then

$$\frac{\partial}{\partial z}\left[k\left(\psi\right)\frac{\partial\left(\psi+z\right)}{\partial z}\right]=0, \quad \text{i.e. } k\left(\psi\right)\frac{\partial\left(\psi+z\right)}{\partial z}=\text{constant} \tag{3.90}$$

The analytical solution for the steady-state infiltration problem with the upper flux boundary (q) can be obtained as,

$$\frac{\partial\psi}{\partial z}+1=\frac{q}{k(\psi)}, \quad \text{i.e. } \psi=-\int_0^z\left(1-\frac{q}{k\left(\psi\right)}\right)dz \tag{3.91}$$

From Equations (3.88) and (3.89), it can be seen that the pore water pressure response of a steady state infiltration system depends only on the water coefficient of permeability ($k(\psi)$) and the applied infiltration rate (q). As shown in Equation (3.79b), $k(\psi)$ can be expressed in terms of k_s and α. Thus, the pore water pressure response of a steady-state infiltration system is a function of k_s, α and q. However, for a transient infiltration, apart from $k(\psi)$ and q, the pore water pressure response also depends on the storage term, $(\partial\theta/\partial\psi)$, [see Equation (3.89)]. As shown in Equation (3.79a), the storage term, $(\partial\theta/\partial\psi)$, can be described in terms of α and $(\theta_s-\theta_r)$. Hence, the transient pore water pressure response depends on k_s, α, $(\theta_s-\theta_r)$ and q. For the infiltration rate (q) on the upper boundary, antecedent and subsequent infiltrate rates (i.e. q_A and q_B) are considered in the initial and subsequent stages, respectively. Therefore, the governing parameters of the one-dimensional transient infiltration system are the five variables; namely α, k_s, $(\theta_s-\theta_r)$, q_A and q_B.

Analysis scheme and input

A homogeneous soil layer ($L = 10\,\text{m}$) and a two-layered soil system ($L_1 = L_u = 5\,\text{m}$) were chosen for this study (see Figures 3.63 and 3.64). The experiments shown in Table 3.12 were run to investigate the influence of hydraulic parameters and rainfall infiltration rates – namely α, k_s, $(\theta_s-\theta_r)$, q_A and q_B – on the pore water pressure profiles in unsaturated soil strata. Series

Table 3.12 Analysis scheme for the parametric study of infiltration (Zhan and Ng, 2004)

Series	Case number	k_s ($\times 10^{-6}$ m/s)	α (kPa^{-1})	$(\theta_s - \theta_r)$	q_A ($\times 10^{-6}$ m/s)	q_B ($\times 10^{-6}$ m/s)
1	1	0.3				
	2	3.0	0.05	0.3	0.0	2.7
	3	30.0				
2a	1		0.01			
	2	3.0	0.05	0.3	0.0	2.7
	3		0.1			
2b	1		0.01			
	2	3.0	0.05	0.3	0.3	2.7
	3		0.1			
3	1			0.20		
	2	3.0	0.05	0.30	0.3	2.7
	3			0.40		
4	1	15.0	0.01	0.5		
	2	15.0	0.05	0.5	0.0	2.7
	3	3.0	0.01	0.5		
	4	15.0	0.01	0.1		
5a	1	3.0	0.01	0.3	0.0	2.7
	2	30.0	0.1			
5b	1	3.0	0.01			15.0
	2	30.0	0.1			
6	1				0.0	2.4
	2	3.0	0.05	0.3	0.3	2.4
	3				0.0	2.7
	4				0.6	3.0
7	1	0.3			0.03	0.27
	2	3.0	0.05	0.3	0.3	2.7
	3	30.0			3.0	27.0
8	1	$k_{sl} = 3.0$ $k_{su} = 30.0$	$\alpha_l = \alpha_u = 0.1$	$(\theta_s - \theta_r)_l$ $= (\theta_s - \theta_r)_u$ $= 0.3$	0.3	2.7
	2	$k_{sl} = 30.0$ $k_{su} = 3.0$				

1–3 are considered to investigate individual influences of the three hydraulic parameters (α, k_s and $(\theta_s–\theta_r)$), respectively. Series 4 and 5 are designed for studying the relative sensitivity of the three hydraulic parameters, whereas Series 6 and 7 are primarily for investigating the effects of q_A and q_B and the last Series is for studying the effect of soil profile heterogeneity. To be simple, a uniform rainfall infiltration lasting for 24 h was simulated for all the series.

Infiltration: a parametric study

Influence of saturated permeability (k_s)

Two extreme cases of the infiltration problem can be first considered: the water coefficient of permeability tending to zero and that tending to infinity. At the former extreme, there will be no infiltration of water into the soil stratum, and hence little drop in the initial negative pore water pressure. At the latter extreme, water will infiltrate into the soil stratum easily but will immediately drain away through the boundaries, so again there is little decrease in the initial negative pore water pressure. However, at intermediate values of water coefficient of permeability, rainwater will infiltrate to a certain degree and will not entirely drain away, resulting in a reduction of the negative pore water pressure. This implies that there exists a critical saturated permeability that may result in the greatest reduction of negative pore water pressure (Ng and Shi, 1998).

Figure 3.65 shows a series of pore water pressure profiles with respect to three different values of saturated permeability (i.e. $k_s = 3 \times 10^{-7}$, 3×10^{-6} and 3×10^{-5} m/s). The embedded small chart illustrates the relationship between water coefficient of permeability (k) and negative pore water pressure (ψ) for different values of k_s [Equation (3.62b)]. As aforementioned, the initial pore water pressure profile ($t = 0$) is deduced from the steady-state flow analysis associated with the antecedent infiltration rate q_A imposed on the ground surface. If q_A is zero, which means zero flux at the ground surface, the hydrostatic condition is thus achieved. For the three cases of Series 1, the antecedent infiltration rate remains zero, so the three initial pore water pressure profiles are identical, regardless of the different values of k_s. Certainly, if the value of q_A is larger than zero, the initial pore water pressure profile will be affected by the k_s value.

As shown in Figure 3.65, the saturated permeability has a significant influence on the pore water pressure redistribution due to subsequent rainfall infiltration. First, the greater the k_s value, the deeper the wetting front. This is consistent with Green and Ampt's model (1911). On the other hand, if the k_s value is relatively small with respect to the applied infiltration rate, the negative pore water pressure near the ground surface decreases greatly, however the wetting front is relatively shallow after a certain duration of rainfall (e.g. 24 h). This is because the infiltrated rainwater accumulates in the sallow soil layer due to the low k_s value, and then gradually wets the soil below. When the k_s value is smaller than the infiltration rate, i.e. $q_B/k_s > 1$ (as for the case with $k_s = 3 \times 10^{-7}$ m/s), a positive pore water pressure appears at the ground surface after a certain duration and the shallow soil will become saturated. This is a result of the small k_s, which prevents some of the rainwater from infiltrating into the soil and results in the development of ponding at the ground surface. On the contrast, for a relatively large

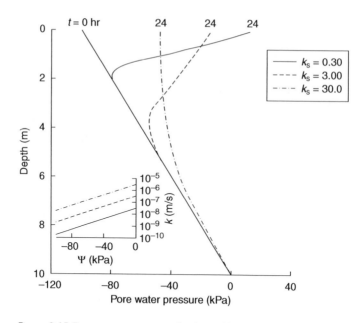

Figure 3.65 Pore water pressure-depth profiles for different values of permeability, k (Zhan and Ng, 2004).

value of k_s, the downward flow will be easier, and hence the infiltrated water are distributed more uniformly over the depth, so a smaller reduction in negative pore water pressure will be observed in the shallow soil layer. This result is similar to Pradel and Raad's (1993) conclusion that the lower the permeability of the soil the higher probability that saturation may develop in the shallow soil of a slope.

As mentioned previously, there is a critical saturated permeability, which leads to the greatest reduction in negative pore water pressure in an infiltration system. A further observation of Figure 3.65 indicates that the area between the initial profile and the one at $t = 24$ h tends be largest for $k_s = 3.0 \times 10^{-6}$ m/s, which means that this value brings about the greatest total reduction in pore water pressure among the three cases considered. This value is closest to the imposed rainfall infiltration rate. This indicates that the critical value may be close to the saturated permeability. Kasim *et al.* (1998) reported a similar result that the steady-state matric suctions in a soil profile decrease with increasing q/k_s ratio, and that the matric suctions reduce to zero throughout the entire depth when the steady-state rainfall infiltration rate approaches the value of the saturated permeability of the soil.

Influence of desaturation coefficient (α)

The desaturation coefficient (α) governs the rate of decrease in water content or water coefficient of permeability with an increasing matric suction or negative pore water pressure [see Equation (6.48b)]. The value of α is related to the grain size distribution of a soil (Philip, 1969). Generally speaking, the greater the clay content, the lower the desaturation rate [(i.e. the smaller the value of α) (Figure 3.61)] and hence the more gradually the water coefficient of permeability decreases with an increase in matric suction (Figure 3.62) (Hillel, 1998). In addition, for a given soil the α value tends to increase with the uniformity of the grain sizes. Typically, α is about $0.1 \, \text{kPa}^{-1}$, and the range of values $0.02–0.5 \, \text{kPa}^{-1}$ seems likely to cover most applications (Philip, 1969). According to Morrison and Szecsody (1985), the range $0.01–10 \, \text{kPa}^{-1}$ covers most of the published α values.

Figure 3.66 shows three pore water pressure profiles at $t = 24 \, \text{h}$ for different values of α (i.e. 0.01, 0.05, and $0.1 \, \text{kPa}^{-1}$, see Series 2a in Table 3.12). A small chart is also embedded to illustrate the relationship between water coefficient of permeability (k) and negative pore water pressure (ψ) for different values of α [Equation (3.62b)]. As shown in Figure 3.66, the initial profiles are identical for the three cases due to zero value of q_A. It can be seen that the smaller the α value, the deeper the wetting front advances. In addition, the three profiles intersect. In other words, the negative pore

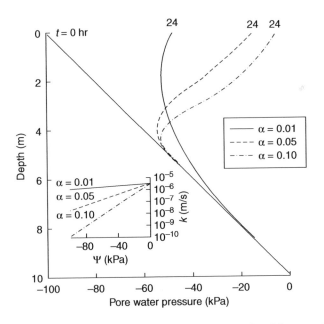

Figure 3.66 Pore water pressure-depth profiles for different values of α ($q_A = 0$) (Zhan and Ng, 2004).

water pressure for a smaller α value is larger above but smaller below the intersect point than that for a larger α value. This may be explained by considering the embedded small chart: if the value of k_s is kept constant, a soil with a larger α value possesses a lower water coefficient of permeability at a given negative pore water pressure than a soil with a smaller α. Because of the lower water coefficient of permeability, more infiltrated water tends to be retained in the shallow soil, resulting in a larger reduction in the negative pore water pressure near the ground surface but less reduction in the deep soil layer. On the other hand, when the value of α is small, the water coefficient of permeability of the soil is relatively large with respect to the same negative pore water pressure, the downward flow is easier and the infiltrated water is more uniformly distributed, so a smaller reduction of negative pore water pressure is observed in the shallow soil layer.

Figure 3.67 shows the results of Series 2b, in which the effect of antecedent infiltration rate, $q_A = 3.0 \times 10^{-7}$ m/s, is also included. First, it can be seen that the three initial (steady-state) pore water pressure profiles ($t = 0$ h) differ significantly from one another. The greater the value of α, on the whole, the lower the initial negative pore water pressure. Equation (3.73) indicates that the larger the ratio of $q_A\alpha/k_s$, the smaller the steady-state negative pore water pressure will result, especially at shallow depths. It is

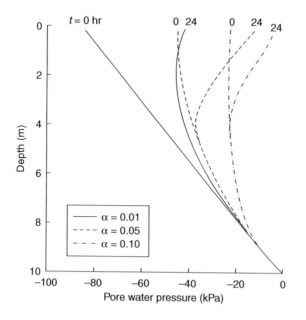

Figure 3.67 Pore water pressure-depth profiles for different values of α ($q_A = 0.3 \times 10^{-6}$ m/s) (Zhan and Ng, 2004).

logical that an unsaturated soil with smaller pore sizes (smaller α) tends to possess higher negative pore water pressure under identical conditions since negative pore water pressure (or matric suction) is inversely proportional to the pore sizes. Because of the discrepancy in the initial pore water pressure distributions, the dependency of the pore water pressure profiles corresponding to 24 h on the value of α is a bit different from Series 2a. On the whole, the smaller the α value, the larger the negative pore water pressure at any depth, and the deeper the wetting front. However, the difference amongst three profiles is not so much as the result of Series 2a.

Here the α value has a significant effect on the initial steady-state pore water pressure distribution, which is apparently contradictory to the conclusion drawn by Kasim *et al.* (1998) – 'the influence of desaturation rate parameter (n) on the steady-state pore water pressure distributions is negligible when compared with that of the air-entry value (a).' This inconsistent conclusion seems to be caused by the narrow range of n value considered in Kasim *et al.*'s study (i.e. from 1.5 to 2.5).

Ng and Pang (2000b) investigated the influence of state dependency on SWCC (SDSWCC) on pore water pressure distributions in a slope subject to rainfall infiltration. Their laboratory test results show that the higher the applied stress on a soil sample, the smaller the desaturation coefficient of the SDSWCC (i.e. α). However, it was found that a SDSWCC with a smaller α value leads to a lower initial steady-state negative pore water pressure in the slope. The difference in the computed results between Ng and Pang's and this study may be attributed to the different water permeability functions adopted.

Influence of water storage capacity, $(\theta_s - \theta_r)$

Water storage capacity of a soil, $(\theta_s - \theta_r)$, is equal to the difference between the saturated and residual volumetric water content. It is a measurement of the maximum amount of water that can be absorbed or desorped by capillary action. It is somewhat different from the soil's water retention ability. Generally, the value of $(\theta_s - \theta_r)$ increases with pore sizes (Vanapalli *et al.*, 1998). To some extent, it is also related to the void ratio.

Figure 3.68 shows three pore water pressure profiles at $t = 24$ h for different $(\theta_s - \theta_r)$ values (i.e. 0.2, 0.3 and 0.4, see Series 3 in Table 3.12). As mentioned before, the initial pore water pressure profile, which is obtained from steady state analysis, depend only on the parameters α, k_s and q_A. In other words, the initial pore water pressure profile is independent of $(\theta_s - \theta_r)$. Certainly, the value of $(\theta_s - \theta_r)$ will affect the volumetric water content profiles because different values of $(\theta_s - \theta_r)$ will lead to different relationships

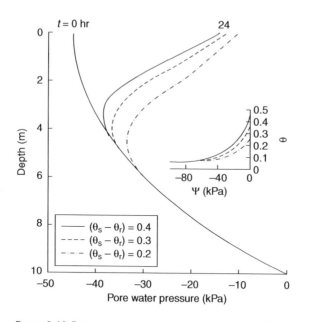

Figure 3.68 Pore water pressure-depth profiles for different values of $(\theta_s - \theta_r)$ (Zhan and Ng, 2004).

between negative pore water pressure and volumetric water content (see the small chart embedded in Figure 3.68).

Figure 3.68 illustrates that $(\theta_s - \theta_r)$ mainly affects the magnitude of the negative pore water pressure but not the shape of the profiles. A larger value of $(\theta_s - \theta_r)$ represents a higher water storage capacity, which results in a lower advance rate of the wetting front as indicated by Green and Ampt's model (1911). The lower advance rate of wetting front is not due to a low water coefficient of permeability, but due to the storage requirement of the shallow soil layer (Wilson, 1997). On the other hand, the larger the $(\theta_s - \theta_r)$ value, the less the reduction in negative pore water pressure is observed. As shown in the small chart embedded in Figure 3.68, for a soil with a larger $(\theta_s - \theta_r)$, an identical amount of increase in volumetric water content will result in a smaller reduction in negative pore water pressure with respect to the same initial value. For this Series, (q_B) remains the same for the three cases, which means the amount of rainwater infiltrated into the soil profiles is identical, so there is a smaller reduction in negative pore water pressure for the soil profile with a larger $(\theta_s - \theta_r)$. In addition, the effect of $(\theta_s - \theta_r)$ appears to be not so great as the effects of α and k_s on the pore water pressure profile.

Relative sensitivity of three hydraulic parameters
$(\alpha, k_s$ and $(\theta_s - \theta_r))$

The effects of the three hydraulic parameters have been investigated separately. For a more comprehensive understanding of their relative effects, a sensitivity analysis is carried out to investigate which parameter is more significant. To meet this objective, Series 4 and 5 were carried out.

In Series 4, the case with $\alpha = 0.01\,\text{kPa}^{-1}$, $k_s = 1.5 \times 10^{-5}\,\text{m/s}$ and $(\theta_s - \theta_r) = 0.5$ was selected as a reference case. Then, each of the three parameters, α, k_s and $(\theta_s - \theta_r)$ was altered five-fold with respect to their individual reference value. For commonly encountered soils, k_s and α can vary over one or even several orders of magnitude, i.e. much larger than that of $(\theta_s - \theta_r)$, which is generally confined to a small range, from 0.1 to 0.5.

Figure 3.69 shows the results of Series 4. As before, all the initial pore water pressure profiles are identical due to the zero value of q_A. Comparisons of the three pore water pressure profiles at 24 h show that the one corresponding to a change in $(\theta_s - \theta_r)$ is closest to the reference profile represented by the solid line. For the two profiles corresponding to changes in α and k_s, both their shape and magnitude are significantly different from the reference profile, especially in the soil near the ground surface. If a wider range of α and k_s values than that of $(\theta_s - \theta_r)$ is also considered, it is obvious

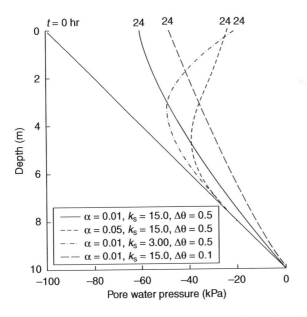

Figure 3.69 Pore water pressure-depth profiles illustrating the relative sensitivity of α, k_s and $(\theta_s - \theta_r)$ (Zhan and Ng, 2004).

that the influence of α and k_s on pore water pressure redistribution will be much more significant than that of $(\theta_s - \theta_r)$.

To investigate the relative sensitivity of α and k_s, Series 5 in Table 3.12, in which the value of k_s/α remains constant but different q_B values are adopted, was analysed. In this Series, two sets of hydraulic parameters were used, as shown in the small chart embedded in Figure 3.70. One set is a small α combined with a small k_s, the other is a relatively large α and k_s. The former generally represents a clayey soil, the latter a sandy soil. For each of these two sets, two different values of rainfall infiltration rate, 2.7×10^{-6} and 15.0×10^{-6} m/s, were selected for the analyses, representing relatively gentle and intense rainfalls, respectively (as compared with the adopted saturated permeability of the soil).

As shown in Figure 3.70, the initial pore water pressure profiles for all the four cases are identical due to the zero value of q_A. For the four profiles at 24 h, it is logical that the two pore water pressure profiles corresponding to a relatively small rainfall are located to the left of the other two, i.e. a relatively small rainfall results in a less reduction of negative pore water pressure.

For the two profiles corresponding to a small rainfall, the reduction of negative pore water pressure in the sandy soil (with a relatively large α and k_s) is larger than that in the clayey soil (with a small α and k_s). However, for the other two profiles corresponding to the relatively intensive rainfall, the

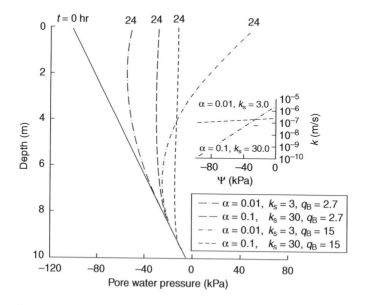

Figure 3.70 Pore water pressure-depth profiles illustrating the relative sensitivity between α and k_s (Zhan and Ng, 2004).

result is different: in the upper part of the layer, the reduction of negative pore water pressure in the clayey soil is significantly larger than that in a sandy soil, and in the lower part, it is slightly smaller than that in the sandy soil.

The above difference may indicate that α and k_s dominate in different situations. As stated above (see Figure 3.61 and the embedded small chart in Figure 3.70), at high negative pore water pressures, a clayey soil possesses a larger water coefficient of permeability than a sandy soil due to its lower desaturation rate α. However, the case is opposite at low negative pore water pressures due to the higher saturated permeability of a sandy soil. Since rainfall infiltration is a wetting process, the soil initially is relatively dry and possesses a relatively high negative pore water pressure. When the rainfall infiltration rate is relatively small as compared with the saturated permeability, the negative pore water pressure in the soil tends to be high, so the clayey soil with a lower α possesses a larger water coefficient of permeability. This leads to more rainwater moving down and draining out of the system, and hence less reduction of negative pore water pressure in the clayey soil than that in the sandy soil. On the other hand, when the given rainfall infiltration rate is relatively large compared with the saturated permeability, the negative pore water pressure in the soil tends to be low, so less reduction of negative pore water pressure is observed in the sandy soil with a larger k_s. These results are similar to a phenomenon described by Hillel (1998), namely that in the case of ponding infiltration due to intensive rainfall, a sandy soil absorbs water more rapidly than a clayey soil due to its larger water coefficient of permeability at or near saturation.

Therefore, within the high negative pore water pressure range, the water coefficient of permeability is controlled primarily by α, and so is the response of the infiltration system. However, at a lower negative pore water pressure, it is primarily k_s that controls the water coefficient of permeability of the soil and hence dominates in the response of a groundwater system. From this analysis, it can be imagined that, in the early period of rainfall infiltration into an initially dry soil, the pore water pressure response is primarily affected by α. Later, however, the saturated permeability tends to control the pore water pressure response as the negative pore water pressure decreases with the duration of rainfall.

Influences of antecedent and subsequent infiltration rate (q_A and q_B)

The rate of rainfall infiltration depends on rainfall intensity as well as the water coefficient of permeability and the hydraulic gradient of the surficial soil. Both the water coefficient of permeability and hydraulic gradient vary throughout the infiltration process due to the variations of negative pore water pressure in the soil, so the rate of infiltration may change with the duration of rainfall even if the rainfall intensity remains constant. If the rainfall intensity is less than the minimum infiltrability (close to the saturated permeability) of the surficial soil, the rate of infiltration is equal to the rainfall intensity, which has been called non-ponding or supply-controlled infiltration (Rubin and Steinhardt, 1963). In this case, if the rainfall intensity is assumed to be constant, then the rate of infiltration will remain constant. This case is for the parametric study in here, i.e. q_A and q_B were assumed to be constant throughout the duration.

The initial pore water distribution is an essential input for transient flow analyses in unsaturated soil. Previous studies indicate that q_A has a significant influence on the initial pore water distribution, and hence on the pore water pressure redistribution as a result of subsequent rainfall infiltration (Ng and Shi, 1998; Mein and Larson, 1973).

Variations in rainfall infiltration rate (q_B) are often restricted by the hydraulic properties of the soil and ground conditions (e.g. the initial pore water pressure distribution). Measurements in initially dry soils show that the infiltration rates vary over a much more narrow range than one might assume in consideration of the wide range in water coefficient of permeability and hydraulic gradient associated with the infiltration process (Houston and Houston, 1995). This is because, though rainfall infiltration in initially unsaturated soils results in an increase in the water coefficient of permeability due to an increase of water content, it simultaneously leads to a decrease of hydraulic gradient due to a decrease of negative pore water pressure. Therefore, it is believed that the sensitivity of an infiltration system to rainfall intensity is not as great as its sensitivity to the initial negative pore water pressure distribution and the hydraulic properties of the soil.

In order to investigate the influence of q_A and q_B and their relative sensitivity, four cases were investigated in Series 6 (Table. 3.12). Case 1 ($q_A =$ 0.0 m/s, $q_B = 2.4 \times 10^{-6}$ m/s) is a reference case. In Cases 2 and 3, q_A and q_B were increased by 10 per cent of the k_s value (3.0×10^{-6} m/s) with respect to the reference case, respectively. In Case 4, both q_A and q_B were increased by 20 per cent of the k_s value.

Figure 3.71 shows the results of Series 6. First, it can be seen that q_A affects the initial (steady-state) pore water pressure profile significantly. If q_A is zero, which represents a zero-flux boundary at the ground surface (e.g. the infiltration rate equal to the evaporation rate), the hydrostatic condition prevails. The initial negative pore water pressure profile for the hydrostatic condition differs significantly from the other cases with non-zero q_A. As anticipated, a smaller q_A value leads to a higher initial negative pore water pressure in the soil (or lower water content).

It is easily imagined that, if all the other parameters are identical, the larger the q_B value, the more reduction of negative pore water pressure will be caused by the q_B. However, comparisons of the results for Case 2 (dashed lines), Case 4 (double-dotted line) and the reference case (solid lines) show that, in the case of different q_A values, the influence of q_B are different from

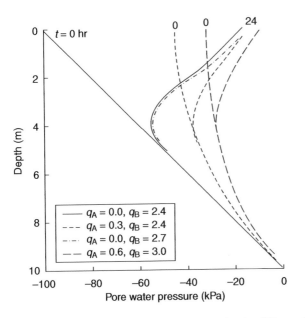

Figure 3.71 Pore water pressure-depth profiles for different infiltration rates (Zhan and Ng, 2004).

the anticipated: a smaller reduction of negative pore-pressure is observed in the shallow soil layer for the case with a larger q_A, regardless of the larger q_B value. This is because the higher antecedent infiltration rate leads to a lower initial negative pore water pressure in the soil, and hence a higher water coefficient of permeability, which facilitates the down-flow and out-discharge of the infiltrated rainwater. This is consistent with the conclusions by Ng and Shi (1998)

> If the duration of antecedent rainfall is relatively short, a subsequent heavy rainfall may greatly reduce the negative pore water pressure in the slope and hence the factor of safety of the slope; however, if antecedent rainfall duration is relatively long, the effect of the subsequent heavy rainfall on the factor of safety is relatively insignificant.

On the other hand, when comparing both the results of Case 2 (dashed lines) and Case 3 (dotted lines) with the results of the reference case, it is found that an increase in q_A has a much more significant effect on the pore water pressure redistribution (24 h) than the same magnitude of increase in q_B.

The discussions above demonstrate that the effect of the antecedent rainfall infiltration rate on pore water pressure redistribution is more significant than the effects of the subsequent rainfall infiltration rate. The antecedent infiltration rate not only controls the initial pore water pressure or water content distribution but also indirectly results in a change of the water coefficient of permeability of the soil. However, as far as the authors are aware, in most of the current analytical and numerical analyses, the determination of the antecedent infiltration rate is largely empirical or semi-empirical, and hence the determined value may not simulate the actual initial conditions reasonably. Therefore, it is of much significance to study the methodology for determining antecedent infiltration rate more realistically.

Certainly, as shown in Figure 3.71, if both the antecedent and subsequent rainfall infiltration rates (q_A and q_B) are highest (Case 4), the change in negative pore water pressure after the rainfall (e.g. 24 h) will be the lowest. Therefore, it can be imagined that the negative pore water pressure in the soil may be greatly reduced or even totally undermined if a prolonged antecedent rainfall is combined with a heavy subsequent rainstorm.

Combined effect of k_s and q_A and q_B

As stated above, the infiltration rate for a soil is strongly related to its saturated permeability, so it is of significance to investigate the combined effects of rainfall infiltration rate and the saturated permeability of the soil.

Figure 3.72 shows the results of Series 7 (in Table 3.12) in which both $q_A\alpha/k_s$ and $q_B\alpha/k_s$ were held constant. Because of the identical values of

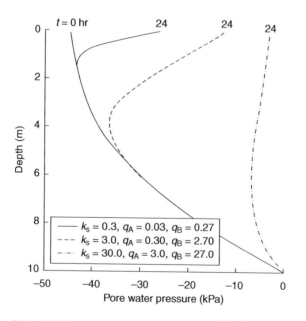

Figure 3.72 Pore water pressure-depth profiles illustrating the combined effects of k_s, q_A and q_B (Zhan and Ng, 2004).

$q_A \alpha / k_s$, the initial pore water pressure profiles from the steady-state analyses are the same for the three cases. However, in spite of the identical values of $q_B \alpha / k_s$ during the subsequent rainfall, the three pore water pressure profiles at 24 h are significantly different from one another. The depth of the wetting front corresponding to a particular duration is proportional to the value of k_s / α. For a low value of k_s / α, the reduction of negative pore water pressure is confined within the shallow soil layer, and the magnitude of reduction is limited. However, provided that the value of k_s / α is relatively large, the wetting front advances to a much greater depth and the negative pore water pressure drops greatly with increasing depth. Therefore, when a relatively permeable soil profile is confronted with a heavy rainfall, the negative pore water pressure may drop significantly throughout the profile.

The comparison of these three cases demonstrates what was stated above: For a steady state flow, the pore water pressure distribution depends only on the value of $q\alpha / k_s$. However, in a transient flow, the pore water pressure response not only depends on the value of $q\alpha / k_s$ but also on the value of k_s / α. When the value of $q\alpha / k_s$ is kept constant, the larger the value of k_s / α, the faster the wetting front advances downward, and the more the negative pore water pressure decreases. This is similar to the conclusion reached by Leach and Herbert (1982)who found that the higher the value of k_s / α, the faster the pressure heads will increase, and the shorter will be the response time of the system to a rainfall event.

Effects of soil heterogeneity

Since the closed-form solution for a two-layer soil system can be obtained only under the condition that the value of α should be identical for the two layers (Srivastava and Yeh, 1991), only soil heterogeneity in terms of k_s is considered here. Two cases were carried out to investigate the effects of the soil heterogeneity on pore water pressure (Series 8 in Table 3.12). In Case 1, the lower layer was relatively impermeable, but it was opposite in Case 2. Figures 3.72 and 3.73 illustrate the results of Cases 1 and 2, respectively. The results show that soil-profile heterogeneity has a great influence on the shape and magnitude of the negative pore water pressure profiles.

As shown in Figure 3.73, in which the saturated permeability of the upper layer is relatively large, the rainwater infiltrates into the upper layer easily, and consequently the negative pore water pressure there decreases quickly. However, a certain negative pore water pressure remains in the soil near the ground surface, because the saturated permeability of the upper soil is greater than the infiltration rate. When the wetting front reaches the top of the lower layer, most of water that has infiltrated tends to accumulate due to the relatively low permeability of the lower layer. So the negative pore water pressure at the interface of the two layers drops greatly. If the

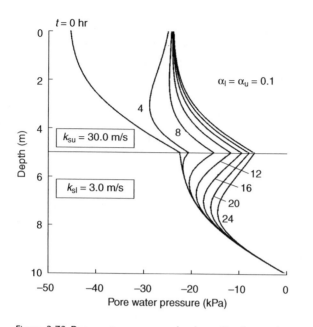

Figure 3.73 Pore water pressure-depth profiles in two layer strata $(k_{su} > k_{st})$ (Zhan and Ng, 2004).

duration and intensity of rainfall are large enough, positive pore water pressure (a perched water table) can develop at the interface (Hillel, 1998). With rainfall infiltration going on, the wetting front in the lower layer advances downwards slowly. Therefore, during the early stage, the infiltration process is controlled by the upper layer (more conductive), while during the late stage, the lower layer (less conductive) controls the infiltration process.

An opposite case is shown in Figure 3.74, in which the upper layer is less permeable. The entire process of infiltration appears to be controlled primarily by this less permeable upper layer. The wetting front advances downwards slowly in upper layer. In other words, most of rainwater that has infiltrated tends to be retained in the upper layer, particularly in the shallow soil layer. Hence the negative pore water pressure in the shallow layer decreases greatly, even changing to a positive value if the rainfall duration and intensity are large enough. However, little rainwater infiltrates into the lower layer during the early stage of rainfall. Even if some water infiltrates into the lower layer later on, the infiltrated water will be quickly discharged away due to the relatively large water coefficient of permeability of the lower soil layer. Therefore, the negative pore water pressure basically remains at the initial value in the lower layer. In any case, the lower layer cannot become saturated, since the restricted outflow

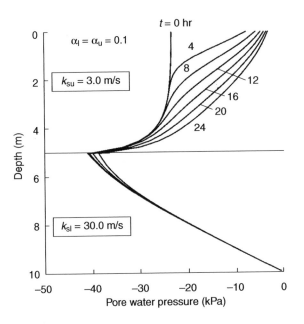

Figure 3.74 Pore water pressure-depth profiles in two layer strata $(k_{su} > k_{sl})$ (Zhan and Ng, 2004).

from the upper less permeable layer cannot sustain flow in the lower layer with a relatively large saturated permeability, except in the case in which a large pond lies on the ground surface. These effects of a sandy sub-layer have been demonstrated and researched by many others as capillary barrier effect (Hillel, 1998). It may be worthwhile to refer to these studies.

Summary

1. The SWCC and the water coefficient of permeability are two essential hydraulic properties for transient flow analyses. For engineering applications, the SWCC can be expressed in terms of desaturation coefficient (α) and water storage capacity $(\theta_s - \theta_r)$; the water coefficient of permeability can be described in terms of the saturated permeability (k_s) and the desaturation coefficient (α). These three hydraulic parameters $(\alpha, k_s$ and $(\theta_s - \theta_r))$, together with two parameters for specifying the initial and boundary conditions $(q_A$ and $q_B)$, are the governing parameters for the one-dimensional infiltration system studied in here.
2. Among the three hydraulic parameters, the effects of α and k_s on pore water pressure response are much more significant than that of $(\theta_s - \theta_r)$.

Furthermore, α and k_s not only affect the initial pore water pressure profiles but also those during the subsequent rainfall. The relative sensitivity of α and k_s depends on the negative pore water pressure range present in the soil. Within a high negative pore water pressure range (relatively dry soils), water coefficient of permeability is primarily controlled by α, and so is the negative pore water pressure response of an infiltration system in initially unsaturated soils. However, within a lower negative pore water pressure range, it is primarily k_s that controls the water coefficient of permeability and hence dominates in an infiltration system.

3. For steady-state infiltration, the pore water pressure response is primarily governed by $q\alpha/k_s$. However, the transient response of an infiltration system not only depends on the value of $q\alpha/k_s$ but also on the value of k_s/α. Generally speaking, the larger the value of $q\alpha/k_s$, the greater will be the reduction of negative pore water pressure as a result of rainfall infiltration. For a given value of α, when the ratio of q/k_s approaches 1, the reduction of negative pore water pressure will be the largest. The larger the value of k_s/α, the faster the wetting front advances downward, and the more the negative pore water pressure decreases.

4. Soil-profile heterogeneity influences the negative pore water pressure profiles greatly. For the case of a relatively permeable layer overlying a less permeable layer, the negative pore water pressure at the interface of the two layers may drop greatly and even change to a positive pore water pressure if the rainfall duration or intensity is large enough. For the opposite case (i.e. a less permeable layer overlying a relatively permeable layer), the upper less permeable layer can impede rainwater flow into the lower layer so that the lower layer may remain unsaturated, and its negative pore water pressure may drop only marginally.

5. For a given soil profile, the larger the antecedent infiltration rate (q_A) value, the lower the initial negative pore water pressure will be. When both hydraulic properties and the initial condition are identical, the larger the subsequent infiltration rate (q_B), the greater will be the reduction of negative pore water pressure due to the rainfall infiltration. The effects of the antecedent infiltration rate on pore water pressure response are more significant than that of the subsequent infiltration rate.

6. The combined effect of q_B and k_s is significant. When a relatively permeable soil profile is subjected to heavy rainfall, the negative pore water pressure may drop greatly throughout the profile. The effect of combining q_A and q_B is also important. When both q_A and q_B are large, i.e. a relatively heavy or prolonged antecedent rain followed by a heavy rainstorm, the negative pore water pressure may drop greatly or even disappear.

Part 2

Collapse, swelling, strength and stiffness of unsaturated soils

Part 2   Frontispiece

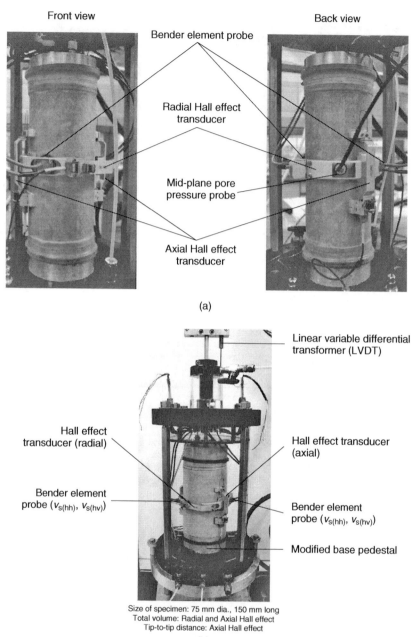

(a)

(b)

Size of specimen: 75 mm dia., 150 mm long
Total volume: Radial and Axial Hall effect
Tip-to-tip distance: Axial Hall effect

Images that represent the content of this Part 2: (a) Arrangement of mid-plane bender element probes and Hall effect transducers (Ng et al., 2004b); (b) Modified triaxial apparatus for measuring unsaturated anisotropic soil stiffness (Ng and Yung, 2007); (c) Measurement of total volume change (double wall system) using a differential pressure transducer (Ng et al., 2002a); (d) Bender elements.

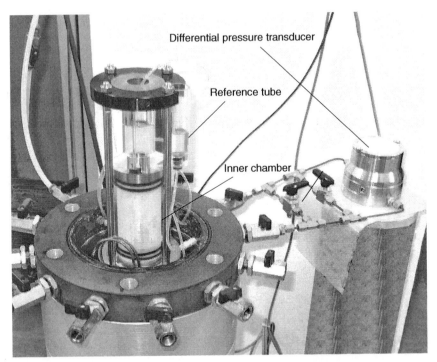

Differential pressure transducer

Reference tube

Inner chamber

(c)

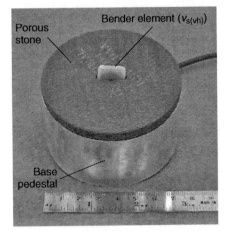

Porous stone

Bender element ($v_{s(vh)}$)

Base pedestal

Bender element embedded in base platen
(12 mm wide, 6 mm out)

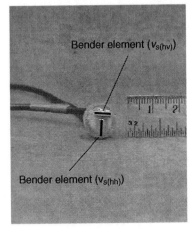

Bender element ($v_{s(hv)}$)

Bender element ($v_{s(hh)}$)

Bender element probe
(9 mm wide, 3 mm out)

(d)

(Continued).

Chapter synopses

Chapter 4: Collapse and swelling caused by wetting

Collapsible and swelling soils are introduced and their common characteristic of being unsaturated soils which react to wetting is explained in terms of stable and meta-stable soil structures. Laboratory testing methods are reviewed. The concept of virgin and non-virgin fills is introduced and their capacity for collapse studied. Mitigation measures against collapsible soils are considered and the behaviour of compacted expansive soils is reviewed along with the design of foundations on expansive swelling soils. This approach based on accumulated experience allied with modern testing is the basis of up-to-date geotechnical design of structures constructed in unsaturated soils as well as engineered unsaturated soil slopes.

Chapter 5: Measurement of shear strength and shear behaviour of unsaturated soils

Shear strength of unsaturated soils is introduced, and triaxial and direct shear test methods of measurement are reviewed. The importance of total volume change measurement (including pore air and pore water) in the triaxial test is explained and a new method of measurement is described in detail. The axis-translation and osmotic techniques for shear testing are compared. The extended Mohr–Coulomb failure criterion is reviewed, and various methods to estimate unsaturated soil shear strength from a soil–water characteristic curve are introduced and explored. The engineering behaviour of unsaturated soils is revealed by a series of testing programmes simulating field stress paths in the laboratory and covering the effects of soil suction on dilatancy. Experimental data revealing the engineering performance of both a loosely compacted volcanic soil and a loose saturated and unsaturated decomposed granitic soil are interpreted and explained using critical state soil mechanics extended for unsaturated soils. Key engineering results include suction-induced dilatancy plus the state-dependency of volume changes and stress–strain responses.

Chapter 6: Measurement of shear stiffness

The engineering importance of small strain shear stiffness of soils is emphasized. The influence of confining pressure, void ratio, grain size distribution and grain shape are explained. Testing equipment including triaxial and resonant column apparatus are described and the role of bender elements highlighted. Theoretical equations are derived to explain the influence of stress conditions and soil suction on stiffness anisotropy. The laboratory measurement of anisotropic shear stiffness of a completely decomposed tuff (CDT) is provided and explained. This combined approach of theory and testing reveal the role of soil suction in both stress-induced stiffness anisotropy as well as inherent stiffness anisotropy.

Chapter 4

Collapse and swelling caused by wetting

Sources

This chapter is made up of verbatim extracts from the following sources for which copyright permissions have been obtained as listed in the Acknowledgements.

- Bell, F.G. and Culshaw, M.G. (2001). Problem soils: a review from a British perspective. *Proc. Symp. on Problematic Soils*, Nottingham Trent University, I. Jefferson, E.J. Murray, E. Faragher and P.R. Fleming, eds, Thomas Telford, London, pp. 1–36.
- Fredlund, D.G. (1996). *The Emergence of Unsaturated Soil Mechanics*. The Fourth Spencer J. Buchanan Lecture, College Station, Texas, A & M University Press, p. 39.
- Houston, S.L. (1995). Foundations and pavements on unsaturated soils – Part one: Collapsible soils. *Proc. Conf. on Unsaturated Soils*, Paris, E.E. Alonso and P. Delage, eds, Balkema, Rotterdam, pp. 1421–1439.
- Jiménez-Salas, J.A. (1995). Foundations and pavements on unsaturated soils – Part two: Expansive clays. *Proc. Conf. on Unsaturated Soils*, Paris, Eds Alonso and Delage, Balkema, Rotterdam, pp. 1441–1464.
- Ng, C.W.W., Chiu, C.F. and Shen, C.K. (1998). Effects of wetting history on the volumetric deformations of unsaturated loose fill. *Proc. 13th Southeast Asian Geotech. Conf.*, Taipei, Taiwan, ROC, pp. 141–146.
- Tripathy, K.S., Subba Rao, K.S. and Fredlund, D.G. (2002). Water content – void ratio swell-shrink paths of compacted expensive soils. *Can. Geotech. J.*, 39, 938–959.

Overview

A major consideration for the technical professional dealing with unsaturated soils relates to the effect of wetting on engineering performance (Houston, 1995). In general, wetting results in volume change and reduction of shear strength (refer to Chapter 5) and stiffness (see Chapter 6). Two often cited unsaturated soil problems confronting the geotechnical and foundation engineers, collapsible soils and expansive soils, are associated with wetting-induced volume change (Fookes and Parry, 1994; Bell and Culshaw, 2001). These problem soils are of greatest concern in arid regions; however, soil collapse or expansion can be of concern in essentially any climate.

Naturally occurring deposits as well as compacted fills may exhibit significant volume change in response to wetting. Generally, when any soil is wetted it may decrease in volume, increase in volume or experience no significant volume change. The amount and type of volume change depend on the soil type and structure, the initial soil density, the imposed stress state and the degree of wetting. Soils that exhibit significant compression or shrink/swell response to moisture content changes are often referred to as moisture-sensitive soils. There are several challenging aspects of the moisture-sensitive soil condition to be considered by the geotechnical engineer as listed as follows (Houston, 1995):

- Identification of moisture-sensitive soils.
- Identification of mechanisms involved in volume change.
- Identification of circumstances leading to building/foundation distress.
- Estimation of volume change and differential movements.
- Development of mitigation and foundation design alternatives.

The discussion on unsaturated soils in this chapter is divided into two distinct categories of volume moisture-sensitivity: part one, collapsible soils, and part two, expansive soils. Although any given soil has the potential to compress or swell when wetted, depending on the conditions at the time of wetting, certain soils and conditions tend to be strongly associated with collapse, and other soils and conditions tend to be strongly associated with swell. Therefore, the subdivision into collapsible and expansive soils has been adopted for the purpose of addressing geotechnical engineering problems in terms of considerations of volume change resulting from wetting of unsaturated soils. The approaches adopted are mainly empirical and semi-empirical in nature. Much of research work is still needed to be done.

Collapsible soils (Bell and Culshaw, 2001; Houston, 1995; Fredlund, 1996)

Occurrence and microstructure of collapsible soils (Fredlund, 1996)

Collapsible soils are often encountered in many parts of the world. Although collapsible soils are often considered a natural hazard, man-made compacted collapsible soil are also imposing challenges to engineers. Most natural occurrence of collapsible soils is comprised primarily of silt and fine sand-sized particles with small amounts of clay, and under some circumstances may contain gravel and cobbles. These natural soils are typically

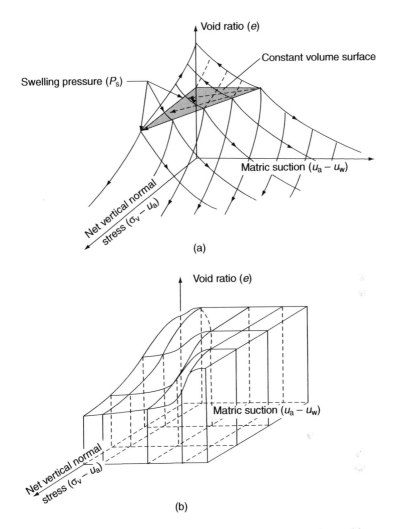

Figure 4.1 Loading and unloading constitutive surfaces for (a) a stable-structured, unsaturated soil; (b) a metastable-structured, unsaturated soil (after Fredlund, 1996).

lightly to heavily cemented by various salts, oxides, dried clay and soil suction, and exist in a loose, metastable condition at relatively low density. Figure 4.1a and b show three-dimensional graphic representations of stable-structured and metastable-structured unsaturated soils in three-dimensional space of void ratio-matric suction-net vertical normal stress, respectively (Fredlund, 1996). For stable-structured unsaturated soils (see Figure 4.1a), a reduction of matric suction due to wetting will induce an increase in void ratio. In contrast, a metastable-structured unsaturated soil will result in a reduction in void ratio (i.e. collapse) due to a decrease in suction from wetting. These soils such as loess, brickearth and certain wind blown silts generally consist of 50 to 90 per cent silt particles and sandy, silty and clayey types have been recognized by Clevenger (1958), with most falling into the silty category (Bell and Culshaw, 2001). They possess porous textures with high void ratios and relatively low densities and they often have sufficient void space in their natural state to hold their liquid limit moisture at saturation. At their natural low moisture content these soils possess high apparent strength but they are susceptible to large reductions in void ratio upon wetting. In other words, the metastable texture collapses as the bonds between the grains break down when the soil is wetted. Hence, the collapse process represents a rearrangement of soil particles into a denser state of packing. Collapse on saturation normally only takes a short period of time, although the more clay such a soil contains, the longer the period tends to be (Bell and Culshaw, 2001). In terms of elasto-plastic constitutive modelling, the collapse of an unsaturated soil may be described as reaching the Loading–Collapse (LC) yield curve (Alonso et al., 1990; Chiu and Ng, 2003). More details are given in Chapter 7.

The fabric of collapsible soils generally takes the form of a loose skeleton of grains (generally quartz) and microaggregates (assemblages of clay or clay and silty clay particles). These tend to be separate from each other, being connected by bonds and bridges, with uniformly distributed pores. The bridges are formed of clay-sized minerals, consisting of clay minerals, fine quartz, feldspar or calcite. Surface coatings of clay minerals may be present on coarser grains. Silica and iron oxide may be concentrated as cement at grain contacts and amorphous overgrowths of silica occur on grains. As grains are not in contact, mechanical behaviour is governed by the structure and quality of bonds and bridges. The structural stability of collapsible soils is not only related to the origin of the material, to its mode of transport and depositional environment but also to the amount of weathering undergone.

Collapsible soil deposits are common in arid regions where evaporation rates exceed rainfall. Arid-region deposits often associated with collapsible soils include alluvium, colluvium and loess (Liu, 1988; Beckwith, 1995; Zhang and Zhang, 1995; Derbyshire et al., 1995; Jefferson et al., 2001). At their normal in situ low moisture contents these soils may exhibit high apparent strength and stiffness. However, collapsible soils densify significantly and quickly upon wetting. Numerous studies have shown collapsible soil deposits to exhibit a high degree of heterogeneity with respect to cementation, collapse potential and other engineering characteristics (Houston and El-Ehwany, 1991; Beckwith, 1995; Zhang and Zhang, 1995; El Nimr, et al., 1995).

Conditions leading to collapse

Under normal circumstances naturally occurring collapsible soils are not wetted to significant depth by precipitation. Rainfall either runs off or infiltrates only a short distance and then evaporates to the surface, particularly in arid regions. Problems with collapsible soils are almost always associated with man-induced changes in the surface water and groundwater regime. This is often coupled with a failure to identify the collapse condition before construction and/or a lack of knowledge of potential wetting sources. Collapse, triggered by increased water content, can typically be linked to engineering modifications and alteration of natural flow patterns. In some cases the overburden stresses alone are sufficient to drive the collapse process when wetting occurs. When cementation is very strong or the collapsible soils are very shallow, additional stress due to a structure or foundation may be necessary for collapse to happen.

Often collapsible soils go undetected until a structure is already in place. Even if collapsible soils are identified prior to construction, lack of knowledge of potential sources of wetting can lead to incomplete mitigation of the problem. For example, due consideration of rising groundwater table may not be given, or infiltration may extend to greater depth than assumed. Moderate-weight to light-weight structures are particularly vulnerable to damage because funds for site exploration and pre-construction mitigation are frequently limited due to highly competitive budget considerations.

Identification of collapsible soils

Geological reconnaissance

Geomorphological considerations are quite valuable as a first step towards identification of a naturally occurring collapsible soil deposit. For example, Beckwith (1995) recommends that Holocene alluvial fan deposits in the southwestern regions of the United States should be assumed to be collapsible unless a comprehensive testing program demonstrates otherwise. Loessial deposits in northwestern China are known to exhibit significant collapse potential. Chinese researchers have found that geographical and geomorphological information is strongly correlated with collapsibility and collapse potential (Liu, 1988; Lin, 1995). Thus, geological reconnaissance, coupled with experience with similar depositional environments, forms an important part of the process of collapsible soil identification.

Laboratory testing

COLLAPSE INDICATORS (BELL AND CULSHAW, 2001)

Several collapse criteria have been proposed for predicting whether a soil is liable to collapse upon saturation. For instance, Clevenger (1958) suggested a criterion for collapsibility based on dry density, that is, if the dry density is less than $1.28\,Mg/m^3$, then the soil is liable to undergo significant settlement. On the other hand, if the dry density is greater than $1.44\,Mg/m^3$, then the amount of collapse should be small, while at intermediate densities the settlements are transitional. Gibbs and Bara (1962) suggested the use of dry unit weight and liquid limit as criteria to distinguish between collapsible and non-collapsible soil types. Their method is based on the premise that a soil, which has enough void space to hold its liquid limit moisture content at saturation, is susceptible to collapse on wetting. This criterion only applies if the soil is uncemented and the liquid limit is above 20 per cent. When the liquidity index in such soils approaches or exceeds unity, then collapse may be imminent. As the clay content of a collapsible soil increases, the saturation moisture content becomes less than the liquid limit so that such deposits are relatively stable. However, Northmore et al. (1996) concluded that this method did not provide a satisfactory means of identifying the potential metastability of brickearth. More simply, Handy (1973) suggested that collapsibility could be determined either by the percentage clay content or from the ratio of liquid limit to saturation moisture content. He maintained that soils with a:

- clay content of less than 16 per cent had a high probability for collapse;
- clay content of between 16 and 24 per cent were probably collapsible;
- clay content between 25 and 32 per cent had a probability of collapse of less than 50 per cent;
- clay content which exceeded 32 per cent were non-collapsible.

Soils in which the ratio of liquid limit to saturation moisture content was less than unity were collapsible, while if it was greater than unity they were safe. After an investigation of loess in Poland, Grabowska-Olszewsla (1988) suggested that loess with a natural moisture content less than 6 per cent was potentially unstable, that in which the natural moisture content exceeds 19 per cent could be regarded as stable, while that with values between these two figures exhibited intermediate behaviour.

Although, empirical indices of collapse may provide a relatively rapid and inexpensive means of determining collapsibility (Denisov, 1963; Feda, 1966; Fookes and Best, 1969; Derbyshire and Mellors, 1988; Northmore et al., 1996), it must be borne in mind that some of the parameters, which are used for assessment are derived from remoulded soil samples. Such samples do not take account of the initial soil fabric. Consequently, the results should be regarded as only general indicators of collapsibility and indeed some indices need to be used with caution.

THE DOUBLE OEDOMETER TEST (JENNINGS AND KNIGHT, 1975)

Jennings and Knight (1975) developed the double oedometer test for assessing the response of a soil to wetting and loading at different stress levels (i.e. two oedometer tests are carried out on identical samples, one being tested at its natural moisture content, whilst the other is tested under saturated conditions, the same loading sequence being used in both cases). They subsequently modified the test so that it involved loading an undisturbed specimen at natural moisture content in the oedometer up to a given load. At this point the specimen is flooded and the resulting collapse strain, if any, is recorded (Figure 4.2). Then the specimen is subjected to further loading.

Table 4.1 provides an indication of the potential severity of collapse. This table indicates that those soils which undergo more than 1 per cent collapse can be regarded as metastable. However, in China a figure of 1.5 per cent is taken (Lin and Wang, 1988) and in the United States values exceeding

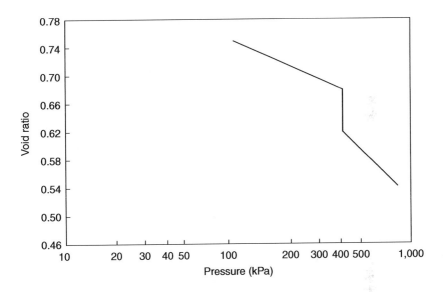

Figure 4.2 Void ratio–log pressure curve of metastable brickearth from south Essex tested in an oedometer (specimen flooded after 24 h loading at 429 kPa), (from Northmore *et al.*, 1996; after Bell and Culshaw, 2001).

Table 4.1 Collapse percentage (defined as $\Delta e/(1 + e)$, where e is void ratio) as an indication of potential severity of soil collapse (from Jennings and Knight, 1975; after Bell and Culshaw, 2001)

Collapse (%)	Severity of problem
0–1	No problem
1–5	Moderate trouble
5–10	Trouble
10–20	Severe trouble
Above 20	Very severe trouble

2 per cent are regarded as indicative of soils susceptible to collapse (Luteneg-ger and Hallberg, 1988).

There are other single specimen collapse tests as described by Houston *et al.* (1988) and ASTM procedure D5333 (ASTM, 1993) and a single point, multiple specimen test procedure (Noorany, 1992). The results of a simple oedometer test will indicate whether a soil is collapsible and at the same time give a direct measure of collapse strain potential (Houston, 1995).

Potential problems associated with the direct sampling method include sample disturbance, field non-homogeneities, and the possibility that the degree of saturation achieved in the field will be less than that achieved in the laboratory. However, performance of laboratory response to wetting tests on undisturbed specimens likely represents the most widely used and economical method for identification of collapsible soils. Of course, the quality of the nature specimen used in the laboratory oedometer test is of concern in making quantitative estimates of collapse potential.

Compacted fills may exhibit moisture sensitivity depending on the soil type, compactive effort, compaction water content and stress level at the time of wetting. Identification of existing collapsible fills is best accomplished by direct measurements of response to wetting, either in the laboratory or in the field. However, the best engineering approach is to identify the conditions leading to compacted collapsible fills through a laboratory testing program before fill construction and then to avoid these conditions for fill placement (Houston, 1995).

Interpretation of laboratory response to wetting tests (Houston, 1995)

Effects of sampling, loading and wetting

The proper interpretation and use of the data from the laboratory oedometer test depends on the history of the test specimen prior to wetting. Three fairly common scenarios can be postulated as follows.

- Case I: A relatively undisturbed sample of naturally occurring collapsible soil is taken from the field and prepared for testing in the laboratory. It is loaded dry up to the overburden pressure in the field and the strain is measured. Then it is wetted while under overburden pressure and this strain is measured.
- Case II: Exactly the same as Case I except that the soil specimen is from an existing compacted fill.
- Case III: A test specimen is compacted directly in the oedometer to the density and water content at which a new fill is to be constructed (Alternatively, the fill to be represented already exists and its density and water content are known. A laboratory test specimen is prepared in

the oedometer to match the fill material as it was first compacted, before the weight of the overlying soil was imposed.) Then the laboratory specimen is loaded to the expected overburden and wetted as in Cases I and II.

A schematic representation of the laboratory test results for all three cases can be constructed, as shown in Figure 4.3. In this figure, it is assumed that the response to wetting for all three cases is collapse.

First, consider Case I which applies to a naturally occurring collapsible soil. For the test result depicted in Figure 4.3, no significant swell would occur for wetting at any confining stress. For more clayey soils more swell might be observed for light confining pressure. The curves AC, AD and AE represent dry loading of specimens with successively greater degrees of sample disturbance. This dependence on sample disturbance is routinely observed (Houston and El-Ehwany, 1991). Greater degrees of disturbance break bonds between particles and decrease resistance to dry loading. Thus more dry strain occurs. If a perfectly undisturbed sample were taken and its density and structure were maintained by cementation and soil suction, it is logical to assume that almost no strain would occur when the in situ overburden stress was reapplied. This is supported by the fact that medium-to-large model footings tests conducted on cemented collapsible soil deposits have exhibited essentially no measurable settlement upon dry loading (Houston *et al.*, 1988, El-Ehwany and Houston, 1990; Mahmoud, 1992). Thus the curve AB is representative of dry loading of truly undisturbed material.

For the curves AC, AD and AE, wetting under overburden stress results in strain to point F. If the stress on the wetted specimen is increased, point G is obtained. If wetting occurs at lower than overburden stress, say point

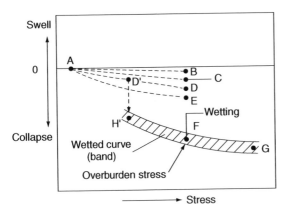

Figure 4.3 Typical response curves of volume change with vertical stress to wetting of a collapsible soil (after Houston, 1995).

D′, then strain to H′ would occur. If a specimen at H′ is subjected to higher stress, the wetted curve passes more or less through points F and G. The important issue is that, regardless of sample disturbance, the specimen ends up on essentially the same wetted curve, H′FG. The experimental data show that the wetted curve is a band, as indicated in Figure 4.3. The width of the band derives in part from sample heterogeneity and in part from path dependency (stress at which wetting occurs). The data obtained to date indicate that most of the width arises from heterogeneity, i.e. sample variability from point to point in the soil profile. However, as a percentage of the total strain ordinate, the width of the band is usually narrow enough that it can be treated as an average wetted curve for engineering applications.

The consequence of this simplifying assumption is that one specimen can be used to obtain the wetted curve, whether wetting occurs at point D′ or D, for example. The greater the sample disturbance the greater is the dry strain and the less is the strain upon wetting. However, an appropriate total of the dry strain plus the wetting strain is always given by the wetted curve. For application to the field the most reasonable interpretation is to use the strain from the origin to the wetted curve as the expected strain upon wetting for the field. This interpretation is based on the position that the particle bonds which are broken by disturbance would have been broken by wetting under load, had the wetting occurred without prior disturbance. This corresponds to neglecting the small strain given by curve AB – the ideal perfect sample curve. By using this interpretation it is possible to use one test specimen to get the wetted curve and then estimate the strain for other points in the soil deposit. Stresses due to structural loads may be added to the overburden stress to obtain the total stress with which to enter Figure 4.3.

Even though a wetted curve can be obtained from only one test specimen, it would not be good engineering practice to try to characterize a site with just one test. In order to capture the effects of spatial variability it is necessary to test several specimens. A band of finite width could then be constructed to represent the wetted curve for the range of test results obtained for a particular layer or zone.

An implied assumption in the foregoing discussion is that, when an ideal perfect sample is unloaded from overburden stress, the strain which occurs is negligible compared to the total strain from the origin to the wetted curve. This assumption is believed to be generally valid for most collapsible soils. If the soil is cemented then cementation helps to prevent volume change during unloading (and reloading back to the same stress). Even if the soil is not cemented, sands and silts exhibit negligible elastic strain upon unloading, while saturated cohesive soil would be expected to exhibit no measurable volume change upon removal and reapplication of overburden stress. If the cohesive soil were unsaturated, however, it is possible that these elastic strains would be measurable and perhaps not negligible. It is possible to estimate these elastic strains as follows.

The specimen should be loaded to overburden stress and then unloaded to some small percentage of overburden and finally reloaded back to overburden. The slopes of the unload/reload curves compared to the primary loading curve can be used to separate elastic and plastic strains. It is reasonable to assume that the plastic strains were due to disturbance effects, but the elastic strains were not. The slope of the unload/reload curve can then be used to construct a curve from the origin out to the overburden stress. This curve can be thought of as a 'disturbance-free' dry loading curve, and corresponds to curve AB in Figure 4.3. If necessary, this curve can be extrapolated a moderate distance beyond the overburden stress to represent the effect of additional structural loads. If extrapolated, the slope should be gradually steepened to account for the plastic strains associated with loading beyond overburden stress. However, the slope of the primary loading curve cited above probably represents a reasonable upper bound for the slope of this extrapolated disturbance-free dry loading curve. By following this procedure, an improved estimate of the dry loading curve is obtained. Then a more nearly unbiased estimate of in situ wetting strain is obtained as the difference in ordinate between the estimated disturbance-free dry loading curve, AB, and the wetted curve H'FG. The refinement suggested above is likely to be rarely justified. For most applications, a satisfactory estimate of field wetting strain can be obtained by simply using the strain ordinate from the origin to the wetted curve.

Consider Case II next, which is identical to Case I except that a sample from an existing compacted fill is taken and tested. The recommended interpretation for Case II is identical to that recommended above for Case I. Although compacted fills that are collapsible may have somewhat less chemical cementation on average than naturally occurring collapsing soils, fill materials are typically cemented with at least some clay and always by soil suction. Thus they are subject to weakening of particle bonds by sampling. If somehow a perfect sample of the fill material were obtained for testing and the overburden stresses were reapplied to it, the strains would be negligible for most cases Therefore, if substantial strains are observed for dry loading of a sample from a fill, these strains are very likely to result from sample disturbance. It follows that the most appropriate test interpretation for Case II is the interpretation recommended for Case I.

Case III differs from Cases I and II in that the soil specimen has been compacted to the initial in situ density and water content but has never experienced the overburden stress corresponding to the point in the field being represented. Furthermore, it is assumed that the laboratory specimen has been compacted in the oedometer similar to the corresponding field element of soil. No sampling occurred, so sample disturbance is not an issue. Therefore, strains which occur upon dry loading correspond to strains which were probably experienced by the field element when the remainder of the embankment was built above it. However, the strains and deformations up to overburden stress occurred in the prototype before infrastructure was constructed on the surface of the fill. Therefore, the dry strains are not particularly important and do not cause damage to structures. The strains due to wetting under overburden (and weight of structures) are important and do cause damage. Referring to Figure 4.3, the strain from the origin to point D (assuming the dry loading curve is AD) can be neglected and the strain from point D to F is the strain of importance. An equally acceptable testing procedure is to re-zero the displacement gauges after the dry strain has occurred and measure only the strain due to wetting, DF.

Summary of recommendations relative to Cases I, II and III (Houston, 1995)

In comparing Cases I, II and III, there is one common factor. In all three cases the dry strain is not extremely important and is not used directly in design exercises. However, in Cases I and II it is included in the total collapse strain, which is taken as the ordinate from the origin to the wetted curve in Figure 4.3. In Case III, however, the dry strain is not included in the wetting under load strain postulated for the prototype. For Case III, the wetting strain postulated for the field is the ordinate from point D to F in Figure 4.3, for example.

Comparison of commonly used laboratory test methods (Houston, 1995)

With the preceding discussion of Cases I, II and III as a context, it is now possible to address the relative merits of three test procedures that have been debated in the literature (Lawton *et al.*, 1991; Noorany, 1992):

- single specimen test
- double oedometer test
- single point, multiple specimen test

The single specimen test is the test that was described under Cases I, II and III above and depicted in Figure 4.3. It is called a single specimen test because one specimen is used to get the wetted curve. The path to the wetted curve is determined by the stress at which wetting occurs. Path dependency refers to the extent to which the position of the wetted curve depends on the path taken to get to it. Although numerous tests conducted as a part of the author's studies of collapsible soils indicate that path dependency is moderately small, particularly for silty and sandy soils which exhibit collapse, it is not zero. The path dependency is greatest when wetting occurs at very low stress and the wetted curve is obtained by continued loading of the wetted specimen. Therefore, this practice is not recommended. Instead it is recommended that wetting occurs at a stress level at least 1/3 to 1/2 of the overburden stress and that the maximum stress expected in the field be achieved in one or two additional load increments. The primary advantage of the single specimen test is that the maximum possible amount of useful data is obtained from one specimen and the remaining testing budget can be used to perform more tests and help capture the spatial variability in the natural deposits and compacted fills.

The double oedometer test (Jennings and Knight, 1956) involves preparing two specimens, which are hopefully identical, for testing in the oedometer. One is tested dry, tracing out the dry compression curve, and the other is tested wet, tracing out the wetted curve. The collapse strain for the field is taken as the ordinate between the wet and dry curves. Recent studies indicate that most of the dry strain (particularly at stresses less than or equal to overburden stress) is due to disturbance and would have occurred under wetting if the disturbance had been absent Therefore, the best interpretation for application to the field is to assume that strain due to wetting in the field will be from the origin to the wetted curve. Using this interpretation, the dry curve adds no useful information and the corresponding testing budget should be used to test another specimen from a different location in the soil profile or fill.

The single point, multiple specimen procedure provides one point on the wetted curve for each specimen. This point is obtained by loading dry up to a particular stress, then wetting. No additional load increments are added after wetting. To obtain another point on the wetted curve, another specimen must be prepared and loaded to a different stress, then wetted. Thus multiple specimens are required. If this procedure is used, it is recommended that the test interpretation advocated for Case I, II or III above be adopted, depending on the actual scenario employed in bringing the test specimen to the wetted curve.

The relative merits of the single point, multiple specimen procedure are as follows (Houston, 1995). First, an advantage of the procedure is that it does not suffer from path dependency. Whether or not path dependency is significant, the test specimen is made to follow the actual path postulated for the field. Thus no error in the results arises from path dependency. Disadvantages of the procedure include inefficiency and inadequate characterization of spatial variability. If a whole suite of tests is performed for each point or small zone within an embankment or soil profile, the test program would be inefficient compared to the single specimen procedure. It should be noted that the author does not advocate using a single specimen to trace out a wetted curve that extends into the expansive region. If a curve covering the full spectrum from expansion to collapse is required, then testing of at least two or three specimens is recommended by the author. Nevertheless, the general procedure associated with the single specimen test is inherently more efficient than the single point, multiple specimen procedure. The price that is paid for this efficiency is the introduction of some error in the position of the wetted curve due to path dependency. It is believed that this error is typically fairly small and is more than compensated for by using the additional testing budget to more nearly capture spatial variability in natural soil deposits and fills. Thus the practice of using the single specimen procedure to trace out significant segments of the wetted curve is considered superior to the single point, multiple specimen procedure, provided the savings in testing budget is used to test more samples from different locations to more fully capture spatial variability.

Field testing

In situ methods of testing for collapsible soils include shallow plate load tests and down-hole plate load tests. In situ measurements appear promising for identification of collapsible soils, particularly for difficult-to-sample materials (Rollins *et al.*, 1994; Zhang and Zhang, 1995; El Nimr *et al.*, 1995; Houston *et al.*, 1995; Souza *et al.*, 1995). In general sample disturbance may be greatly reduced, and a quantitative measure of collapse potential can be obtained. A quantitative measure of collapse potential in terms of strain is somewhat more difficult to obtain from in situ tests because the degree and extent of wetting, and therefore the zone of influence, are not routinely determined (Houston *et al.*, 1995).

In spite of difficulties such as sample disturbance and soil heterogeneity over very short distances, it is desirable to obtain some direct, quantitative measure of the collapse potential. Ideally, an adequate number of points within the profile can be tested so that a reasonable engineering estimate of collapse settlement potential can be made. The cost of conducting laboratory and/or in situ response to wetting tests should not be significantly greater than the costs associated with less-quantitative identification methods.

Concept of virgin and non-virgin fills (Ng et al., 1998)

Theoretical background

Experimental data demonstrate that wetting-induced collapse behaviour of unsaturated soil is an irrecoverable process. This has led to develop elasto-plastic models by some researchers for unsaturated soils. Alonso *et al.* (1990) proposed an elasto-plastic model under the critical state framework for unsaturated soils based on experimental data from suction-controlled tri-axial tests on compacted clay. The model is defined in terms of four state variables:

$$\text{Mean net stress, } p = \frac{1}{3}(\sigma_1 + \sigma_2 + \sigma_3) - u_a \tag{4.1}$$

Deviator stress, $q = \sigma_1 - \sigma_3$ (4.2)

Suction, $s = u_a - u_w$ (4.3)

Specific volume, $v = 1 + e$ (4.4)

This model assumes elastic behaviour if the soil remains inside a yield surface defined in q–p–s space, with plastic strains commencing once the yield surface is reached. The general shape of the yield surface is shown in Figure 4.4. For soil initially at a stress state A inside the yield surface, elasticity is assumed. An increase of p (isotropic loading path AB) causes elastic compression and a reduction of s (wetting path AF) causes elastic swelling. As the stress state reaches the current position of the yield surface (e.g. B, D or F), an increase of p (isotropic loading path BC), an increase of q (shearing path DE) or a reduction of s (wetting path FG) can expand the yield surface further. For isotropic stress states, Alonso et al. (1990) defines a LC yield curve as the intersection of the yield surface with $q = 0$ plane (see Figure 4.4). The existence of the LC yield curve is further supported by experimental evidence provided by Wheeler and Sivakumar (1995) and Cui and Delage (1996).

Testing program

An experimental program was carried out to study the effects of wetting history on collapse potential of unsaturated and loosely compacted virgin and non-virgin fills. In this chapter, collapse potential is defined as the ratio

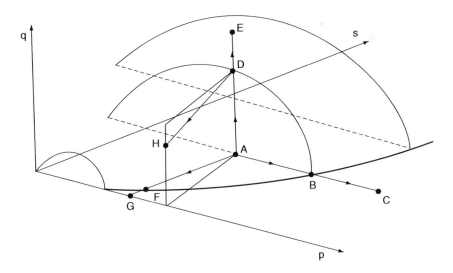

Figure 4.4 Yield surface in q–p–s space (after Alonso et al., 1990).

between changes in specific volume and initial specific volume of soil sample. Virgin fill is refers to a soil which has never been subjected to a suction smaller than its current value, whereas non-virgin fill is defined as a soil which has been subjected to a suction smaller than its current value. One-dimensional wetting tests were conducted on statically compacted specimens of completely decomposed volcanic taken from a slope in Shatin, Hong Kong. The index properties of the soil are summarized in Table 4.2. All the tests were performed in a modified oedometer ring, which is fitted with a flexible tube type tensiometer to measure negative pore water pressure simultaneously with vertical settlement during a wetting stage of tests. The diameter and height of the soil sample are 25 and 100 mm, respectively. Only soil particles passing through a 2 mm sieve were used for testing because of the size limitations of the oedometer. Details of the testing conditions are summarized in Table 4.3. The advantage of using an oedometer is that tests can be conducted relatively quickly due to the short drainage path. During

Table 4.2 Summary of index properties for sandy silt (after Ng et al., 1998)

Grain size distribution	
Percentage of gravel (%)	35
Percentage of sand (%)	18
Percentage of silt (%)	21
Percentage of clay (%)	26
Specific gravity	2.66
Standard Proctor test	
Maximum dry density (kg/m^3)	1,460
Optimum moisture content (%)	26

Table 4.3 Wetting history testing program (after Ng et al., 1998)

Test No.	Initial relative compaction (%)	Initial compaction stress (kPa)	Initial moisture content (%)	First cycle of wetting		Second cycle of wetting	
				Suction (kPa)	Applied vertical stress (kPa)	Suction (kPa)	Applied vertical stress (kPa)
T1	68	99	22.5	77	99	77	99
T2	81	271	22.5	77	271	77	271
T3	90	540	22.5	77	540	77	540
T4	67	50	26	60	50	60	50
T5	79	167	26	60	167	60	167
T6	86	406	26	60	406	60	406
T7	66	26	30	20	26	20	26
T8	78	63	30	20	63	20	63
T9	91	246	30	22	246	22	246
T10	65	50	26	58	50	58	167
T11	65	26	30	20	69	N.A.	N.A.

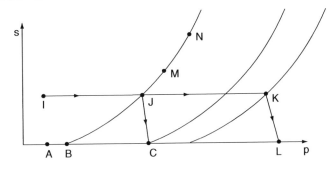

Figure 4.5 Load collapse yield surface (after Ng *et al.*, 1998).

the tests soil specimens are subjected to one-dimensional stress conditions with unknown horizontal stress. The deviator stress is likely non-zero for most situations. The stress state of a virgin fill loaded in the oedometer may be represented by D in Figure 4.4 which lies on its current yield surface. It is most likely that the horizontal stress could increase during wetting and DH is a possible wetting path under a constant vertical stress. Under these stress conditions, it is expected that the form of yield curve in the space of net vertical stress–matric suction stress is likely to be similar to the LC yield curve shown in Figure 4.5. The concept of the proposed elasto-plastic model may be used to explain the results of the oedometer tests.

Testing procedures

The soil specimens were statically compacted to various initial conditions (moisture content and dry density) as summarized in Table 4.3. The dry densities are expressed in terms of relative compaction which is the percentage of maximum dry density obtained from the standard Proctor test. A tensiometer was installed into the soil specimen before application of vertical loads. A 5.5-mm diameter hole was drilled into each compacted soil sample through a circular opening provided in the mid-height of the modified oedometer ring. The drilled hole is slightly smaller than the tensiometer tip (6.3 mm in diameter) to ensure a good contact between the tip and the surrounding soils. The soil samples were subjected to different stress paths after assembling into the oedometer.

A total of nine tests (T1–T9) were carried out for soil samples compacted to different initial conditions but they all followed the same stress path as illustrated in Figure 4.6. After assembling into the oedometer, soil samples were loaded to the stresses equal to those applied in compacting the specimens previously (loading path IT). Then they were covered by plastic wrap to minimize loss of moisture content and left them for approximately 24 h to make sure that 99 per cent of vertical settlement due to the applied

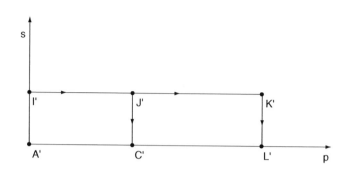

Figure 4.6 Stress paths of repeated cycles of wetting (after Ng et al., 1998).

vertical load was reached. Subsequently, the soil samples were wetted with distilled water from the bottom of the oedometer (wetting path J'C'). The corresponding vertical settlements due to wetting were recorded by a dial gauge and matric suction at the mid-height of the samples was measured simultaneously by a tensiometer. After completing the first cycle of wetting, water remained in the oedometer was drained away and plastic wrap on the soil sample was taken away. Then the applied loads was removed (unloading path C'A') and the soil samples were left for drying until matric suction returned to their initial values (drying path AT). During this stage, small vertical movement was recorded. The soil samples were then reloaded to the previous applied stress (reloading path IT), ready for the second cycle of wetting (wetting path J'C').

For studying the effects of different stress paths on the collapse behaviour of virgin and non-virgin fills, two more tests were carried out. The soil sample for test no. T10 was compacted to the same initial conditions of test no. T4 and it followed the same stress path as test no. T4 to J' (path I'J'C'ATJ'). Instead of wetting, it was loaded to a higher vertical stress to K' and then it was subjected to the second cycle of wetting (wetting path K'L'). Similarly, the soil sample of test no. T11 was compacted to the same initial conditions of test no. T4. After loading the sample to a higher vertical stress than that used in compacting the specimen previously (loading path I'K'), it was subsequently subjected to a wetting path K'L'.

Collapse behaviour of virgin fill

Typical results of a two-cycle wetting test are shown in Figure 4.7. It is shown that there is a substantial amount of deformation when the soil sample subjected to the first cycle of wetting (virgin fill). The behaviour of the second wetting cycle is discussed later. Figure 4.8 shows the effect of relative compaction on collapse potential for the virgin fill (i.e. results from

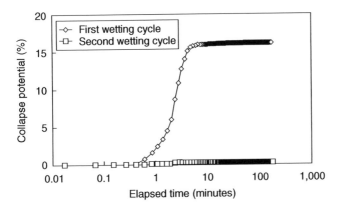

Figure 4.7 Collapse potential relationship with time for a two cycle wetting test (after Ng *et al.*, 1998).

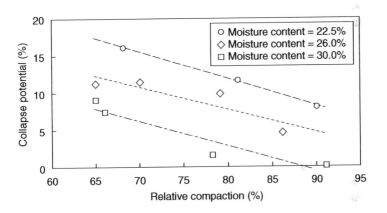

Figure 4.8 Collapse potential relationship with relative compaction (after Ng *et al.*, 1998).

first wetting cycle only). For a given initial moisture content, the amount of collapse caused by wetting decreases with an increase in relative compaction or initial dry density. Similar relationship between relative compaction and collapse potential was observed and reported by Chiu *et al.* (1998) on a sandy silt compacted at moisture contents ranging from 5 to 30 per cent. Tadepalli and Fredlund (1991) conducted some tests on Indian Head silt in Canada at moisture contents ranging from 7 to 13 per cent and they reported an inverse linear relationship between dry density and collapse potential. It can also be seen from the figure that for a given compaction effort, the collapse potential decreases as moisture content increases. Since matric

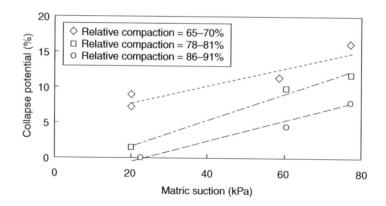

Figure 4.9 Collapse potential relationship with matric suction (after Ng *et al.,* 1998).

suction is related to moisture content, this observed collapse behaviour can be regarded as controlled by matric suction. Conceptual explanation of this observed collapse behaviour is given in the following paragraph.

Figure 4.9 shows the effect of initial matric suction (i.e. moisture content) on collapse potential for the virgin fill (i.e. results from first wetting cycle only). For a given range of relative compaction, the amount of collapse caused by wetting increases with an increase of initial matric suction (i.e. a decrease in moisture content). As illustrated in Figure 4.4, the yield surface can be expanded by an increase of p or q, or a reduction of s. DH of Figure 4.4 is a possible stress path for wetting tests conducted in the oedometer under constant vertical stress. The expansion of the yield curve is mainly governed by the reduction of s. For soil samples of virgin fill compacted to the same initial dry density, they would lie on the same yield curve. The higher the initial suction of the sample is, the further away from zero suction axis (see J, M and N in Figure 4.5). As the wetting process brings the suction in the samples to zero, larger expansion of the yield surface is expected from samples with higher suction and this means that change of volume is greater.

The applied vertical stress also influences the collapse behaviour of virgin fill. Samples of tests T7 and T11 were compacted to the same initial conditions, but the sample of T11 was loaded to a higher vertical stress than that of T4. Their stress states before wetting are shown in Figure 4.10. It is shown in Table 4.4 that T11 has a larger collapse potential for first wetting cycle than T4. Upon wetting, both samples exhibit the same amount of reduction in matric suction, but they laid on different yield curves and followed different stress paths. Different amounts of plastic strains are expected in these two samples.

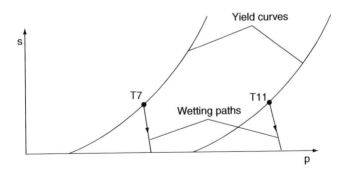

Figure 4.10 Stress states of tests T7 and T11 before first wetting cycle (after Ng et al., 1998).

Table 4.4 Collapse potential observed in test T4, T7, T10 and T11 (after Ng et al., 1998)

Test No.	Collapse potential for first wetting path (%)	Collapse potential for second wetting path (%)
T4	11.7	0.3
T10	11.3	6.9
T7	7.3	0.1
T11	9	N.A.

Collapse behaviour of non-virgin fill

The influence of wetting history on collapse potential of non-virgin fill can be demonstrated by subjecting the soil sample to two cycles of wetting and drying under constant vertical load (see Figure 4.7). The soil sample was initially compacted at a moisture content of 22.5 per cent with relative compaction of 65 per cent. The initial matric suction was 77 kPa. The test results indicate a large collapse potential during the first cycle of wetting, however, a significant reduction of collapse potential during the second cycle of wetting. The observed collapse potentials at the first and second cycles are 16.2 and 0.22 per cent, respectively. It is important to distinguish the nature of a virgin fill and a non-virgin fill as both of them exhibit significant difference in response to wetting. The volumetric deformation induced by first cycle of wetting can be explained by the yield curve shown in Figure 4.4. After application of the vertical load, the stress state of virgin fill lies on the yield curve at point J in Figure 4.4. The first wetting path JC takes place in the plastic region and causes expansion of the yield curve to point C. Large amount of collapse is therefore expected from the first wetting path. However, the results observed from the second cycle of wetting are different from those would be predicted by an elasto-plastic model (Alonso et al.,

1990). Before the commencement of second wetting cycle, the stress state of non-virgin fill would return to point J in Figure 4.4 and lies inside the expanded yield curve (according to the model). The second wetting path JC takes place within the elastic region. Elastic swelling would be predicted by Alonso's model. However, a relatively small amount of elastic compression instead of swelling is observed in Figure 4.4. The wetting paths conducted in the oedometer were not under the conditions of constant mean stress and zero deviator stress. These conditions may influence the amount of elastic strains recorded during the wetting tests. Further experiments are required to study the behaviour of soil inside the yield curve.

The applied vertical stress also affects the amount of collapse exhibited by the non-virgin fill. Samples of tests T4 and T10 were compacted to the same initial conditions and they followed the same stress path for the first wetting–drying cycle. Before the commencement of second wetting cycle, T10 was reloaded to a higher vertical stress than T4 (i.e. T10′ in Figure 4.11). The collapse potentials of second wetting cycle for no and T4 are 6.9 and 0.3 per cent, respectively (see Table 4.4). This significant difference in the observed collapse potential could be explained by the model proposed by Alonso *et al.* (1990). The stress states of the two samples just before the second wetting cycle are shown in Figure 4.11. T4 lies inside the yield curve, while T10′ lies on the yield curve due to the additional applied stress. The substantial amount of collapse observed for T10′ during the second wetting cycle is the result of expansion of the yield curve caused by reduction of matric suction.

Summary

1. Unsaturated virgin fill is referred to a soil which has never been subjected to a suction smaller than its current value. For virgin fill, the collapse potential decreases with an increase in initial dry density but increases with an increase in initial matric suction.

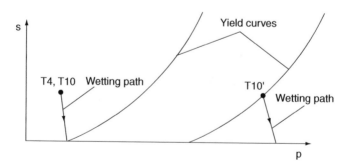

Figure 4.11 Stress states of tests T4 and T10 before second wetting cycle (after Ng *et al.*, 1998).

2. While substantial amount of collapse is observed for virgin fill, a significant reduction of collapse is recorded for non-virgin fill, which is defined as a soil which has been subjected to a suction smaller than its current value. It is important to distinguish the nature of a virgin fill and a non-virgin fill as both of them exhibit significant difference in response to wetting.

3. Existing elasto-plastic models may be used to predict the collapse behaviour of virgin fill subjected to wetting. However, these models do not seems to be able to predict further collapse of non-virgin fill during repeated wetting and drying cycles. On the contrary, they would predict swelling.

Estimations of foundation settlements (Houston, 1995)

Although lateral strains and two- or three-dimensional deformations induced by wetting can be very important for certain field geometries such as slopes, estimating the collapse settlement potential using one-dimensional test results and one-dimensional analyses may be adequate for making engineering design and mitigation decisions (Houston, 1995). The general procedure followed for estimating potential collapse settlements involves evaluation, at a series of representative points of:

(1) the vertical stress state,
(2) the change in stress,
(3) the probable degree of wetting and
(4) the probable strain due to wetting.

A laboratory technique for estimating strain in the field due to thorough wetting was described in previous sections. The collapse settlement is computed by integration of a strain profile over the depth of anticipated wetting.

Difficulties associated with estimating collapse settlement include the typical existence of variable cementation and gradation leading to soil non-homogeneities over distances of only a few centimetres both laterally and vertically, and lack of knowledge of sources of water (Beckwith, 1995; Zhang and Zhang, 1995). The greatest uncertainty in estimating collapse settlement is linked to the uncertainty of the extent and degree of wetting (Houston et al., 1993; Andrei and Manea, 1995). In fact, numerous studies have shown that when the extent and degree of wetting is known, the simple procedure for computing collapse settlement outlined above can result in very good agreement between observed and computed settlements (Houston et al., 1988; Walsh et al., 1993).

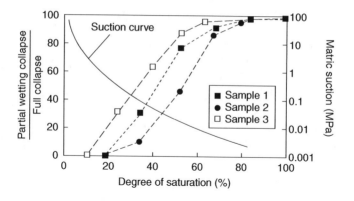

Figure 4.12 Partial collapse due to partial wetting curves for three ML soils (after Houston, 1995).

The degree of saturation achieved during the conduct of conventional laboratory response to wetting tests is quite high, typically 85 to 95 per cent. Estimated collapse settlements based on fun-wetting collapse potential may not be realized in-situ. The experience in China reported by Zhang and Zhang (1995) is that actual field collapse settlements are commonly only about one- seventh of estimated full-wetting collapse settlements. Walsh *et al.* (1993) report a case history for which collapse settlements for the prototype were only about one-tenth (1/10) of collapse settlements estimated from full-wetting laboratory response to wetting tests. The most likely explanation for the common occurrence of lesser field collapse settlement compared to estimated collapse settlement potential is that when a soil is only partially wetted, only a portion of the full collapse potential is realized. Further, there are many field situations for which only partial wetting occurs. Results of partial-wetting tests on several collapsible silts from the Phoenix, Arizona, USA, area are depicted in Figure 4.12. Partial-wetting collapse curves such as those in this figure have been used to make corrections to collapse settlement estimates based on full-wetting response to wetting tests, leading to very good agreement between observed and estimated wetting-induced settlements (Mahmoud, 1992, Walsh *et al.*, 1993). Partial wetting considerations are particularly helpful for forensic studies and for assessment of pre-wetting and controlled wetting mitigation alternatives.

Mitigation measures against collapsible soils (Houston, 1995)

Several mitigation alternatives are available for dealing with collapse phenomena (Turnbull, 1968; Clemence and Finbarr, 1981; Houston and Houston, 1989; Rollins and Rogers, 1994; Beckwith, 1995;

Evstatiev, 1995). Mitigation measures which have been used in the past can generally be fit into one of the categories listed below:

1. Removal of volume moisture-sensitive soil (Anayev and Volyanick, 1986)
2. Removal and replacement or compaction (Abelev, 1975; Souza *et al.*, 1995)
3. Avoidance of wetting (Royster and Rowan, 1968)
4. Chemical stabilization or grouting (Sokolovich and Semkin, 1984)
5. Pre-wetting (Holtz and Hilf, 1961)
6. Controlled wetting (Bally and Oltulescu, 1980)
7. Dynamic compaction (Lutenegger, 1986)
8. Pile or pier foundations (Gao and Wu, 1995)
9. Differential settlement resistant foundations (Cintra *et al.*, 1986).

The most appropriate mitigation method or combination of methods for a given collapsible soil site depends on several factors. The bases upon which the decision is made should include (1) the time at which the soil collapsibility is discovered, (2) the geometry and source of stress and (3) the source (type) of wetting. The collapse problem is often not detected until after the structure is built or the embankment or fill is placed. In this case the choices for mitigation become more limited and typically more expensive. The maximum flexibility for selection of mitigation and foundation options exists when a thorough site investigation has been performed and the regions of potential wetting-induced volume change have been identified prior to construction.

It is common practice to distinguish between shallow and deep deposits of collapsible soils. However, there is more than one basis for choosing between shallow and deep as a description. For example, one basis is the absolute depth to which significantly collapsible soils exist. If the collapsible soils extend to a few metres only, most engineers would classify the deposit as shallow. If the absolute depth of collapsible soils is a few tens of metres, most engineers would select deep as the descriptor. It is also useful to relate the depth classification to the percentage of the vertical stress on the soil due to overburden. Depth classification might also take into consideration the anticipated sources of wetting and the expected extent of wetting. For example, a site wetted by rising groundwater table might tend to be classified as deep, whereas the same site might be classified as shallow if the wetting is expected to extend only a few metres and to come from surface infiltration. For deep collapsible soil deposits the overburden stress is typically large compared to the stress generated by a structure. By contrast, for shallow deposits the stress from the structure can be quite significant.

Development of profiles of the percent vertical stress due to overburden, such as that shown in Figure 4.13, can be very useful in selection of appropriate mitigation measures. The example shown in the figure is a mat

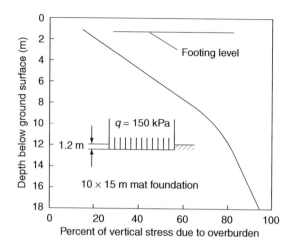

Figure 4.13 Typical variation with depth of overburden stress as a fraction of total stress (after Houston, 1995).

foundation on a collapsible soil deposit. The overburden stress is at least 80 percent of the total stress for depths greater than 12 m. Mitigation alternatives which are effective for deep collapsible soil deposits would likely be appropriate for this situation for mitigation of collapse strains below a depth of about 12 m. It is unlikely that the stresses due to structural load will have a significant influence on the total collapse strain if they only contribute 15–25 per cent of the total stress. With respect to mitigation methods, soil layers for which the overburden stress contributes most significantly might be termed deep, whereas soil layers for which structural loads are quite significant would be termed shallow. For many situations the collapsible soil deposit will consist of both shallow and deep collapsible soil layers. In these cases, it is often appropriate to select a combination of at least one mitigation alternative directed towards resolving the shallow collapsible soil problems and one directed towards the deep layers.

Expansive soils (Bell and Culshaw, 2001; Tripathy et al., 2002)

Introduction (Tripathy et al., 2002)

Engineering problems related to expansive soils have been reported in many countries of the world (e.g. Bao and Ng, 2000) but are generally most serious in arid and semi-arid regions. As a result, highly reactive soils undergo substantial volume changes associated with the shrinkage and swelling processes. Consequently, many engineered structures suffer severe distress and damage. The swelling potential of an expansive soil is commonly assessed in the laboratory by utilizing a conventional oedometer apparatus. The specimens are generally allowed to swell under an arbitrary surcharge pressure for one cycle to determine the total magnitude of swell.

Laboratory cyclic swell–shrink tests on expansive soils have shown that vertical swell potential may reduce or even increase by a factor of two when compared with the first cycle of swelling. Therefore, the assessment of expansive soil behaviour without considering cyclic seasonal fluctuations may underestimate the swelling potential of the soil.

A number of research studies are available on laboratory cyclic swell–shrink tests on compacted–remoulded expansive soils. The studies have shown that after about 3–5 swell–shrink cycles, the soils reach an equilibrium condition, where the vertical deformations during swelling and shrinkage are the same (Warkentin and Bozozuk, 1961; Popescu, 1986; Subba Rao and Satyadas, 1987; Day, 1994; Al-Homoud et al., 1995; Songyu et al., 1998). This reversible volume change between fixed limits has been observed for surface horizons of agricultural soils that have undergone numerous drying and wetting cycles (Yong and Warkentin, 1975).

The nature of expansive soils (Bell and Culshaw, 2001)

Some clayey soils undergo slow volume changes that occur independently of loading and are attributable to swelling or shrinkage. These volume changes can give rise to ground movements which can cause damage to buildings. Low-rise buildings are particularly vulnerable to such ground movements since they generally do not have sufficient weight or strength to resist. In addition, shrinkage settlement of embankments can lead to cracking and break up of the roads they support.

The principal cause of expansive clays is the presence of swelling clay minerals such as montmorillonite. Differences in the period and amount of precipitation and evapotranspiration are the principal factors influencing the swell–shrink response of a clay soil beneath a building. Poor surface drainage or leakage from underground pipes also can produce concentrations of moisture in clay. Trees with high water demand and uninsulated hot process foundations may dry out clay causing shrinkage.

The depth of the active zone in expansive clays (i.e. the zone in which swelling and shrinkage occurs in wet and dry seasons, respectively) varies. Many soils in temperate regions such as Britain, especially in the south, southeast and south Midlands of England, possess the potential for significant volume change due to changes in moisture content. However, owing to the damp climate in most years volume changes are restricted to the upper 1.0 to 1.5 m in clay soils. The susceptible soils include most of the clay formations of Mesozoic and Tertiary age such as the Edwalton Formation of the Mercia Mudstone, Lias Clay, Oxford Clay, Kimmeridge Clay, Weald Clay, Gault Clay, clays of the Lambeth Group. Generally speaking, the older formations are less susceptible than the younger ones (Bell and Culshaw, 2001).

The potential for volume change in clay soil is governed by its initial moisture content, initial density or void ratio, its microstructure and the vertical stress, as well as the type and amount of clay minerals present. These clay minerals are responsible primarily for the intrinsic expansiveness, whilst the change in moisture content or suction (where the pore water pressure in the soil is negative, that is, there is a water deficit in the soil) controls the actual amount of volume change, which a soil undergoes at a given applied pressure. The rate of heave depends upon the rate of accumulation of moisture in the soil.

Grim (1962) distinguished two modes of swelling in clay soils, namely inter-crystalline and intra-crystalline swelling. Inter-particle swelling takes place in any type of clay deposit irrespective of its mineralogical composition, and the process is reversible. In relatively dry clays the particles are held together by relict water under tension from capillary forces. On wetting, the capillary force is relaxed and the clay expands. In other words, intercrystalline swelling takes place when the uptake of moisture is restricted to the external crystal surfaces and the void spaces between the crystals. Intracrystalline swelling, on the other hand, is characteristic of the smectite family of clay minerals, and montmorillonite in particular. The individual molecular layers, which make up a crystal of montmorillonite are weakly bonded so that on wetting water enters not only between the crystals but also between the unit layers which comprise the crystals. Generally kaolinite has the smallest swelling capacity of the clay minerals and nearly all of its swelling is of the interparticle type. Illite may swell by up to 15 per cent but intermixed illite and montmorillonite may swell some 60 to 100 per cent. Swelling in calcium montmorillonite is very much less than in the sodium variety, ranging from about 50 to 100 per cent. Swelling in sodium montmorillonite can amount to 2000 per cent of the original volume, the clay then having formed a gel (Bell and Culshaw, 2001).

Cemented and undisturbed expansive clay soils often have a high resistance to deformation and may be able to absorb significant amounts of swelling pressure. Therefore, remoulded expansive clays tend to swell more than their undisturbed counterparts. In less-dense soils expansion initially takes place into zones of looser soil before volume increase occurs. However, in densely packed soil with low void space, the soil mass has to swell more or less immediately to accommodate the volume change. Therefore, clay soils with a flocculated fabric swell more than those that possess a preferred orientation. In the latter, the maximum swelling occurs normal to the direction of clay particle orientation. Because expansive clays normally possess extremely low permeabilities, moisture movement is slow and a significant period of time may be involved in the swelling–shrinking process. Accordingly, moderately expansive clays with a smaller potential to swell but with higher permeabilities than clays having a greater swell potential may swell more during a single wet season than more expansive clays (Bell and Culshaw, 2001).

Determinations of potential swell (Bell and Culshaw, 2001)

The swell–shrink behaviour of a clay soil under a given state of applied stress in the ground is controlled by changes in soil suction. The relationship between soil suction and water content depends on the proportion and type of clay minerals present, their microstructural arrangement and the chemistry of the pore water. Changes in soil suction are brought about by moisture movement through the soil due to evaporation from its surface in dry weather, by transpiration from plants or alternatively by recharge consequent upon precipitation. The climate governs the amount of moisture available to counteract that which is removed by evapotranspiration (i.e. the soil moisture deficit). The volume changes that occur due to evapotranspiration from clay soils can be conservatively predicted by assuming the lower limit of the soil moisture content to be the shrinkage limit. Desiccation beyond this value cannot bring about further volume change.

Methods of predicting volume changes in soils can be grouped into empirical methods, soil suction methods and oedometer methods (Bell and Maud, 1995; Bell and Culshaw, 2001). Empirical methods make use of the swelling potential as determined from void ratio, natural moisture content, liquid and plastic limits, and activity. For example, Driscoll (1983) proposed that the moisture content, In, at the onset of desiccation (pF = 2) and when it becomes significant (pF = 3) could be approximately related to the liquid limit, LL, in the first instance m = Q.5 LL and in the second m = 0.4 LL. The Building Research Establishment (BRE) (Anon, 1980) suggested that the plasticity index provided an indication of volume change potential as shown in Table 4.5. A degree of overlap was allowed. The activity chart, proposed by Van Der Merwe (1964), frequently has been used to assess the expansiveness of clay soils (Figure 4.14).

However, because the determination of plasticity is carried out on remoulded soil, it does not consider the influence of soil texture, moisture

Table 4.5 Soil activity related to swelling (after Bell and Culshaw, 2001)

Plasticity index (%)	Potential for volume change
Over 35	Very high
22–48	High
12–32	Medium
Less than 18	Low

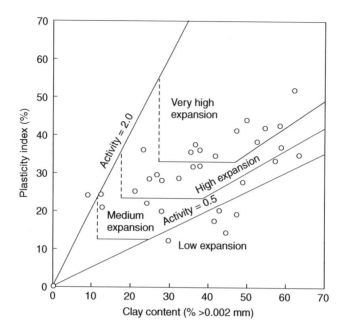

Figure 4.14 Activity chart of Van Der Merwe (1964) for estimation of the degree of expansiveness of clay soil. Some expansive clays from Natal, South Africa, are shown (from Bell and Maud, 1995; after Bell and Culshaw, 2001).

content, soil suction or pore water chemistry, which are important factors in relation to volume change potential. Therefore, over-reliance on the results of such tests must be avoided. Consequently, empirical methods should be regarded as simple swelling indicator methods and nothing more. As such, it is wise to carry out another type of test and to compare the results before drawing any conclusions.

Soil suction methods use the change in suction from initial to final conditions to obtain the degree of volume change. Soil suction is the stress which, when removed allows the soil to swell, so that the value of soil suction in a saturated fully swollen soil is zero. O'Neill and Poorymoayed

Table 4.6 USAEWES classification of swell potential (from O'Neill and Poorymoayed, 1980; after Bell and Culshaw, 2001)

Liquid limit (%)	Plastic limit (%)	Initial (in situ) suction (kPa)	Potential swell (%)	Classification
Less than 50	Less than 25	Less than 145	Less than 0.5	Low
50–60	25–35	145–385	0.5–1.5	Marginal
Over 60	Over 35	Over 385	Over 1.5	High

(1980) quoted the United States Army Engineers Waterways Experimental Station (USAEWES) classification of potential swell (Table 4.6) which is based on the liquid limit, plasticity index and initial (in situ) suction. The latter is measured in the field by a psychrometer. Soil suction is not easy to measure accurately. Filter paper has been used for this purpose (McQueen and Miller, 1968). According to Chandler *et al.* (1992), measurements of soil suction obtained by the filter paper method compare favourably with measurements obtained using psychrometers or pressure plates.

The oedometer methods of determining the potential expansiveness of clay soils represent more direct methods (Jennings and Knight, 1957). In the oedometer methods, natural (undisturbed) samples are placed in the oedometer and a wide range of testing procedures are used to estimate the likely vertical strain due to wetting under vertical applied pressures. The latter may be equated to overburden pressure plus that of the structure which is to be erected. In reality most expansive clays are fissured, which means that lateral and vertical strains develop locally within the ground. Even when the soil is intact, swelling or shrinkage is not truly one-dimensional. The effect of imposing zero lateral strain in the oedometer is likely to give rise to overpredictions of heave and the greater the degree of fissuring, the greater the overprediction. The values of heave predicted using oedometer methods correspond to specific values of natural moisture content and void ratio of the sample. Therefore, any change in these affects the amount of heave predicted. Gourley *et al.* (1994) mentioned the use of a stress path oedometer to determine volume change characteristics of expansive soils. Such a method provides data on vertical and radial total stresses, suction and void ratio.

Water content-void ratio swell–shrink paths of compacted expansive soils (Tripathy et al., 2002)

Overview

The swelling of desiccated soils occurs in three different phases (Day, 1994), namely primary swelling, secondary swelling and no swelling. During primary swelling, the cracks developed during drying close. Primary swelling occurs at a very rapid rate. Secondary swelling includes closure of microcracks and reduction of entrapped air. During the third phase (i.e. no swelling), no further void ratio changes occur. The three phases of swelling of desiccated Otay Mesa clay from Day (1999) are shown in Figure 4.15b.

Hanafy (1991) proposed a characteristic S-shaped curve to describe the potential volume change of an expansive clayey soil for the change in void ratio relative to changes in water content resulting from desiccation and water absorption. The S-shaped curve (Figure 4.16) can be determined in

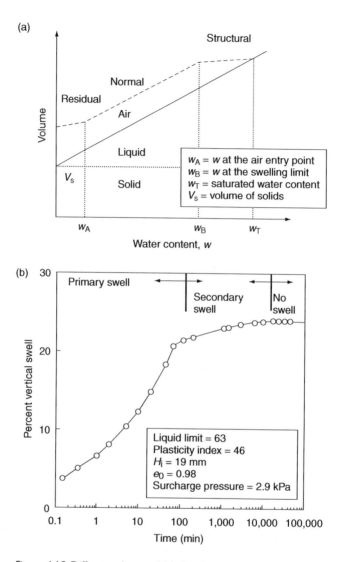

Figure 4.15 Different phases of (a) shrinkage and (b) swelling for a compacted expansive soil where e_0 is the initial void ratio, H_0 is the initial height of the test specimen (after Tripathy *et al.*, 2002).

the laboratory using conventional consolidation test equipment by carrying out one complete swelling–shrinkage test with two additional partial tests. Expansive soil specimens, desiccated or partially desiccated with an initial water content, w_n, and a corresponding natural void ratio, e_o, are used to trace the path during testing. The S-shaped curve can be used to classify the swelling potential of desiccated expansive clayey soils.

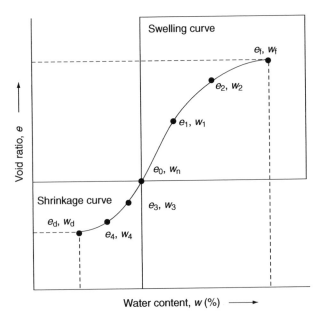

Figure 4.16 Illustration of the S-shaped curve of the water content/void ratio relationship (after Tripathy *et al.*, 2002).

Soils used and testing procedures

Two expansive soils collected from the Northern Karnataka State of India were selected for the study. The natural soils were processed to be finer than $425\,\mu$m with resulting liquid limits of 100 and 74 per cent. These soils are called soils A and B, respectively. The properties of the soils are presented in Table 4.7. Laboratory test procedures, following ASTM standards, were performed to evaluate the physical properties of the soils. The free swell properties of the soils were determined as per the procedure given by Holtz and Gibbs (1956). Figure 4.17 shows the standard Proctor (ASTM test method D698; ASTM, 1998b) and modified Proctor (ASTM test method D1557; ASTM, 1998a) compaction curves for the soils. The shrinkage limit and liquid limit of the soils indicate that both soils are highly plastic and have high moisture absorption capacities. X-ray diffraction patterns of magnesium-saturated and glycerol-solvated specimens of soils A and B show the existence of montmorillonite as the dominating clay mineral (Figure 4.18). The Indian standard classification system (IS 1498, Bureau of Indian Standards, 1970) classifies the 'degree of expansion' and 'danger of severity' of both the soils as *very high* and *severe*, respectively. The classification is based on the liquid limit (w_l), plasticity index ($I_p = w_l - w_p$) and

Table 4.7 Properties of the soils used in the study of swell–shrink paths of expansive soils (after Tripathy et al., 2002)

	Soil A	Soil B
Liquid limit, w_l (%)	100	74
Plastic limit, w_p (%)	42	32
Plasticity index, I_p (%)	58	42
Shrinkage limit, w_s (%)	10.6	13.5
Specific gravity, G_s	2.68	2.73
% passing sieve No. 200 (425 μm)	98	80
Clay content (<0.002 mm; %)	62	52
Silt content (%)	36	28
Fine sand content (%)	2	20
Free swell (%)	340	225

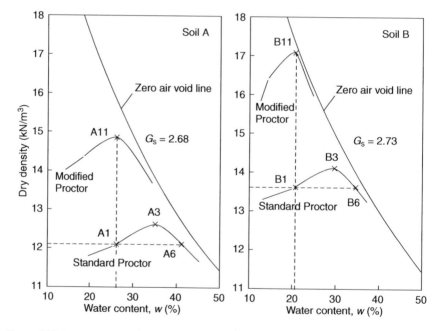

Figure 4.17 Standard Proctor and modified Proctor curves for soils A and B used in the study of swell-shrink paths of expansive soils (after Tripathy et al., 2002).

shrinkage index ($I_s = w_l - w_s$) values, where w_p is the plastic limit of the soil, and w_s is the shrinkage limit of the soil.

The specimen conditions selected for the study are marked on the compaction curves for both soils (Figure 4.17). Specimens were selected such that the dry density could be varied while the water content remained constant

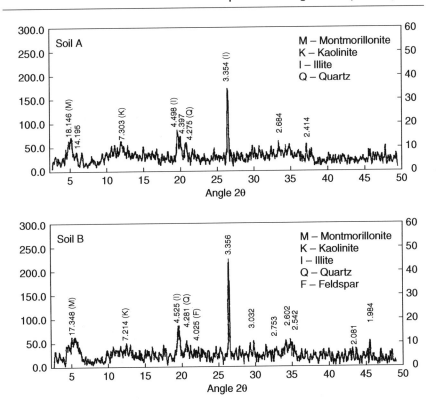

Figure 4.18 X-ray diffraction patterns of the soils used in the study of swell–shrink paths of expansive soils (after Tripathy *et al.*, 2002).

(specimens A1 and A11 for soil A, and specimens B1 and B11 for soil B) and the dry density was maintained constant while the water content was varied (specimens A1 and A6 for soil A, and specimens B1 and B6 for soil B). In addition, specimens A3 for soil A and B3 for soil B were chosen to represent the standard Proctor optimum conditions of the respective soils. Several mixes were prepared using the air-dried soil mixed with predetermined quantities of water. The mixes were kept in closed plastic bags and cured for 7 days. After ensuring the desired water content for the mix, the moist soil was compacted to a height of 13 ± 0.5 mm directly into a stainless steel oedometer ring, 76.2 mm in diameter and 38 mm in height. Laboratory static compaction was used to achieve the desired dry densities.

Cyclic swell–shrink tests

Cyclic swell–shrink tests were conducted in a fixed-ring oedometer with a modification to shrink the specimens under a controlled surcharge pressure

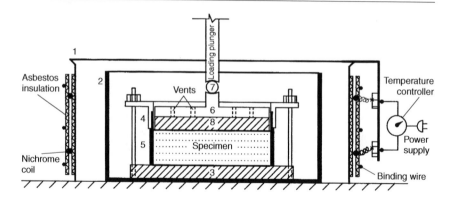

Figure 4.19 Schematic diagram of the fixed-ring oedometer cell arrangement used in the study of swell–shrink paths of expansive soils. Note: 1, outer stainless steel jacket; 2, water jacket; 3, bottom porous stone; 4, outer ring; 5, specimen ring; 6, pressure pad; 7, pressure ball; 8, top porous stone (after Tripathy *et al.*, 2002).

and temperature (i.e. $40 \pm 5\,°C$). A schematic diagram of the set-up is shown in Figure 4.19. The set-up consists of a fixed-ring oedometer cell placed inside a stainless steel container (outer jacket). The outer face of the outer jacket holds a 1 kW capacity coil tightly sandwiched between two flexible asbestos sheets. The flexible asbestos sheets serve as an insulator. The two ends of the coil were connected to porcelain connectors to which power was provided through a temperature controller. A known surcharge pressure was applied to each specimen by means of the lever arm of action of the oedometer frame. The specimens were then allowed to swell after being inundated with water under an ambient temperature. After completion of the full swelling process, the process of shrinkage was commenced. To start the shrinkage process, water was removed from the inner cell (i.e. water jacket) and then the temperature controller was switched on to maintain a constant temperature throughout the shrinkage process.

Observation of vertical movements of the specimens during swelling and shrinkage was made by using a dial gauge with a minimum reading of ± 0.002 mm and a travel of 25 mm. The combination of one swelling and shrinkage cycle is designated as one swell–shrink cycle. At the end of each cycle, the temperature controller was switched off and the specimen temperature was returned to room temperature in about 2–3 h. The specimen was again inundated with water for the next swelling and then shrinkage process to complete the second cycle. Thus, several cycles of swelling and shrinkage were performed until the specimen reached equilibrium conditions. Equilibrium was defined as the condition where swelling and shrinkage were of constant magnitude for each cycle. In the case of full swelling followed by full shrinkage tests, specimens in each cycle were allowed to fully swell and then shrink to a stage where there was no further change in height. The time allowed for each swelling process was about 3 days, and the full shrinkage process required about 6 days to complete at a temperature of 40 °C. In the case of full swelling followed by partial shrinkage, specimens in each cycle were allowed to fully swell and then shrink to a height corresponding to 50 per cent of the first swollen height in each cycle. The time taken for partial shrinkage was about 2 days.

The vertical deformation of the specimens was represented as the change in height (ΔH) of the specimen (during either swelling or shrinkage) and is expressed as a percentage of the initial height of the specimen (H_i) at the beginning of the first swell–shrink process. By plotting the vertical deformations of a specimen for several swell–shrink cycles, the per cent change in height of the specimen during any of the swelling or shrinkage cycles can be observed.

Table 4.8 shows the number of specimens used to trace the swell–shrink path under each surcharge pressure for soils A and B. The surcharge pressures applied to the specimens were 6.25, 50.00 and 100.00 kPa for specimen A3 of soil A and 6.25 and 50.00 kPa for specimen B3 of soil B. The effect of initial placement conditions on the equilibrium swell–shrink path was studied under a surcharge pressure of 6.25 kPa. The effect of full swelling– partial shrinkage was studied for specimen A3 of soil A under a surcharge pressure of 6.25 kPa.

WATER CONTENT–VOID RATIO PATHS

The void ratio changes for changes in water content in specimen A3 under surcharge pressures of 6.25 and 50.00 kPa and specimen B3 under a surcharge pressure of 6.25 kPa from the as-compacted state to the fourth cycle are shown in Figures 4.20–4.22, respectively, for swelling and shrinkage.

Table 4.8 Testing program used in the study of swell–shrink paths of expansive soils (after Tripathy et al., 2002)

Surcharge pressure (kPa)	Type of test	No. of specimens
Soil A		
6.25	Full swelling – full shrinkage	92[a]
6.25	Full swelling – partial shrinkage	32
50.00	Full swelling – full shrinkage	45
100.00	Full swelling – full shrinkage	45
Soil B		
6.25	Full swelling – full shrinkage	62[a]
50.00	Full swelling – full shrinkage	40

Note
a Includes specimens from all the initial placement conditions.

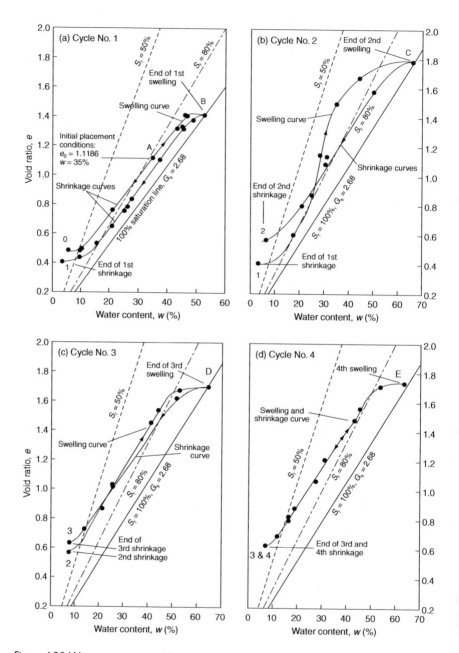

Figure 4.20 Water content – void ratio swell–shrink paths of specimen A3 at 6.25 kPa for full swelling – full shrinkage cycles: (a) cycle 1, (b) cycle 2, (c) cycle 3 and (d) cycle 4 (after Tripathy *et al.*, 2002).

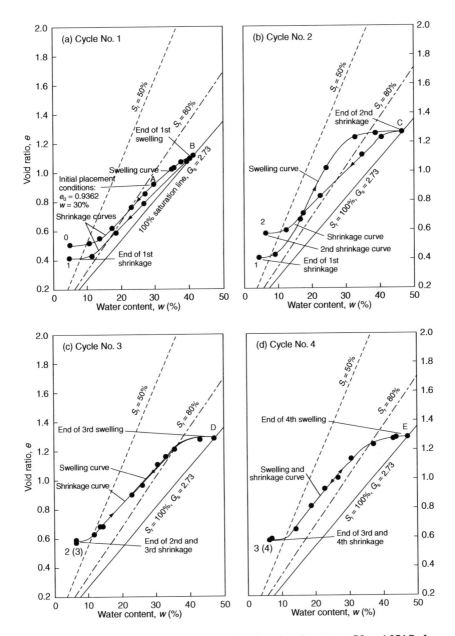

Figure 4.21 Water content – void ratio swell–shrink paths of specimen B3 at 6.25 kPa for full swelling – full shrinkage cycles: (a) cycle 1, (b) cycle 2, (c) cycle 3, and (d) cycle 4 (after Tripathy et al., 2002).

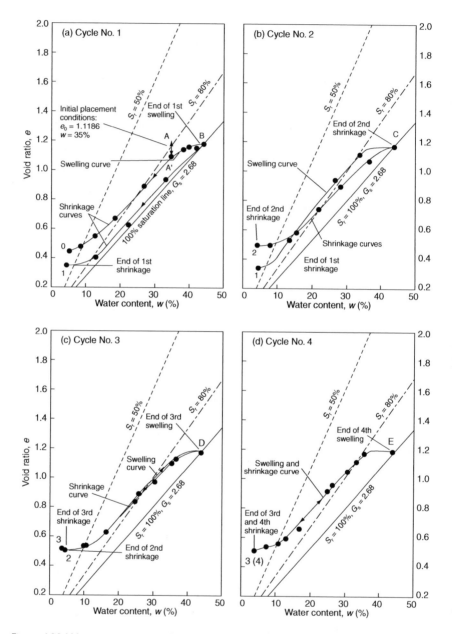

Figure 4.22 Water content – void ratio swell–shrink paths of specimen A3 at 50.00kPa for full swelling – full shrinkage cycles: (a) cycle 1, (b) cycle 2, (c) cycle 3, and (d) cycle 4 (after Tripathy *et al.*, 2002).

Test results for specimen B3 under a surcharge pressure of 50.00 kPa and A3 under a surcharge pressure of 100.00 kPa exhibited similar trends.

In Figure 4.20, points 0, 1, 2, 3 and 4 represent the positions of specimens at full shrinkage stages in different cycles, and points B, C, D and E represent the positions of the specimen at full swelling for different cycles. The initial void ratio and water content of the specimens are e_0 and w_0 (i.e. point A). The sequence of wetting and drying of specimen A3 in Figure 4.20 is as follows: (i) swelling took place from point A to point B and shrinkage from point B to point 1 (Figure 4.20a), (ii) swelling took place from point 1 to point C and shrinkage from point C to point 2 (Figure 4.20b), (iii) swelling took place from point 2 to point D and shrinkage from point D to point 3 (Figure 4.20e) and (iv) swelling took place from point 3 to point E and shrinkage from point E to point 4. Point 4 becomes the same as point 3 as shown in Figure 4.20d. The shrinkage curve of the specimen from position A was also traced (i.e. A to 0 in Figure 10a). Similar explanations hold for Figures 4.21 and 4.22. In Figure 4.22, the specimen showed compression from A to A' after being loaded. The swell–shrink cycle in this case started from A'.

Figures 4.20–4.22 show that, subsequent to swelling, all specimens attained almost full saturation. All swelling curves, except curve AB (i.e. specimen subjected to swelling after application of surcharge pressure), and all shrinkage curves, except curve AO [i.e. specimen shrunk from point A (A' in Figure 4.22)], are S-shaped. This indicates that the paths from the end of the shrinkage to the end of the swelling and from the end of the swelling to the end of the shrinkage consist of two curvilinear portions and a linear portion. It can also be seen that for a change in water content, the void ratio changes are insignificant at the top and bottom curvilinear portions of the S-shaped curves. Except for the second swelling path (i.e. curve 1C), the linear portions of all other swelling and shrinkage paths are essentially parallel to the 100 per cent saturation line. The slope of the linear portion of the second swelling curve is slightly steeper. This indicates that a large volume change occurred in the second swelling cycle as compared with any other swelling cycle.

The shrinkage paths shifted away from the saturation line and the specimens became more unsaturated with minimal changes in void ratio for a decrease in water content. This tendency essentially ends at a water content corresponding to a degree of saturation, Sr, of 80 per cent in cycles 3 and 4. In earlier cycles, the transition from full saturation to unsaturated conditions occurred at still higher degrees of saturation, indicating that the distance from the shrinkage curve to the saturation line increased with an increase in the number of swell–shrink cycles. Subsequently, the shrinkage curves took on a slope of about 45° until the specimen approached the shrinkage limit water content (i.e. Ws = 10.6 per cent for soil A and Ws = 13.5 per cent for soil B). Similar trends can be seen in Figures 4.21 and 4.22.

The three phases of shrinkage can be identified in the entire shrinkage curves (i.e. structural shrinkage, normal shrinkage and residual shrinkage). Similarly, in all the swelling curves traced from the full shrinkage stage, the phases can be identified as primary, secondary and no swelling. The shrinkage curves obtained by drying the specimen from position A (A' in Figure 4.22) showed only the normal and residual shrinkage phases. The trends are similar to those observed by Ho et al. (1992) and Sitharam et al. (1995) for compacted specimens subjected to one cycle of shrinkage.

The void ratios and water contents attained by the specimens were at a minimum at the end of the first shrinkage cycle. At the end of other cycles, however, higher void ratio and water content were observed. Similarly, the void ratio and water content attained by the specimen at the end of the first swelling cycle were lower than the corresponding values at all other swelling cycles. The water content and void ratio at the end of the second swelling cycle were higher than at any other cycle. Beyond the second cycle, the water content and void ratio decreased slightly and then stabilized.

Hysteresis in swelling and shrinkage paths was observed for the first three cycles. When each specimen was allowed to shrink, the void ratio at the end of the cycle changed. Similarly, the void ratio and water content of the specimen at the end of the swelling cycle changed when the specimen was rewetted. This occurred until about the third cycle. In the fourth swell–shrink cycle, the path traced by the specimen during both swelling and shrinkage was found to be the same, signifying reversibility in the swelling and shrinkage paths and the elimination of hysteresis.

The results indicate that hysteresis in the wetting–drying path of the soil–water characteristic curves (i.e. matric suction versus water content curves) of clayey soils is due to changes in both void ratio and water content during the initial cycles but is due to changes in water content alone with an increasing number of cycles.

EFFECT OF INITIAL PLACEMENT CONDITIONS

Specimens A1, A6 and A11 for soil A and B1, B6 and B11 for soil B were subjected to swell–shrink cycles under a surcharge pressure of 6.25 kPa. The water content and void ratio of specimens A1, A3, A6 and A11 for soil A and B1, B3, B6 and B11 for soil B at the end of the fourth swelling and fourth shrinkage cycles are shown in Table 4.9. The values are nearly the same for all the specimens at the respective full swelling and full shrinkage states.

Furthermore, a few of the specimens corresponding to placement conditions A1, A3, A6 and A11 for soil A and B1, B3, B6 and B11 for soil B were run for five swell–shrink cycles under a surcharge pressure of 6.25 kPa. The water content versus void ratio paths during swelling and shrinkage were traced. An S-shaped curve was established for all the specimens of each soil as shown in Figure 4.23 (curve XYX_1) and 14 (curve $X'Y'X_1'$). The range

Table 4.9 Water content and void ratio of the specimens at the fourth swell–shrink cycle (after Tripathy et al., 2002)

Specimen	Water content, w (%)			Void ratio, e		
	Initial	Fourth shrinkage	Fourth swelling	Initial	Fourth shrinkage	Fourth swelling
A1	26.0	6.2	63.20	1.2058	0.6242	1.727
A3	35.0	6.5	63.00	1.1186	0.6242	1.733
A6	41.0	6.8	62.90	1.2058	0.6240	1.735
A11	26.0	7.3	63.30	0.8023	0.6236	1.742
B1	21.0	6.5	46.20	1.0040	0.5918	1.289
B3	30.0	6.5	46.50	0.9362	0.5826	1.295
B6	34.5	6.3	46.65	1.0040	0.5937	1.293
B11	21.0	6.8	46.35	0.6012	0.6012	1.288

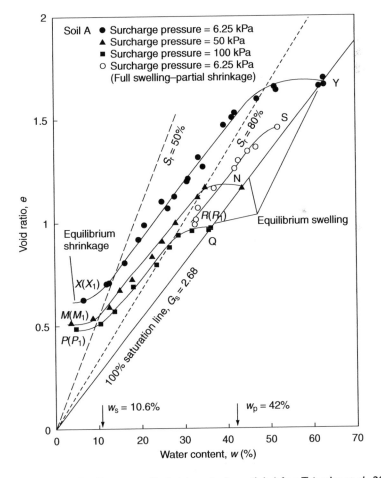

Figure 4.23 Equilibrium swell–shrink paths for soil A (after Tripathy et al., 2002).

of void ratio and water content over which the swell–shrink path occurred remained almost the same as that observed for specimens A3 and B3 in the fourth swell–shrink cycle (Figures 4.20d, 4.21d).

The range of water content and void ratio over which the swelling and shrinkage occurred is less for soil B. This can be attributed to the lower liquid limit of soil B. The water content variation in the fifth cycle was from 6.5 to 63.0 per cent for soil A and 6.5 to 44.0 per cent for soil B.

EFFECT OF SURCHARGE PRESSURES

Figures 4.23 and 4.24 show the equilibrium swell–shrink paths traced at the fifth swell–shrink cycle for specimen A3 under surcharge pressures of

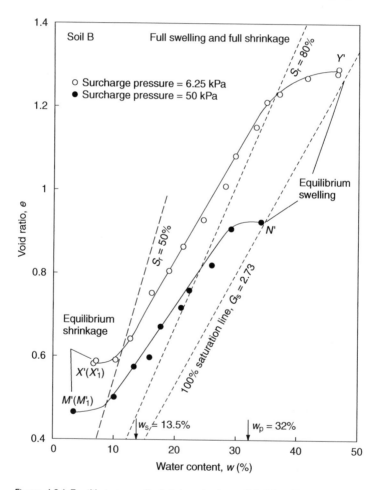

Figure 4.24 Equilibrium swell–shrink paths for soil B (after Tripathy et al., 2002).

50.00 kPa (curve MNM_1) and 100.00 kPa (curve PQP_1) and for specimen B3 under a surcharge pressure of 50.00 kPa ($M'N'M'_1$). All the equilibrium swell–shrink curves have a similar S shape. However, the range of water content and void ratio change over which the curves occurred becomes less with an increase in surcharge pressure.

It is evident from the results that the central straight-line portions of all the equilibrium swell–shrink curves are within the degree of saturation range from about 50 to 80 per cent and are almost parallel to the 100 per cent saturation line. As the surcharge pressure increases, the curves are shifted towards the 100 per cent saturation line. The equilibrium swell–shrink curves are also similar to the characteristic volume change S-shaped curve proposed by Hanafy (1991), as shown in Figure 4.16.

The ratio of the change in void ratio to the change in water content (i.e. $\Delta e / \Delta w$) in the range of degrees of saturation from 50 to 80 per cent is about 0.025 for soil A and about 0.020 for soil B. This is true for all equilibrium swell–shrink paths studied. Although these ratios are not strictly comparable with CLOD index values, the values are of similar magnitude.

The plastic limit, w_p, and shrinkage limit, w_s, of the soils are also indicated in Figures 4.23 and 4.24. It can be noted that w_p and w_s, which have a specific meaning for an initially slurried soil subjected to shrinkage, have no specific meaning for the volume change of a compacted expansive soil. The water contents at the end of all the swelling cycles were well below the liquid limit of the soils and fall below or above the plastic limit of the soils.

VOLUMETRIC CHANGE AND VERTICAL DEFORMATION AT EQUILIBRIUM CYCLE

The volumetric change and vertical deformation for changes in degree of saturation during equilibrium swell–shrink cycles for the full swelling–full shrinkage tests for both soils are shown in Figure 4.25. These plots also show three distinct phases. The plots indicate that at any degree of saturation, as the surcharge pressure increases, the volumetric and vertical deformation decrease. Similarly, at any deformation, the degree of saturation increases for the specimens with a higher surcharge pressure.

An attempt was made to distinguish the three phases and find the percentage of the changes (i.e. both volumetric and vertical) that occurred in each phase. The first phase of swelling or the last phase of shrinkage is designated as phase I ($S_r < 50$ per cent), the last phase of swelling or first phase of shrinkage as phase III ($S_r > 80$ per cent) and the middle phase as phase II ($S_r = 50$–80 per cent).

Table 4.10 shows the changes that occurred in each phase for both soils. About 80–90 per cent of the total volumetric change for soil A and about 70–80 per cent of the total volumetric change for soil B occurred in phase

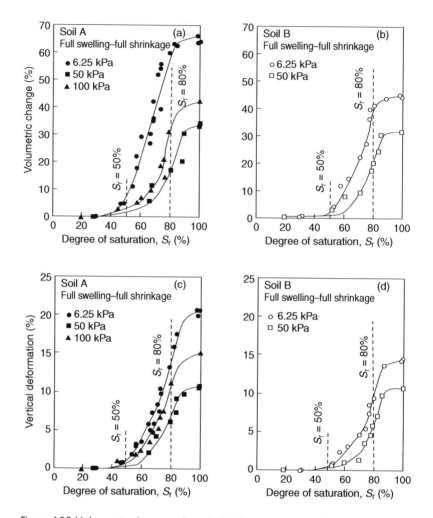

Figure 4.25 Volumetric change and vertical deformation at equilibrium swell–shrink cycle (after Tripathy *et al.*, 2002).

II. This indicates that for a change in degree of saturation of about 30 per cent, most of the volume change occurred in phase II. Similarly, about 50–60 per cent of the total vertical deformation for soil A and about 50–55 per cent for soil B occurred in the same phase.

Figure 4.26 shows the vertical deformation versus volumetric change plots for soils A and B under the surcharge pressures studied at the fifth swell–shrink cycle. The relationship between volumetric change and vertical deformation is linear in the three different phases. In phase II, where most of

Table 4.10 Volumetric change and vertical deformation for soils A and B (after Tripathy *et al.*, 2002)

Phase	Surcharge pressure (kPa)	Volumetric change (%)	Vertical deformation (%)
Soil A			
I	6.25	0–8.48	0–1.20
I	50.00	0–0.44	0–0.63
I	100.00	0–0.33	0–0.50
II	6.25	8.48–59.50	1.20–13.00
II	50.00	0.44–39.10	0.63–10.00
II	100.00	0.33–26.11	0.50–7.00
III	6.25	59.50–66.33	13.00–18.52
III	50.00	39.10–43.30	10.00–15.08
III	100.00	26.11–32.94	7.00–12.20
Soil B			
I	6.25	0–1.05	0–0.50
I	50.00	0–0.44	0–0.50
II	6.25	1.05–39.83	0.50–8.00
II	50.00	0.44–23.17	0.50–6.50
III	6.25	39.83–45.00	8.00–13.44
III	50.00	23.17–31.26	6.50–11.63

Note
$S_r < 50\%$ for phase I, $50 < S_r < 80\%$ for phase II, and $S_r > 80\%$ for phase III.

the deformation occurred, the slope is steeper than those for the other two phases. The average ratios of the volumetric change to vertical deformation in phase II are presented in Table 4.11. The ratio decreases with an increase in surcharge pressure and varies from about 3.9 to 4.5 for soil A and from 3.8 to 5.2 for soil B for the surcharge pressures considered in the investigations.

Design of foundations on expansive swelling soils (Jimenez-Salas, 1995)

Design philosophy

The design of foundations on expansive clays takes two approaches as follows:

- By action on the ground by substitution or by stabilization, or
- By action on the supported structure which may be flexible, rigid or isolated ('palaphite').

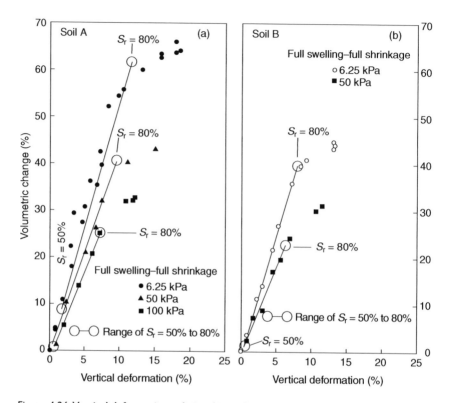

Figure 4.26 Vertical deformation relationship with volumetric change at equilibrium swell–shrink cycle (after Tripathy *et al.*, 2002).

Table 4.11 Ratio of volumetric change to vertical deformation at the equilibrium cycle for surcharge pressures of 6.25 kPa, 50.00 kPa and 100.00 kPa (after Tripathy *et al.*, 2002)

	Soil A			Soil B	
	6.25	50.00	100.00	6.25	50.00
Ratio of volumetric change to vertical deformation	4.50	4.10	3.90	5.17	3.78

Substitution of the ground

In some cases, the active depth is only 1 or 2 m. This could be due, for instance, to the relative proximity of the phreatic table. The removal of the active material and substitution by an inert one will eliminate the problem.

More delicate is the case in which a certain thickness of active clay is found over an inactive shale, according to a very usual disposition in the south of Spain. The active clay is a residual soil or a vertisol coming from the gentle slopes, in any case consisting of the same formation. The weathered material has characteristics which are related to the black cotton soils, and also with the North African tirs.

It is then necessary to take into account that, if the substituted material is more permeable than the clay, an underground reservoir for the rain water has been prepared. Formerly, the shale was protected by the clay, but it should be recalled (Morgenstern, 1979; Morgenstern and Balasubramanian, 1980) that some shales or heavily overconsolidated clays are inert if in contact with pore water with a salt content similar to that of their own pores, but they become expansive if in contact with pure water (rain water for example) with a much lower osmotic potential.

In India, and after research carried out over several years (Katti, 1987), it has been found that many of the problems of the expansive black cotton soils can be solved if these clays are covered with a layer of cohesive non-swelling soil (CNSS). This must be some inert material, but with enough clay content to give it a low permeability and significant cohesion. In the referenced research it has been found that a modest thickness of this kind of material produces a great improvement on the behaviour of the remaining expansive material. This is attributed to the fact that the cohesive non-swelling material impedes the initiation of cracks, 'welding' them together. Without cracks, the vertical diffusivity is much lower, and the same is true of active depth.

Mechanical stabilization of the ground

The term 'stabilization' can have many different meanings. Stabilization of the material can be by mechanical means or by use of chemical additives like, mainly, lime.

Mechanical stabilization is the mixing of the active soil with a coarser material. This concept can be extrapolated to other processes, including those of reinforcement with membranes or geogrids.

Portland cement has also been used, but the most interesting method is, without doubt, lime stabilization. Its use is well known as a constructive aid in the case of ponded areas on clayey or silty soils, providing a stable surface for the traffic of the equipment. But in the case of a permanent stabilization of an active clay, its action is complicated and still not well understood. Many of the most active clays have sodium as the exchangeable base, and in these cases it is clear that their saturation with calcium must produce a reduction in their plasticity indexes, volume change, etc. But this cannot be the explanation of its beneficial effect on some of the already calcium-saturated clays.

Several mechanisms of action have been invoked, such as lime carbonation, pozzolanic reaction, cementation, flocculation, etc.

Lime carbonation must have a place, as it has one in the lime mortars of so many old buildings, but it cannot explain the short-term effects. Probably, there are some pozzolanic effects in clays with alophane, so frequently found in tropical environments. But in the black earths of the Spanish south, the examination of lime-stabilized soil over more than 5 years found only very light signs of attack on the borders of the clay platelets. Cementation must also have an effect for medium or long-term periods. Flocculation can explain the observed effects, at least in an important proportion of cases, but not in all of them, as there are many soils that are already heavily flocculated.

These and other causes, some of them of a purely technological character, in many circumstances make the use of lime stabilization impractical, and it is necessary to consider the probabilities of success before undertaking it. Some laboratory tests have been proposed to check the 'reactivity' of the clay, and to estimate the proportion of lime necessary for the object.

Nelson and Miller (1992) present a very good summary of these tests. These authors also describe the different modalities of the technological processes. They classified the lime stabilization procedures in three groups:

1. Mixed in place and recompacted
2. Drill hole lime
3. Pressure injected lime

Mixing in place procedures have been used mainly for areas to be paved. The soil must be previously prepared by ploughing, harrowing, etc. breaking it into clods of small size.

For greater depths the drill hole method has been employed. It consists of pouring quicklime or hydrated lime slurry into holes of 150–300 mm diameter. The holes are distributed in a net of 1.2–1.5 m of mesh size.

The last group of procedures quoted by Nelson and Miller involves pressure injected lime (Cothren, 1984). It is essentially similar to the preceding one, but the slurry is injected with a pressure up to 0.6 MPa. This does not improve the penetration of the liquid into the fine pores, but opens the cracks and fissures, even creating many new ones. The data about the efficiency of this method presents contradictions, but it has proved to be beneficial in many cases. Local experience in a particular formation and circumstances must be taken into account before making a decision.

Stabilization of the moisture content

Stabilization of the moisture content can take place by passive means such as protection and encapsulation, or by active means such as pre-heaving or by controlled water injection.

There are two ways of avoiding the changes of moisture content, which in turn, are the origin of volume change. First, it is possible to establish barriers against the moisture transfer or even to encapsulate soil masses, and in this manner preserve their volumetric stability (Ramaswany and Aziz, 1987).

Second, one can avoid the desiccation of the soil by introducing water at the points and in the quantities required to prevent drying. There seems to be no record of active stabilization forcing the drying of the soil.

Passive stabilization

Passive stabilization has been the usual practice in most cases. The beneficial effect of an ample sidewalk has been known for a long time, but sidewalks usually crack, allowing some infiltration. Today, they have been advantageously substituted by membranes. The sidewalk constituted a horizontal barrier, and its breadth is determined by the extent of the possible edge effect. The membranes can also be arranged horizontally, with 200–400 mm of earth cover, either just for mechanical protection or sometimes to allow some seeding of the cover.

Active defence against the moisture changes

In arid climates, the problem is the progressive increase in water content under the building, due to its protective effect against the intense potential evaporation. An obvious idea is to moisten initially the soil until it reaches its final equilibrium water content.

This ideal is difficult to achieve. Aside from the uncertainties relating to the final equilibrium moisture content, it is nearly impossible to produce in a reasonable time a moisture pattern similar to the equilibrium one, since the expansive clay has a low coefficient of permeability.

The previous flooding of foundation areas is used from time to time, with variable results (Blight et al., 1992).

Another technique of active moisture stabilization is the controlled addition of water, generally by means of conveniently distributed borings. Another method is by conventional irrigation, by sprinklers or by any other means, of which an important case can be found in (Williams, 1980).

Measures on the structure

Flexible structures

There are extensive areas of swelling clays where the problem is not felt, as the traditional architecture uses wood or bamboo providing enough flexibility to absorb the swelling without sensible damage. But the same principle can be applied in many other constructions, especially of the industrial type, and this can be particularly useful when the movements to be expected are only of a few centimetres. Isostatic trusses for the cover and walls of pre-fabricated plates are able to follow those deformations.

Rigid structures

The attention of authors dealing with the problems of swelling clays has been caught, primarily, by solutions consisting of rigid structures, capable of 'floating' over the deformed soil without experiencing damage.

The pioneers in this line were the South Africans. Dealing with economical houses for the workers in the new mining districts, they found it possible to construct houses rigid enough to resist the expected bending moments, using the walls as beams of great height. To obtain these results, the walls were composed of reinforced brick masonry, with sometimes reinforced concrete frames around the openings (Rosenhaupt and Mueller, 1963).

The problem that has retained the attention of the authors has been the estimation of the forces to be resisted by the structure, and it is convenient to make the point in today's somewhat confusing panorama (Pidgeon, 1980).

Deep foundations

The last group of solutions for swelling clay problems includes methods of making the structures independent from the expansive layers. The typical features of these solutions are those of a 'palaphite' which is a structure resting on piles going down until they reach deep and stable layers. A gap is provided between the structure and the swelling surface of the soil. Usually the piles are bored and cast in place, as the consistency of these clays makes the driving difficult.

The piles pass across the expansive layers, which produces a negative side resistance which tends to raise the piles (Blight, 1984), annulling the desired soil-structure independence. Two different procedures exist against this possibility: the first one is to make the piles also independent from the raising layers, which can be made with several different kinds of shuttering, leaving a gap between pile and the ground in the upper part of the shaft. Although this seems to be a very rational solution, the fact is that today the second possibility is preferred, which is to anchor the piles into the deep stable layers and to reinforce the shaft to resist the tensile forces which are produced in it.

There are, in turn, two different ways of forming the anchorage: boring a supplementary length of pile, or under-reaming the base to form an enlarged end of the pile shaft. It is difficult to decide between the two possibilities, and it is necessary to consider many aspects, chief among them, of course, being cost.

Chapter 5

Measurement of shear strength and shear behaviour of unsaturated soils

Sources

This chapter is made up of verbatim extracts from the following sources for which copyright permissions have been obtained as listed in the Acknowledgements.

- Fredlund, D.G. and Rahardjo, H. (1993). *Soil Mechanics for Unsaturated Soils*. John Wiley and Sons, Inc, New York, 517.
- Fredlund, D.G., Xing, A., Fredlund, M.D. and Barbour, S.L. (1996). The relationship of the unsaturated soil shear strength to the soil–water characteristic curve. *Can. Geotech. J.*, 33, 440–445.
- Gan, J.K.M. and Fredlund, D.G. (1996). Shear strength characteristics of two saprolitic soils. *Can. Geotech. J.*, 33, 595–609.
- Ng, C.W.W. and Chiu, A.C.F. (2001). Behaviour of a loosely compacted unsaturated volcanic soil. *Proc. J. Geotech. Geoenviron. Eng.*, ASCE, December, 1027–1036.
- Ng, C.W.W. and Chiu, A.C.F. (2003a). Laboratory study of loose saturated and unsaturated decomposed granitic soil. *Proc. J. Geotech. Geoenviron. Eng.*, ASCE, June, 550–559.
- Ng, C.W.W. and Zhou, R.Z.B. (2005). Effects of soil suction on dilatancy of an unsaturated soil. *Proc. 16th Int. Conf. Soil Mech. Geotech. Eng.*, Osaka, Japan, Vol. 2, 559–562.
- Ng, C.W.W., Zhan, L.T. and Cui, Y.J. (2002a). A new simple system for measuring volume changes in unsaturated soils. *Can. Geotch. J.* 39, 757–764.

- Ng, C.W.W., Cui, Y.J., Chen, R. and Delage, P. (2007a). The axis-translation and osmotic techniques in shear testing of unsaturated soils: a comparison. *Soils and Foundations.* Vol. 47, No. 4, 657–684.

Introduction to shear strength (Fredlund and Rahardjo, 1993)

Overview

Many geotechnical problems such as bearing capacity, lateral earth pressures, and slope stability are related to the shear strength of a soil. The shear strength of a soil can be related to the stress state in the soil. The stress state variables generally used for an unsaturated soil are the net normal stress, $(\sigma - u_a)$, and the matric suction, $(u_a - u_w)$.

The two commonly performed shear strength tests are the triaxial test and the direct shear test. The theory associated with various types of triaxial tests and direct shear tests for unsaturated soils are compared and discussed here.

Soil specimens which are 'identical' in their initial conditions are required for the determination of the shear strength parameters in the laboratory. If the strength parameters of natural soil are to be measured, the tests should be performed on specimens with the same geological and stress history. On the other hand, if strength parameters for a compacted soil are being measured, the specimens should be compacted at the same initial water content and with the same compactive effort. The soil can then be allowed to equalize under a wide range of applied stress conditions. It is most important to realize that soils compacted at different water contents, to different densities, are 'different' soils. In addition, the laboratory test should closely simulate the loading conditions that are likely to occur in the field.

Unsaturated soil specimens are sometimes prepared by compaction. In this case, the soil specimens must be compacted at the same initial water content to produce the same dry density in order to qualify as an 'identical' soil. Specimens compacted at the same water content but at different dry densities, or vice versa, cannot be considered as 'identical' soils, even though their classification properties are the same. Soils with differing density and water content conditions can yield different shear strength parameters, and should be considered as different soils as illustrated in Figure 5.1.

Triaxial tests on unsaturated soils

One of the most common tests used to measure the shear strength of a soil in the laboratory is the triaxial test. The theoretical concepts behind the measurement of shear strength are outlined in this section. There are various procedures available for triaxial testing, and these methods are explained and compared in this section. However, there are basic principles used in the

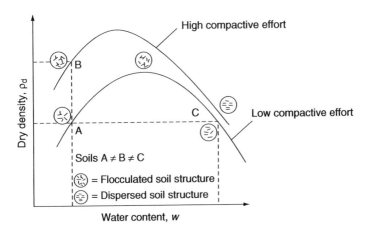

Figure 5.1 The particle structure of clay specimens compacted at various dry densities and water contents (from Lambe, 1958; after Fredlund and Rahardjo, 1993).

triaxial test that are common to all test procedures. The triaxial test is usually performed on a cylindrical soil specimen enclosed in a rubber membrane, placed in the triaxial cell. The cell is filled with water or other suitable fluids and pressurized in order to apply a constant all-around pressure or confining pressure. The soil specimen can be subjected to an axial stress through a loading ram in contact with the top of the specimen.

Stage one

The application of the confining pressure is considered as the first stage in a triaxial test. The soil specimen can either be allowed to drain (i.e. consolidate) during the application of the confining pressure or drainage can be prevented. The term consolidation is used to describe the process whereby excess pore pressures due to the applied stress are allowed to dissipate, resulting in volume change. The consolidation process occurs subsequent to the application of the confining pressure if the pore fluids are allowed to drain. On the other hand, the consolidation process will *not* occur if the pore fluids are maintained in an undrained condition. The *consolidated* and *unconsolidated* conditions are used as the first criterion in categorizing triaxial tests.

Stage two

The application of the axial stress is considered as the second stage or the shearing stage in the triaxial test. In a conventional triaxial test, the soil specimen is sheared by applying a compressive stress. The total confining pressure generally remains constant during shear. The axial stress is continuously increased until a failure condition is reached. The axial stress generally acts as the total major principal stress, σ_1, in the axial direction, while the isotropic confining pressure acts as the total minor principal stress, σ_3, in the lateral direction. The total intermediate principal stress, σ_2, is equal to the total minor principal stress, σ_3 (i.e. $\sigma_2 = \sigma_3$). The stress conditions associated with a consolidated drained triaxial test at pre-failure and failure are illustrated in Figure 5.2.

The pore fluid drainage conditions during the shearing process are used as the second criterion in categorizing triaxial tests. When the pore fluid is allowed to flow in and out of the soil specimen during shear, the test is referred to as a *drained* test. On the other hand, a test is called an *undrained* test if the flow of pore fluid is prevented. The pore air and pore water phases can have different drainage conditions during shear.

The shear strength test is performed by loading a soil specimen with increasing applied loads until a condition of failure is reached. There are several ways to perform the test, and there are several criteria for defining failure. Consider a consolidated drained triaxial compression test where the pore pressures in the soil specimen are maintained constant (Figure 5.3a). The soil specimen is subjected to constant matric suction and is surrounded by a constant net confining pressure, i.e. the net minor normal stress is $(\sigma_3 - u_a)$. The specimen is failed by increasing the net axial pressure (i.e. the net major normal stress), $(\sigma_1 - u_a)$. The difference between the major and

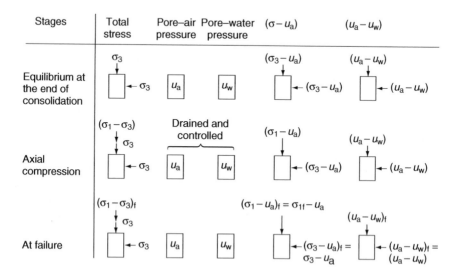

Figure 5.2 Stress conditions during a consolidated drained triaxial compression test (after Fredlund and Rahardjo, 1993).

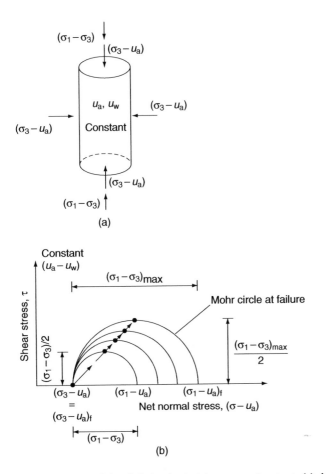

Figure 5.3 Consolidated-drained triaxial compression test. (a) Applied stresses; (b) Mohr's circles for changes in stress state during shear (after Fredlund and Rahardjo, 1993).

minor normal stresses, commonly referred to as the deviator stress, $(\sigma_1 - \sigma_3)$, is a measure of the shear stress developed in the soil (see Figure 5.3b). As the soil is compressed, the deviator stress increases gradually until a maximum value is obtained, as illustrated in Figure 5.3b.

The maximum deviator stress $(\sigma_1 - \sigma_3)_{max}$ is an indicator of the shear strength of the soil and has been used as a failure criterion. The net principal stresses corresponding to failure conditions are called the net major and net minor normal stresses at failure (i.e. $(\sigma_1 - u_a)_f$ and $(\sigma_3 - u_a)_f$, respectively, as indicated in Figure 5.3b.

Types of tests

Various triaxial test procedures are used for unsaturated soils based upon the drainage conditions adhered to during the first and second stages of the triaxial test. The triaxial test methods are usually given a two-word designation or abbreviated to a two-letter symbol. The designations are

1) consolidated drained or CD test,
2) constant water content or CW test,
3) consolidated undrained or CU test with pore pressure measurements,
4) unconsolidated undrained or UU test, and
5) unconfined compression or UC test.

In the case of CD and CU tests, the first letter refers to the drainage condition prior to shear, while the second letter refers to the drainage condition during shear. The constant water content test is a special case where only the pore air is kept in a drained mode, while the pore water phases is kept undrained during shear (i.e. constant water content). The pore air and pore water are not allowed to drain throughout the test for the undrained triaxial test. The unconfined compression test is a special loading condition of the undrained triaxial test. These five testing procedures are explained in the following sections. The air, water, or total volume changes may or may not be measured during shear. A summary of conventional triaxial testing conditions used, together with the measurements performed, is given in Table 5.1.

The shear strength data obtained from triaxial tests can be analysed using the stress state variables at failure or using the total stresses at failure

Table 5.1 Various triaxial tests for unsaturated soils (after Fredlund and Rahardjo, 1993)

Test methods	Consolidation prior to shearing process	Drainage		Shearing process		
		Pore–air	Pore–water	Pore–air pressure, u_a	Pore–water pressure, u_w	Soil volume change, ΔV
Consolidated drained (CD)	Yes	Yes	Yes	C	C	M
Constant water content (CW)	Yes	Yes	No	C	M	M
Consolidated undrained (CU)	Yes	No	No	M	M	–
Undrained	No	No	No	–	–	–
Unconfined compression (UC)	No	No	No	–	–	–

Notes
M = measurement, C = controlled.

when the pore pressures are not known. This concept is similar to the effective stress approach and the total stress approach used in saturated soil mechanics. In a drained test, the pore pressure is controlled at a desired value during shear. Any excess pore pressures caused by the applied load are dissipated by allowing the pore fluids to flow in or out of the soil specimen. The pore pressure at failure is known since it is controlled, and the stress state variables at failure can be used to analyse the shear strength data. In an undrained test, the excess pore pressure due to the applied load can build up because pore fluid flow is prevented during shear. If the changing pore pressures during shear are measured, the pore pressures at failure are known, and the stress state variables can be computed. However, if pore pressure measurements are not made during undrained shear, the stress state variables are unknown. In this case, the shear strength can only be related to the total stress at failure.

The total stress approach should be applied in the field only for the case where it can be assumed that the strength measured in the laboratory has relevance to the drainage conditions being simulated in the field. In other words, the applied total stress that causes failure in the soil specimen is assumed to be the same as the applied total stress that will cause failure in the field. The above simulation basically assumes that the stress state variables control the shear strength of the soil; however, it is possible to perform the analysis using total stresses. It is difficult, however, to closely simulate field loading conditions with an undrained test in the laboratory. Rapid loading of a fine-grained soil may be assumed to be an undrained loading condition.

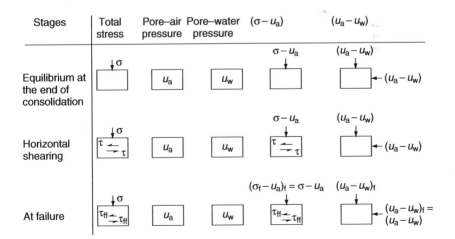

Figure 5.4 Stress conditions during a consolidated-drained direct shear test (after Fredlund and Rahardjo, 1993).

Direct shear tests on unsaturated soils

A direct shear test apparatus basically consists of a split box, with a top and bottom portion. The test is generally performed using a consolidated drained procedure, as shown in Figure 5.4.

A soil specimen is placed in the direct shear box and consolidated under a vertical normal stress, σ. During consolidation, the pore air and pore water pressures must be controlled at selected pressures. The axis-translation technique can be used to impose a matric suction greater than 1 atm. The direct shear test can be conducted in an air-pressurized chamber in order to elevate the pore air pressure to a magnitude above atmospheric pressure (i.e. 101.3 kPa). The pore water pressure can be controlled below the soil specimen using a high air-entry disk. At the end of the consolidation process, the soil specimen has a net vertical normal stress of $(\sigma_n - u_a)$ and a matric suction of $(u_a - u_w)$. Shearing is achieved by horizontally displacing the top half of the direct shear box relative to the bottom half. The soil specimen is sheared along a horizontal plane between the top and bottom halves of the direct shear box. The horizontal load required to shear the specimen, divided by the nominal area of the specimen, gives the shear stress on the shear plane. During shear, the pore air and pore water pressures are controlled at constant values. Shear stress is increased until the soil specimen fails. The failure plane has a shear stress designated as τ_f, corresponding to a net vertical normal stress of $(\sigma_f - u_a)_f$ [i.e. equal to $(\sigma - u_a)$ at failure] and a matric suction of $\underline{(u_a - u_w)_f}$ [i.e. equal to $(u_a - u_w)$ at failure], as illustrated in Figure 5.4. A typical plot of shear stress versus horizontal displacement for a direct shear test is shown in Figure 5.5.

The failure envelope can be obtained from the results of direct shear tests without constructing the Mohr circles. The shear stress at failure, τ_{ff}, is plotted as the ordinate, and $(\sigma_f - u_a)_f$ and $(u_a - u_w)_f$ are plotted as the abscissas to give a point on the failure envelope, as shown in Figure 5.6.

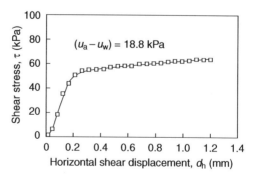

Figure 5.5 A typical shear stress versus displacement curve from a direct shear test (from Gan, 1986; after Fredlund and Rahardjo, 1993).

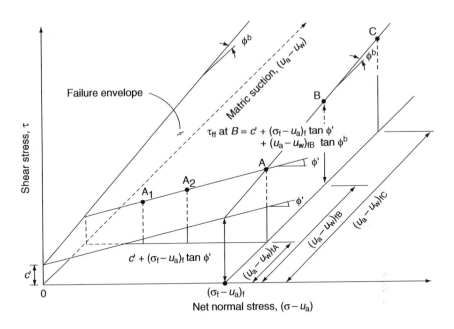

Figure 5.6 Extended Mohr–Coulomb failure envelope established from direct shear test results (after Fredlund and Rahardjo, 1993).

A line joining points of equal magnitude of $(\sigma_f - u_a)_f$ determines the ϕ^b angle (e.g. a line joining points A, B and C in Figure 5.6). Similarly, a line can be drawn through the points of equal $(u_a - u_w)_f$ to give the angle of internal friction, ϕ' (e.g. a line drawn through points A_1, A_2 and A). Details of the extended Mohr–Coulomb failure theory [see Equation (5.1)] will be discussed later.

The direct shear test is particularly useful for testing unsaturated soils due to the short drainage path in the specimen. The low coefficient of permeability of unsaturated soils results in 'times to failure' in triaxial tests which can be excessive. Other problems associated with testing unsaturated soils in a direct shear apparatus are similar to those common to saturated soils (e.g. stress concentrations, definition of the failure plane, and the rotation of principal stresses).

A new simple system for measuring volume changes of unsaturated soils in the triaxial cell (Ng et al., 2002a)

Introduction

Accurate measurement of total volume changes in an unsaturated soil specimen is vital but it is much more difficult and complicated than the same measurement on a saturated soil specimen. In a saturated soil, the total volume change of the soil specimen is generally assumed to be equal to the change in the water volume (assuming water is incompressible), and it can be relatively easy to measure with a volume gauge. In an unsaturated soil, however, the total volume change generally comprises two components: changes in the volume of the air and changes in the volume of the water in the void spaces. Geiser *et al.* (2000) summarized some existing methods for measuring volume change in unsaturated soil specimens and classified them into the following three broad categories:

1. measurement of the cell fluid,
2. measurement of the air and water volumes separately, and
3. direct measurement of the soil specimen.

For the first category, a volume change in a soil specimen is recorded by measuring the volume change in the confining cell fluid. This method is simple to set up and use. A standard triaxial apparatus can be used if it is carefully calibrated. However, several problems are usually encountered using this method such as expansion–contraction of the cell wall, connecting tubes, and cell fluid because of pressure and temperature variations, creep under pressure, and possible water leakage. The accuracy of the cell fluid measuring method depends on the quality of the calibrations, which are extremely difficult to perform accurately for this category of the measuring method.

To minimize the effects of expansion–compression of the cell and volume change in the cell fluid, Bishop and Donald (1961) developed a modified triaxial cell, which included an additional inner cylindrical cell sealed to the outer cell base (i.e. a double cell). Mercury was used as the cell fluid in the lower part of the inner cell.

Another double-wall triaxial apparatus developed by Wheeler (1986) differed from the set-up of Bishop and Donald (1961) in that the inner cell was sealed at both ends and both the inner and outer cells were filled completely with water. Equal cell pressures were applied to the inner and outer cells to avoid expansion of inner acrylic cells. The volume change in the soil specimen was measured by monitoring the flow of the water into or out of the inner cell with a burette system connected to a differential pressure transducer (DPT) or a volume-change device. The electrical readouts of the volume change were transformed for automatic data acquisition and feedback. Careful calibrations were carried out to correct any apparent volume change caused by cell pressure application (as a result of compression of the cell fluid, slight flexure of the top plate, and expansion of the connecting tubes), water absorption by the acrylic cell wall, and ram displacement (Sivakumar, 1993). This system has been used in many triaxial tests of unsaturated soils. There are some limitations of this method, however, caused by:

- possible time delay between the application of pressure to the inner and outer cell,
- slight flexure in the top plate of the inner cell that is not completely surrounded by the outer cell, and
- the inner cell has to be assembled in a large water reservoir to ensure complete de-airing of the cell.

For the second category of measuring systems, it is generally accepted that accurate measurement of air volume change is rather difficult because the volume of air is very sensitive to variations in the atmospheric pressure and ambient temperature, and undetectable air leakage through the tubes and connections and air diffusion through the rubber membrane cannot be avoided (Geiser, 1999). Therefore, the second method classified by Geiser *et al.* (2000) (i.e. direct measurement of air and water volume changes) would be troublesome and impractical in the triaxial testing of unsaturated soils.

In the third category of measuring systems is the measurement of the total volume change using local displacement transducers, which are directly mounted on the specimen (Chiu, 2001; Ng and Chiu, 2001), to measure the local vertical and radial deformations of soil specimens. This approach is best suited to small deformation tests and requires a fairly rigid specimen.

In this section, a new simple, open-ended, bottle-shaped inner cell is introduced, explained, and calibrated for measuring total volume changes in unsaturated soil specimens. Various measures have been taken to improve the accuracy of the system. The calibrated results of this system are compared with those of other existing devices and discussed.

The design of a total volume change measuring system

The basic principle of the measuring system introduced in this paper is that the overall volume change in an unsaturated–saturated specimen is measured by recording the differential pressures between the water inside the open-ended, bottle-shaped inner cell and the water inside a reference tube using a high-accuracy DPT. Details of the measuring system are shown in Figures 5.7 and 5.8. In the outer cell of the triaxial apparatus, the aluminum inner cell is sealed onto a pedestal. The high-accuracy DPT is connected to the inner cell and to a reference tube to record changes in differential pressures between the water-pressure change inside the inner cell due to a volume change in the specimen and the constant water pressure in the reference tube. The system is developed in a large Bishop and Wesley (1975) type triaxial stress-path cell, which normally accommodates 100-mm diameter

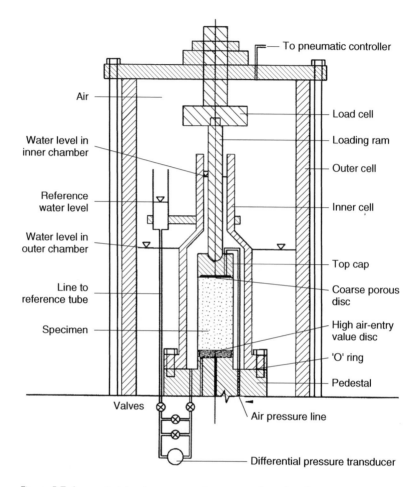

Figure 5.7 A new total volume measuring system for triaxial testing of unsaturated soils (after Ng *et al.*, 2002a).

specimens, so that enough space is provided for the inner cell and reference tube. After the installation of the inner cell and the reference tube, the triaxial apparatus can accommodate only 38-mm diameter specimens.

The system has been licensed by Hong Kong University of Science and Technology (HKUST) to GDS Instruments Ltd. This commercially available system is shown in Figure 5.9.

To improve the accuracy and sensitivity of the total volume change measuring system, the following important steps were taken:

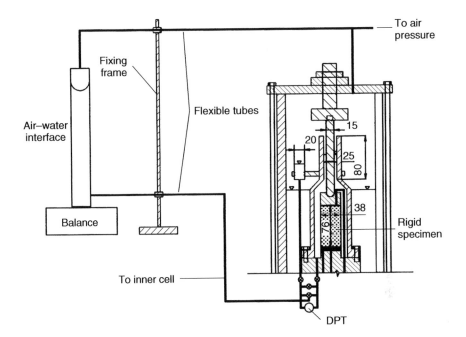

Figure 5.8 Calibration system for the volume change measuring system. All dimensions are in millimetres. DPT is the differential pressure transducer (after Ng et al., 2002a).

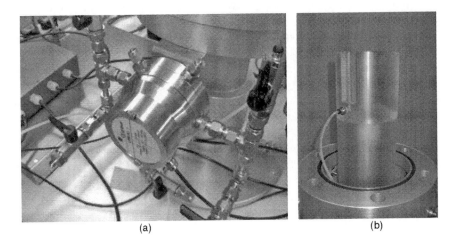

Figure 5.9 (a) The HKUST volume change system fitted to a stress path cell, (b) inner cell machined from a single block of acrylic polymer (permission of GDS Instruments Ltd).

(1) The inner cell is designed to be open-ended and bottle-shaped. At the bottle neck, the inner diameter is slightly (10 mm) larger than the diameter of the loading ram. During a test, a change in water level take places only within the bottle neck. Thus, the measurement of the water level inside the inner cell due to any volume change in the specimen becomes more sensitive because of its small cross-sectional area (314 mm^2) at the bottle neck.

(2) The open-ended design at the top of the inner cell ensures that an identical pressure can be applied to both the inner and outer cells, hence eliminating any expansion–compression of the inner cell caused by pressure differences between the inner and outer cells. This is different from the system described by Wheeler (1986) and Sivakumar (1993).

(3) The material used in the inner cell wall is aluminum, which generally exhibits relatively negligible creep and hysteretic effects compared with acrylic materials. It is also generally recognized that water absorption by an aluminum cell wall is much less than that of an acrylic cell wall.

(4) A high-accuracy DPT (Druck LPM9381) is adopted. This DPT gives bidirectional measurements ranging from −1 to 1 kPa (−100 to 100 mm of water head at 20 °C), an accuracy (including non-linearity, hysteresis, and repeatability) better than 0.1 per cent full-scale (i.e. readable to 0.1 mm), and a long-term stability of less than 0.1 per cent full-scale per annum. Therefore, the estimated accuracy of the volume change measuring system is on the order of 31.4 mm^3 (i.e. 314 mm^2 × 0.1 mm) if other potential errors are not accounted for (discussed later in the paper). For a specimen 38 mm in diameter and 76 mm in height, this estimated accuracy is equivalent to 0.04 per cent of the volumetric strain.

(5) To minimize the potential expansion–compression of various connecting tubes due to application of cell pressure, bronze tubes are used.

(6) De-aired water is used inside the inner cell and the reference tube and a thin layer of paraffin is added to the surface of both the inner cell and the reference tube to minimize evaporation of the water and slow down the rate of air diffusion into the water as suggested by Sivakumar (1993).

(7) To reduce diffusion, two layers of membrane sandwiching a layer of greased tin foil are used to seal the specimen. A number of slots are cut in the foil to reduce any reinforcing effects on the soil specimen.

(8) The reference tube is fixed onto the inner cell wall so that the tube moves together with the inner cell, which is fixed on the bottom pedestal, during tests. Thus, no calibration is needed for the effects of the pedestal movement on the volume-change measurement. Of course, the apparent volume change due to the relative movement of the loading ram inside the inner cell is calibrated.

(9) The internal diameter of the reference tube is enlarged at the top so that its cross-sectional area becomes the same as that at the neck part of the

inner cell, thus the evaporation from the water in the reference tube is identical to that in the inner cell. Also, the enlarged area can reduce the change in the reference water level resulting from a change of cell pressure.

Calibration of the volume change measuring system

Calibration for the differential pressure transducer

The DPT in the system can detect only changes in water level (i.e. water pressure) inside the inner cell relative to the reference water level. A calibration factor (CF) is needed to relate the measured differential water pressure by the DPT to the overall volume change in the specimen. This calibrated factor is associated only with the cross-sectional area at the bottle neck of the inner cell and thus it can be calibrated directly in conjunction with the entire system.

Calibration for apparent volume change

Regardless of how precise the design of the total volume change measuring system, several practical issues remain problematic, such as changes in cell pressure, fluctuations in the ambient temperature, possible creep in the inner cell wall, the hysteresis effect of water flow and possible water leakage through the membranes. All these issues could result in the measurement of an apparent volume change rather than of the actual volume change in a specimen. Therefore, calibrations were performed to investigate any possible apparent volume change of a specimen.

Calibration for relative movement of the loading ram in the inner cell

Since the inner cell is fixed onto the bottom pedestal (see Figure 5.7), the inner cell moves up vertically together with the bottom pedestal during a shearing test. Thus, a relative movement occurs between the inner cell and the fixed upper loading ram. Since the loading ram is partially submerged in the inner cell water, any relative movement between the inner cell and the loading ram will displace cell water and hence result in a rise of water level in the inner cell, which is not caused by the actual volume change in a specimen and must be accounted for in the total volume measurements.

Summary and conclusions

(1) A new and simple total volume measuring system was developed for triaxial tests on unsaturated and saturated soils. The overall volume

change of a specimen is measured by recording the change of differential pressures between the pressure inside an open-ended, bottle-shaped inner cell and a constant pressure inside a reference tube using a high-accuracy differential pressure transducer. Any volume change in a soil specimen will result in a change in the water level inside the inner cell.

(2) To improve the accuracy of the measuring system, an open-ended inner cell was designed to ensure that identical pressure can be applied to both the outer and the inner cells and hence to eliminate expansion–compression of the inner cell caused by pressure differences between the outer and inner cells.

(3) To increase the sensitivity of the measuring system, a bottle-shaped inner cell was designed. At the bottle neck, the inner diameter is only a slightly (10 mm) larger than the diameter of the loading ram. This greatly increases the sensitivity of the system because of the small cross-sectional area (314 mm^2) at the bottle neck.

(4) In the newly developed system, all the apparent volumetric strains due to changes in the cell pressure, fluctuations in the ambient temperature, and creep of the membranes and the inner cell wall are smaller than those in other similar measuring systems described in the literature.

(5) It has been verified that the CF of the system is fairly linear, reversible, and slightly dependent on the cell pressures once the apparent volume changes due to changes in cell pressure, fluctuations in the ambient temperature, creep of membranes and inner cell wall, and relative movements of the loading ram and the inner cell are properly calibrated and corrected. It is believed that the estimated resolution of the measuring system is on the order of 32 mm^3 (or 0.04 per cent volumetric strain for a triaxial specimen 38 mm in diameter and 76 mm in height).

Comparisons of axis-translation and osmotic techniques for shear testing of unsaturated soils (Ng et al., 2007)

Introduction

Suction plays a significant role in unsaturated soils. To study unsaturated soil behaviour in laboratory and field tests, suction control or measurement is an essential issue. Generally suction in unsaturated soils can be divided into matric suction and osmotic suction. The matric suction, $(u_a - u_w)$, is defined as the difference between pore air pressure, u_a, and pore water pressure, u_w. The effect of matric suction on unsaturated soil behaviour is mostly concerned. Therefore, the term 'suction' is simply taken to mean matric suction henceforth. The axis-translation technique (Hilf, 1956) and the osmotic technique (Zur, 1966) are generally used to control suction in laboratory testing of unsaturated soils.

Hilf (1956) introduced the axis-translation technique of elevating values of total stress and pore air pressure to increase pore water pressure in order to prevent cavitation in drainage system. This technique provides a method of measuring pore water pressure in laboratory without altering soil structure. Subsequently, this method has been widely used to measure or control matric suction in unsaturated soil testing.

The majority of experimental results about unsaturated soils have been obtained by the application of axis-translation technique. The validity of axis-translation technique lays on an assumption of two independent stress state variables for unsaturated soils (Fredlund and Morgenstern, 1977), whereas only limited experimental evidence in literature is available to support this validity. Bishop and Blight (1963) investigated the effect of axis-translation technique on measured shear strength by conducting unconfined triaxial compression tests on a compacted Selset clay and a compacted Talybont clay. The experiments showed that the measured shear strength was not affected by the application of axis-translation technique. However, the authors did not provide complete information about the specimen characteristics, especially about degree of saturation, which is regarded as an indicator of continuity of air phase in unsaturated soils. Therefore, consideration should be given to the state of air phase in unsaturated soils when using these experiments to verify the axis-translation technique. The null tests of Tarantino and Mongiovi (2000a) verified the axis-translation technique for the case where air phase is continuous. The null tests of Fredlund and Morgenstern (1977) confirmed the validity of axis-translation technique at high degree of saturation (the range from 0.833 to 0.95 with one exception of 0.759), where air phase is believed to be occluded (Juca and Frydman, 1995). The analysis of Bocking and Fredlund (1980), however, suggested that the axis-translation technique is no longer valid when air phase in unsaturated soil is occluded. Thus, the validity of axis-translation technique is controversial for the case where air phase is occluded. Furthermore, since cavitation of pore water is hindered in the axis-translation technique, elevation of pore air pressure may alter desaturation mechanism of soils (Dineen and Burland, 1995). It is then fundamental to understand whether experimental results obtained by using the axis-translation technique can be extrapolated to unsaturated soils under atmospheric conditions in the field.

An alternative suction control method is the osmotic technique. Delage et al. (1998) reported that this technique was initially developed by biologists (Lagerwerff et al., 1961) and then adopted by soil scientists (Zur, 1966). In geotechnical testing, this application was successfully adapted in oedometer (Kassif and Ben Shalom, 1971; Dineen and Burland, 1995), hollow cylinder triaxial apparatus (Komornik et al., 1980), and standard triaxial apparatus (Cui and Delage, 1996).

In the osmotic technique, suction in an unsaturated soil specimen is controlled by a difference of solute concentration between the specimen and an osmotic solution, which are separated by a semi-permeable membrane. The membrane is permeable to water and ions in the soil but impermeable to large solute molecules and soil particles (Zur, 1966). Energy analysis of water on both sides of the membrane by Zur (1966) indicated that matric suction in the soil specimen should be equal to osmotic pressure of the solution. Polyethylene glycol (PEG) is the most commonly used solute in biological, agricultural, and geotechnical testing for its safety and simplicity. The value of osmotic pressure depends on the concentration of solution: the higher the concentration, the higher is the osmotic pressure. The maximum value of osmotic pressure for PEG solution was reported to be above 10 MPa (Delage et al., 1998).

Tarantino and Mongiovi (2000a) believed that osmotic technique is more suitable for investigating the behaviour of clays and silty clays especially in the range of transition from saturation to the state where air and water are both continuous. In addition, the air pressure around the sample remains atmospheric, similar to actual conditions in the field.

Since the applied suction in osmotic technique is equal to osmotic pressure of PEG solution, calibration between osmotic pressure and concentration of PEG solution is essential. However, the relationship between osmotic pressure and concentration of PEG solution is founded to be affected significantly by calibration method (Dineen and Burland, 1995; Slatter et al., 2000). Thus, Dineen and Burland (1995) suggested the need for direct measuring negative pore water pressure in soil specimen when using osmotic technique. Furthermore, an evaluation on the performance of three different semi-permeable membranes by Tarantino and Mongiovi (2000b) indicated that all membranes experienced a chemical breakdown as the osmotic pressure of PEG solution exceeded a threshold value, which was found to depend on the type of membrane. Beyond this value, solute molecules were no longer retained by the semi-permeable membrane and passed into the soil specimen, with a reduction of concentration gradient and resulting decay of soil suction.

If unsaturated soil behaviour is governed by two independent stress variables, i.e. net normal stress $(\sigma - u_a)$ and suction $(u_a - u_w)$, not by component stresses, there should be no difference in test results obtained by using axis-translation and osmotic techniques when the applied net stress and suction are the same. However, sparse experimental comparison between these two techniques is available especially in terms of shear testing. Zur (1966) found that for a sandy loam, gravimetric soil–water characteristic curve (SWCC) obtained by using these two techniques showed a good agreement, however for a clay, equilibrium water content obtained by using axis-translation technique was higher than that obtained by using osmotic technique under the same applied nominal suction. Zur (1966) related this

observed difference to test period, suggesting that 2 days were not sufficient for suction equilibrium in axis-translation technique. When the test
period for suction equilibrium was extended to 3 days, better agreement
was achieved. However, Williams and Shaykewich (1969) found that even
when the suction equilibrium period was extended to more than 10 days
in axis-translation technique, equilibrium water content obtained by using
axis-translation technique was still slightly higher than that obtained by
using osmotic technique for a clay. As far as the authors are aware, the comparison between these two techniques in terms of shear testing has not been
investigated, despite the importance of shear strength in engineering practice. Therefore, it is necessary to investigate any difference in axis-translation
and osmotic techniques for shear testing of unsaturated soils.

To investigate any difference in axis-translation and osmotic techniques
for testing of unsaturated soils, a collaborative research was carried out
at the HKUST and the Ecole Nationale des Ponts et Chaussees (ENPC).
In this research, three series of unsaturated triaxial shear tests were conducted on a recompacted expansive soil with an initial degree of saturation
of 82 per cent. Two of these series were performed at HKUST using the
axis-translation technique and the osmotic technique. Another series was
performed at ENPC using the osmotic technique and the results were firstly
presented by Mao et al. (2002). In the series using the axis-translation
technique at HKUST, single-stage testing procedure and multi-stage testing
procedure were compared to investigate the influence of these two testing
methods on shear testing. In both series using osmotic technique at HKUST
and ENPC, test results were compared to study the reliability of osmotic
technique. The test results using axis-translation technique and osmotic
technique were compared in an attempt to investigate any difference in
axis-translation and osmotic techniques for shear testing of the expansive
soil at high degree of saturation. It should be noted that the test results
were interpreted by using the extended Mohr–Coulomb failure theory [see
Equation (5.1)] which will be discussed later.

Testing equipment

At HKUST, two different testing systems with the application of axis-
translation technique and osmotic technique were used to conduct triaxial
shear tests on the unsaturated expansive soil sample.

In the system using axis-translation technique, a triaxial apparatus
equipped with two water pressure controllers and two pneumatic controllers
was used. Suction was applied to a specimen through one water pressure
controller and one air pressure controller. Pore water pressure was applied
at the base of the specimen through a ceramic disc with an air-entry value of
500 kPa. Pore air pressure was applied at the top of the specimen through a

sintered copper filter. A double-cell measurement system was used to measure total volume change of the specimen. Details of this testing system are described by Ng et al. (2002a).

The system using the osmotic technique at HKUST is similar to that used at ENPC except that there was no direct measurement of total volume changes for specimens at HKUST during testing. However, dimensions of the specimens tested at HKUST were measured after shearing and then volume changes were back-calculated. In the osmotic technique at HKUST and ENPC, suction was applied to specimens through two pieces of semi-permeable membrane, which were kept in contact with the top and bottom surfaces of the specimens. Details of the triaxial testing system using the osmotic technique are described by Cui and Delage (1996).

Osmotic technique in this study involved a Spectra/Por@ 2 Regenerated Cellulose dialysis membrane with a value of 12,000–14,000 Da MWCO (Molecular Weight Cut Off). The corresponding PEG has a value of 20,000 Da MWCO.

Testing material and specimen preparation

An expansive soil from Zao-Yang (ZY) at about 400 km far from Wuhan in China was used in this study. It is composed of 49 per cent clay, 44 per cent silt and 7 per cent sand. It has a liquid limit of 68 per cent and a plastic limit of 29 per cent. The clay fraction is of 60 per cent kaolinite, 20–30 per cent montmorillonite and 10–20 per cent illite. According to the USCS classification, the soil can be described as a clay of high plasticity (Wang, 2000).

A drying SWCC of this expansive clay is shown in Figure 5.10, which was obtained from four identical specimens under zero vertical stress in a volumetric pressure plate extractor (Wang, 2000). The details about these specimens can be found in Table 5.2. As the solid line in Figure 5.10, the equation of Fredlund and Xing (1994) is used to fit the measured data. The air-entry value is obtained by extending the constant slope portion of the curve to intersect suction axis at saturated state. The corresponding value of suction is taken as the air-entry suction value of this soil. This estimated air entry value is 60 kPa. The gentle gradient of the SWCC implies a high water storage potential of this expansive soil, which enables the soil to maintain a high degree of saturation within a large range of suction.

For triaxial tests, the soil was first dried in an oven of a temperature about 40–50 °C, then ground in a mortar with a pestle and passed through a 2-mm sieve. De-aired water was carefully sprayed layer by layer to reach a final water content of 30.3 per cent. After keeping the soil–water mixture in a sealed plastic bag for moisture equalization for 24 h, the soil was statically recompacted in a cylindrical mould of 38 mm in diameter and 76 mm in height. Compaction was performed at a rate of 0.3 mm/min in three layers

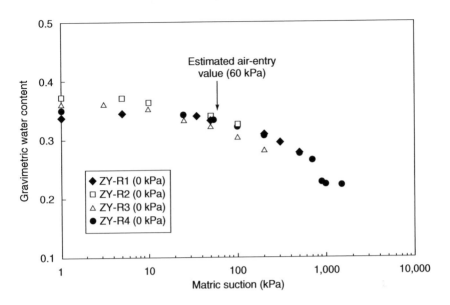

Figure 5.10 SWCC of recompacted ZY expansive soil (after Wang, 2000; Ng *et al.*, 2007).

of 25.3 mm high to ensure good homogeneity (Cui and Delage, 1996). The final dry density was about 1.36 g/cm^3. Soil suction was measured by a tensiometer on an identical specimen, showing that the initial suction of these specimens was 62 kPa. The initial characteristics of the specimens are

Table 5.2 The testing conditions for the soil test specimens using the AS technique (Ng *et al.*, 2007)

Type	Specimen	$u_a - u_w$ (kPa)	$p - u_a$ (kPa)
Multi-stage			25
	AM1	50	50
			100
			25
	AM2	100	50
			100
Single-stage	AS1	50	100
	AS2		50
	AS3	100	100

Notes
A = axis-translation; M = multi-stage; S = single-stage.
$u_a - u_w$: Matric suction after equalisation.
$p - u_a$: Net mean stress after consolidation.

Table 5.3 The testing conditions for the soil test specimens using the OS technique (Ng et al., 2007)

Type	Specimen	$u_a - u_w$ (kPa)	$p - u_a$ (kPa)
ENPC	OE1	0	50
	OE2		100
	OE3	100	100
	OE4		25
	OE5	200	50
	OE6		100
HKUST	OH1	50	50
	OH2		100
	OH3	100	25
	OH4		100
	OH5	200	100

Notes
O = Osmotic; E = ENPC; H = HKUST.
$u_a - u_w$: Matric suction after equalisation.
$p - u_a$: Net mean stress after consolidation.

presented in Tables 5.2 and 5.3 for the triaxial tests using axis-translation technique and osmotic technique, respectively.

Testing programme

All triaxial shear tests were consolidated type under a constant applied suction. The test series using the axis-translation technique include five single-stage tests and two multi-stage tests (refer to Table 5.2). For specimen identification used in Table 5.2, letter 'A' denotes axis-translation technique, 'S' denotes single-stage test, and 'M' denotes multi-stage test. The applied suction ranged from 0 to 100 kPa and net confining pressure ranged from 25 to 100 kPa. The multi-stage tests can maximize shear strength information from one specimen and assist in eliminating the effect of soil variability between specimens (Ho and Fredlund, 1982). Both series using the osmotic technique at HKUST and ENPC are single-stage tests, and the testing conditions are presented in Table 5.3. For specimen identification used in Table 5.3, letter 'o' denotes osmotic technique, 'H' denotes tests performed at HKUST, and 'E' denotes tests performed at ENPC. The applied suction ranged from 0 to 165 kPa and net confining pressure ranged from 25 to 100 kPa.

Testing procedures

In the test series using the axis-translation technique, a specimen was firstly equalized at a desired suction under a mean net stress of 10 kPa. Suction equalization was terminated when the variation of water content was within 0.05 per cent/day. The duration for this suction equilibrium procedure ranged from 4 to 9 days while the applied suction ranged from 0 to 100 kPa (Table 5.2). Thereafter, the specimen was subjected to isotropic consolidation at the constant suction. This stage of consolidation was terminated when the rate of change of water content was less than 0.05 per cent/day. This stage lasted from 3 to 9 days while the net confining pressure ranged from 25 to 100 kPa. After consolidation, the specimen was sheared at a constant rate under the constant net confining pressure and suction. Two types of shearing procedures, i.e. single-stage shearing and multi-stage shearing, were carried out. In a single-stage shearing procedure, the specimen was sheared directly to failure after initial consolidation. The multi-stage shearing procedure consisted of three stages. The net confining pressure was varied from one stage to another while the suction was maintained constant. In each stage of the multi-stage shearing procedure, each loading stage commenced from a condition of zero deviator stress, which is called cyclic loading procedure (Ho and Fredlund, 1982). To ensure drained condition, shearing was performed at a constant rate of 3.07×10^{-5} per cent/s. This rate was lower than the shearing rate of 4.39×10^{-5} per cent/s used at ENPC (Mao et al., 2002), considering that longer drainage path (Le. single drainage) was used in the application of axis-translation technique. For the single-stage shearing, the duration was about 8 days and for multi-stage shearing, the duration was about 16 days.

In the series using the osmotic technique at HKUST, suction was equalized by water interchange between the specimen and a PEG solution with a desired concentration under a mean net stress of 10 kPa. After suction equalization, the specimen was consolidated under a constant net confining pressure. The criteria to terminate suction equalization and consolidation were the same with those adopted in the series using axis-translation technique. The duration for suction equilibrium ranged from 5 to 11 days for suction range from 0 to 165 kPa, whereas duration for consolidation ranged from 3 to 7 days for net confining pressure range from 25 to 100 kPa (Table 5.3). When the consolidation stage was completed, the specimen was sheared to failure at a constant rate of 3.07×10^{-5} per cent/s under the same constant net confining pressure and suction. The shearing rate was the same as the one adopted in the tests using the axis-translation technique but it was slower than that adopted in the series using osmotic technique in ENPC (Mao et al., 2002). In the application of osmotic technique, pore air pressure was maintained at the atmospheric pressure during the whole triaxial testing. To ensure performance of the semi-permeable membrane, inspections of the membrane were carried out before and after test following a method described by Tarantino and Mongiovi (2000a). In the inspections, the semi-permeable membranes were first clamped on top cap and base in triaxial apparatus, and then a negative water pressure was applied below the membranes. Therefore, possible breaks in the membranes can be detected by ingress of air through the membranes. All the triaxial tests performed at HKUST were under a temperature of $20 \pm 1°C$.

In the test series using the osmotic technique at ENPC, specimens were first wrapped by a semi-permeable membrane and kept in a PEG solutions for 5–6 days to reach desired suctions. Then they were put in triaxial cell for 3 days to reach equilibrium under desired suctions and confining pressures. Shearing was performed at a constant rate of 4.39×10^{-5} per cent/s. Details of the test procedures are described by Mao *et al.* (2002).

Test results from using axis-translation technique

Although multi-stage tests can maximize shear strength information from one specimen, it may be useful to verify this shearing method with a single-stage test result. Figure 5.11 presents the relationships between deviator stress, $q = (\sigma_1 - \sigma_3)$, and axial strain, ε_a, during shearing for one multi-stage test (AM1) and two single-stage tests (AS2 and AS3). These three tests were performed under a suctions of 100 kPa. For the multi-stage test AM1 in three stages, net confining pressures, p', were 25, 50 and 100 kPa. For single-stage tests AS2 and AS3, the net confining pressures were 50 and 100 kPa.

As shown in Figure 5.11, the first stage of test AM1 shows a stiffer behaviour than tests AS2 and AS3 at the start of shearing. However, stress–strain relationship at the second stage of test AMI is consistent with that of

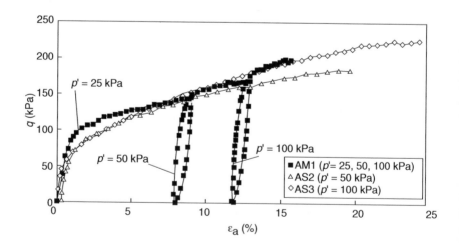

Figure 5.11 Stress–strain relationships of multi-stage and single-stage triaxial tests using the axis-translation technique with s = 100 kPa (Ng et al., 2007).

the single-stage test AS2, which is under the same net confining pressure of 50 kPa. The stress–strain relationship at the third stage of multi-stage test is almost identical to that of the single-stage test AS3, which is under the same net confining pressure of 100 kPa. This comparison illustrates that the stress–strain relationships for multi-stage tests at different stages coincide to those of the corresponding single-stage tests, which are under the same suction and net confining pressure.

The stress points at failure are plotted in Figure 5.12 where

$$s' = \left(\frac{\sigma_1 + \sigma_3}{2}\right) - u_a \text{ and } t = \left(\frac{\sigma_1 - \sigma_3}{2}\right),$$

This figure shows that the shear strengths obtained from single-stage tests are consistent with those obtained from multi-stage tests under the same suction and net confining pressure (e.g. the variations of shear strength for the specimens AS1, AS2 and AS3 are 2.6, 4 and 2.2 per cent respectively, as compared with the corresponding multi-stage tests).

For this triaxial test series, the internal friction angle, ϕ', ranges from 19.3 to 20.8° for suction range from 0 to 100 kPa, and the average value is 20.0°. By assuming a constant friction angle of 20°, the failure lines for this series fit well with the test results (see Figure 5.12). The increasing intercepts on the t axis indicate that apparent cohesion increases with suction.

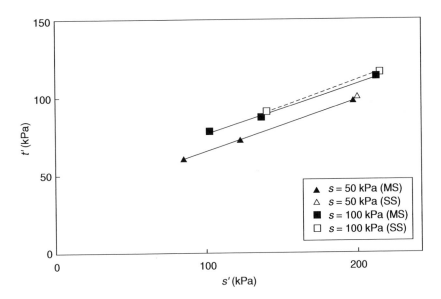

Figure 5.12 Comparison of strength between multi-stage and single-stage triaxial tests using the axis-translation technique (Ng et al., 2007).

At saturated state (i.e. zero suction), there appears to be a small intercept of 19.6 kPa on the t axis when the failure envelop is extended linearly to zero value of s. To demonstrate whether this apparent intercept is attributed to a true cohesion, a soaking test without any stress applied was performed on a specimen compacted to the same initial state as the specimens used in the triaxial tests. It was found that the specimen collapsed completely while assessing to water. It indicates that no true cohesion is present in the recompacted specimen. Hence, the apparent intercept is likely due to experimental error or the failure envelope is curved within the range of small stress (see Figure 5.12).

Test results from using osmotic technique

The stress points at failure from using the osmotic technique at HKUST and ENPC are plotted in Figure 5.13. Under the same net confining pressure and suction, tests OH4 and OE3 exhibit similar shear strength. The comparison between tests OHS and OE6 shows the same conclusion. The consistency of the results from HKUST and ENPC confirms the reliability of osmotic technique. The tests at HKUST and ENPC are regarded as one test series using the osmotic technique henceforth.

For this test series, the internal friction angle, ϕ', ranges from 19.5° to 20.8° for suction ranges from 0 to 165 kPa, and the average value is 20.2°. Again, by assuming a constant friction angle of 20°, the failure lines for this series fit well with the test results, as shown in Figure 5.13. The

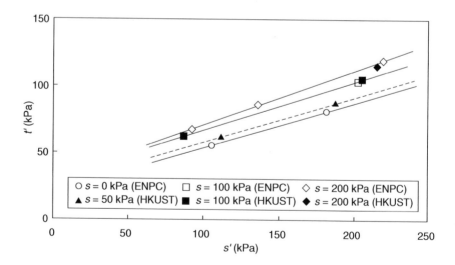

Figure 5.13 Comparison of strength of the triaxial tests performed at ENPC and HKUST using the osmotic technique (Ng et al., 2007).

increasing intercepts on the t axis indicate that apparent cohesion increases with suction. Similar to the results from using the axis-translation technique, at saturated state, there also appears a small intercept of 19.4 kPa on the t axis when the failure envelop is extended linearly to zero value of s. Previous analysis exhibited that this apparent intercept is likely due to experimental error or the failure envelope is curved within the range of small stress (see Figure 5.13).

Comparison between test results from axis-translation and osmotic techniques

Both test series using the axis-translation and osmotic techniques can be approximated by an internal friction angle of 20°. It indicates that there is no difference in determining internal friction angle when using the axis-translation technique and the osmotic technique for testing this material.

In the view of the results of soaking test, it is believed that no true cohesion is present in the recompacted specimens. Thus, apparent cohesions due to suction can be calculated from intersects on the t axis at different suctions subtracting the intersect at saturated state. The data of apparent cohesion at different suctions are plotted in Figure 5.14a. For the test series using the axis-translation technique, the relation curve shows a linear increase of shear strength with suction within the suction range from 0 to 100 kPa. For the test series using the osmotic technique, a non-linear increase of shear strength with suction is observed within the suction range from 0 to 165 kPa. In the test series using the axis-translation technique, apparent cohesion due to suction is consistently larger than that in the test series using the osmotic technique. The value of ϕ^b at different suctions can be calculated from the cohesion versus suction curve in Figure 5.14a and its variation with suction is shown in Figure 5.14b. At zero suction, ϕ^b may be assumed equal to the saturated friction angle, ϕ' (i.e. 20°) (Gan and Fredlund, 1996). As shown in Figure 5.14b, in the test series using the axis-translation technique, ϕ^b is slightly larger than that in the test series using the osmotic technique by 2°–3° when suction is larger than zero. The larger ϕ^b indicates that the tests using the axis-translation technique exhibit a larger suction effect on shear strength as compared with that using the osmotic technique. Inspecting specific volume at failure in the tests under net confining pressure of 100 kPa (refer to Figure 5.15), it is found that the specimens in the test series using the axis-translation technique have lower values than those using the osmotic technique at different suctions. Perhaps the smaller specific volume at failure induced a larger shear strength and then the tests using the axis-translation technique exhibited a larger values of ϕ^b. However, reason for smaller specific volume in the tests using the axis-translation technique is not clear yet.

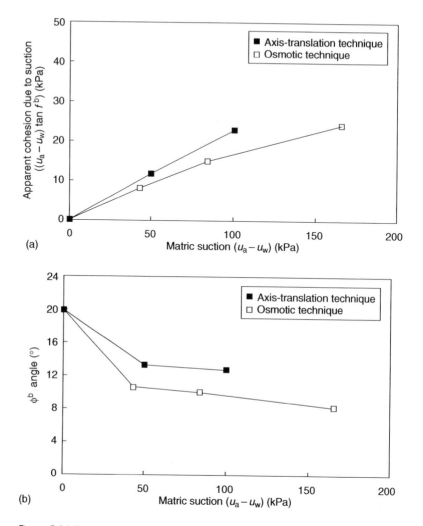

Figure 5.14 Suction versus shear strength: (a) apparent cohesion due to suction; (b) variation in ϕ^b angle with suction (Ng *et al.*, 2007).

Summary

In the application of osmotic technique, relationship between osmotic pressure and concentration of PEG solution is of importance. Different calibration methods lead to different calibrated relationships between osmotic pressure and concentration of PEG solution. The main difference comes from whether a semi-permeable membrane is used in calibration.

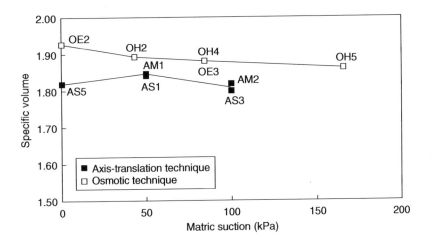

Figure 5.15 Specific volume at failure in the tests with a net confining pressure of 100 kPa (Ng et al., 2007).

For the recompacted expansive clay used in this research, comparison of the results of single-stage and multi-stage tests indicated that these two testing methods provide consistent shear strength information. The consistency of the results from using the osmotic technique at HKUST and ENPC showed the reliability of osmotic technique.

Based on the extended Mohr–Coulomb shear strength formulation, comparison of the results using the axis-translation technique and the osmotic technique showed that there is no difference in determining internal friction angle ϕ'. However, the values of ϕ^b in the test series using the axis-translation technique are slightly larger than those in the test series using the osmotic technique by 2°–3°. Perhaps the larger values of ϕ^b are due to smaller values of specific volume at failure in the tests using the axis-translation technique. However, reason for the smaller specific volume in the tests using the axis-translation technique is not clear yet.

Extended Mohr–Coulomb failure criterion (Fredlund and Rahardjo, 1993)

Introduction

The shear strength of an unsaturated soil can be formulated in terms of independent stress state variables (Fredlund et al., 1978). Any two of the three possible stress state variables can be used for the shear strength equation. The stress state variables, $(\sigma - u_a)$ and $(u_a - u_w)$, have been shown to be the most advantageous combination for practice. Using these stress variables, the shear strength equation is written as follows:

$$\tau_f = c' + (\sigma - u_a)_f \tan \phi' + (u_a - u_w)_f \tan \phi^b \tag{5.1}$$

where

c' = intercept of the 'extended' Mohr–Coulomb failure envelope on the shear stress axis where the net normal stress and the matric suction at failure are equal to zero; it is also referred to as 'effective cohesion'

$(\sigma - u_a)_f$ = net normal stress

u_a = pore air pressure

u_w = pore water pressure

σ = normal total stress

ϕ' = angle of internal friction associated with the net normal stress state variable, $(\sigma - u_a)_f$

$(u_a - u_w)_f$ = matric suction at failure

ϕ^b = angle indicating the rate of increase in shear strength relative to the matric suction, $(u_a - u_w)_f$.

The shear strength equation for an unsaturated soil exhibits a smooth transition to the shear strength equation for a saturated soil. As the soil approaches saturation, the pore water pressure, u_w, approaches the pore air pressure, u_a, and the matric suction, $(u_a - u_w)$, goes to zero. The matric suction component vanishes, and Equation (5.1) reverts to the equation for a saturated soil:

$$\tau_f = c' + (\sigma - u_w)_f \tan \phi' \tag{5.2}$$

The failure envelope for a saturated soil is obtained by plotting a series of Mohr circles corresponding to failure conditions on a two-dimensional plot, as shown previously. The line tangent to the Mohr circles is called the failure envelope, as described by Equation (5.2). In the case of an unsaturated soil, the Mohr circles corresponding to failure conditions can be plotted in a three-dimensional manner, as illustrated in Figure 5.16.

The three-dimensional plot has the shear stress, τ, as the ordinate and the two stress state variables, $(\sigma - u_a)$ and $(u_a - u_w)$, as abscissas. The frontal plane represents a saturated soil where the matric suction is zero. On the frontal plane, the $(\sigma - u_a)$ axis reverts to the $(\sigma - u_w)$ axis since the pore air pressure becomes equal to the pore water pressure at saturation. Thus it can be seen that a saturated soil is just a special case of an unsaturated soil.

The Mohr circles for an unsaturated soil are plotted with respect to the net normal stress axis, $(\sigma - u_a)$, in the same manner as the Mohr circles are plotted for saturated soils with respect to effective stress axis, $(\sigma - u_w)$. However, the location of the Mohr circle plot in the third dimension

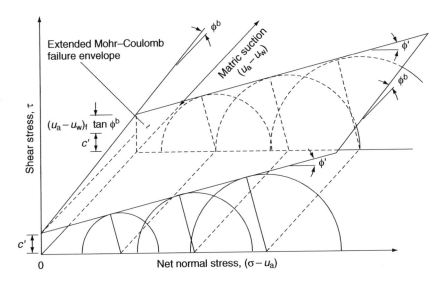

Figure 5.16 Extended Mohr–Coulomb failure envelope for unsaturated soils (after Fredlund and Rahardjo, 1993).

is a function of the matric suction (Figure 5.16). The surface tangent to
the Mohr circles at failure is referred to as the extended Mohr–Coulomb
failure envelope for unsaturated soils. The extended Mohr–Coulomb failure
envelope defines the shear strength of an unsaturated soil. The intersection
line between the extended Mohr–Coulomb failure envelope and the frontal
plane is the failure envelope for the saturated condition.

The inclination of the theoretical failure plane is defined by joining the
tangent point on the Mohr circle to the pole point. The tangent point on
the Mohr circle at failure represents the stress state on the failure plane at
failure.

Clearly, the extended Mohr–Coulomb failure envelope is idealized as a
planar surface. In reality, the failure envelope is somewhat curved (Gan and
Fredlund, 1996).

Figure 5.16 shows a planar failure envelope that intersects the shear stress
axis, giving a cohesion intercept, c'. The envelope has slope angles of ϕ' and
ϕ^b with respect to the $(\sigma - u_a)$ and $(u_a - u_w)$ axes, respectively. Both angles
are assumed to be constants. The cohesion intercept, c', and the slope angles,
ϕ' and ϕ^b, are the strength parameters used to relate the shear strength to the
stress state variables. The shear strength parameters represent many factors
are density, void ratio, degree of saturation, mineral composition, stress
history and strain rate. In other words, these factors have been combined
and expressed mathematically in the strength parameters.

The mechanical behaviour of an unsaturated soil is affected differently by changes in net normal stress than by changes in matric suction (Jennings and Burland, 1962). The increase in shear strength due to an increase in net normal stress is characterized by the friction angle, ϕ'. On the other hand, the increase in shear strength caused by an increase in matric suction is described by the angle, ϕ^b. The value of ϕ^b is consistently equal to or less than ϕ', as indicated for soils from various geographic locations as shown in Table 5.4.

The failure envelope intersects the shear stress versus matric suction plane along a line of intercepts, as illustrated in Figure 5.17.

The line of intercepts indicates an increase in strength as matric suction increases. In other words, the shear strength increase with respect to an increase in matric suction is defined by the angle, ϕ^b. The equation for the line of intercepts is as follows:

$$c = c' + (u_a - u_w)_f \tan \phi^b \tag{5.3}$$

Table 5.4 Experimental values of ϕ^b (after Fredlund and Rahardjo, 1993)

Soil type	c' (kPa)	ϕ' (degrees)	ϕ^b (degrees)	Test procedure	Reference
Compacted shale; $w = 18.6\%$	15.8	24.8	18.1	Constant water content triaxial	Bishop et al. (1960)
Boulder clay; $w = 11.6\%$	9.6	27.3	21.7	Constant water content triaxial	Bishop et al. (1960)
Dhanauri clay; $w = 22.2\%$, $\rho_d = 1580\,\mathrm{kg/m^3}$	37.3	28.5	16.2	Consolidated drained triaxial	Satija (1978)
Dhanauri clay; $w = 22.2\%$, $\rho_d = 1478\,\mathrm{kg/m^3}$	20.3	29.0	12.6	Constant drained triaxial	Satija (1978)
Dhanauri clay; $w = 22.2\%$, $\rho_d = 1580\,\mathrm{kg/m^3}$	15.5	28.5	22.6	Consolidated water content triaxial	Satija (1978)
Dhanauri clay; $w = 22.2\%$, $\rho_d = 1478\,\mathrm{kg/m^3}$	11.3	29.0	16.5	Constant water content triaxial	Satija (1978)
Madrid grey clay; $w = 29\%$	23.7	22.5[a]	16.1	Consolidated drained direct shear	Escario (1980)
Undisturbed decomposed granite; Hong Kong	28.9	33.4	15.3	Consolidated drained multistage triaxial	Ho and Fredlund (1982a)
Undisturbed decomposed rhyolite; Hong Kong	7.4	35.3	13.8	Consolidated drained multistage triaxial	Ho and Fredlund (1982a)
Tappen–Notch Hill silt; $w = 21.5\%$, $\rho_d = 1590\,\mathrm{kg/m^3}$	0.0	35.0	16.0	Consolidated drained multistage triaxial	Krahn et al. (1989)
Compacted glacial till; $w = 12.2\%$, $\rho_d = 1810\,\mathrm{kg/m^3}$	10	25.3	7–25.5	Consolidated drained multistage direct shear	Gan et al. (1988)

Note
a Average value.

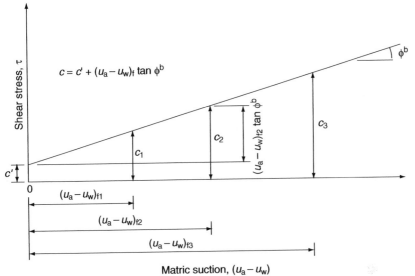

Figure 5.17 Line of intercepts along the failure plane on the τ versus $(u_a - u_w)$ plane (after Fredlund and Rahardjo, 1993).

where
c' = intercept of the extended Mohr–Coulomb failure envelope with the shear stress axis at a specific matric suction, $(u_a - u_w)_f$, and zero net normal stress; it can be referred to as the 'total cohesion intercept'.

Relationships between ϕ^b and χ

Bishop (1959) proposed a shear strength equation for unsaturated soils which had the following form:

$$\tau_f = c' + [(\sigma - u_a)_f - \chi (u_a - u_w)_f] \tan \phi' \tag{5.4}$$

where
χ = a parameter related to the degree of saturation of the soil.

Let us assume that the shear strength computed using Equation (5.1) can be made to be equal to the shear strength given by Equation (5.4). Then it is possible to illustrate the relationship between $\tan \phi^b$ and χ as:

$$(u_a - u_w)_f \tan \phi^b = \chi (u_a - u_w)_f \tan \phi' \tag{5.5}$$

It is then possible to solve for the parameter, χ as follows:

$$\chi = \frac{\tan \phi^b}{\tan \phi'} \tag{5.6}$$

A graphical comparison between the ϕ^b representation of strength and the χ representation of strength [i.e. Equation (5.4)] is shown in Figure 5.18.

Using the ϕ^b method, the increase in shear strength due to matric suction is represented as an upward translation from the saturated failure envelope. The magnitude of the upward translation is equal to $\left[(u_a - u_w)_f \tan \phi^b\right]_f$ (i.e. point A in Figure 5.18). In this case, the failure envelope for the unsaturated soil is viewed as a third-dimension extension of the failure envelope for the saturated soil. On the other hand, the χ parameter method uses the same failure envelope for saturated and unsaturated conditions. Matric suction is assumed to produce an increase in the net normal stress. This increase is a fraction of the matric suction at failure (i.e. $\chi (u_a - u_w)_f$). The shear strength at point A using the ϕ^b method is equivalent to the shear strength at point A' in the χ parameter method, as depicted in Figure 5.18.

Theoretically, only one χ value is obtained from Equation (5.6) for a particular soil when the failure envelope is planar. A planar failure envelope uses one value of ϕ' and one value of ϕ^b. If the failure envelope is bilinear with respect to the ϕ^b, there will be two values for χ. A χ value equal to 1.0 corresponds to the condition where ϕ^b is equal to ϕ'. A χ value less

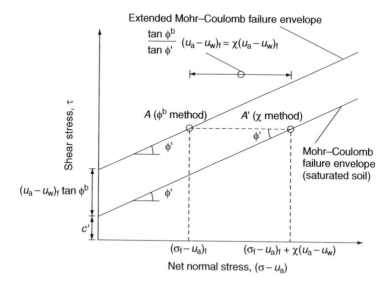

Figure 5.18 Comparisons of the ϕ^b and χ methods of designating shear strength (after Fredlund and Rahardjo, 1993).

than 1.0 corresponds to the condition when ϕ^b is less than ϕ'. For envelopes which are highly curved with respect to matric suction, there will be various χ values corresponding to different matric suctions.

The χ parameter has commonly been correlated with the degree of saturation of the soil. Unfortunately, the χ value has sometimes been obtained from shear strength tests on soil specimens compacted at different water contents. The different initial compacted water contents may have been used to give varying initial matric suctions. However, the soil specimens compacted at different water contents do not represent an 'identical' soil. As a result, the χ values obtained from specimens compacted at different water contents are essentially obtained from different soils. The ϕ^b and χ relationship in Equation (5.6) applies only to initially 'identical' soils. These may be soils compacted at the same water content to the same dry density. The χ parameter need not be unique for both shear strength and volume change problems. The χ relationship given in Equation (5.6) is only applicable to the evaluation of the shear strength of the soil. From a practical engineering standpoint, it would appear to be better to use the $(\sigma - u_a)$ and $(u_a - u_w)$ stress state variables in an independent manner for designating the shear strength of an unsaturated soil.

The relationship of the unsaturated soil shear strength to the soil–water characteristic curve (Fredlund et al., 1996)

Introduction

Laboratory studies have shown that there is a relationship between the soil–water characteristic curve and the unsaturated soil properties (Fredlund and Rahardjo, 1993). Several models have been proposed to empirically predict the permeability function for an unsaturated soil from the soil–water characteristic curve by using the saturated coefficient of permeability as the starting value (Fredlund et al., 1994). The following provides engineers with a means of estimating the shear strength function for an unsaturated soil from a soil–water characteristic curve by using the saturated shear strength parameters as the starting values.

Soil–water characteristic curve

The soil–water characteristic curve for a soil is defined as the relationship between water content and suction. The water content variable (i.e. volumetric water content, gravimetric water content, or degree of saturation) defines the amount of water contained in the pores of the soil. The variable has often been used in a dimensionless form where the water content is referenced to a residual or zero water content and is defined as follows:

$$\Theta = \frac{\theta - \theta_r}{\theta_s - \theta_r} \tag{5.7}$$

where

θ = volumetric water content at any suction (or $\theta(u_a - u_w)$),
θ_s = volumetric water content at saturation,
θ_r = volumetric water content at residual conditions, and
Θ = normalized volumetric water content. When the reference volumetric water content, θ_r is taken as being zero, $\Theta = \theta/\theta_s$.

The suction may be either the matric suction, $(u_a - u_w)$, or total suction (matric plus osmotic suction) of the soil. At high suctions (e.g. > 3,000 kPa), matric suction and total suction are generally assumed to be essentially the same.

The total suction corresponding to zero water content appears to be essentially the same for all type of soils. A value slightly below 1,000,000 kPa has been experimentally supported for a variety of soils (Croney and Coleman, 1961; Russam and Coleman, 1961; Fredlund, 1964). The value is also supported by thermodynamic considerations (Richards, 1965). In other words, there is a maximum total suction value corresponding to a zero relative humidity in any porous medium. A general equation describing the soil–water characteristic curve over the entire suction range (i.e. 0–1,000,000 kPa) is given by Fredlund and Xing (1994):

$$\theta = \theta_s \left[1 - \frac{\ln\left(1 + \dfrac{\psi}{\psi_r}\right)}{\ln\left(1 + \dfrac{1,000,000}{\psi_r}\right)} \right] \left\{ \frac{1}{\ln\left[e + \left(\dfrac{\psi}{a}\right)^n\right]} \right\}^m \tag{5.8}$$

where

ψ = total soil suction (kPa),
e = natural number, 2.71825..,
ψ_r = total suction (kPa) corresponding to the residual water content, θ_r,
a = a soil parameter that is related to the air-entry value of the soil (kPa),
n = a soil parameter that controls the slope at the inflection point in the soil–water characteristic curve, and
m = a soil parameter that is related to the residual water content of the soil.

The parameters, a, n and m, in Equation (5.8) can be determined using a non-linear regression procedure outlined by Fredlund and Xing (1994). The residual water content, θ_r, is assumed to be zero. The normalized (volumetric or gravimetric) water content when referenced to zero water content is equal to the degree of saturation, S, provided the total volume change is negligible (Fredlund *et al.*, 1994).

The shear strength of a soil is a function of matric suction, as it goes from the saturated condition to an unsaturated condition. In turn, the water content is a function of matric suction. Equation (5.8) can be expressed in terms of the matric suction of the soil:

$$\theta = \theta_s \left\{ 1 - \frac{\ln\left[1 + \frac{(u_a - u_w)}{(u_a - u_w)_r}\right]}{\ln\left[1 + \frac{1,000,000}{(u_a - u_w)_r}\right]} \right\} \times \left\{ \frac{1}{\ln\left[e + \left(\frac{(u_a - u_w)}{a}\right)^n\right]} \right\}^m \tag{5.9}$$

where $(u_a - u_w)_r$ = matric suction corresponding to the residual water content, θ_r.

A model for the shear strength function for unsaturated soils

The contribution of matric suction to the shear strength of an unsaturated soil can be assumed to be proportional to the product of matric suction, $(u_a - u_w)$, and the normalized area of water, a_w, at a particular stress state (Fredlund *et al.*, 1995a):

$$\tau = a_w (u_a - u_w) \tan \phi' \tag{5.10}$$

where

$a_w = A_{dw}/A_{tw}$,
A_{dw} = area of water corresponding to any degree of saturation and
A_{tw} = total area of water at saturation.

The normalized area of water, a_w, decreases as the matric suction increases. The chain rule of differentiation on Equation (5.10) shows that there are two components of shear strength change associated with a change in matric suction.

$$d\tau = \tan \phi' [a_w d(u_a - u_w) + (u_a - u_w) da_w] \tag{5.11}$$

The normalized area of water in the soil, a_w, may be assumed to be proportional to the normalized volumetric water content at a particular suction value by applying Green's theorem (Fung, 1977) (i.e. $\Theta(u_a - u_w)$, which is equal to $\theta(u_a - u_w)/\theta_s$. The normalized area of water can be defined by the following equation:

$$a_w = [\Theta (u_a - u_w)]^\kappa \tag{5.12}$$

where

$\Theta(u_a - u_w)$ = normalized water content as a function of matric suction, and
κ = a soil parameter dependent upon the soil type.

Then, substituting Equation (5.12) into Equation (5.11) gives

$$d\tau = \tan \phi' \left\{ [\Theta(u_a - u_w)]^\kappa + \kappa(u_a - u_w) \right.$$

$$\left. \times \left[\Theta(u_a - u_w)^{\kappa - 1} d\Theta(u_a - u_w) \right] \right\} d(u_a - u_w) \tag{5.13}$$

The normalized volumetric water content, $\Theta(u_a - u_w)$, is defined by the soil–water characteristic function and can be obtained from Equation (5.9). In other words, Equation (5.10) can be used to predict the shear strength function of an unsaturated soil using the soil–water characteristic curve and the saturated shear strength parameters.

Referring now to the fundamental equation for shear strength of unsaturated soil:

$$\tau = c' + (\sigma_n - u_a) \tan \phi' + (u_a - u_w) \tan \phi^b \tag{5.14}$$

Equation (5.14) can be written in a different form as follows:

$$\tau = c' + (\sigma_n - u_a) \tan \phi' + (u_a - u_w)[\Theta(u_a - u_w)]^\kappa \tan \phi \tag{5.15}$$

This equation is found by substituting Equation (5.12) into Equations (5.10) and (5.14). The simple form of Equation (5.15) now allows for the easy substitution of a normalized soil–water characteristic curve.

Comparison of model to example data

Consider two different soils, soil 1 and soil 2, shown in Figure 5.19. The soil–water characteristic curves are typical of a medium-grained sand and a fine-grained sand, respectively. Soil 1 has an effective cohesion of 0 kPa, an effective angle of internal friction of 32.0°, and an air-entry value of 20 kPa. Soil 2 has an effective cohesion of 0 kPa, an effective angle of internal friction of 25.0°, and an air-entry value of 60 kPa. The predicted shear strength curves for soil 1 and soil 2 using Equation (5.15) are shown in Figure 5.20. It can be seen that the shear strength of both soils increases linearly at the rate of $\tan \phi'$ up to the air-entry values of the soils. Beyond the air entry values, the rate of change of shear strength with matric suction decreases. The change in shear strength with respect to suction is in accordance with Equation (5.14).

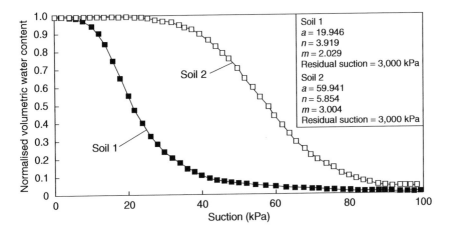

Figure 5.19 Two sample soil–water characteristic curves from Equation (5.7) (after Fredlund et al., 1996).

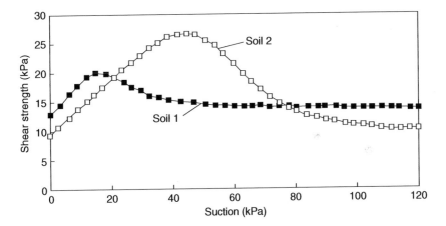

Figure 5.20 Predicted shear strength curves using Equation (5.15) and the soil–water characteristic curves in Figure 5.19 (after Fredlund et al., 1996).

The shapes of the shear strength curves with respect to matric suction are similar to those measured by Donald (1956). Donald's test results for several sands are shown in Figure 5.21. In each case, the shear strength increases with suction and then drops off to a lower value. A similar behaviour was observed in the testing of a fine- to medium-grained decomposed tuff from Hong Kong (Figure 5.22). The results indicate that at low confining pressures the shear strength may increase and then start to fall with increasing

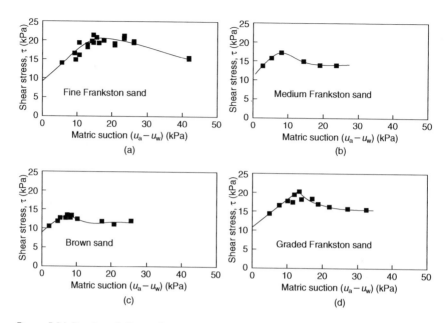

Figure 5.21 Results of direct shear tests on sands under low matric suctions (modified from Donald, 1956; after Fredlund *et al.*, 1996).

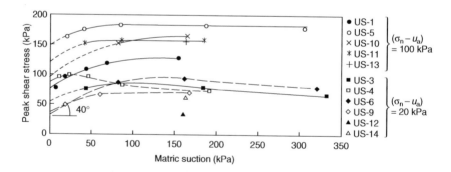

Figure 5.22 Peak shear stress versus matric suction envelope for the completely decomposed fine ash tuff (after Fredlund *et al.*, 1996; Gan and Fredlund, 1996).

suction. At higher confining pressures the shear strength shows a continuing rise in strength with increasing suction. These results also illustrate the importance of applying an appropriate confining pressure to the soil when measuring the soil–water characteristic curve.

It should be noted that the unsaturated ϕ^b term may be expressed as:

$$\tan \phi^b = \frac{d\tau}{d\,(u_a - u_w)} = \{[\Theta\,(u_a - u_w)]^\kappa + \kappa\,(u_a - u_w)$$
$$\times [\Theta\,(u_a - u_w)]^{\kappa-1}\,d\Theta\,(u_a - u_w)\}\tan \phi' \qquad (5.16)$$

The ratio between the two parameters (i.e. ϕ' and ϕ^b can be shown as a function of the normalized water content as follows:

$$\beta = \frac{\tan \phi^b}{\tan \phi'} = \left\{[\Theta\,(u_a - u_w)]^\kappa + \kappa\,(u_a - u_w)\times[\Theta\,(u_a - u_w)]^{\kappa-1}\,d\Theta\,(u_a - u_w)\right\}$$

$$(5.17)$$

Equations (5.16) and (5.17) show that the angle ϕ^b is equal to the effective angle of internal friction, ϕ', up to the air-entry value of the soil (i.e. $\Theta(u_a - u_w)$ equal to 1). Beyond the air-entry value, ϕ^b decreases as the matric suction increases (Figure 5.22). This is in agreement with experimental observations (Gan et al., 1988).

Equation (5.15) has been tested using experimental data from a completely decomposed tuff from Hong Kong (Gan and Fredlund, 1992). A best-fit soil–water characteristic curve using Equation (5.9) is found by using a curve-fitting program to match the measured water contents at various matric suction values (Figure 5.23). The shear strength function

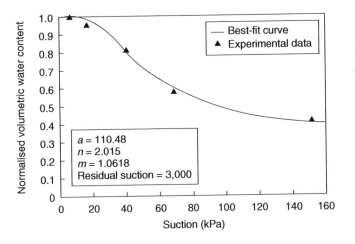

Figure 5.23 Soil–water characteristic curve for a completely decomposed tuff (specimen US-1) from Hong Kong; experimental values (from Gan and Fredlund, 1992) and calculations using Equation (5.9) (after Fredlund et al., 1996).

is calculated from Equation (5.14). The predicted shear strength values, along with the measured shear strength values are shown in Figure 5.24. The parameters used in the model are listed in Table 5.5. The value of the soil parameter, κ, was set to 1 for the prediction. The model with κ equal to 1, appears to give satisfactory predictions for sandy soils. The value of κ generally increases with the plasticity of the soil and can be greater than 1.0.

The value of κ affects the rate at which the angle decreases as the matric suction exceeds the air-entry value of the soil. The effect of κ on the shear strength function of a soil with a soil–water characteristic curve defined in Figures 5.25 and 5.26 is shown in Figure 5.27. The values of κ are in the range from 1.0 to 3.0. The influence of κ on the shape of the shear strength function occurs once the air-entry value of the soil is exceeded. In the example shown, a value of κ equal to 2.0 shows that the shear

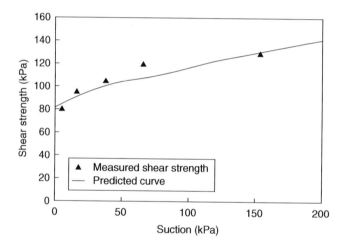

Figure 5.24 Comparisons of the predicted shear strength curve with the experimental shear strength data for the completely decomposed tuff (specimen US-1) from Hong Kong (from Gan and Fredlund, 1992; after Fredlund et al., 1996).

Table 5.5 Soil properties and fitting parameters for the Hong Kong soil US-1 (after Fredlund et al., 1996)

Equation [13]			Equation [7]			
tan ϕ'	c' (kPa)	κ	a (kPa)	n	m	ψ_r (kPa)
0.8012	0	1	110.48	2.015	10.618	3,000

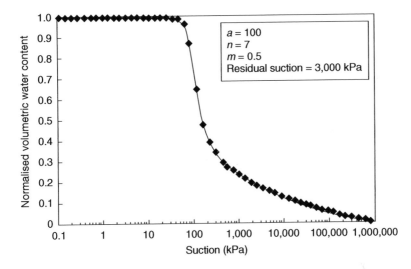

Figure 5.25 The normalized soil–water characteristic curve over the entire range of suction values (after Fredlund *et al.*, 1996).

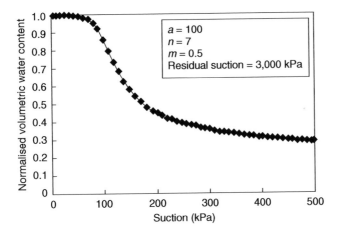

Figure 5.26 Effect of the parameter, κ, on the shear strength function of a soil; the soil–water characteristic curve of the soil (after Fredlund *et al.*, 1996).

strength envelope becomes essentially horizontal shortly after the air-entry value is exceeded. The variable κ can be visualized as an indication of the relationship between the volumetric representation of water in the voids and the area representation of water in the voids, as represented by an unbiased plane passed through the soil mass.

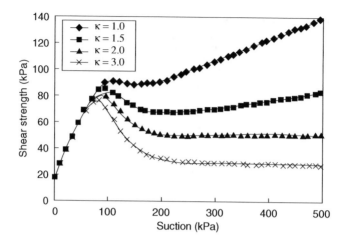

Figure 5.27 Effect of the parameter, κ, on the shear strength function of a soil; the shear strength function showing the effect of varying κ (after Fredlund et al., 1996).

Closed-form solutions

Closed-form solutions for shear strength functions are developed for two cases, using the empirical equations proposed by McKee and Bumb (1984) and Brooks and Corey (1964), respectively.

The following exponential relationship for the soil–water characteristic curve given in Equation (5.18) is suggested by McKee and Bumb (1984) for the case where the suction is greater than the air-entry value (i.e. $(u_a - u_w) \geq (u_a - u_w)_b$). A sample plot of the equation proposed by McKee and Bumb (1984) can be seen in Figure 5.28.

$$\Theta = e^{-[(u_a - u_w) - (u_a - u_w)_b]/f} \tag{5.18}$$

where

$(u_a - u_w)_b$ = air-entry value (also known as the bubbling pressure), and
f = fitting parameter.

Equation (5.18) describes the soil–water characteristic curve for suction values greater than the air-entry value. The normalized volumetric water content, Θ, is assumed to be constant in the range from zero soil suction to the air-entry value of the soil. For simplicity, the soil parameter, κ, was assumed to be equal to 1. Substituting Equation (5.18) into Equation (5.14) gives the closed-form equation:

$$\tau = c' + (\sigma_n - u_a) \tan \phi' + \left(e^{-[(u_a - u_w) - (u_a - u_w)_b]/f}\right)^{\kappa} (u_a - u_w) \tan \phi' \tag{5.19}$$

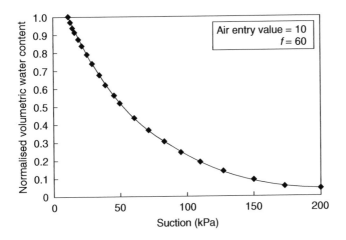

Figure 5.28 A sample plot of the McKee and Bumb equation (1984) (after Fredlund et al., 1996).

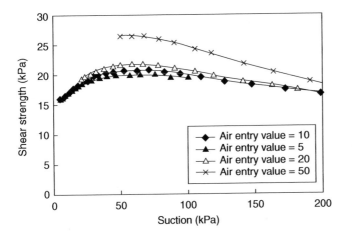

Figure 5.29 Shear strength function predicted using the McKee and Bumb equation (1984); illustrating the effect of varying the air-entry value (after Fredlund et al., 1996).

Sample plots of Equation (5.19) showing the effect of varying the air-entry value and f parameter are shown in Figures 5.29 and 5.30, respectively.

The soil–water characteristic curve given by Brooks and Corey (1964) can be expressed in the following form in Equation (5.20) for the case where the suction is greater than the air-entry value (i.e. $(u_a - u_w) \geq (u_a - u_w)_{AEV}$).

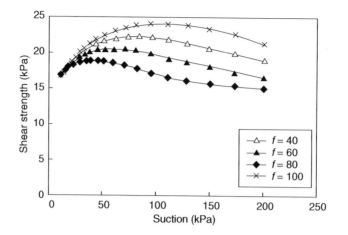

Figure 5.30 Shear strength function predicted using the McKee and Bumb equation (1984); illustrating the effect of varying the f parameter (after Fredlund et al., 1996).

$$\Theta = \left[\frac{(u_a - u_w)_{AEV}}{(u_a - u_w)} \right]^{f'} \tag{5.20}$$

where
$(u_a - u_w)_{AEV}$ = air-entry value, and
f' = fitting parameter.

Equation (5.20) is valid for matric suctions greater than the air-entry value (i.e. the value of Θ is assumed to be a constant up to the air-entry value). A sample plot of Equation (5.20) can be seen in Figure 5.31.

Substituting Equation (5.20) into Equation (5.14) gives

$$\tau = c' + (\sigma_n - u_a) \tan \phi' + \left[\frac{(u_a - u_w)_{AEV}}{(u_a - u_w)} \right]^{f'} (u_a - u_w) \tan \phi' \tag{5.21}$$

Sample plots using Equation (5.21) with varying f' and air-entry values are shown in Figures. 5.32 and 5.33, respectively. The parameters f and f' appear to have similar effects on the shear strength function (Figures 5.30, 5.33) as the parameter κ (Figure 5.27). The parameters f and f' can therefore be expressed in terms of the parameter κ, thus eliminating one additional parameter from Equations (5.19) and (5.21).

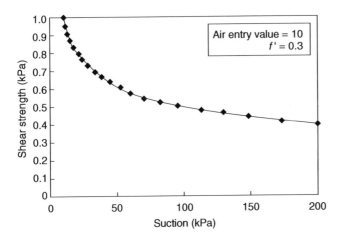

Figure 5.31 A sample soil–water characteristic curve using the Brooks and Cory equation (1964).

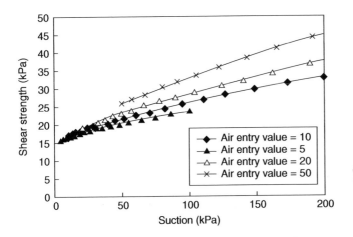

Figure 5.32 Shear strength equation using the Books and Corey equation (1964) for the soil–water characteristic curve and illustrating the effects of varying the air-entry value (after Fredlund *et al.*, 1996).

Alternate solution to the general shear strength equation for an unsaturated soil

The experimental data used to illustrate the use of Equation (5.15) were from a sandy soil (a decomposed tuff from Hong Kong). The data set

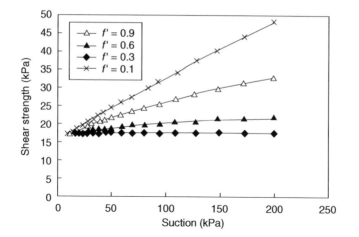

Figure 5.33 Shear strength equation using the Books and Corey equation (1964) for the soil–water characteristic curve and illustrating the effects of varying the *f* parameter (after Fredlund *et al.*, 1996).

showed a good fit with the parameter κ set equal to 1. However, for highly plastic soils, the parameter κ is greater than 1. At present, its magnitude is an unknown variable.

Attempts to best-fit other data sets have shown that it is possible to always leave the variable K at 1.0 but change the upper limit of integration to reflect the soil suction near residual conditions. Unfortunately, the best-fit of the shear strength data often occurs when the residual conditions vary from those used in the best-fit of the soil–water characteristic curve. In other words, there may not be a common residual suction value for both the soil–water characteristic data and the shear strength data.

More data sets are required, along with further best-fit regression analyses, in order to better understand how best to predict the shear strength of an unsaturated soil.

Effects of soil suction on dilatancy of an unsaturated soil (Ng and Zhou, 2005)

Introduction

Soil behaviour is strongly influenced by dilatancy and so it is an important and essential component in elasto-plastic modelling of soils (refer to Chapter 7). Dilatancy in saturated soils has been studied for many years. It is fairly well understood. On the other hand, due to the complexity and long duration of testing unsaturated soils, fundamental understanding of the role of dilatancy in unsaturated soils is rather limited. Most of existing unsaturated constitutive models treat dilatancy using an assumption or deriving it from assumed yield surfaces with associated or non-associated flow rules (Alonso *et al.*, 1990; Wheeler and Sivakumar, 1995). Recently Chiu and Ng (2003) have introduced state-dependent dilatancy in their elasto-plastic model. However, relatively very limited and reliable laboratory data are available for verifying assumptions and calibrating constitutive relationships for unsaturated soils.

Cui and Delage (1996) performed three series of unsaturated triaxial tests on compacted silt under an osmotically controlled-suction technique. Their test results obtained at mean constant stress ratio $(\eta = q/p = 1)$ at constant cell pressure under various suctions are shown in Figures 5.34a and b, respectively. A summary of the notation used in this book is given in Appendix B. Based on the results shown in Figure 5.34a, they drew two

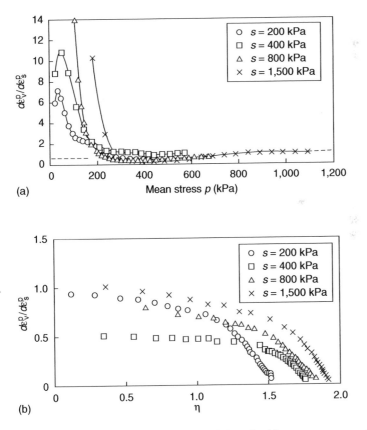

Figure 5.34 Effect of suction on dilatancy of silt under (a) constant stress ratio $(\eta = 1)$ and (b) constant cell pressure $(\sigma_3 = 200\,\text{kPa})$ (after Cui and Delage, 1996; Ng and Zhou, 2005).

conclusions: first, the ratio between plastic volumetric strain to plastic shear strain $d\varepsilon_v^p/d\varepsilon_s^p$ (also called the 'flow rule') is independent of suction; second, the final value of $d\varepsilon_v^p/d\varepsilon_s^p$ is equal to 1. For constant cell pressure cases (Figure 5.34b), they interpreted their data by separating these curves into two segments, i.e. one segment with small slope gradients for low η values and the other one with large slope gradients for higher η values. Based on a volumetric criterion, they claimed that the η values at intersection points of these two segments corresponded to yield points. They also claimed that in the plastic zone beyond the intersection points, the gradient of these slopes are independent of suction, thus a linear and suction-independent stress–dilatancy relationship is derived.

Recently Chiu (2001) and Chiu and Ng (2003) have investigated stress–dilatancy of a decomposed volcanic soil and a decomposed granitic soil using a computer-controlled traixial apparatus. Figure 5.35 shows experimental relationships of dilatancy $(d\varepsilon_v^p/d\varepsilon_s^p)$ and normalized stress ratio (η/M) obtained by shearing decomposed volcanic soil specimens using constant water content stress paths where M is the gradient of the critical state line (CSL). Details of the extended critical state framework for unsaturated soils are given in Chapter 7.

Following the same approach proposed by Cui and Delage (1996), no distinct yield point can be identified. At high stress ratios, the experimental data formed two groups according to their suction values. If an average line is drawn through these two groups of data at $\eta/M = 0.8$ or higher,

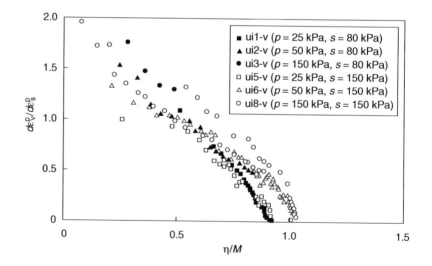

Figure 5.35 Stress dilatancy relationship for specimens of decomposed volcanic soil under constant water content stress paths (after Chiu, 2001; Ng and Zhou, 2005).

the gradients of these two fitted lines may be assumed to be parallel. This implies that the dilatancy rate with respect to η/M is independent of suction.

Here, a series of five laboratory tests on a compacted completely decomposed granite (CDG) were carried out in a suction-controlled direct shear box. Five suctions varying from 0 to 400 kPa were considered. Effects of soil suction on the evolution of dilatancy, stress dilatancy relationship, and maximum dilatancy are investigated.

Soil type and test procedures

The soil used in the experiment was sieved CDG from Beacon Hill (BH), Hong Kong. Based on visual inspection, the BH soil was a coarse-grained sand with 22 per cent silt and 2 per cent clay. The soil was sieved into its constituent particle sizes. Dry sieving was performed and particles larger than 2 mm were discarded in the tests. Index properties were determined in accordance with BS1337 (BSI, 1990). The material has a liquid limit, $w_L = 44$ per cent, and a plastic limit, $w_P = 16$ per cent. According to Proctor compaction test results, the maximum dry density was 1845 kg/m^3, and the optimum water content was 14.2 per cent. The drying SWCC for the CDG is shown in Figure 5.36. Also included in this figure is a drying SWCC of a silty soil from Nishimura and

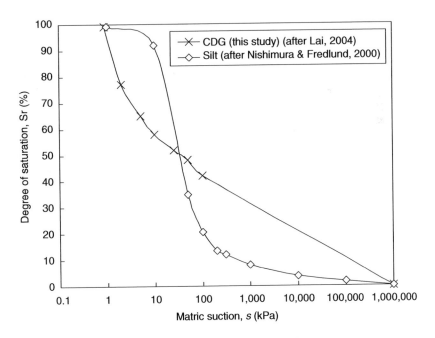

Figure 5.36 SWCCs of two soils (after Ng and Zhou, 2005).

Table 5.6 Air entry values and residual suction values of two soils
(after Ng and Zhou, 2005)

Soil type	Air entry value (kPa)	Residual suction value (kPa)
CDG	1	1,258
Silt	10	200

Fredlund (2000). Air entry and residual suction values were determined using an equation proposed by Fredlund and Xing (1994) and are summarized in Table 5.6. The major difference between the two SWCCs is that the silt has a steeper SWCC than CDG does. This indicates that the silt has a more uniform particle size distribution than that of CDG. Any relationship between the SWCCs and shear test results are discussed later.

An unsaturated direct shear apparatus was used to carry out a series of unsaturated soil tests. The apparatus was originally developed by Gan et al. (1988) and modified by Zhan (2003) in order to facilitate the measurements of water volume change automatically and accurately. The axis translation technique was utilized to control pore air and pore water pressures applied to a soil specimen.

A series of five unsaturated direct shear box tests on recompacted CDG were conducted at various controlled suctions. Specimen with $50 \times 50 \times 21 \, mm^3$ (length×width×height) was used in the tests. Each specimen was prepared by static compaction technique and compacted at the optimum water content of 14.2 per cent. The compacted initial dry density was $1.53 \, g/cm^3$, corresponding to 82.7 per cent of the maximum dry density.

After static compaction, each specimen was placed in the chamber of the unsaturated shear box equipped with a high air-entry value of 500 kPa ceramic disk, which was saturated according to the procedure proposed by Fredlund and Rahardjo (1993). Then the specimen was soaked under zero total vertical stress over 24 h as illustrated in the path $A \rightarrow O$ in Figure 5.37. Then the specimen was loaded along path $O \rightarrow B$ to a constant vertical normal stress $(\sigma_v - u_a)$ of 50 kPa, where σ_v is total vertical stress and u_a is pore air pressure. Subsequently, pore air pressure was applied at the top of the specimen step up step while maintaining zero pore water pressure at the base of the specimen till a target air pressure was reached. Once the target air pressure was reached, suction equalization was monitored by recoding changes of water volume. When the change of water volume was less than 0.01 per cent of initial water content, this suction equalization stage was terminated. Then shearing was carried out at a constant suction (i.e. along paths $(B \rightarrow B', \ C \rightarrow C', \ D \rightarrow D', \ E \rightarrow E'$ and $F \rightarrow F')$.

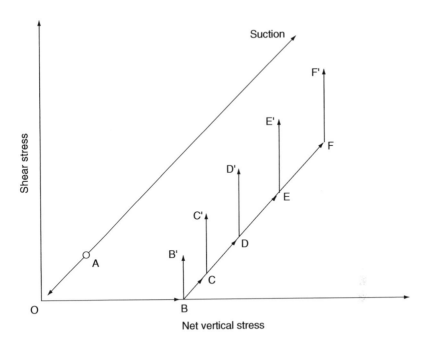

Figure 5.37 Stress paths in direct shear box tests (after Ng and Zhou, 2005).

Discussion of test results

Evolution of dilatancy during shear

Figures 5.38a and b show the relationships of stress ratio $(\tau/(\sigma_v - u_a))$ and dilatancy versus horizontal displacement (δx) respectively. In the figure, dilatancy is defined as the ratio $(\delta y/\delta x)$ of incremental vertical displacement (δy) to incremental horizontal displacement. Negative sign (or negative dilatancy) means expansive behaviour. In Figure 5.38a, it can be seen that at zero suction and suctions of 10 and 50 kPa, the stress ratio–displacement curve displayed strain hardening behaviour. With an increase in suction, strain softening behaviour was observed at suctions of 200 and 400 kPa. Generally, measured peak and ultimate stress ratios increased with suction, except the ultimate stress ratio measured at suction of 200 kPa. These test results are consistent with existing elasto-plastic models (e.g. Alonso et al., 1990; Wheeler and Sivakumar, 1995; Chiu and Ng, 2003) for unsaturated soils.

Figure 5.38b shows the effects of suction on dilatancy of CDG in the direct shear box tests. Under the saturated conditions, the soil specimen showed contractive behaviour (i.e. positive dilatancy). On the other hand, under unsaturated conditions, all soil specimens displayed contractive behaviour initially

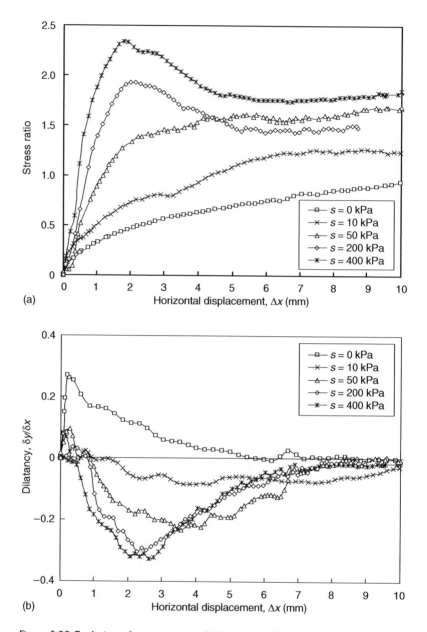

Figure 5.38 Evolution of stress ratio and dilatancy of CDG subjected to shear under different controlled suctions (after Ng and Zhou, 2005).

but then dilative behaviour as horizontal displacement continued to increase. The measured maximum negative dilatancy was enhanced by an increase in suction. This measured trend was consistent with test results on a compacted silt reported by Cui and Delage (1996). The increase in maximum negative dilatancy was likely attributed to a closer particle packing (i.e. a smaller void ratio) under a higher suction. At the end of each test, dilatancy of all specimens approached zero, indicating the attainment of the critical state.

An interesting phenomenon observed in Figure 5.38 was that strain hardening was recorded at suctions of 10 and 50 kPa as soil dilated. A maximum negative dilatancy at the latter test did not lead to strain softening behaviour. Another phenomenon observed was that when controlled suctions were equal to 200 and 400 kPa, strain softening was observed (see Figure 5.38a). However, the measured peak stress ratio in each test did not correspond with its maximum negative dilatancy (see Figure 5.38b). This feature was not consistent with a common assumption in constitutive modelling that the point of peak strength was usually associated with the peak negative dilatancy in saturated granular materials (Bolton, 1986).

Stress–dilatancy relationship

Figure 5.39 shows the measured stress–dilatancy relationships in the five tests. At low stress ratios (i.e. 0.2 or smaller), positive dilatancy was observed for all specimens. The soil contracted initially during shear. As the stress

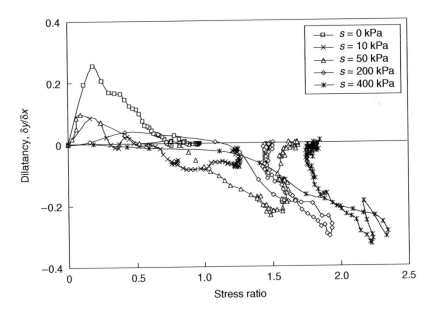

Figure 5.39 Experimental stress–dilatancy relationship (after Ng and Zhou, 2005).

ratios increased, negative dilatancy was measured in all unsaturated soil specimens. Obviously there was a phase transformation from positive to negative dilatancy as stress ratio increased. It is evident that the stress ratio corresponding to a maximum negative dilatancy increased with soil suction. As shearing continued, all the unsaturated soil specimens reached or approached zero dilation (i.e. critical state) at the end of each test. At the critical state, the measured stress ratio increased with suction, except at suction equal to 200 kPa. This measured trend is consistent with test results published by Cui and Delage (1996) and Chiu (2001) as shown in Figures 5.34 and 5.35, respectively. It should be pointed out in Figure 5.39 that a distinct loop was observed in tests with suctions equal to 50 kPa or higher. This implies that a state-dependent dilatancy soil model (Chiu and Ng, 2003) is necessary to capture this type of soil behaviour.

Maximum dilatancy

Figure 5.40 shows relationships of maximum dilatancy and controlled suctions for the CDG specimens. From comparisons, experimental data from unsaturated shear box tests on a silt from Nishimura (2000) are reinterpreted and included in the figure. Also included in this figure are the air-entry values and the residual suction values of these two soils. Their soil–water characteristic curves (SWCCs) are given in Figure 5.36.

As illustrated in Figure 5.40, maximum dilatancy of CDG decreases (i.e. more negative or dilative) with an increase in suction, i.e. the soil dilates

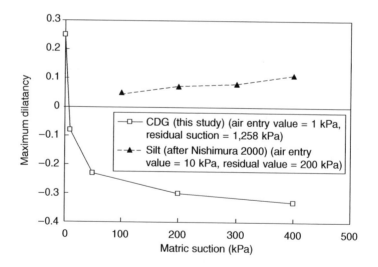

Figure 5.40 Relationship between maximum dilatancy and suction (after Ng and Zhou, 2005).

more at a higher suction. The relationship between maximum dilatancy and suction is highly non-linear. On the contrary, maximum dilatancy of the silt published by Nishimura (2000) increases (i.e. more contractive) linearly with suction. The major difference in the observed maximum dilatancy–suction relationships of these two soils is likely attributed to the difference in their initial soil densities (i.e. void ratios). The CDG and the silty soil have an initial dry density of 1.53 and $1.27\,g/cm^3$, respectively. Therefore, it is not surprising to observe that CDG is more dilative than the silt for a given suction. By comparing the absolute values of the maximum dilatancy of these two soils, variations in the magnitude of the maximum dilatancy are relatively larger in CDG than that in the silt for a given change of soil suctions. In addition to the difference in the initial soil density, the compressibility of CDG is relatively large as compared with other silty soils because of the presence of crushable feldspar (Ng et al., 2004). Therefore, for a given increase in soil suctions, it is believed that CDG has undergone a larger reduction in void ratio than that in the silt, resulting in the measured larger absolute maximum dilatancy.

SWCC of a soil has been proposed and used to predict shear strength and many other properties of unsaturated soils (Vanapalli et al., 1996). Since the shear strength of a soil is closely related to its dilatancy, it is interesting to explore any correlation between the SWCCs of CDG and the silt and their corresponding maximum dilatancy. Based on the limited data shown in Figures 5.34 and 5.38, it is very difficult, if not impossible, to draw any conclusion between measured maximum dilatancy and the SWCCs.

Summary and conclusions

A series of consolidated drained shear tests were carried out on compacted CDG to investigate the effects of suction on shear behaviour and dilatancy. Measured stress ratio corresponding to a maximum negative dilatancy increased with soil suction. The measured peak stress ratio in each test did not correspond with its maximum negative dilatancy. Maximum dilatancy of CDG was strongly dependent on suction and soil density. The maximum dilatancy decreased (i.e. more negative or dilative) with an increase in suction, i.e. the soil dilated more at a higher suction. The relationship between maximum dilatancy and suction was highly non-linear. State-dependent flow rules are essential and vital for modelling unsaturated soils properly and correctly.

Behaviour of a loosely compacted unsaturated volcanic soil (Ng and Chiu, 2001)

Introduction

Many studies of loose materials reported in the literature have been focused on clean sands. These studies show that saturated loose clean sands can exhibit a significant strain softening response during undrained shear. This may lead to so-called 'static liquefaction' (Sladen *et al.*, 1985; Ishihara, 1993; Sasitharan *et al.*, 1993; Yamamuro and Lade, 1997). Most fill materials are not clean sands. They are mainly decomposed granitic and volcanic soils that contain some quantity of fines. Recently, Yamamuro and Lade (1998) have shown that sand specimens with a certain amount of fines exhibit the reverse behaviour as compared with that from clean sands. This reverse behaviour is interpreted as the static liquefaction of the loose silty sands at low confining pressures with an increasing tendency of dilatancy with increasing confining pressure.

Testing program and procedures

Introduction

Three series of stress path triaxial tests were conducted in this study, and they are summarized in Table 5.7. The first series entailed consolidated undrained tests on saturated specimens. These tests established reference points from which to interpret the test results of the unsaturated specimens. The second series of tests were constant water content tests on the unsaturated specimens with suction measurements. In the constant water content test, the soil specimen is sheared under a drained condition in the pore air phase and under an undrained condition in the pore water phase. As the permeability of air is much greater than that of water in unsaturated soils, the drainage conditions of the constant water content tests may represent those of a slope element subjected to a fast shearing rate, like the conditions in static liquefaction. The last series of tests were shear tests under a constant deviator stress with a decreasing suction on the unsaturated specimens. They were used to simulate the stress path of a slope element subjected to rainfall infiltration (Brand, 1981). The testing conditions are given in Table 5.8.

Table 5.7 Testing programme (after Ng and Chiu, 2001)

Specimen identity	Type of test	Consolidation path	Degree of saturation during shear	Drainage condition
si1–si5	Consolidated undrained	Isotropic	Saturated	Undrained
ui1–ui8	Constant water content	Isotropic	Unsaturated	Air phase – drained, water phase – undrained
ua1–ua4	Field stress path that simulates rainfall infiltration (reduce suction at constant net stress)	Anisotropic	Unsaturated	Air phase – drained, water phase – drained

Table 5.8 Testing conditions (after Ng and Chiu, 2001)

Specimen identity	Initial specific volume	Applied suction (kPa)	Specific volume at applied suction	Volumetric strain at applied suction (%)[a]	Net mean stress before shearing (kPa)	Specific volume before shearing	Stress ratio before shearing (q/p)	Specific volume after shearing	Final suction (kPa)
si1	2.422	0	2.097	13.4	25	2.065	1	2.065	0
si2	2.419	0	2.101	13.1	50	2.005	1	2.005	0
si3	2.430	0	2.099	13.6	100	1.945	1	1.945	0
si4	2.464	0	2.145	12.9	200	1.882	1	1.882	0
si5	2.435	0	2.091	14.1	400	1.797	1	1.797	0
ui1	2.486	80	2.476	0.40	25	2.462	1	2.305	74
ui2	2.503	80	2.508	−0.20	50	2.416	1	2.136	72
ui3	2.526	80	2.523	0.12	150	2.215	1	1.917	79
ui4	2.504	80	2.507	−0.12	310	2.068	1	1.815	76
ui5	2.505	150	2.504	0.04	25	2.467	1	2.291	144
ui6	2.493	150	2.492	0.04	50	2.408	1	2.162	140
ui7	2.490	150	2.488	0.08	100	2.256	1	1.978	142
ui8	2.489	150	2.489	0.00	150	2.178	1	1.885	141
ua1	2.511	150	2.509	0.08	25	2.469	1.5	2.338	17
ua2	2.507	150	2.508	−0.04	50	2.327	1.5	2.240	40
ua3	2.489	150	2.489	0.00	100	2.148	1.5	2.154	62
ua4	2.507	150	2.504	0.12	140	2.071	1.5	2.108	60

Note
a Positive means contraction and negative means dilation.

The soil used in this study was a colluvium taken from a fill slope on Victoria Peak, Hong Kong. The colluvium was created from decomposed volcanic soil. The index properties were determined in accordance with the procedures given in BS1337 [British Standards Institution (BSI) 1990]. The particle size distribution of this decomposed volcanic soil is shown in Table 5.9 along with the distributions of four other soils for comparison. The decomposed volcanic soil can be described as slightly sandy silt (GCO, 1988). Proctor compaction tests determined the maximum dry density of $1,540 \, kg/m^3$ at the optimal moisture content of 21 per cent. Its liquid and plastic limits are 48 and 35 per cent, respectively.

All the triaxial specimens were 76 mm in diameter and 152 mm high and were prepared by wet tamping at a moisture content of 20 per cent to ensure that each one had the same soil fabric and structure. All specimens were compacted to a dry density of 70 per cent of the Proctor maximum, which corresponds to a specific volume v of 2.468 and a degree of saturation S_r of 36 per cent.

In the saturated specimen tests, the specimens were saturated by creating a vacuum of 15 kPa to evacuate air bubbles, followed by a carbon dioxide

Table 5.9 Particle size distributions (after Ng and Chiu, 2001)

Type of soil	Percentage				
	Decomposed volcanic soil	*Collapsible silt (Maatouk et al. 1995)*	*Toyoura sand (Ishihara 1995)*	*Completely decomposed granite (Lee and Coop 1995)*	*Completely decomposed fine ash tuff (Gan and Fredlund 1996)*
Gravel	0	0	0	44	8
Sand	25	16	100	51	26
Silt	65	66	0		61
Clay	10	18	0	5[a]	5

Note
a 5% for silt and clay combined.

flush for at least 1 h, after which, de-aired water was slowly introduced from the bottom of the specimens. Back pressure was used until the 5-value achieved a minimum value of 0.97. After saturation, consolidation was started and followed by shearing at a constant rate of strain. The stress path of an isotropically consolidated undrained test on a saturated specimen is denoted as *O–A–B–C–D* in Figure 5.41, where *O–A–B* is the wetting path during saturation of an initially unsaturated specimen, and *B–C* and

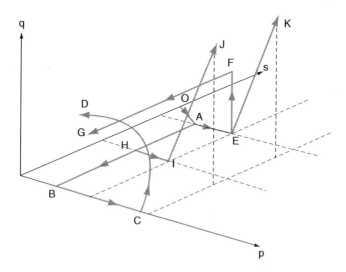

Figure 5.41 Various stress paths for tests on decomposed volcanic soil (after Ng and Chiu, 2001).

C–D are the stress paths for isotropic consolidation and undrained shear, respectively.

In the unsaturated specimen tests, a computer-controlled triaxial stress path apparatus equipped with three water pressure controllers and one air pressure controller was used. Suction was applied to the specimens through one pore water and one pore air pressure controllers. Pore water pressure was applied or measured at the base of the specimen through a porous filter, which had an air-entry value of 500 kPa. Positive pore water pressures were maintained by using the axis translation principle proposed by Hilf (1956). Pore air pressure was applied at the top of the specimen through a filter with a low air-entry value. During the constant water content test, no drainage of the pore water was permitted. As a result, the variation of suction during shear was measured from the difference between the positive pore water pressure at the bottom of the specimen and the constantly applied pore air pressure at the top of the specimen. The volume change of the unsaturated specimen was determined by two methods. Under small strains, the volume change was calculated from the axial and radial strains measured by a Hall effect local transducer (Clayton et al., 1989). When the volumetric strain exceeded the measuring limit of the local transducer, the volume change was measured by monitoring the flow of water into or out of the triaxial cell. A metallic cell wall was used instead of a Perspex one in order to minimize the deformation of the cell wall. The calibration of compliance for the metallic cell wall and the experimental set-up are given by Chiu (2001).

In the beginning, unsaturated specimens were controlled at initial suctions of 80 and 150 kPa by following wetting paths O–A–H and O–A (Figure 5.41), respectively. Then, eight soil specimens (ui1 to ui8) were isotropically consolidated to the net mean stresses given in Table 5.8 at a constant suction (H–I and A–E in Figure 5.41). After consolidation, these specimens were sheared at a constant water content (I–J and E–K). Four other soil specimens (ua1 to ua4) were anisotropically consolidated at a constant stress ratio (q/p) of 1.5 under constant suction of 150 kPa (O–A–E–F) where

$$p = \frac{\sigma_1 + \sigma_2 + \sigma_3}{3} - u_a \text{ and } q = \sigma_1 - \sigma_3.$$

After consolidation, these specimens were sheared under a constant deviator stress with a decreasing suction (F–G).

Behaviour of isotropically consolidated saturated decomposed volcanic soil

Initially unsaturated specimens during saturation

During saturation of the initially unsaturated specimens of the decomposed volcanic soil (wetting path O–A–B in Figure 5.41), substantial contractive volumetric strains were observed. The measured volumetric strains were between 12.9 and 14.1 per cent (Table 5.8) for initial specific volumes ranging from 2.419 to 2.464 (si1—si5). The approximate initial suctions of the compacted specimens were measured using the principle of axis translation. It was found that they were on the order of 150 kPa. Similar results were obtained in one-dimensional inundation tests (wetting tests) conducted with an oedometer and a Rowe cell on the same material (Chiu et al., 1998). It was found that volumetric strains of 13.5 and 12.5 per cent were recorded by the oedometer and the Rowe cell, respectively, for soil specimens compacted to the same initial conditions adopted in this study.

Wheeler and Karube (1995) postulated that the capillary effect arising from suction increases the normal forces at inter-particle contacts, from which a very loose state of the soil specimen could be maintained. When these forces vanish as the suction disappears during saturation, the inter-particle friction cannot hold together such a loose structure, and the soil particles start to slide over each other and lead to a large and irrecoverable change of volume.

The contractive volumetric strain induced by saturation (or wetting) may also be postulated by using the Loading–Collapse yield curve as suggested by Alonso et al. (1990). Before saturation, the decomposed volcanic soil may be considered as a virgin fill. In this instance, the term virgin fill refers to a soil that has never been subjected to suction greater than its current value (Chiu et al., 1998). Hence, the initial state (X in Figure 5.42) of this virgin fill should lie inside its current LC yield surface (LCI). For the saturation of an initially unsaturated soil specimen, XYZ can be the wetting path. According to Alonso et al. (1990), as the state of the soil lies inside its current yield surface, a reduction of suction (wetting) causes elastic swelling (path XY). As the stress state reaches the current position of the yield curve

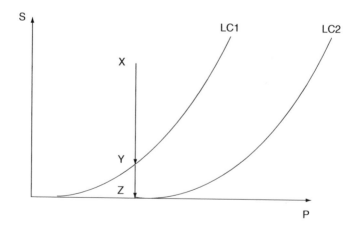

Figure 5.42 Postulated collapse behaviour of virgin decomposed volcanic soil during saturation (after Ng and Chiu, 2001).

at Y, the yield can be produced by a further reduction of suction (path YZ). As a result, plastic and irrecoverable strains would be expected, and LC2 becomes the new yield curve. The shape of the LC yield curve and the collapse postulation for the decomposed volcanic soil will be discussed later.

Isotropic compression of saturated specimens

After saturation, isotropic consolidation was carried out on specimens si1 to si5. Isotropic compression curves of the saturated specimens are given in Figure 5.43 along with the curves of four other soils for comparison. For the range of effective mean stress considered, the compressibility of the saturated decomposed volcanic soil is similar to a compacted kaolin (Wheeler and Sivakumar, 1995). It is more compressible than a compacted decomposed granitic gravelly sand (Lee and Coop, 1995), a collapsible silt (Maatouk et al., 1995), and a Toyoura sand (Ishihara, 1993). The high compressibility of decomposed volcanic soil reflects the well-graded nature of the soil. A yield stress of about 20 kPa is estimated for the decomposed volcanic soil from the Casagrande graphical method. For the range of applied effective mean stress, an isotropic normal compression line (NCL) can be identified. In addition, the decomposed volcanic soil resembles the behaviour of clay because of the occurrence of a distinct yield point and a post-yield NCL from the isotropic compression tests.

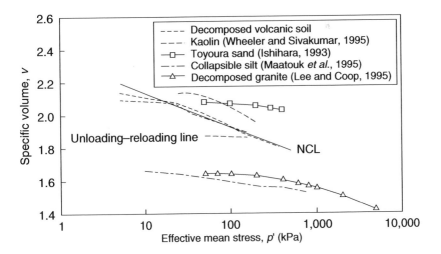

Figure 5.43 Isotropic compression curves for saturated specimens (after Ng and Chiu, 2001).

Shear behaviour of saturated specimens under undrained conditions

Figure 5.44a shows the stress–strain relationship of consolidated undrained tests on the saturated specimens. As expected, the deviator stress increases monotonically with the axial strain. The deviator stress also increases with the applied effective mean stress p'. The stress–strain curves indicate there is

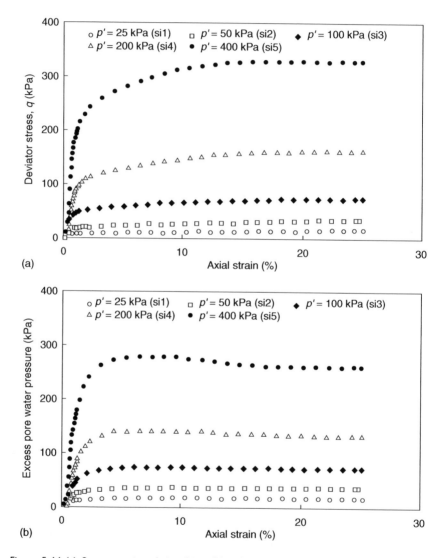

Figure 5.44 (a) Stress–strain relationships; (b) relationships between excess pore water pressure and axial strain for consolidated undrained triaxial tests on saturated test specimens (after Ng and Chiu, 2001).

ductile behaviour without any evidence of strain softening. Bulging failure is observed for all specimens at the end of the tests. It can be seen in the figure that a critical state is reached for all five tests at an axial strain in excess of 25 per cent, and hence a CSL can be determined later.

Figure 5.44b shows the relationship between the excess pore water pressure generated and the axial strain of consolidated undrained tests on the saturated specimens. Positive excess pore water pressure was recorded in all five undrained tests. In soil specimens consolidated to an effective mean stress smaller than 100 kPa (si1 and si2), the positive excess pore water pressure increases steadily with the axial strain and finally reaches a plateau at the end of the tests. In the other soil specimens (si3 to si5), the positive excess pore water pressure increases to a peak at a small strain, reduces from its maximum value as the strain increases continuously, and finally reaches a plateau at the end of the tests. The initial build-up of the positive excess pore water pressure suggests that the specimens exhibit contractive behaviour. The reduction in the positive excess pore water pressure after the peak indicates that the specimens change from contractive to dilative behaviour during shear, as illustrated by the effective stress paths shown in Figure 5.45a. All the effective stress paths show a similar trend. Each path moves towards the left-hand side initially until reaching a turning point, after which it turns right and finally reaches the CSL at the end of the test. This turning point is termed the point of phase transformation (Ishihara, 1993), which is defined as a temporary state of transition from contractive to dilative behaviour. This observed behaviour resembles the behaviour of typical clean sands. For the range of effective mean stress considered, the behaviour of the decomposed volcanic soil is different from the reverse behaviour of the loose silty sand reported by Yamamuro and Lade (1998), where static liquefaction is observed at low effective mean stresses and an increasing tendency of dilatancy with increasing effective mean stress is found. As no strain softening behaviour is observed, no static liquefaction would be expected from the decomposed volcanic soil compacted to an initial dry density higher than or equal to 70 per cent of relative compaction for the considered range of effective mean stress.

The critical states of saturated specimens can be represented by the CSL in the stress plane q–p' and the compression plane v–$\log p'$, as shown in Figure 5.45. The gradient of the CSL $M(s)$ at zero suction in the q–p'-plane is 1.317. This corresponds to a critical state angle of internal friction ϕ' of 33° in compression. The gradient of the CSL, $\psi(s)$, at zero suction in the v–$\log p'$ plane is 0.087, which is very close to that of NCL [$\lambda(s) = 0.089$ at zero suction]. Hence, the NCL is approximately parallel with the CSL for the range of effective mean stress considered.

Based on the test results, the saturated decomposed volcanic soil resembles clay in isotropic compression. On the other hand, it behaves like clean

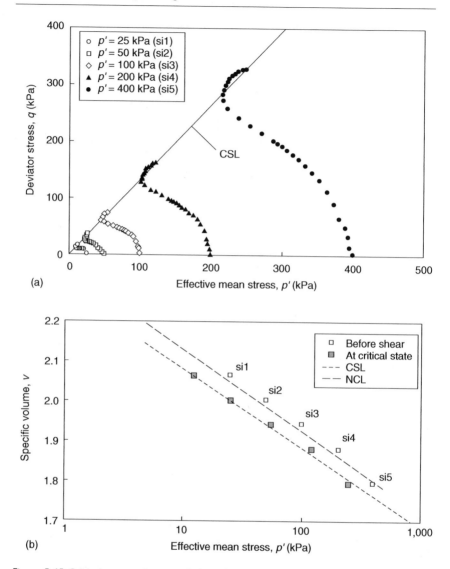

Figure 5.45 Critical states of saturated test specimens: (a) plane of deviator stress against effective mean stress; (b) plane of specific volume against effective mean stress (after Ng and Chiu, 2001).

sand during undrained shear. For the limited range of effective mean stress considered, although its states before shear lie on the loose side of the critical state, it contracts initially and then dilates without softening as strain increases.

Behaviour of isotropically consolidated unsaturated decomposed volcanic soil

Isotropic compression of unsaturated specimens

The isotropic compression curves of the unsaturated specimens at suctions of 80 and 150 kPa are shown in Figure 5.46, along with the curves of the saturated specimens. It can be seen from the figure that there is a clear yield point in each isotropic compression curve, identified by a marked change in the slope of the plot of specific volume against the logarithm of the net mean stress. They show similar behaviour to that observed in saturated clays, where a yield stress and a post-yield NCL are found. Further inspection of the data from the figure reveals that the isotropic NCLs may be represented by the following:

$$v = N(s) - \lambda(s) \ln\left(\frac{p}{p_{at}}\right) \tag{5.22}$$

where $\lambda(s)$ = gradient of the isotropic normal compression hyperline (Wheeler and Sivakumar, 1995); and $N(s)$ = specific volume p_{at}, which is a reference pressure taken as 100 kPa. The isotropic normal compression hyperline describes a locus of isotropic compression states defined in a three-dimensional space of $\{v : p : s\}$. The gradients of the isotropic normal compression hyperline $\lambda(s)$ are found to be a function of suction and are shown in Figure 5.47, together with $\lambda(s)$ of the collapsible silt (Maatouk

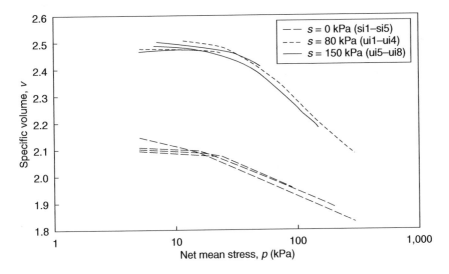

Figure 5.46 Isotropic compression curves for saturated and unsaturated test specimens (after Ng and Chiu, 2001).

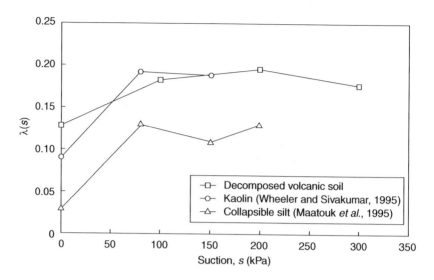

Figure 5.47 Relationships between gradients of isotropic normal compression hyperline and suction for some soils (after Ng and Chiu, 2001).

et al., 1995) and the kaolin (Wheeler and Sivakumar, 1995) for comparison. For suction ranging from 80 to 150 kPa, the value of $\lambda(s)$ of the decomposed volcanic soil shows relatively little variation with suction. However, a significant drop in the value of $\lambda(s)$ is observed as the suction reduces to zero. This variation of $\lambda(s)$ is consistent with the test results from the collapsible silt and the kaolin for the range of suction considered. Since there is a substantial reduction in volume due to saturation of an initially unsaturated specimen, the saturated specimen is much denser than the unsaturated one even though they are compacted to the same initial specific volume. As a result, the denser saturated specimen exhibits a stiffer response to compression and has a lower gradient of the isotropic NCL than the gradient of an unsaturated specimen. In addition, the degree of saturation for the decomposed volcanic soil lies between 35 and 51 per cent for suction ranging from 80 to 150 kPa. At such a low degree of saturation, the contribution of suction to stiffen the micro-structure of the unsaturated soil may be limited because of a small contact area of the menisci (air–water interface). Hence, the suction has a relatively limited effect on the compressibility of the unsaturated decomposed volcanic soil for the range of suction considered.

Yielding behaviour

The initial LC yield curve exhibited by the unsaturated specimens can be estimated by the yield stresses observed from the isotropic compression curves at suctions of 80 and 150 kPa only (Figure 5.46). This is because there was relatively small change in volume (Table 5.8, column 5) when the unsaturated specimens (ui1—ui8) were controlled to the above values of suction. This indicates that the initial LC yield curve remains unchanged. The yield stresses of the unsaturated specimens can be estimated from Casagrande's graphical method, and they are 35 kPa (ui2—ui4) and 40 kPa (ui6—ui8) for suctions of 80 and 150 kPa, respectively. On the other hand, collapse was recorded during the saturation of unsaturated specimens (si1—si5) where suction was reduced to zero. As a result, the initial LC yield curve appears to have zero yield stress at zero suction, as shown in Figure 5.48.

The experimental yield points measured from the isotropic consolidation tests and the corresponding LC yield curves for the collapsible silt (Maatouk et al., 1995) and the kaolin (Wheeler and Sivakumar, 1995) are shown in the figure for comparison. For decomposed volcanic soil, it appears that there is a relatively small increase in the yield stress for a change of suction from 80 to 150 kPa. This is consistent with the finding of the collapsible silt, from which the yield stress only increases from 30 to 60 kPa as the suction increases from 80 to 600 kPa. When compared with kaolin, the rate of increase in the yield stress with respect to suction is much lower for the decomposed volcanic soil at the range of suction considered. As kaolin is finer than the other two soils, the rate of desaturation is slower than the other two soils for the range of suction considered. As a result, the rate of losing the contact area of the menisci in kaolin is slower and the suction is more effective to stiffen the soil. Hence, kaolin exhibits a higher rate of

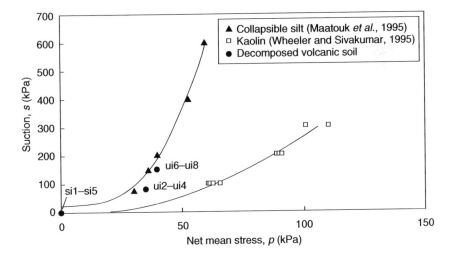

Figure 5.48 Experimental yield stresses and predicted LC yield curves (after Ng and Chiu, 2001).

increase in the yield stress with respect to suction compared with the other two soils. The yield stresses observed from kaolin are higher than those stresses of the other two soils at the same suction. This is attributed to the higher initial densities of the compacted kaolin. Based on the tests on a limited range of suction, it seems that the shape of the LC yield curve for the loosely compacted, decomposed volcanic soil may be approximated from the one predicted by Maatouk *et al.* (1995) for collapsible silt.

The shape of the postulated LC yield curve for the decomposed volcanic soil provides evidence to support the collapse postulation discussed previously for the observed volumetric contraction during saturation of the initially unsaturated specimens. The saturation was conducted at a low confining pressure, for example, 10 kPa. As the suction reduces to zero, the wetting path XYZ (Figure 5.42) should cross over the postulated initial LC yield curve, and yield is produced. As a result, plastic and irrecoverable volumetric strains are expected.

Unsaturated shear behaviour under constant water content

To study the 'undrained' behaviour of unsaturated loosely compacted specimens, the constant water content test is considered to be an appropriate one for resembling their rapid collapse behaviour. Figures 5.49 and 5.50 show the results of the constant water content tests obtained at initial suctions of 80 and 150 kPa, respectively. Similar behaviour is observed from the tests conducted at these values of suction. The deviator stress increases steadily

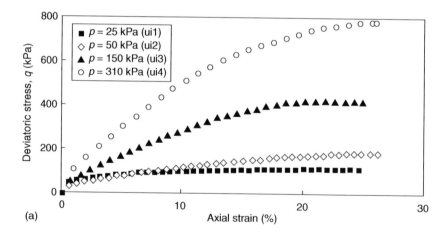

(a)

Figure 5.49 (a) Stress–strain relationships; (b) volumetric strain–axial strain relationships; (c) relationships between change of suction and axial strain; for constant water content tests on unsaturated specimens at initial suction of 80 kPa (after Ng and Chiu, 2001).

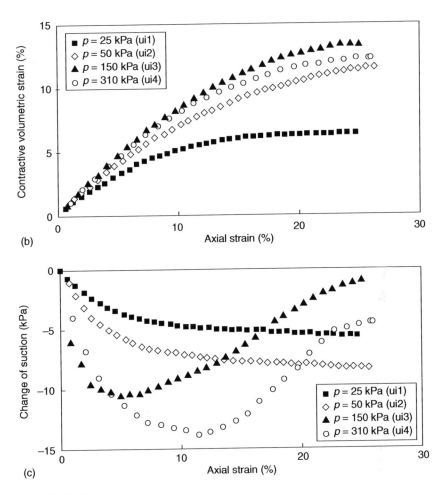

(b)

(c)

Figure 5.49 (Continued).

with axial strain and approaches almost a plateau at an axial strain over 25 per cent, with no evidence of strain softening (Figures 5.49a and 5.50a). The deviator stress also increases with the applied net mean stress. Only contractive volumetric behaviour is observed for all specimens (Figures 5.49b and 5.50b). For net mean stresses not greater than 150 kPa (ui1–ui3 and ui5–ui8), the amount of contraction increases with the net mean stress. As the net mean stress increases beyond 150 kPa (ui4), the amount of contraction decreases with an increase in the net mean stress. All the tests show a reduction in suction and the amount of reduction does not exceed 15 kPa (Figures 5.49c and 5.50c), which may reflect the low initial degree

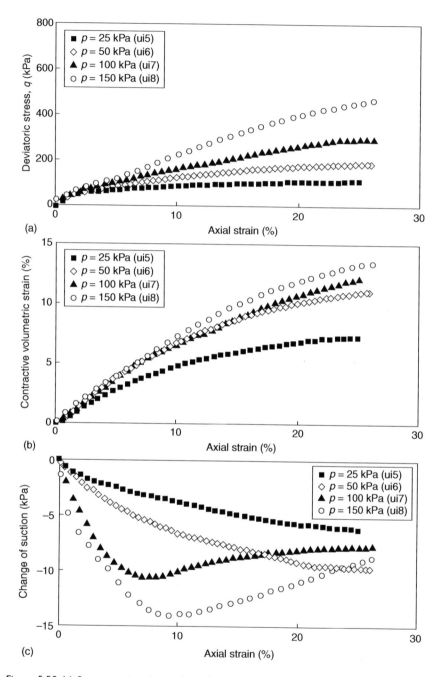

Figure 5.50 (a) Stress–strain relationships; (b) volumetric strain–axial strain relationships; (c) relationships between change of suction and axial strain; for constant water content tests on unsaturated specimens at initial suction of 150 kPa (after Ng and Chiu, 2001).

of saturation of the unsaturated specimens ($S_r = 35$–51 per cent). The measured suctions at 25 per cent axial strain range from 72 to 79 kPa and from 140 to 144 kPa for suctions of 80 and 150 kPa, respectively (Table 5.8). Two distinct patterns for the variation of suction with the axial strain can be observed. For net mean stresses not greater than 50 kPa (ui1, ui2, ui5 and ui6), the suction reduces steadily with axial strain and approaches a steady value at the end of the test. In other specimens (ui3, ui4, ui7 and ui8), initially the suction decreases to a minimum at small strains. As the axial strain increases further, the suction increases from its minimum value. It continues to increase towards the end of the test at axial strain of about 25 per cent.

The observed contractive volumetric behaviour may be explained by considering the fabric of an unsaturated compacted soil. Croney et al. (1958) suggested that the clay compacted on the dry side of optimum exists as saturated packets of clay, which contain small and water-filled intra-packet voids, separated by large inter-packet air voids. Similarly, the loosely compacted soil may possess an open fabric that is dominated by the comparatively large inter-packet air voids. In an unsaturated specimen consolidated to a low net mean stress (ui1–ui3 and ui5–ui8), its volumetric behaviour is mainly governed by the large inter-packet air voids. Although change in water content is not allowed during shear, it could be possible for the specimen to change in volume by expelling the pore air and rearranging the soil packets. As the net mean stress increases, the specimen contracts more by expelling more pore air (Figures 5.49b and 5.50b). On the other hand, an unsaturated specimen consolidated to a high net mean stress (ui4) exhibits a more densely packed structure after consolidation. Even though its overall volume contracts during shear, it has an increasing tendency to dilate as the net mean stress increases because of the interlocking effect between the more densely packed soil packets during shear. As a result, the overall volumetric contraction decreases with an increase in net mean stress (see ui4 in Figure 5.49b).

The interplay between inter-packet air voids and the saturated soil packets may also be used to explain the observed change in suction. Since the tests were carried out under constant water content and volumetric contraction was observed (Figures 5.49b and 5.50b), it may be expected that suction could decrease (ui1, ui2, ui5 and ui6) during shear. For other specimens (ui3, ui4, ui7 and ui8), the initial decrease in suction is also associated with the overall volumetric contraction during shear at the constant water content. These specimens were consolidated to higher net mean stress. As a result, they are denser than those consolidated to lower net mean stress. As the volumetric contraction continues further, these specimens could reach a state in which the saturated soil packets play a more dominant role than the inter-packet air voids. Even though the overall volume contracts, the individual soil packets may dilate. As a result, the suction may not decrease further but commence to increase from the minimum.

Figures 5.49 and 5.50 show that most of the tests only approach the critical state but do not actually reach it even at strains in excess of 25 per cent. However, by inspecting the stress–strain and the volumetric–axial strain relationships in both figures, the rates of change in deviator stress and volume reduce towards the end of the test, and the state at the end of each test may be used to approximate the critical state. Figure 11 shows the approximated critical states plotted in the planes of q–p and v–log p for the isotropically consolidated specimens. The critical state values may be represented by the following two equations for the range of suction considered:

$$q = M(s) + \mu(s) \tag{5.23}$$

$$v = \Gamma(s) = \psi(s) \ln \left(\frac{p}{p_{at}} \right) \tag{5.24}$$

where $M(s)$ and $\mu(s) =$ gradient and intercept of the critical state hyperline in the q–p plane, respectively; $\psi(s) =$ gradient of the same hyperline in the $v - \ln p$ plane; and $\Gamma(s) =$ specific volume at p_{at}. Equations (5.23) and (5.24) are called the critical state hyperline by Wheeler and Sivakumar (1995), which defines the critical states in a four-dimensional space of $\{v : p : q : s\}$. It can be found that the critical state for all unsaturated specimens may be represented by a single CSL despite their final suction ranging from 72 to 144 kPa (Table 5.8, column 10).

In Figure 5.51a, the approximated CSL for the unsaturated specimens appears to intersect the one for the saturated specimens at a net mean stress of about 720 kPa. Maatouk *et al.* (1995) reported that CSLs of unsaturated specimens at suction ranging from 150 to 600 kPa also converge towards the CSL of saturated specimens at a net mean stress of about 700 kPa in the collapsible silt. The figure also shows that the states of all unsaturated specimens before shear lie either close to or on the loose side of the corresponding CSL. According to the extended critical state framework, contractive behaviour is expected during shear if the initial state of the unsaturated specimen lies on the loose side of the corresponding critical state and the results of the constant water content tests agree with the predictions (Figures 5.49b and 5.50b).

The CSL for unsaturated specimens in the q–p plane has a gradient $M(s)$ of 1.327 and an intercept $\mu(s)$ of 33 kPa. In addition, this line is approximately parallel to the one for saturated specimens. This may suggest that the gradient (i.e. the angle of internal friction ϕ') of the CSL in the q–p plane is a constant for the limited range of applied suction and suction does not appear to affect ϕ'. Gan and Fredlund (1996) also reported that ϕ' is independent of the suction for a CDG and a completely decomposed fine

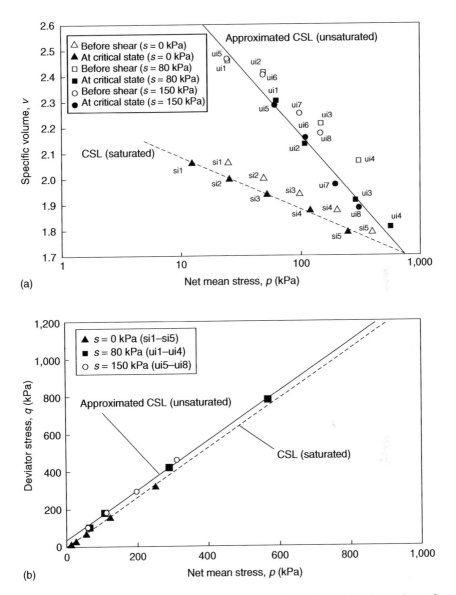

(a)

(b)

Figure 5.51 Critical states of isotropically consolidated specimens: (a) plane of specific volume against net mean stress; (b) plane of deviator stress against net mean stress (after Ng and Chiu, 2001).

ash tuff. As ϕ' is a strength parameter related to the factional characteristic of the inter-particle contacts, which is an intrinsic property of the soil, it may be independent of the stress state variable, such as suction. However, other researchers (Maatouk *et al.*, 1995; Wheeler and Sivakumar, 1995) have shown that ϕ' is a function of the suction by considering a wider range of suction.

The intercept $\mu(s)$ of 33 kPa reflects the contribution of the suction on the shear strength. For the decomposed volcanic soil, the suction makes a large contribution to its shear strength for suctions ranging from 0 to 79 kPa but it does not have a significant effect for suctions ranging from 72 to 144 kPa. The relationship between $\mu(s)$ and suction is shown in Figure 5.52, along with those relationships of kaolin (Wheeler and Sivakumar, 1995), and an undistributed decomposed fine ash tuff (Gan and Fredlund, 1996) for comparison. The figure shows that the decomposed volcanic soil and the decomposed fine ash tuff exhibit a non-linear $\mu(s)$–suction relationship, except kaolin. There is relatively little increase in the $\mu(s)$ when suction exceeds about 80 kPa for the decomposed volcanic soil and the decomposed fine ash tuff. On the contrary, there is a reduction in the $\mu(s)$ for the decomposed fine ash tuff as the suction increases beyond around 150 kPa. A similar non-linear relationship was also observed by other researchers (Maatouk *et al.*, 1995; Gan and Fredlund, 1996). Maatouk *et al.* (1995) postulated that the effect of suction on shear strength is limited to a range of net stress for the collapsible silt.

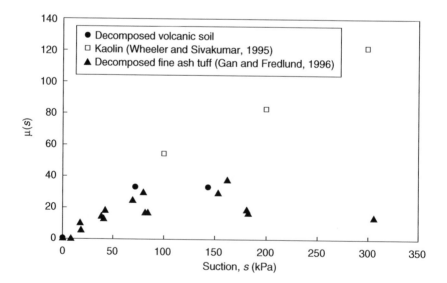

Figure 5.52 Relationship of suction and $\mu(s)$ for some soils (after Ng and Chiu, 2001).

Behaviour of an isotropically consolidated unsaturated decomposed volcanic soil

Figure 5.53a shows the variation of axial strain with suction for the field stress path tests, which simulate rainfall infiltration, conducted on anisotropically consolidated unsaturated specimens. The deviator stress and the net mean stress were kept constant while the suction was decreased. As the suction decreases from an initial value of 150 to about 80 kPa, a small axial strain (< 4 per cent) is mobilized for specimens consolidated to a net mean stress > 25 kPa (ua2–ua4). As the suction continues to decrease, the rate of increase in axial strain accelerates towards the end of the test. For the other specimen (ua1), the mobilized axial strain gradually increases as the suction decreases. It may be seen from the figure that the specimen approaches failure at higher suction for a higher applied net mean stress. This observation could be explained by the concept of state boundary surface and will be discussed later.

Figure 5.53b shows the variation of volumetric strain with suction for the field stress path tests. Relatively small volumetric strain is mobilized as the suction decreases from an initial value of 150 to about 80 kPa. As the suction continues to decrease, contractive behaviour is observed for soil specimens consolidated to net mean stress smaller than 100 kPa (ua1 and ua2) but dilative behaviour is observed for the other two specimens (ua3 and ua4). The anisotropically consolidated unsaturated specimens change from contractive to dilative behaviour as the applied net mean stress increases.

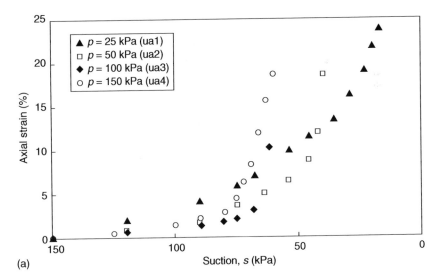

(a)

Figure 5.53 Relationship between (a) axial strain and suction (b) volumetric strain and suction for stress path tests that simulate rainfall infiltration conducted on unsaturated specimens (after Ng and Chiu, 2001).

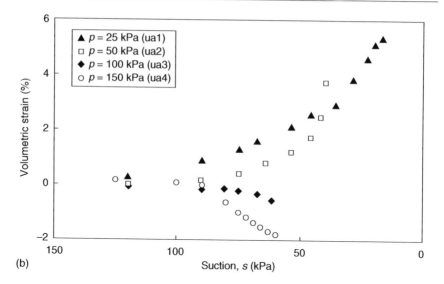

(b)

Figure 5.53 (Continued).

A similar trend can also be seen in Figures 9c and 10c for isotropically consolidated specimens sheared under constant water content. This type of observed behaviour cannot be readily explained by existing elasto-plastic constitutive models extended for unsaturated soils (Alonso *et al.*, 1990; Wheeler and Sivakumar, 1995).

Figure 5.54 illustrates the states of unsaturated specimens in the plane of v–$\log p$ after anisotropic consolidation and at the end of the field stress path test. The CSLs for saturated specimens ($s = 0\,\mathrm{kPa}$) and unsaturated specimens ($s = 72$–$144\,\mathrm{kPa}$) are also shown in the figure. Two CSLs for $s = 15$–40 and 60–$65\,\mathrm{kPa}$ may be assumed to lie between the measured CSLs for $s = 0$ and 72–$144\,\mathrm{kPa}$. It is also assumed that the state at the end of the field stress path test can be approximated as the critical state. For specimens ua1 and ua2, the measured suctions at the approximated critical state are 17 and 40 kPa (Table 5.8, column 10), respectively. Since these two suctions are much lower than 72 kPa, it seems that their corresponding CSLs should lie close to the one for saturated specimens ($s = 0\,\mathrm{kPa}$). As a result, the stress states after anisotropic consolidation for specimens ua1 and ua2 may lie on the loose side of their corresponding CSLs and contractive behaviour would be expected. On the other hand, the measured suctions at the approximated critical state for specimens ua3 and ua4 are 62 and 60 kPa (Table 5.8), respectively. Since these two suctions are quite close to 72 kPa, it might be possible that their corresponding CSLs lie very close to the one for unsaturated specimens ($s = 72$–$144\,\mathrm{kPa}$). As a result, the stress states after anisotropic consolidation for specimens ua3 and ua4 might fall on the dense side of their corresponding CSLs and dilative behaviour would be expected. This postulation should be further verified by other experimental data.

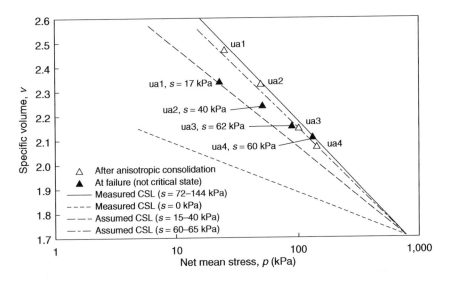

Figure 5.54 Relationship between specific volume and net mean stress for anisotropically consolidated unsaturated specimens subjected to stress paths that simulate rainfall infiltration (after Ng and Chiu, 2001).

Summary and conclusions

Three series of triaxial stress path shear tests were conducted on a loose decomposed volcanic soil (slightly sandy silt) compacted to an initial dry density of 70 per cent of the Proctor maximum ($v = 2.468$). The tests included consolidated undrained, constant water content and a reducing suction path under a constant deviator stress. The test results have provided insights into the fundamental behaviour of the loose fill.

An initially unsaturated specimen of the decomposed volcanic soil exhibits volumetric contraction of around 14 per cent during saturation. This substantial change in volume can be explained by the concept of the loading collapse yield surface.

In an isotropically consolidated saturated specimen lying on the loose side of the CSL, it appears that the specimen behaves like clay under compression as the occurrence of a yield point, and the post-yield NCL can be readily identified. The observed high compressibility is similar to kaolin. The NCL is parallel to the CSL. On the other hand, the saturated specimen resembles clean sand behaviour during undrained shear. For the limited range of effective mean stress considered, the specimen initially contracts and then dilates without strain softening as shear strain increases.

In an isotropically consolidated unsaturated specimen, it exhibits strain-hardening behaviour and possesses a non-linear shear strength–suction relationship. Suction makes a large contribution to the apparent cohesion and hence the shear strength of the fill for suctions ranging from 0 to 79 kPa, but it does not have any significant effect for suctions ranging from 72 to 144 kPa. The angle of internal friction appears to be independent of suction. Regarding volumetric behaviour, only contraction is observed during shearing at a constant water content. This amount of contraction increases with the net mean stress up to 150 kPa. As the net mean stress increases further, the amount of contraction decreases. Two distinct patterns are observed for the variation of suction during shearing at a constant water content. In specimens consolidated to a low net mean stress, the suction steadily reduces with axial strain and approaches a steady value at failure. In other specimens, the suction decreases to a minimum at small strains and increases from its minimum as the strain continues to increase. This seems to imply that there was a tendency of dilation in individual soil packets.

In an anisotropically consolidated unsaturated specimen subjected to a decreasing suction under a constant deviator stress, the specimen changes from contractive to dilative behaviour with increasing applied net mean stress. A similar trend can also be observed in isotropically consolidated specimens sheared under constant water content. This type of observed behaviour cannot be explained by existing elasto-plastic constitutive models extended for unsaturated soils.

Since dilative and ductile stress–strain behaviour is observed in the saturated fill, static liquefaction of the fill is very unlikely to occur. Passive stabilizing measures such as soil nails may be used to improve the stability of slopes, provided that sufficient soil movement is allowed to mobilize the friction between the fill and the reinforcement. On the other hand, the observed volumetric behaviour of the unsaturated fill is strongly influenced by the stress paths and the stress conditions. Volumetric contraction was observed when it was sheared under constant water contents for the ranges of net mean stress and suction considered. This seems to imply that the possibility of static liquefaction cannot be completely ruled out if the rate of air escape from voids in the soil is slower than the rate of volumetric deformation of the soil having a high degree of saturation.

Laboratory study of a loose saturated and unsaturated decomposed granitic soil (Ng and Chiu, 2003a)

Soil type and specimen preparation

The soil used in this study was a decomposed granitic soil taken from a quarry located in Cha Kwo Ling, Hong Kong. The in situ water content of the soil was 15 per cent. The soil tended to form aggregates of larger than 20 mm in diameter in the natural state. In order to break up the aggregates through constant energy, the free fall method used by the Japanese Society of Soil Mechanics and Foundation Engineering (JSSMFE, 1982; Lee and Coop, 1995) was adopted. Soil particles over 5 mm were discarded to avoid the need for an excessively large specimen size (Head, 1992). The discarded coarser particles accounted for about 2 per cent of the total specimen.

The classification tests on the soil were conducted in accordance with the procedures described in BS1337 (BSI, 1990). The test results are summarized in Table 5.10 together with the particle size distribution shown in Figure 5.55a. The soil comprises 21 per cent gravel, 53 per cent sand, 12 per cent silt and 14 per cent clay, which can be described as a silty, clayey, gravelly SAND (GCO, 1988) and classified as a coarse-grained and gap-graded soil. The SWCC of the decomposed granitic soil may be estimated from a similar coarse-grained decomposed soil from Butterfly Valley, Hong Kong (Gan and Fredlund, 1997) because both soils have similar particle size distributions (refer to Figure 5.55a). The SWCC is shown in Figure 1b and the residual degree of saturation (S_r) is around 47 per cent.

Triaxial specimens 76 mm in diameter and 152 mm in height were prepared by the wet tamping method (dynamic compaction). The specimens were compacted with water contents (w) of 19 per cent and dry densities ranging from 74 to 84 per cent of the Proctor maximum, which corresponded to the specific volume ($v = 1 + e$) ranging from 1.852 to 2.099 and S_r ranging from 45 to 58 per cent. Thus, the initial S_r of the triaxial specimens of decomposed granitic soil were very close to the residual value (see Figure 5.55b).

The specimen was compacted directly on the base pedestal of the triaxial cell in order to avoid any disturbance caused by transportation to

Table 5.4 Characteristics of decomposed granitic soil (after Ng and Chiu, 2003a)

Index test	Decomposed granitic soil
Standard compaction tests	
Maximum dry density (kg/m^3)	1,670
Optimum water content	20
Grain size distribution	
Percentage of gravel	21
Percentage of sand	53
Percentage of silt	12
Percentage of clay	14
D_{10} (mm)	0.0015
D_{30} (mm)	0.220
D_{60} (mm)	1.400
Coefficient of uniformity (D_{60}/D_{10})	933
Coefficient of curvature [$(D_{30})^2/(D_{60} \times D_{10})$]	23
Specific gravity	2.61
Maximum void ratio	1.19

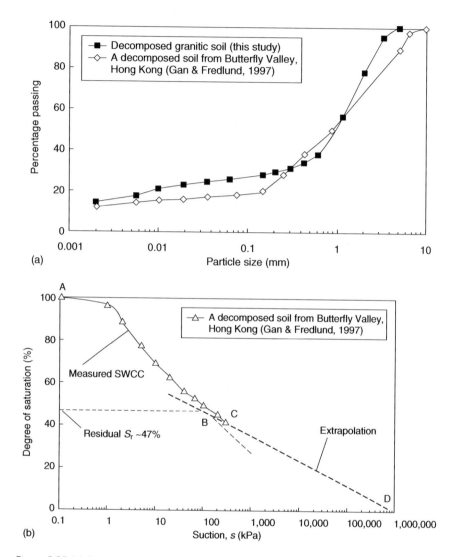

Figure 5.55 (a) Particle size distribution curves of decomposed granitic soil (this study) and coarse grained decomposed soil from Butterfly Valley, Hong Kong (Gan and Fredlund, 1997); (b) soil–water characteristic curve for coarse grained soil from Butterfly Valley, Hong Kong (Gan and Fredlund, 1997) (after Ng and Chiu, 2003a).

the triaxial cell. Each soil specimen was compacted in 10 layers using a compaction apparatus, which consists of a supporting frame and a falling hammer (Chiu, 2001). The weight and falling height of the hammer were controlled to ensure soil uniformity within each layer. The undercompaction method proposed by Ladd (1978) was adopted to prepare a uniform specimen. Specimen uniformity was verified by measuring the dry densities of three sections obtained from the centre, middle and bottom of a frozen compacted specimen. According to the results of the uniformity tests, the maximum difference in the dry densities was about ±0.9 per cent. This corresponded to a difference in specific volume of ±0.019.

Testing program and procedures

Three series of triaxial tests were conducted in this study. The tests consisted of undrained tests on saturated specimens, unsaturated constant water content tests and continuous wetting tests at constant deviator stress on unsaturated specimens. The undrained tests were carried out to establish reference states from which the test results from the unsaturated specimens were interpreted. The constant water content tests were used to study how pore pressures are generated during the collapse of a loose soil structure due to shear, in which the undrained condition of the pore water phase represented one of the critical scenarios. The wetting tests at constant deviator stress were used to simulate the stress path of a slope element subjected to rainfall infiltration (Brand, 1981). The stress paths of these tests are shown in Figure 5.56. The testing conditions are summarized in Tables 5.11 and 5.12.

All the specimens exhibited a suction of 20 kPa after compaction (refer to point O' in Figure 5.56). The stress path of the undrained tests (si1-g to si13-g) is shown as $O'C'D'E'F'$. First, an unsaturated specimen was saturated $(O'C'D')$ until the B-value achieved a minimum of 0.97. Subsequently, the specimen was subjected to isotropic compression $(D'E')$ ranging from 25 to 400 kPa. Then, the shearing tests were conducted under the undrained conditions $(E'F')$ for about 16 h.

The stress path for wetting tests at a constant deviator stress (ua1-g to ua4-g) is shown as $O'X'Y'A'B'$. Similar to constant water content tests, the unsaturated specimen was initially controlled to an initial suction of 40 kPa by following drying path $O'X'$. The specimen was subjected subsequently to isotropic compression $(X'Y')$ ranging from 25 to 150 kPa. Then, the deviator stress $(q = \sigma_1 - \sigma_3$, where σ is principal total normal stress) was increased (i.e. $Y'A'$) at a constant net mean stress, $p = ((\sigma_1 + \sigma_2 + \sigma_3)/3 - u_a)$, where u_a is the pore air pressure, until reaching a net stress ratio $(\eta = q/p)$ of 1.4. After applying the deviator stress, the specimen was wetted at the constant deviator stress by reducing the suction $(A'B')$. It took around 10–14 days to complete the wetting process.

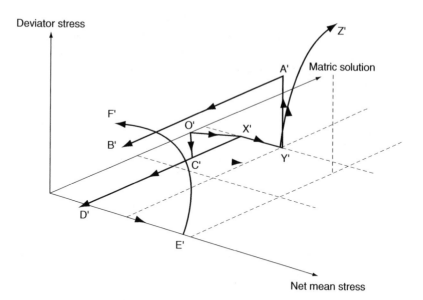

Figure 5.56 Various stress paths for triaxial tests conducted on loosely compacted decomposed granites soil (after Ng and Chiu, 2003a).

Table 5.11 Testing conditions for undrained tests on saturated specimens of decomposed granitic soil (after Ng and Chiu, 2003a)

Specimen identity	Initial percentage of Proctor maximum	Initial specific volume	Effective mean stress before shearing (kPa)	Specific volume before shearing	State parameter before shearing
si1-g	75	2.094	25.0	2.038	0.060
si2-g	80	1.949	25.0	1.928	−0.050
si3-g	75	2.080	49.4	1.961	0.059
si4-g	77	2.032	50.0	1.944	0.042
si5-g	81	1.933	50.0	1.879	−0.023
si6-g	84	1.852	50.0	1.813	−0.089
si7-g	75	2.093	100.0	1.927	0.101
si8-g	81	1.918	100.0	1.872	0.046
si9-g	75	2.083	200.0	1.857	0.107
si10-g	74	2.097	200.0	1.840	0.090
si11-g	81	1.931	200.0	1.820	0.070
si12-g	74	2.099	300.0	1.820	0.115
si13-g	80	1.942	400.0	1.755	0.081

Table 5.12 Testing conditions for undrained tests on saturated specimens of decomposed granitic soil (after Ng and Chiu, 2003a)

Specimen identity[a]	Specific volume at applied suction[d]	Degree of saturation at applied suction (%)	Net mean stress before shearing or wetting (kPa)	Specific volume before shearing or wetting	State parameter before shearing or wetting	Final specific volume	Final suction (kPa)	Final degree of saturation (%)
ui1-g[b]	2.087	43.5	25[e]	2.076	−0.143	2.067	33	44.3
ui2-g[b]	2.091	42.9	50[e]	2.053	−0.041	1.967	30	48.1
ui3-g[b]	2.083	43.1	100[e]	1.983	0.014	1.838	27	55.2
ui4-g[b]	2.089	43.1	150[e]	1.916	0.020	1.753	25	61.2
ua1-g[c]	2.085	43.3	25[f]	2.080	−0.139	2.084	15	45.7
ua2-g[c]	2.087	42.9	50[f]	2.037	−0.057	2.011	17	49.0
ua3-g[c]	2.096	42.7	100[f]	1.923	−0.046	1.894	10	56.5
ua4-g[c]	2.093	43.1	150[f]	1.829	−0.067	1.789	5	65.5

Notes
a All specimens were compacted to an initial dry density of 75% Proctor maximum.
b Constant water content test.
c Wetting test at constant deviator stress.
d Initial applied suction was 40 kPa.
e Shearing.
f Net stress ratio ($\eta = q/p$) of 1.4 was applied to the specimens before wetting.

Shear behaviour of saturated specimens

After saturation, the saturated specimens (si1-g to si13-g) were compressed isotropically to confining pressures ranging from 25 to 400 kPa and then sheared in undrained conditions. The conditions of the 13 specimens before shearing are summarized in Table 5.11.

Figure 5.57a shows the typical stress–strain relationship of the undrained tests on five representative saturated specimens. For clarity, not all the 13 test results are illustrated. To compare the influence of different initial effective confining pressures before shearing, three of the specimens (si5-g, si8-g and si9-g) were compressed to three different initial effective confining pressures ranging from 50 to 200 kPa but similar specific volumes ($v = 1.857$–1.879). For the specimen compressed to low effective confining pressures, p' (si5-g), or lying on the dense side of the critical state (refer to Figure 5.59a; the determination of the critical state and the CSL for the saturated specimens will be discussed later), the deviator stress increases monotonically with the axial strain and approaches a steady value at the end of the test. No significant strain-hardening behaviour is observed for saturated specimens compacted to an initial dry density of around 75 per cent of the Proctor maximum and compressed to an effective confining pressure of 50 kPa (si5-g) on the dense side (dry) of the critical state. For specimens compressed to high effective

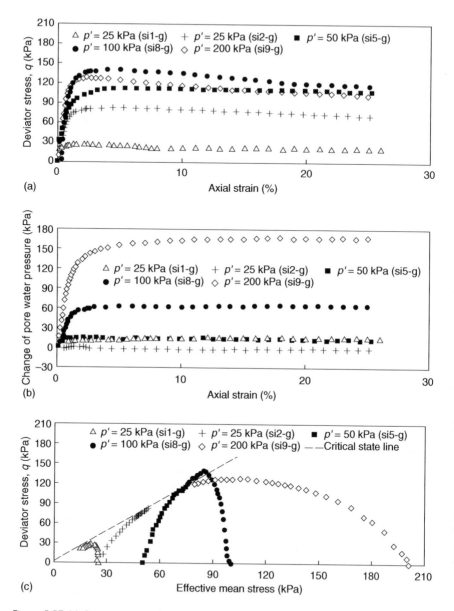

Figure 5.57 (a) Stress–strain relationships; (b) relationships between excess pore water pressure and axial strain; (c) effective stress paths for undrained tests on saturated decomposed granitic soil specimens (after Ng and Chiu, 2003a).

confining pressure (si8-g and si9-g) or lying on the wet (loose) side of the critical state (refer to Figure 5a), strain-softening behaviour is observed. The deviator stress reaches a peak value at a low axial strain. It then reduces from the maximum as the strain further increases and continues to drop but at a much reduced rate towards the end of the test. The ratio between the drop in the shear strength at the end of the tests and the peak strength for specimens si8-g and si9-g are 16 and 20 per cent, respectively. The observed strain-softening behaviour implies that a potential static liquefaction may occur in loose fill slopes formed with decomposed granitic soil in Hong Kong during rainstorms (Cheuk *et al.*, 2001).

To investigate the dependence of stress–strain behaviour on the initial state (i.e. void ratio) of soil with respect to the dry (dense) and wet (loose) side of the CSL in v–log p' space at the same initial confining pressure $p' = 25\,\text{kPa}$ (see Figure 5.59a), the observed results from two specimens (si1-g and si2-g), which were compressed to different initial specific volumes, are also shown in Figure 5.57. As expected, the soil specimen located at the dry side (si2-g) of the CSL exhibits a higher shear resistance than that located at the wet side (si1-g) of the critical state, due to the tendency of dilation during undrained shear.

Figure 5.57b shows the typical relationship between the generated excess pore water pressure generated and the axial strain. Positive excess pore water pressure is generated for all saturated specimens. For specimens lying on the dense side of the CSL (si2-g and si5-g), the positive excess pore water pressure increases to a peak value at a low axial strain and then drops to an approximate constant value as the strain increases continuously. For specimens lying on the loose side of the CSL (si1-g, si8-g and si9-g), the positive excess pore water pressure increases steadily with the axial strain and finally reaches a plateau at the end of the tests.

Figure 5.57c shows typical effective stress paths of the undrained tests in the $p' : q$ stress plane. The volumetric behaviour of the specimens can be indirectly reflected from the shape of the effective stress paths. The volumetric behaviour of the specimen is influenced by the state. For specimens lying on the dense side of the critical state (si2-g, si5-g), the effective stress path moves towards the right-hand side. This is an indication of the tendency to dilate. For specimens located on the loose side of the critical state (si1-g, si8-g and si9-g), the effective stress path moves towards the left-hand side, which is a sign of the tendency to contract. Even though the decomposed granitic soil contains some quantity of fines, the shear behaviour of the 13 tests on saturated specimens is similar to that of the clean sands for the ranges of density and pressure considered in this paper. Recently, Cheuk (2001) and Fung (2001) carried out a significant number of tests on saturated granitic soils obtained from two different sites, and they observed similar shear behaviour. Although the undrained tests do not exhibit phase transformation, phase transformation was observed in undrained tests on decomposed granitic soil taken from another site and compacted with a higher degree of relative compactions ranging from 87 to 93 per cent (Fung, 2001).

The dilatancy of the saturated soil can be examined by studying the direction of the plastic strain increment, $d\varepsilon_v^p/d\varepsilon_q^p$, from the drained test, where $d\varepsilon_v^p$ and $d\varepsilon_q^p$ are plastic components of the volumetric strain increment $(d\varepsilon_v)$ and the shear strain increment $(d\varepsilon_q)$, respectively. In the undrained test, there is no volume change and the volumetric strain increment equals zero. Consequently, the magnitude of $d\varepsilon_v^p$ equals that of the elastic component of the volumetric strain increment $(d\varepsilon_v^e)$. The latter can be related to the increment of effective mean stress (dp') by Hooke's law:

$$d\varepsilon_v^p = -d\varepsilon_v^e = -\frac{\kappa}{vp'}dp' \qquad (5.25)$$

where κ is the elastic stiffness parameter for saturated soil and v is the initial specific volume. Based on the unloading path of isotropic compression tests, κ is estimated to be 0.11. The elastic component of the shear strain increment $(d\varepsilon_q^e)$ is relatively small and is assumed to be negligible. Therefore, $d\varepsilon_q^p$ can be approximated by

$$d\varepsilon_q^p \approx d\varepsilon_q \qquad (5.26)$$

Now, Equations (5.25) and (5.26) can be used to evaluate the dilatancy measured from the undrained test of a saturated specimen.

Different stress–dilatancy relationships have been proposed in the literature for saturated granular materials. Rowe (1962) postulated that dilatancy is related only to the effective stress ratio and the angle of soil friction. However, experimental evidence revealed that the dilatancy of granular materials also depends on the material state. Bolton (1986) proposed a parameter called the relative dilatancy index, which depends on both the relative density and the confining pressure. Ishihara (1993) proposed the use of the state index, which depends on the quasi steady state line and the upper reference line. Recently, several state-dependent dilatancy relationships have been proposed (Manzari and Dafalias, 1997; Wan and Guo, 1998; Li and Dafalias, 2000). Li and Dafalias (2000) postulated the following state-dependent dilatancy for saturated sands:

$$\frac{d\varepsilon_v^p}{d\varepsilon_q^p} = d_0\left(e^{m\psi} - \frac{\eta'}{M}\right) \qquad (5.27)$$

where η' is the effective stress ratio (q/p'), M is the gradient of the CSL in the $p' : q$ stress plane, ψ is state parameter defined by Been and Jefferies (1985), and d_0 and m are material parameters.

Following Equation (5.27), the dilatancy is plotted as a function of effective stress normalized by M (i.e. η'/M) and shown in Figure 5.58. The

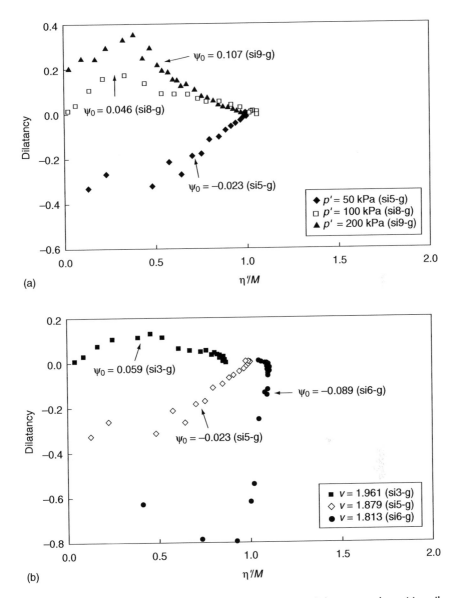

Figure 5.58 Dilatancy versus η'/M for saturated specimens of decomposed granitic soil compressed to (a) different confining pressures; (b) different specific volumes (after Ng and Chiu, 2003a).

saturated specimens are compressed to different states (or state parameter, ψ) before shearing. It is apparent that the dilatancy of the saturated specimens depends on the state. For specimens lying on the loose side of the critical state (or a positive ψ, e.g. si3-g, si8-g and si9-g), the dilatancy increases with the effective stress ratio until reaching a peak. Thereafter the dilatancy decreases as the effective stress ratio further increases. For specimens located on the dense side of the critical state (or a negative ψ, e.g. si5-g and si6-g), the dilatancy increases monotonically with the effective stress ratio.

Most of the undrained shear tests approach the critical state but do not actually reach it even at strains of about 25 per cent (Figure 5.57). However, by inspecting the stress–strain relationship and the change of pore water pressure with respect to axial strain, the rate of change in the deviator stress reduces, and the excess pore water pressure remains constant towards the end of the tests. Hence, the state at the end of the tests may be used to approximate the critical state. The approximated critical states of the saturated specimens can be represented by a CSL in the $q : p'$ stress plane and the $v : \ln p'$ compression plane as shown in Figure 5.59. The gradient of the CSL of the saturated specimens in the $q : p'$ plane is 1.549. This

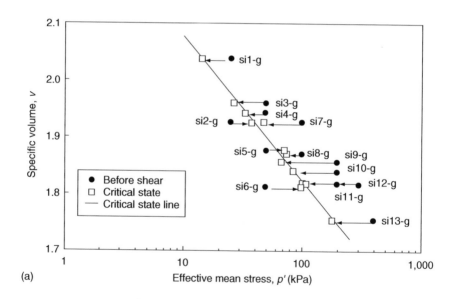

(a)

Figure 5.59 Critical states of saturated specimens of decomposed granitic soil in plane of (a) specific volumes and effective mean stresses; (b) deviator stresses and effective mean stresses (after Ng and Chiu, 2003a).

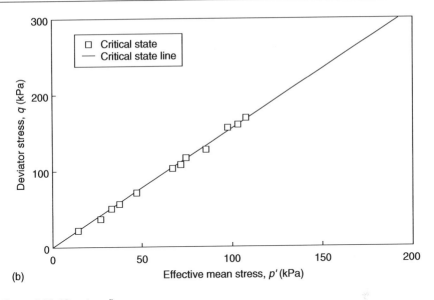

Figure 5.59 (Continued).

corresponds to a critical state angle of internal friction (ϕ') of 38° under compression. The gradient of the CSL in the $v : \ln p'$ plane is 0.11.

Based on the 13 undrained tests on saturated soil specimens as shown in Figure 5.59, it seems that the critical states of the specimens lie very close to a single straight line in both the v–log p' and q–p' planes and they seem to illustrate the uniqueness of a CSL for the decomposed granite.

Shear behaviour of unsaturated specimens

After applying an initial suction of 40 kPa, four unsaturated specimens (ui1-g to ui4-g) were isotropically compressed and then sheared at a constant water content. The conditions of the specimens before shearing are summarized in Table 5.12. Figure 5.60 shows that the deviator stress increases steadily with axial strain and almost approaches a plateau at an axial strain over 25 per cent, with no evidence of strain softening. The deviator stress also increases with the applied net mean stress.

Figure 5.60b shows that only contractive volumetric behaviour is observed for all specimens and the amount of contraction increases with the applied net mean stress. For specimens compressed to a net mean stress of 25 kPa (ui1-g) or located on the dense side of the critical state (refer to Figure 5.62a; the determination of the critical state and the CSL for unsaturated specimens will be discussed later), the amount of contraction increases to a peak value at a low axial strain and then reduces from the maximum to an approximated constant value with increasing axial strain. This is a sign of the tendency to dilate after reaching the peak value. For other specimens (ui2-g to ui4-g), the amount of contraction increases monotonically with the axial strain to a steady value.

The volumetric behaviour may be governed by the compression of inter-particle air voids and the interlocking effect between individual soil particles during shearing. For specimens compressed at a low net mean stress (ui1-g), the state before shearing lies below and far away from the critical state line (refer to Figure 5.62a). This specimen behaves like a dense material. The volumetric compression of the inter-particle air voids may be compensated by the volumetric expansion due to the interlocking between the soil particles. As a result, the amount of the overall volumetric contraction is relatively small. For specimens compressed to higher net mean stress (ui4-g), the state before shearing lies above the CSL (refer to Figure 5.59a). The specimen behaves like a loose material. During shearing, the relative large inter-particle air voids can be compressed, and individual soil particles may roll into these inter-particle air voids. Thus, the specimen exhibits substantial amounts of volumetric contraction. This implies that the design of any stabilization measures in loose granitic fill slopes and foundation structures resting on this type of material should take into account the significant shear-induced volume changes in the soil leading to potential serviceability problems.

Figure 5.60c shows that all the tests exhibit a reduction in suction, and the amount of reduction increases with the applied net mean stress. The suction reduces steadily with axial strain and approaches a steady value at the end of the test. The suctions at 25 per cent axial strain range from 25 to 33 kPa (refer to Table 5.12). As the applied net mean stress increases, the volumetric contraction of the specimen increases (refer to Figure 5.60b). As a result, the degree of saturation also increases and a lower value of suction is expected. Hence, a larger reduction in the suction is observed for a specimen compressed at a higher net mean stress.

Regarding the dilatancy for unsaturated specimens, it is assumed that the elastic components of the volumetric and shear strain increments are relatively small and are negligible. The dilatancy can thus be approximated by $d\varepsilon_v/d\varepsilon_q$. The dilatancy of unsaturated specimens from constant water content tests (ui1-g to ui3-g) is plotted as a function of net mean stress normalized by M (i.e. η/M, where η is the net stress ratio (q/p)) and shown in Figure 5.61. The dilatancy of three saturated specimens (si1-g, si3-g and si7-g), which were compressed to similar confining pressures, are also shown in the figure for comparison. Generally, the dilatancy of all soil specimens decreases as the normalized stress ratio increases. At zero suction, the magnitudes of η and η' are identical. It is apparent that the dilatancy observed at a suction of 40 kPa is greater than that at a suction of 0 kPa. For a given stress ratio and shear strain increment, unsaturated specimens sheared at suction of 40 kPa exhibit more contractive volumetric strain than saturated specimens sheared at zero suction. Similar results were also

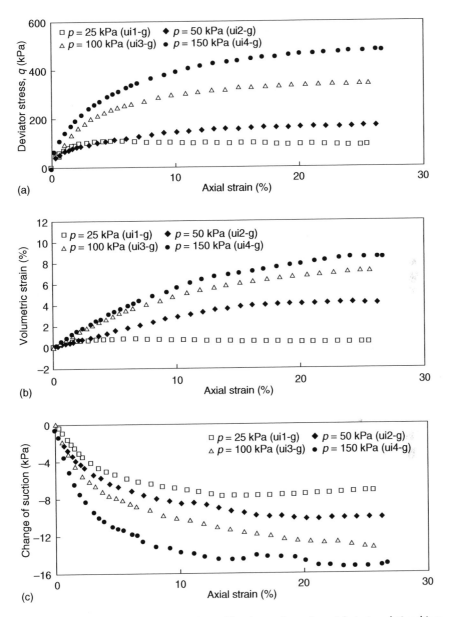

Figure 5.60 (a) Stress–strain relationships; (b) volumetric strain–axial strain relationships; (c) relationship between change of suction and axial strain for constant water content tests on unsaturated specimens of decomposed granitic soils (after Ng and Chiu, 2003a).

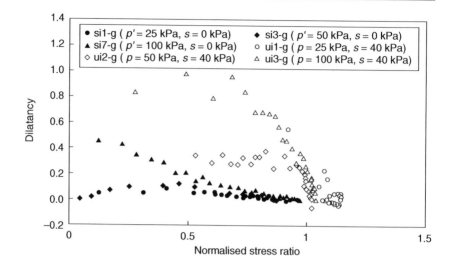

Figure 5.61 Dilatancy versus normalized stress ratio for specimens of decomposed granitic soils at suctions of 0 and 40 kPa (after Ng and Chiu, 2003a).

observed in a loose decomposed volcanic soil from Hong Kong (Ng and Chiu, 2001). Besides, unsaturated specimens compressed to different initial states (refer to the sixth column of Table 5.12) exhibit different amounts of dilatancy, in agreement with the test results on saturated specimens. An expression of a form similar to Equation (5.27) may be used to evaluate the dilatancy of unsaturated specimens, except that the material parameters d_0, and m may be suction dependent.

Most shear tests only approach the critical state but do not actually reach it even under strains in excess of 25 per cent. However, by inspecting the stress–strain and the volumetric–axial strain relationships in both Figure 5.60a and b, respectively, the rates of change in deviator stress and volume reduce towards the end of the test. The state at the end of each test may therefore be used to approximate the critical state. Figure 5.62 shows the approximated critical states plotted in the planes of $q : p$ and $v : \ln p$ planes for the unsaturated specimens. It can be seen that the critical state values may be represented by the following two equations for the range of suction considered:

$$q = M(s) + \mu(s) \tag{5.28}$$

$$v = \Gamma(s) - \omega(s) \ln\left(\frac{p}{p_{at}}\right) \tag{5.29}$$

Equations (5.28) and (5.29) are called the critical state hyperline by Wheeler and Sivakumar (1995). This hyperline defines the critical states in

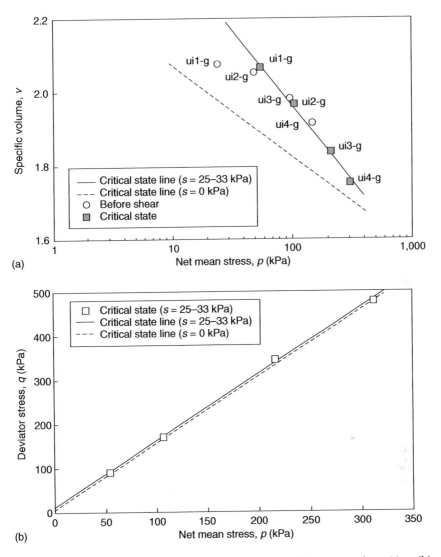

Figure 5.62 Critical state lines for unsaturated specimens of decomposed granitic soil in plane of (a) specific volume and net mean stress; (b) deviator stress and net mean stress (after Ng and Chiu, 2003a).

a four-dimensional space of $\{v : p : q : s\}$. $M(s)$ and $\mu(s)$ are the gradient and the intercept of the critical state hyperline in the $q : p$ plane, respectively. $\omega(s)$ is the gradient of the same hyperline in the $v : \ln p$ plane, and $\Gamma(s)$ is the specific volume at p_{at}, which is a reference pressure taken as 100 kPa. It is found that the critical state for all unsaturated specimens with an initial suction of 40 kPa may be represented by a single CSL (see Figure 5.62a) despite the final suctions ranging from 25 to 33 kPa (second to the last column of Table 5.12). The CSLs for $s = 0$ and the $s = 25$–33 kPa are not parallel, but they tend to converge as the net mean stress increases. Similar behaviour has also been observed by Wheeler and Sivakumar (1995, 2000) in their tests. A possible explanation of the difference between the saturated ($s = 0$) and the unsaturated (initial $s = 40$ kPa) soil specimens may be attributed to the difference in the soil fabric including the macro-structural arrangement of fines packets and inter-packet voids and the micro-structure within the individual packets. Note that the decomposed granite consists of 26 per cent of fines (see Table 5.9). The difference in the soil fabric (i.e. both the macro-structure and micro-structure) may be erased to different extents during shearing to the critical state. The higher the net mean stress during shearing, the greater the extents of the erasing effects, leading to the convergence of the two CSLs.

Figure 5.62b shows that the CSL for unsaturated specimens in the $q : p$ plane has a gradient ($M(s)$) of 1.535 and an intercept ($\mu(s)$) of 7 kPa. In addition, this line is approximately parallel to the one for saturated specimens. This may suggest that the gradient of the CSL (i.e. the angle of internal friction, ϕ', at the critical state) in the $q : p$ plane is a constant for the range of applied suctions. Hence, the suction does not appear to affect the angle of internal friction. Similar observations were obtained from other weathered soils including a Hong Kong decomposed volcanic soil (Ng and Chiu, 2001), a Korean decomposed granite (Lee and Coop, 1995), a Hong Kong decomposed fine ash tuff and granite (Gan and Fredlund, 1996). As ϕ' at the critical state is a strength parameter related to the frictional characteristic of the inter-particle contacts, which is an intrinsic property of the soil, it should be independent of the stress–state variables including as suction. The intercept ($\mu(s)$) of 7 kPa reflects the contribution of the suction on the shear strength. As the initial S_r of the unsaturated specimens is very close to the residual value (refer to Figure 5.55b), the suction has little effect on the inter-particle forces due to the low contact area of the meniscus. Thus, the contribution of suction to the shear strength is small for the decomposed granitic soil for the range of suction considered. This agrees with the results reported by Gan and Fredlund (1996). They investigated the effect of suction on the behaviour of an unsaturated decomposed granitic soils from Hong Kong, and their results showed that the shear strength increases initially with the suction and reaches a peak value. As the suction increases further (or reaches a residual value), the shear strength begins to decrease from the peak.

Wetting behaviour of unsaturated specimens

After isotropic compression, four unsaturated specimens (ua1-g to ua4-g) were sheared at constant net mean stress and suction ($s = 40$ kPa) until reaching a net stress ratio (η) of 1.4. Then, the specimens were wetted by

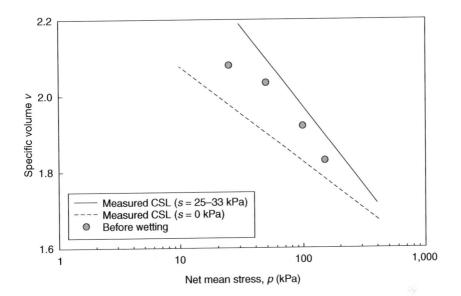

Figure 5.63 States of unsaturated specimens of decomposed granitic soil before wetting (after Ng and Chiu, 2003a).

reducing the soil suction while the deviator stress and the net mean stress were kept constant. The stress states of these specimens (ua1-g to ua4-g) before wetting are summarized in Table 5.12 and shown in Figure 5.63. It is noted that all the states lie below the CSL obtained from the constant water content tests ($s = 25$–$33 \, \text{kPa}$).

Figure 5.64a shows the variation of the axial strain with suction for the wetting tests at a constant deviator stress. As the suction decreases from an initial value of 40 to about 20 kPa, limited axial strain (less than 4 per cent) is mobilized. As the suction continues to decrease, the rate of increase in the axial strain accelerates towards the end of the test. There is a substantial increase in axial strain for the last decrement of suction for all specimens. All the tests were stopped because the travel limit of the bottom piston of the stress path apparatus was reached.

Sasitharan *et al.* (1993) conducted similar stress path tests (increasing pore water pressure under a constant deviator stress) but on loose saturated Ottawa sand. These stress path test results showed that after the pore water pressure had increased to a critical value (equivalent to a decrease in matric suction), a slight increase in the pore water pressure could initiate a catastrophic collapse of the saturated specimen. Sasitharan *et al.* (1993) also reported that it was not possible to record the test data due to the rapid collapse of the saturated specimen. However, such catastrophic collapse of the soil skeleton was not observed in the wetting tests of this study because

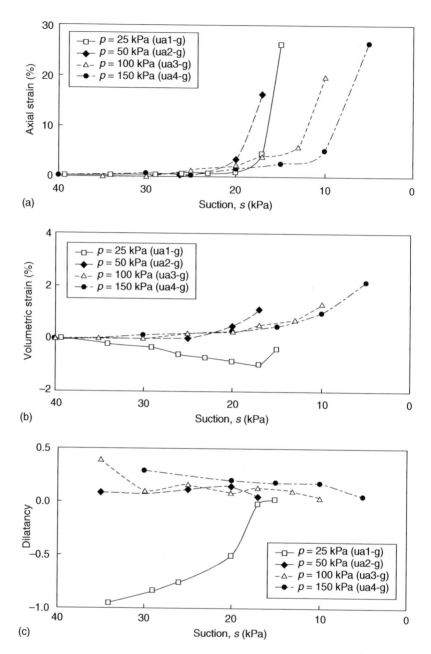

Figure 5.64 Relationships between (a) axial strain and suction; (b) dilatancy and suction for wetting tests at constant deviator stress conducted on unsaturated specimens of decomposed granitic soil (after Ng and Chiu, 2003a).

the time taken between failure and the last decrement of suction was about 12 h. This may due to the low S_r, which was about 46–66 per cent. At such a low S_r, the collapse of the soil skeleton of the loose unsaturated soil generates only a limited amount of excess pore water pressure due to the high compressibility of the air–water mixture. It takes this limited excess pore water pressure longer to affect the whole soil skeleton due to low permeability of the pore water phase. As a result, the volume reduces progressively and the soil becomes denser before reaching failure.

It is noted that the unsaturated specimens approach failure at lower suctions for higher applied net mean stresses. This agrees with the observation from the constant water content tests. As the confining pressure increases, the volumetric contraction of the specimen increases. As a result, the degree of saturation increases, which corresponds to a smaller value of suction.

Figure 5.64b shows the variations of volumetric strain with suction. For unsaturated specimens compressed to a net mean stress higher than 50 kPa (ua2-g and ua3-g), relatively small volumetric contraction is mobilized as the suction decreases from an initial value of 40 to about 25 kPa. As the suction continues to decrease, the rate of increase in volumetric contraction increases towards the end of the test. For other specimens (ua1-g), dilative behaviour is observed and the volume reaches a maximum at a suction of 17 kPa. As the suction further decreases, the specimen commences to contract and the amount of volumetric dilation reduces. All four specimens exhibit an increase in the rate of volumetric contraction upon the application of the last decrement of suction, no matter if the specimen exhibits an overall volumetric contraction or dilation. It seems that this increase in the rate of volumetric contraction can be attributed to the collapse of the soil skeleton. The figure also shows that the unsaturated specimens exhibit an increasing tendency towards volumetric contraction as the applied net mean stress increases. This agrees with the results of the constant water content tests on unsaturated specimens and of the undrained tests on saturated specimens. However, the magnitude of induced volumetric strains at collapse during the wetting test path is considerably smaller than that induced during the constant water content tests (see Figure 5.60b). This is because the wetting process (or reducing of soil suction) was conducted under a high constant net stress ratio (i.e. $\eta = 1.4$), which suppressed the amount of negative dilation (i.e. volumetric contraction) as observed in other results shown in Figure 5.61. This implies that anisotropic pre-shearing of a soil to a high net stress ratio appears to reduce the amount of contractive strains during the subsequent wetting. In other words, it seems that steep slopes may be less vulnerable than gentle slopes to static liquefactions.

Figure 5.64c shows the variation of dilatancy with suction. For specimens compressed to net mean stress higher than or equal to 100 kPa (ua3-g and ua-4g), the dilatancy reduces with the suction. This agrees with the results from the constant water tests. For other specimens (ua1-g and ua2-g), the dilatancy increases with the suction except for the last suction decrement of specimen ua2-g.

Summary and conclusions

Three series of triaxial stress path shear tests were conducted on a loose decomposed granitic soil (silty, clayey, gravely sand) compacted to initial dry densities ranging from 74 to 84 per cent of the Proctor maximum ($v = 1.852$–2.099). The tests included undrained tests on saturated specimens, constant water content tests and wetting tests at a constant deviator stress on unsaturated specimens. The test results have provided insights into the fundamental behaviour of the loose decomposed soil.

Saturated loose decomposed granitic soil behaves like clean sands during undrained shearing. Strain-softening behaviour is observed in a saturated specimen compacted to an initial dry density of around 75 per cent of the Proctor maximum and compressed to an effective confining pressure as low as 25 kPa. This corresponds to the range of specific volumes commonly found in the loose fill slopes in Hong Kong and the range of overburden pressure relevant to a shallow slope failure. The observed strain-softening behaviour implies that a potential static liquefaction may occur in loose fill slopes formed with decomposed granites during rainstorms in Hong Kong.

In unsaturated loose decomposed granitic soil sheared at a constant water content, a hardening stress–strain relationship and a relatively significant volumetric contraction behaviour are observed for the considered range of net mean stresses. This implies that the design of any stabilization measures in loose granitic fill slopes and of geo-structures should take into account the shear-induced volume changes in the soil, which may lead to potential serviceability problems. For the range of suction considered, the degree of saturation is close to the residual value. The suction therefore contributes little to the apparent cohesion of the unsaturated soil. The angle of internal friction at the critical state appears to be independent of the soil suction.

For unsaturated loose decomposed granitic soil wetted at constant deviator stress, the behaviour of the unsaturated soil changes from dilative to contractive with increasing net mean stress. The same wetting tests also show that the unsaturated soil fails at a degree of saturation far below full saturation, and the suction at failure decreases with an increase in the net mean stress. The magnitude of induced volumetric strains at collapse during the wetting test path is considerably smaller than that induced during the constant water content tests. This is because the wetting process was conducted under a high constant net stress ratio (i.e. $\eta = 1.4$), which suppressed the amount of volumetric contraction. This implies that anisotropic pre-shearing of a soil to a high net stress ratio appears to reduce the amount of contractive strains during the subsequent wetting. In other words, it seems that steep slopes may be less vulnerable to static liquefactions than gentle slopes.

Based on a limited number of tests, the dilatancy of the unsaturated soil depends on the suction, the state and the stress path. For unsaturated soil compressed to a net mean stress higher than or equal to 100 kPa, the dilatancy increases with the suction in both the constant water content test and the wetting test at a constant deviator stress. In unsaturated soil compressed to a net mean stress lower than 100 kPa, the dilatancy increases with suction in the constant water content tests but decreases with suction in the wetting tests at constant deviator stress.

Chapter 6

Measurement of shear stiffness

Sources

This chapter is made up of verbatim extracts from the following sources for which copyright permissions have been obtained as listed in the Acknowledgements.

- Cabarkapa, Z., Cuccovillo, T. and Gunn, M. (1999). Some aspects of the pre-failure behaviour of unsaturated soil. *Proc. Conf. on Pre-failure Deformation Characteristics of Geomaterials*, Jamiolkowski, Lancellotta and Lo Presti (eds), Balkema, Rotterdam, 1, pp. 159–165.
- Mancuso, C., Vassallo, R. and d'Onofrio, A. (2000). Soil behaviour in suction controlled cyclic and dynamic torsional shear tests. *Proc. Conf. on Unsaturated Soils for Asia*, Rahardjo, Toll and Leong (eds), Balkema, Rotterdam, pp. 539–544.
- Mancuso, C., Vassallo, R. and d'Onofrio, A. (2002). Small strain behaviour of a silty sand in controlled-suction resonant column-torsional shear tests. *Can. Geotech. J.*, 39, 22–31.
- Ng, C.W.W. and Yung, S.Y. (2007). Determinations of anisotropic shear stiffness of an unsaturated decomposed soil. Provisionally accepted by *Géotechnique*.
- Ng, C.W.W., Pun, W.K. and Pang, R.P.L. (2000b). Small strain stiffness of natural granitic saprolite in Hong Kong. *Proc. J. of Geotech. Geoeviron. Eng. ASCE*, 819–833.
- Qian, X., Gray, D.H. and Woods, R.D. (1993). Voids and granulometry: effects on shear modulus of unsaturated sands. *J. Geotech. Eng. ASCE*, 119(2), 295–314.

Introduction (Ng et al., 2000b)

One of the most distinctive features of soils is their non-linear and stress-path dependent deformation characteristics and stiffness properties. Most soils behave non-linearly even at very small to small strains (0.001–1 per cent). The non-linear small strain deformation characteristics and properties of dry and saturated sands and clays are well documented in geotechnical engineering and soil testing literatures. This is due to the recent advances in techniques of in situ and laboratory testing, field monitoring and numerical modelling, which have led to a great improvement in understanding of ground response to stress changes due to underground constructions such as excavations and tunnels (Burland, 1989; Mair, 1993; Ng and Lings, 1995; Atkinson, 2000). It is now widely accepted that the stress–strain relationship of dry and saturated soils is highly non-linear (Atkinson and Sällfors, 1991; Simpson, 1992; Tatsuoka and Kohata, 1995). Construction of retaining walls, foundations and tunnels in saturated stiff soils often results in a range of strains over which there is a large variation of soil stiffness (Mair, 1993; Ng and Lings, 1995). A summary of typical mobilized strain ranges in saturated stiff soils under working load conditions is given in Figure 6.1. Understanding of soil response at low levels of strain is clearly very important, not only for earthquake problems but it also has its practical significance and applications for designs and analyses of earth retaining structures subjected to static loads. For ease of identification, Atkinson

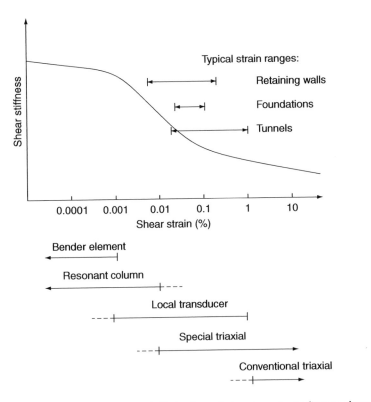

Figure 6.1 Approximate strain limits for soil structures in medium to dense and stiff soils (after Mair, 1993).

and Sällfors (1991) define three ranges of strains: very small strains (0.001 per cent or less), small strains (between 0.001 and 1 per cent) and large strains (larger than 1 per cent). These definitions are adopted in this chapter.

Although substantial research work on soil stiffness has been carried out by using saturated reconstituted and natural sedimentary soils, sands and soft rocks (Jardine et al., 2004; Clayton and Khatrush, 1986; Richardson, 1988; Dobry, 1991; Ng et al., 1995; Stokoe et al., 1995; Tatsuoka and Kohata, 1995; Lo Presti et al., 1997; Salgado et al., 1997); and on natural weathered rocks such as granitic and volcanic saprolites, which are common in Brazil, Portugal and the Far East (Viana Da Fonseca et al., 1997; Ng et al., 2000b, 2004b; Ng and Wang, 2001; Wang and Ng, 2005; Ng and Leung, 2006), relatively only a limited number of research work has been conducted to determine soil stiffness of unsaturated soils. In this chapter, laboratory equipments and instrumentation developed for measuring shear modulus, G_0, at very small strains (0.001 per cent or less) of various unsaturated sands and silts are summarized and reported. In addition, the anisotropic shear modulus, $G_{0(ij)}$, in the shear plane ij a recompacted volcanic soil is discussed and reported below.

Effects of initial water contents on small strain shear stiffness of sands (Qian et al., 1993)

Introduction

Qian et al. (1993) reported the results of an experimental investigation the effects of initial water contents at compaction on the very small strain shear modulus of sands in a resonant column. Their studies included: (i) void ratio; (ii) confining pressure; (iii) grain shape and (iv) grain size distribution on shear modulus. All the tests were conducted under isotropic stress conditions. Thirteen cohesionless soils were selected for testing to investigate shear modulus under unsaturated conditions. Four were natural sands, Glazier Way, mortar, Ottawa F-125 and Agsco. The minus 200 sieve size fractions of these four sands were non-plastic. Glazier Way and Ottawa F-125 sands were selected as the benchmark test materials, and the mortar and Agsco sands were included because their grain shapes are very similar to those of Glazier Way and Ottawa F-125 sands, respectively, but have different grain sizes and distributions. These four sands were dry-sieved into uniform size fractions or, alternatively, into size fractions with specific grain size distributions. These tailored size fractions were tested to determine the influence of gradation in a systematic manner.

Glazier Way sand and mortar sand are angular sands with roundness values (R) of 0.22 and 0.21, respectively. Ottawa F-125 sand and Agsco sand are subrounded sands with roundness values of 0.43 and 0.45, respectively. The roundness (Youd, 1973) is defined as the ratio of the average of the radii of the corners of a sand grain image to the radius of the maximum circle that can be inscribed within the grain image.

The other nine sands were man-made or tailored cohesionless sands derived from the four basic sands. Sands A1 through A6 were angular sands fractionated from Glazier Way sand; sands R1 through R3 were subrounded sands fractionated from Ottawa F-125 sand. Sands through A3 have different size distributions in their coarse size fractions, but have the same content of minus 400 sieve size fraction. The gradation characteristics of A4, A5 and A6 sands, and R1, R2 and R3 sands were similar to those of A1, A2 and A3 sands.

The basic properties of these 13 sands are listed in Table 6.1. The gradation curves for the eight angular sands are shown in Figure 6.2; and the gradation curves for the five subrounded sands are shown in Figure 6.3. Eleven uniform angular sand fractions were used in their study; these sand fractions were designated as 45, 50, 70, 80, 100, 120, 140, 170, 200, 270 and 400. For example, 100 sand is defined as a sand with all grains passing through the No. 80 sieve and retained on the No. 100 sieve; the average diameter of this uniform sand is 164.5 μm. Nine uniform subrounded sand fractions were used; they were designated as 70, 80, 100, 120, 140, 170, 200, 270 and 400.

The test sands were prepared by mixing soil and distilled water to a pre-selected degree of saturation. The mixture was placed in an airtight container and stored for at least 24 h to assure uniform moisture distribution. Specimens were then prepared by compacting the soil–water mix to the

Table 6.1 Basic material properties of 13 test soils (after Qian et al., 1993)

Name of material (1)	Specific gravity (2)	Minimum void ratio (3)	Maximum void ratio (4)	Effective grain size D_{10} (mm) (5)	Uniformity coefficient (C_u) (6)	Roundness (R) (7)
(a) Angular sands						
Glazier way	2.66	0.64	1.15	0.032	2.13	0.22
Mortar	2.69	0.59	1.07	0.058	2.51	0.21
A1	2.66	0.66	1.21	0.039	1.23	0.22
A2	2.66	0.53	0.98	0.040	3.40	0.22
A3	2.66	0.46	0.82	0.045	6.60	0.22
A4	2.66	0.66	1.16	0.022	2.05	0.22
A5	2.66	0.50	0.91	0.022	4.77	0.22
A6	2.66	0.43	0.80	0.022	11.36	0.22
Ottawa F-125	2.65	0.52	0.90	0.068	1.50	0.43
(b) Subrounded sands						
Agsco	2.67	0.50	0.90	0.070	1.37	0.45
R1	2.65	0.54	0.92	0.040	1.18	0.43
R2	2.65	0.45	0.82	0.045	2.38	0.43
R3	2.65	0.42	0.79	0.058	3.17	0.43

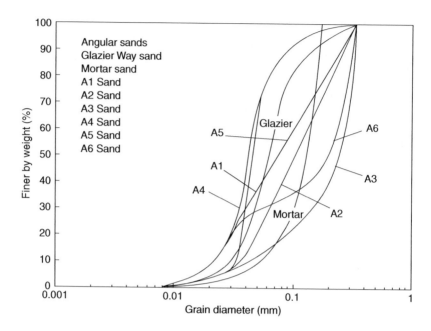

Figure 6.2 Grain size distribution of angular sands (after Qian et al., 1993).

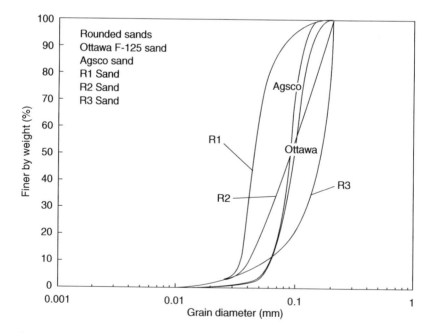

Figure 6.3 Grain size distribution of subrounded sands (after Qian et al., 1993).

required void ratio in a metal mould mounted directly on a resonant column device. Measurements of the first mode resonant frequency were recorded at pre-selected time intervals.

Effects of confining pressure and void ratio on maximum shear modulus ratio and optimum degree of saturation

In order to illustrate the difference in shear modulus between unsaturated sand and completely dry sand, the ratio of the shear modulus in an unsaturated condition to the shear modulus in a completely dry condition, $G_0/G_{0(dry)}$ was selected as a basic representation of shear modulus response. This ratio was plotted as a function of the degree of saturation at various confining pressures for specific void ratios. Figures 6.4 and 6.5 show that

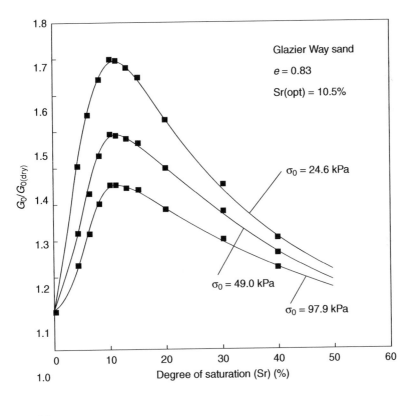

Figure 6.4 $G_0/G_{0(dry)}$ versus degree of saturation at various confining pressures for Glazier Way sand (after Qian et al., 1993).

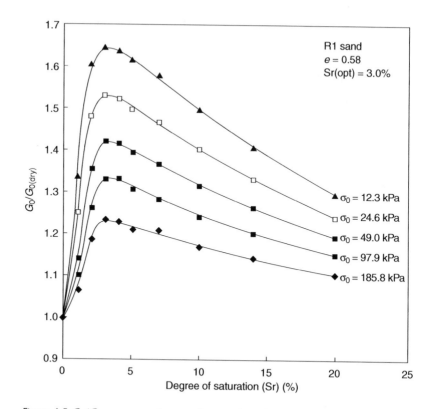

Figure 6.5 $G_0/G_{0(dry)}$ versus degree of saturation at various confining pressures for R1 sand (after Qian *et al.*, 1993).

the shear modulus ratio, $G_0/G_{0(dry)}$, varies significantly with the degree of saturation. In the figures, the content of the minus 400 sieve size fraction, C_s, is defined as a ratio of the weight of the minus 400 sieve size fraction to the total weight. When the degree of saturation increases from a completely dry condition ($S_r = 0$), the ratio increases very rapidly to a peak value, i.e. the maximum shear modulus ratio. The maximum shear modulus ratio of Glazier Way sand was about 1.7 when the confining pressure was 24.6 kPa, and the void ratio was 0.83 as shown in Figure 6.4. On the other hand, the maximum shear modulus ratio of R1 sand was about 1.5 when the confining pressure was 24.6 kPa, and the void ratio was 0.58 as shown in Figure 6.5. The value of the optimum degree of saturation for Glazier Way sand in Figure 6.4 was 10.5 per cent at a void ratio of 0.83, while the optimum degree of saturation for R1 sand in Figure 6.5 was 3.0 per cent at a void ratio of 0.58. When the degree of saturation increased beyond the optimum degree of saturation, the curves slope downwards and gradually flattened out. Both Figures 6.4 and 6.5 show that the shear modulus ratio, $G_0/G_{0(dry)}$ decreased as the confining pressure increased. From a practical standpoint, this finding indicates that the capillary effects on the shear modulus at shallow depths will be more pronounced than those at greater depths.

The relationship between the shear modulus ratio, $G_0/G_{0(dry)}$, and the degree of saturation, at various void ratios and constant confining pressure for A4 and Ottawa F-125 sands, is shown in Figures 6.6 and 6.7. The maximum shear modulus ratio, $G_0/G_{0(dry)}$, of A4 sand was 2.04 at a confining pressure of 24.6 kPa and void ratio of 0.68. The maximum shear modulus ratio decreased to 1.60 when the void ratio increased to 1.13. In a similar manner the maximum shear modulus ratio, $G_0/G_{0(dry)}$, of Ottawa- F-125 sand was 1.32 at a confining pressure of 24.6 kPa and void ratio of 0.58; the ratio dropped to a 1.19 as the void ratio increased to 0.88. As shown in both Figures 6.6 and 6.7, $G_0/G_{0(dry)}$ increased as the void ratio decreased for all degrees of saturation. This means that the value of the maximum

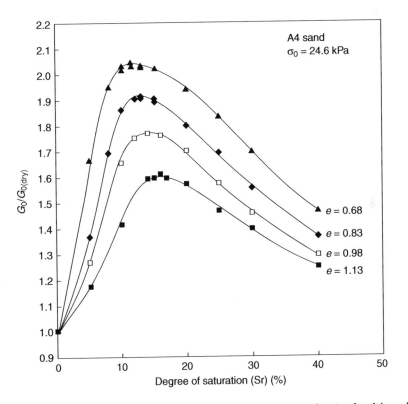

Figure 6.6 $G_0/G_{0(dry)}$ versus degree of saturation at various void ratios for A4 sand (after Qian et al., 1993).

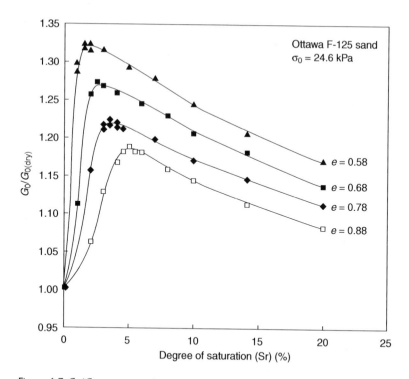

Figure 6.7 $G_0/G_{0(dry)}$ versus degree of saturation at various void ratios for Ottawa F-125 sand (after Qian et al., 1993).

shear modulus ratio should be greatest in a soil that has the lowest void ratio, all other factors constant.

Effects of grain size distribution on maximum shear modulus ratio and optimum degree of saturation

To illustrate the effects of grain size distribution on the shear modulus of unsaturated sands, the different size groups of uniform angular sands fractionated from Glazier Way sand and uniform subrounded sands fractionated from Ottawa F-125 sand were tested in an unsaturated condition. Figure 6.8 shows the relationship between the shear modulus ratio and the degree of saturation for 70, 100, 140, 170, 200, 270 and 400 angular, uniform size fractions at a confining pressure of 24.6 kPa and void ratio of 0.78. It can be seen that uniform sands with the same grain shape from 45 to 400 size fractions have the same optimum degree of saturation. This optimum degree of saturation is equal to 4.5 per cent for angular sand at a void ratio of

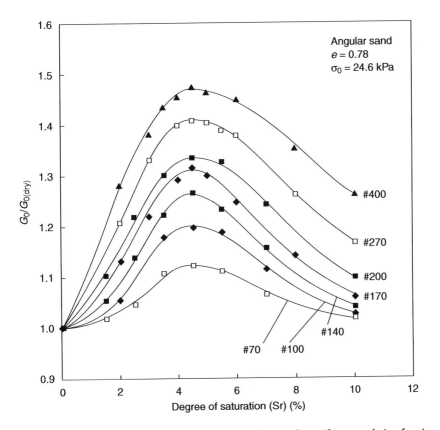

Figure 6.8 $G_0/G_{0(dry)}$ versus degree of saturation for angular uniform sand size fractions (after Qian *et al.*, 1993).

0.78 even though they have different values of the maximum shear modulus ratio. Subrounded sands behaved in a similar manner. The maximum shear modulus ratio increased with decreasing soil grain size as shown in the figure. For example, the maximum shear modulus ratio of 400 angular, uniform sand was about 1.47, and the maximum shear modulus ratio of 70 angular, uniform sand was about 1.12 at a confining pressure of 24.6 kPa and void ratio of 0.78. This finding indicates that the soil grain size is also an important factor affecting the value of the maximum shear modulus ratio for unsaturated cohesionless soils. The higher the content of the fine particles in a sand, the greater will be the maximum shear modulus ratio.

Figure 6.9 shows the relationship between the shear modulus ratio, $G_0/G_{0(dry)}$, and the degree of saturation, for angular and subrounded minus 400 sieve size fractions at a void ratio of 0.78 and confining pressure of

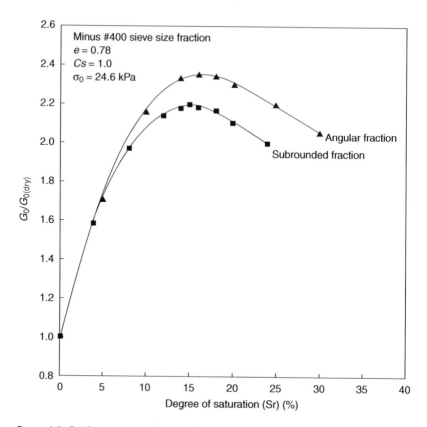

Figure 6.9 $G_0/G_{0(dry)}$ versus degree of saturation for angular and subrounded minus 400 sieve size fractions (after Qian *et al.*, 1993).

24.6 kPa. The content of the minus 400 sieve size fraction, C_s, is defined as a ratio of the weight of the minus 400 sieve size fraction to the total weight. Its value varies from 0 to 1.0. If the smallest grain of a sand is larger than 400 sieve size, the content of the minus 400 sieve size fraction of this soil is 0 ($C_s = 0$). All soils shown in Figure 6.8 fall in this category. If all grains of a cohesionless soil are less than 400 sieve size, the content of the minus 400 sieve size fraction of this soil is 1.0 ($C_s = 1.0$). All the soils shown in Figure 6.9 fall in this category.

A comparison between the curve for the angular minus 400 sieve size fraction in Figure 6.9 and the curves in Figure 6.8 indicates quite different modulus ratios between the minus 400 sieve size fraction and grains that are larger than the 400 sieve size. The value of the optimum degree of saturation of the angular minus 400 sieve size fraction was equal to 16.0 per cent, as shown in Figure 6.6. This was much higher than the value of the optimum

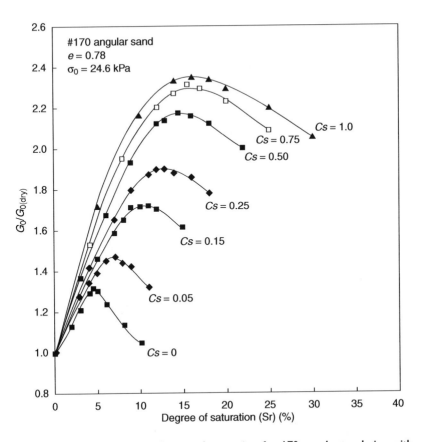

Figure 6.10 $G_0/G_{0(dry)}$ versus degree of saturation for 170 angular sand size with various C_s (after Qian et al., 1993).

degree of saturation of the angular grains that were larger than 400 sieves size, namely, 4.5 per cent as shown in Figure 6.8. The value of the optimum degree of saturation of the subrounded minus 400 sieve size fraction was equal to 15.0 per cent, as shown in Figure 6.10, which was also much higher than the value of the optimum degree of saturation of the subrounded grains that were larger than 400, namely, 3.0 per cent. If an angular sand, whose round-ness value is also 0.22, has grains that are both larger and smaller than 400 sieve size, the optimum degree of saturation should lie between 4.5 and 16.0 per cent. A similar conclusion can be drawn for the subrounded sand. Thus, the content of the minus 400 sieve size fraction of a cohesionless soil will also affect the optimum degree of saturation in a partially saturated condition.

To understand the effect of the content of the minus 400 sieve size fraction on the optimum degree of saturation, 170 angular sand was selected as the common ingredient of artificial mixes having different contents of the minus 400 sieve size fraction. Figure 6.10 shows the relationships between the shear modulus ratio and the degree of saturation for 170 angular sand with various contents of the minus 400 sieve size fraction at a confining pressure of 24.6 kPa and void ratio of 0.78. In the figure, there are seven curves, which represent seven values of the minus 400 sieve size fraction content, namely, 0, 0.05, 0.15, 0.25, 0.50, 0.75 and 1.0. The test results in Figure 6.8 show that soil grains larger than the 400 sieve size only affect the value of the maximum shear modulus ratio and do not affect the value of the optimum degree of saturation. Figure 6.10 shows that the content of the minus 400 sieve size fraction, on the other hand, affects both the maximum shear modulus ratio and the optimum degree of saturation. Thus, grain size distribution is also an important factor affecting dynamic behaviour of unsaturated cohesionless soil.

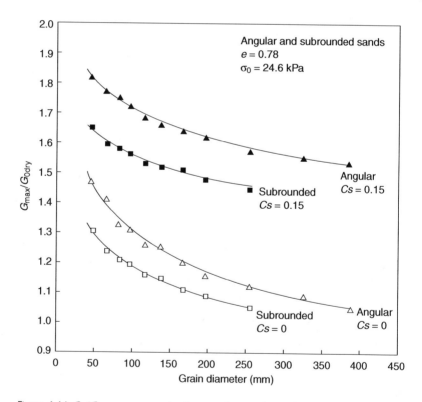

Figure 6.11 $G_0/G_{0(dry)}$ versus grain diameter for angular and subrounded sands (after Qian et al., 1993).

Effects of grain shape on maximum shear modulus ratio and optimum degree of saturation

To observe the effect of grain shape on the shear modulus of unsaturated cohesionless soils, the maximum shear modulus ratio, $G_{0(max)}/G_{0(dry)}$, at the optimum degree of saturation for both angular and subrounded sands with the same minus 400 sieve size fraction content were both plotted as a function of grain diameter. Figure 6.11 shows the relationship between $G_{0(max)}/G_{0(dry)}$ and grain diameter at a confining pressure of 24.6 kPa and void ratio of 0.78. The two curves in the figure through solid points represent angular sand whose C_s values were 0 and 0.15. The two curves through open points represent subrounded sand whose C_s values were also 0 and 0.15.

Grain shape also affected the optimum degree of saturation. Figure 6.12 shows the relationship between the optimum degree of saturation and void ratio for both angular and subrounded sands at three minus 400 sieve size contents, 0, 0.15 and 1.0. These results show that the optimum degree of saturation of angular sand is also higher than that of subrounded sand. Therefore, soil grain shape is an important factor affecting both the optimum degree of saturation and the maximum shear modulus ratio.

Effects of suction on elastic shear modulus of quartz silt and decomposed granite (silty sand)

Shear stiffness of quartz silt (Cabarkapa et al., 1999)

Triaxial equipment and instrumentation

Cabarkapa et al. (1999) modified a computer-controlled hydraulic triaxial cell for unsaturated soil testing (see Figure 6.13). The system used a hydraulic loading system, Bishop and Wesley type, where diaphragms seal a piston in a Bellofram cylinder. The cell was equipped with pressure control units for cell pressure, ram pressure, pore water and pore air pressure. The pressures were controlled either by electropneumatic converters or digital pressure controllers. Water was used as cell and pore fluid.

The apparatus was able to test samples of 38 and 100 mm diameter. The axis-translation technique was used to control matric suction. The pore water pressure (u_w) was applied at the base of the sample via a high air-entry porous disc with an air-entry value of 1.5 MPa. The pore air pressure (u_a) was supplied and controlled at the top of the sample using a digital pressure–volume controller. The deviatoric load acting on the sample was measured with a 4.5 kN internal load cell. Axial displacements were measured externally using a linear variable differential transformer (LVDT) and were corrected for compliance error by the software.

For the internal axial strain measurements two miniature submersible transducers (LVDTs) were mounted diametrically opposite on the sample over a central length of 50 mm to provide measurements down to 0.0001 per cent. For unsaturated soil radial strains need to be determined directly on the sample since an indirect determination from volume gauges would be highly inaccurate. In this system, a radial strain belt was positioned at mid-height of the sample (see Figure 6.14). The radial strain belt used a miniature LVDT mounted tangentially to the belt.

The internal measurements of axial and radial strains were used to calculate the overall volume change. The water volume changes were measured with accuracy of about 0.001 per cent using a standard 50 cc Imperial College volume gauge fitted with a LVDT. Air volume changes were measured using a digital pressure–volume controller (Adams *et al.*, 1996). Even if temperature was controlled within (±0.5 °C) this controller showed to be sensitive to temperature fluctuation with a volumetric drift of about

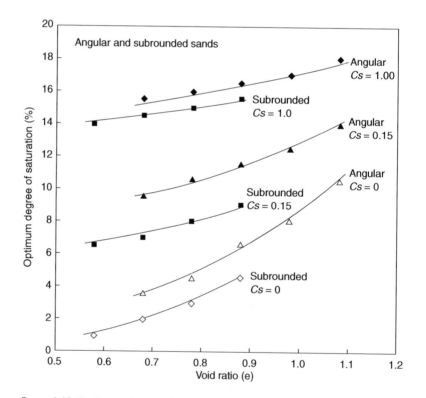

Figure 6.12 Optimum degree of saturation versus void ratio for angular and sub-rounded sands with various C_s (after Qian et al., 1993).

0.3 per cent per degree Celsius so that these measurements could not be used to calculate volumetric changes at small strain levels.

One of the key features in the triaxial system developed for testing unsaturated soils was the ability to measure and control suction as well as measuring the shear modulus G_0 from the local instrumentation during continuous loading and from bender elements. To install the bender elements, the transmitter in the top cap was located in a slot cut in a standard porous stone 3 mm thick. The receiver was located in the pedestal which was designed to accommodate a 6-mm thick and 30-mm diameter disc with a high air-entry value of 1,500 kPa.

Design and assembly techniques described by Dyvick and Madshus (1985) were modified for unsaturated soil testing and particular attention was paid to connecting the receiver bender element with the high air-entry disc since high impedance bender devices cannot be exposed to moisture and require therefore perfect insulation. An epoxy composite with polyurethane coating was used to cover and insulate the bender elements. For this test, a sine wave pulse with a frequency of 9 kHz was chosen to avoid the near field effect and to obtain a clear received signal. Using this frequency the ratio of the distance between the tips of the benders (d) and the wavelength (λ) varied

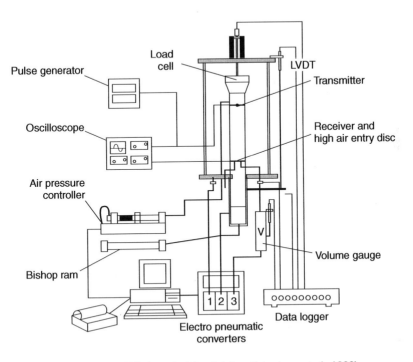

Figure 6.13 Bishop and Wesley triaxial cell (after Cabarkapa *et al.*, 1999).

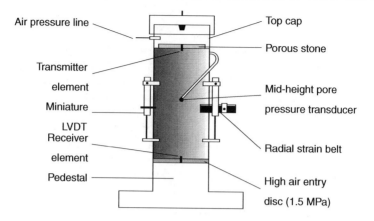

Figure 6.14 Set-up of the test specimen in the apparatus with internal instrumentation (after Cabarkapa *et al.*, 1999).

in the range 4.5–6.5 as recommended by Sanchez-Salinero *et al.* (1986) to minimize any near field effect.

Material tested

The soil tested was a quartz silt with angular particles (mean particle size $D_{50} = 0.02$ mm) with a specific gravity (G_s) of 2.67, a liquid limit (w_L) of 31 per cent and a plastic limit (w_p) of 0 per cent. The grain size distribution is shown in Figure 6.15.

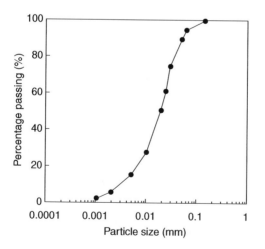

Figure 6.15 Particle grain size distribution (after Cabarkapa *et al.*, 1999).

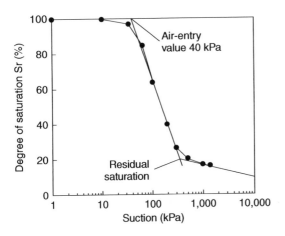

Figure 6.16 Soil–water characteristic curve (after Cabarkapa et al., 1999).

The SWCC of the silt tested was obtained experimentally in a stepped process during which different values of pore air and pore water pressure were maintained. The system was allowed time to come into equilibrium. This procedure was repeated until the final pressure was set and resulting equilibrium attained. The SWCC shown in Figure 6.16 was obtained in the triaxial apparatus able to reach a confining pressure of 1.7 MPa. The estimated air-entry value is around 40 kPa.

Compression and swelling characteristics under constant suctions

Figure 6.17 shows the data for compression and swelling for three samples with suctions of 50, 100 and 200 kPa and a dry sample. The samples with suctions of 50 and 100 kPa show clear evidence of yielding and a value of p for yield can be estimated by extrapolating the slopes of the initial and later parts of the test paths. On the other hand, the evidence for yielding of the sample with a suction of 200 kPa and the dry sample is not so clear.

Variations of G_0 during isotropic compression

Values of measured shear wave velocity and shear moduli (G_0) of an unsaturated soil sample obtained from bender element tests were computed using the following equation:

$$G_0 = \rho V_s^2 \tag{6.1}$$

where V_s is shear wave velocity and ρ mass density of the soil.

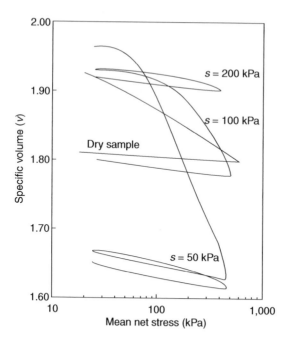

Figure 6.17 Isotropic compression tests results (after Cabarkapa *et al.*, 1999).

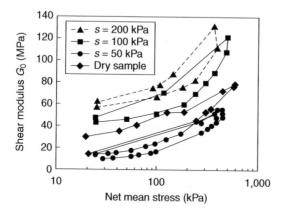

Figure 6.18 Variation of G_0 during isotropic compression tests (after Cabarkapa *et al.*, 1999).

Figure 6.18 shows values of G_0 obtained from bender element tests on the dry sample and the three unsaturated samples. Each sample experiences isotropic loading and unloading as shown in Figure 6.18. The following key points can be made about the measured results:

- The results for each sample are similar. The shear modulus increases with increasing p as the sample is loaded. On unloading the shear modulus decreases but at each value of net mean stress p, G_0 is higher than during the loading (in other words, over-consolidation increases stiffness).
- The effect of suction is to increase stiffness. It should be noted that the results for the dry sample lie between those for suctions of 50 and 100 kPa. It was not expected if the dry sample simply corresponded to the case of zero suction.
- In the post-yield region, the relationship between G_0 and net mean stress p for different suctions is somewhat linear in a double logarithm plot as shown in Figure 6.19.

Shear stiffness of decomposed granite (silty sand) (Mancuso et al., 2000, 2002)

Mancuso *et al.* (2000, 2002) conducted suction-controlled tests on recompacted decomposed granite specimens in a resonant column-torsional shear (RCTS) apparatus to investigate the effects of suction and fabric on unsaturated soil behaviour, shear stiffness at very small strain in particular.

The RCTS and testing procedures

To investigate the small strain behaviour of unsaturated soils, an RCTS apparatus working under controlled-suction conditions was developed. Figure 6.20 shows a general layout of the apparatus. This apparatus allows separate

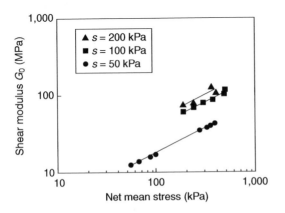

Figure 6.19 Variation of G_0 in the post-yield region (after Cabarkapa *et al.*, 1999).

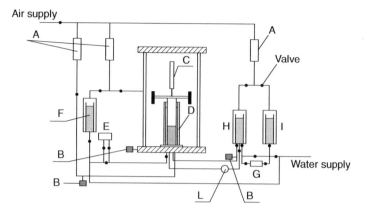

Figure 6.20 General layout of the controlled-suction resonant column-torsional shear device for unsaturated soils developed in Naples. A, electromagnetic regulator; B, transducer; C, linear variable differential transducer; D, aluminium hollow cylinder; E and G, differential pressure transducers; F, reference burette; H and I, double-walled burettes and L, peristaltic pump (after Mancuso *et al.*, 2002).

control of cell, pore air and pore water pressures using three electropneumatic regulators (A in Figure 6.20). A feedback process sets pressures at any value established by the operator, taking advantage of readings from transducers (B in Figure 6.20) connected to the pressure lines. Suction is controlled using the axis-translation technique. Pore air and pore water pressures are applied through the base pedestal. A high air-entry value porous disk protects the drainage line from short-term desaturation. The resulting complex design of the pedestal arose from the desire to not modify the fixed-free torsional constraint condition in the RCTS for unsaturated soils.

The specific volume changes of the specimen are inferred from the separate measurement of axial and radial displacements (Figure 6.20). The former are evaluated using an LVDT coaxial to the specimen, i.e. C, and the latter are obtained indirectly with a differential pressure transducer (i.e. E) monitoring the change of the water level in a bath surrounding the sample. A reference burette (i.e. F) gives the zero reading. To slow down water evaporation, a layer of silicone oil is placed above the water–air interfaces.

The measurement of changes in water content (Δw) is obtained through a system of two double-walled burettes coupled to a differential pressure transducer (DPT; G in Figure 6.20). The first burette (i.e. H) is connected to the specimen drainage line and the second (i.e. I) gives a reference value for the DPT readings. A flushing system, in addition to the high air-entry value disk, protects the drainage line from long-term desaturation, and water evaporation is delayed through the placement of silicone oil on the water–air interfaces.

The testing procedure used with this RCTS device consists of three stages: equalization, compression and cyclic-dynamic torsional shear. Details of the stages are given by Mancuso *et al.* (2002).

The tested material

The tested material is a decomposed granite extracted from a quarry in Calabria, Italy, used to build the core of a zoned earth dam across the Metramo River (Baldovin *et al.*, 1991). The material is a silty sand with a high uniformity coefficient of 400 and a clay fraction of about 16 per cent. The liquid limit w_L measured on the fraction having diameters of soil particles lower than 0.4 mm is around 35 per cent, and the plasticity index PI is 14 per cent, indicating that the finer fraction of the soil can be classified as lean clay in the USCS Plasticity chart (ASTM, 1993). The above and other physical properties of the material are summarized in Table 6.2.

This soil has been tested after compaction by the modified Proctor procedure (ASTM, 1991) at the optimum moulding water content ($w_{OPT} = 6.8$ per cent) and on the wet side of the relevant compaction curve (water content $w = w_{OPT} + 2.5$ per cent). The undercompaction technique (Ladd, 1978) was adopted to improve soil homogeneity along the specimen height.

Three 36-mm diameter specimens were obtained from each Proctor sample. Average specific volumes v and dry unit weights γ_d resulting from compaction are $v = 1.338 \pm 0.009$ and $\gamma_d = 16.35 \pm 0.14 \, \mathrm{kN/m}^3$ for the optimum soil and $v = 1.462 \pm 0.004$ and $\gamma_d = 17.71 \pm 0.005 \, \mathrm{kN/m}^3$ for the wet material. The after-compaction suction measured using the Imperial College tensiometer (Ridley and Burland, 1993) is around 800 and 60 kPa for optimum and wet of optimum compacted specimens, respectively (Vinale *et al.*, 1999).

The tested soil has a non-negligible fine-grained component. It is therefore expected that, depending on compaction water content, the soil may assume a quite different structure due to bulk water, menisci water and soil particles arrangement. This is not supported in the paper by scanning electron microscopy or mercury intrusion porosimetry measurements, and arises

Table 6.2 Main physical properties of Metramo silty sand (after Mancuso et al., 2000, 2002)

G_S	D_{max} (mm)	Sand (%)	Silt (%)	Clay (%)	U_c	w_L (%)	w_P (%)	PI (%)
2.64	2	63	21	16	400	35.4	21.7	13.7

Note
G_S, specific gravity of soil particles; w_P, plastic limit.

from literature studies that show direct observations of compaction-induced microfabric (Barden and Sides, 1970; Wan and Gray, 1995; Delage *et al.*, 1996). From these studies it can be argued that dry compaction generally results in a soil fabric made up of aggregates of varying sizes, and usually with a bimodal pore-size distribution. Wet of optimum compaction tends to produce a more homogeneous, matrix-dominated soil fabric with a single pore-size distribution.

Observed relationships of shear stiffness and suction

At the end of compression stages, several series of resonant column (RC) and torsional shear (TS) tests were performed. To analyse strain-rate effects, each series was performed at constant torque and at loading frequencies of 0.06, 0.1, 0.5 and 2 Hz. Tests were initiated immediately after the end of compression to eliminate time effects and thus obtain homogeneous data.

For the intermediate net stress $(\sigma - u_a)$ values, the RC and TS tests were performed at torques that increased progressively not crossing over the elastic threshold strain of the soil $(G_{(y)}/G_0 = 95$ per cent, where G is the shear stiffness, γ is the shear strain and G_0 is the initial shear stiffness. Once the highest chosen $(\sigma - u_a)$ was reached, torque was increased up to the maximum allowable value (0.43 Nm) to obtain the G–γ curve (not discussed here). Table 6.3 summarizes the investigation suction and mean net stress levels that were used in the tests (Mancuso *et al.*, 2002).

Figures 6.21 and 6.22 show the initial shear stiffness G_0 against matric suction obtained in the suction-controlled RC and in TS tests at a 0.5 Hz loading frequency. It can be seen that most of the suction effects are observed for s ranging from 0 to about 100 kPa. For suction higher than 100 kPa, G_0 tends towards a threshold value that depends on the mean net stress level. The effect of suction is significant. Suction variation in the 0–400 kPa range causes an increase in G_0 ranging from 85 to 50 per cent for the optimum soil and from 165 to 40 per cent for the wet compaction. The stiffness values of both optimum and wet compacted materials are clearly distinct for constant stress levels and show an S-shaped increase with suction. This mechanical response complies with the typical S-shape of the soil–water characteristic

Table 6.3 Investigated suction and mean net stress levels with resonant column and torsional shear tests (after Mancuso *et al.*, 2002)

Material	Suction (kPa)	Mean net stress (kPa)
Optimum	0, 25, 50, 100, 200, 400	100, 200, 400
Wet of optimum	0, 100, 200, 400	100, 200, 400

curve (SWCC) and can be explained by dividing the G_0 versus s curves into three different zones.

1. Zone 1 starts in saturated conditions at null suction and is restricted to low suction values. Herein, bulk-water effects dominate the soil behaviour, since the amount of air present in the specimens is negligible. In this zone, variations in s are practically equivalent to changes in mean effective stress. The initial gradient of the G_0–s relationship is then expected to be the same as that of the G_0–p' function of the saturated material, calculated at a mean effective stress p' equal to $(p-u_a)_c$, where $(p-u_a)_c$ is the mean net stress of the drying path under consideration. This gradient is also expected to decrease less than linearly as $(p-u_a)_c$ increases, according to the typical shape of the G_0-p' function (Rampello *et al.*, 1995).

2. In zone 2, i.e. intermediate suction values, the amount of air present in the pore voids becomes ever more significant as suction increases. A progressive shift of the soil response occurs from bulk-water regulated behaviour to menisci-water regulated behaviour.

3. In zone 3, for suction values high enough to allow menisci water to have a dominant influence, suction-change effects conform to those expected on the basis of the Fisher (1926) model. Thus, the initial shear stiffness increases with suction at an initially fast rate and then tends towards a threshold value.

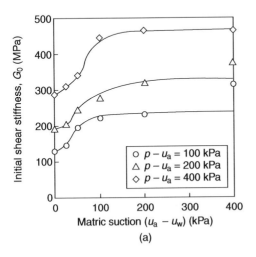

Figure 6.21 Initial shear stiffness versus suction for the optimum compacted soil: (a) RC tests; (b) TS tests (after Mancuso *et al.*, 2000).

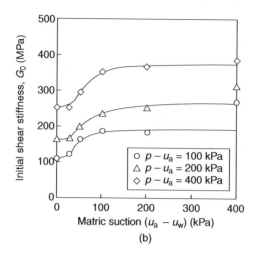

Figure 6.21 (Continued).

Moulding water content seems to cause some significant differences in soil response. This variable affects the ratio between the saturated value and the unsaturated threshold values of shear stiffness, and the suction values (called s^* in the following) characterizing the transition between bulk-water reg-ulated behaviour and menisci-water regulated behaviour. This latter point can be explained on the basis of intuitive physical considerations. In fact,

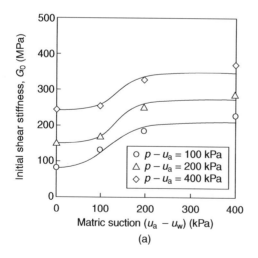

Figure 6.22 Initial shear stiffness versus suction for the wet compacted soil: (a) RC tests; (b) TS tests (after Mancuso *et al.*, 2000).

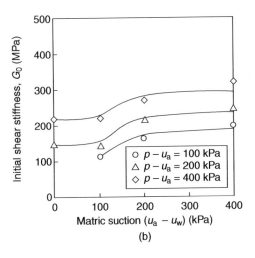

Figure 6.22 (Continued).

changes in soil fabric due to an increase in moulding water content induce smaller pore sizes in the wet material with respect to the optimum soil, even if the overall porosity follows the opposite trend (Barden and Sides, 1970; Lambe and Whitman, 1979; Delage *et al.*, 1996). This is also suggested by the characteristic curves of compacted soils from Vanapalli *et al.* (1996), who observed that optimum compaction leads to a lower air-entry value than wet compaction. This could justify the lower s value of the optimum soil (a few tens of kilopascals in Figure 6.21) with respect to that of the wet soil (around 100 kPa in Figure 6.22). It must be noted, however, that the differences between the two figures could be partly due to hydraulic hysteresis.

Data from Figures 6.21 and 6.22 confirm that wet compaction induces a weaker soil fabric with respect to optimum compaction. In fact, the increase in moulding water content causes a strong reduction in initial shear stiffness, both under RC and TS conditions. For example, the RC data from saturated specimens indicate that the G_0 value at p' of 100 kPa varies from about 80 MPa in the wet case to 130 MPa in the optimum compacted specimen. That is, stiffness increases by more than 60 per cent by decreasing compaction water content from 12.3 (wet) to 6.8 per cent (optimum). Similar effects are observed at all other suction levels.

Besides the observed effect of moulding water content (w_{mld}), the experimental data confirm that the soil fabric resulting from w_{mld} is an important factor regulating soil behaviour, since ageing, stress and suction states or stress histories were not changed in the previous comparison.

Empirical equations for shear stiffness and suction relationships (Mancuso et al., 2002)

For saturated conditions, G_0 values can be expressed, for example, by following the relationship proposed by Rampello *et al.* (1995):

$$\frac{G_0}{p_a} = A\left(\frac{p'}{p_a}\right)^n \text{OCR}^m \tag{6.2}$$

where p_a is the reference pressure (atmospheric); A is the stiffness index and n and m are the stiffness coefficients representing the stiffness of the material under the reference pressure (at an over-consolidation ratio, OCR, of 1) and the sensitivity of the stiffness to stress state and history, respectively.

As suggested by Mancuso *et al.* (2002), the stiffness $(G_0)_{s=0}$ at null suction is

$$\frac{(G_0)_{s=0}}{p_a} = A\left[\frac{(p - u_a)_C}{p_a}\right]^n \text{OCR}^m \tag{6.3}$$

where $(p - u_a)$ is the mean net stress of the drying path under consideration.

Equation (6.3) can be extended up to the air-entry value of the tested soil by simply adding s to mean net stress, giving the following equation for the G_0–s relationship in zone 1:

$$\frac{(G_0)_{s \leq s_{ev}}}{p_a} = A\left[\frac{(p - u_a)_c + s}{p_a}\right]^n \text{OCR}^m \tag{6.4}$$

Further, small increases in suction cause progressive soil desaturation and move the G_0–s relationship from the bulk-water regulated zone to the menisci-water regulated zone. In this suction range (zone 2) it can be reasonably assumed that the effect of suction starts with a gradient equal to that which can be derived from Equation (6.4).

The G_0–s curve in this zone is the link between the functions in zones 1 and 3 and must tend towards the latter for higher suctions. However, both the shape of the curve in zone 2 and the suction level at which menisci water starts prevailing (i.e. the edge of zone 2) depend on the details of the specific desaturation process, and are not straightforward.

In this session, a simplified G_0 versus s relationship is assumed, i.e. both zone 2 and zone 3 behaviour can be described by an equation similar to that proposed by Alonso *et al.* (1990) for the $\lambda(s)$ function. This assumption implies a discontinuity between zones 1 and 3, i.e, zone 2 reduces to zero width. Using the suction value at the transition between bulk-water regulated behaviour and menisci-water regulated behaviour, already indicated as s^*, experimental data can be fitted by the relationship:

$$G_0 = (G_0)_s \{(1-r)e^{-\beta(s-s^*)} + r\} \tag{6.5}$$

where β is the parameter that controls the rate of increase of soil stiffness with increasing suction and is associated with the soil sensitivity to suction changes; and r is the ratio between shear stiffness at s^* and the threshold value of G_0 for increasing suction and is related to the stiffness increase due to menisci water.

It is worth noting that Equation (6.5) does not follow directly from Alonso et al. (1990). As a more reasonable analogy would be expected between shear stiffness and bulk modulus, G_0 should vary inversely with the $\lambda(s)$ function proposed by the authors. The shape of such a relation, however, does not differ from that of Equation (6.5) which is preferred because of its better manageability and straightforward physical meaning of the r and β coefficients.

Figure 6.23 displays the application of the above criteria to the data from the optimum and wet of optimum compacted soils for the data collected at $(p-u_a) = 400\,\text{kPa}$. This figure suggests that equation (6.5) can properly fit experimental data obtained in zone 3, where stiffness values are higher than those predicted by Equation (6.4) obtained extending the saturated soil approach up to s^*.

On the basis of these observations it can be concluded that soil behaviour is dominated by menisci-water effects at relatively high suctions, whereas at low matric suctions (i.e. relatively high degree of saturation) bulk-water effects prevail. In the former case, suction increases the normal forces acting on particles, whereas in the latter case suction is roughly equivalent to a change in mean net stress (having smaller effect on shear stiffness). To confirm this point, however, further experimental evidence is necessary.

Anisotropic shear stiffness of completely decomposed tuff (clayey silt) (Ng and Yung, 2007)

Introduction

With the introduction of an additional horizontally mounted bender element probe consisting of two pairs of orthogonally oriented bender elements by Pennington et al. (1997) in a triaxial apparatus, velocities of horizontally propagated shear waves with horizontal and vertical polarizations, as well as conventional vertically propagated shear wave with horizontal polarization, can be measured within a single soil specimen. The development of this additional horizontally mounted bender element probe enables researchers to evaluate anisotropic shear moduli in different planes of soils (Ng et al., 2004b; Ng and Leung, 2006).

Theoretical considerations

In dry and saturated soils, shear wave velocity is found to be dependent on the effective stresses in the directions of wave propagation and particle motion and is almost independent of the effective stress normal to the plane of shear for saturated soil (Roesler, 1979; Knox *et al.*, 1982; Jamiolkowski *et al.*, 1995; Stokoe *et al.*, 1995; Bellotti *et al.*, 1996; Rampello *et al.*, 1997). In addition to effective stresses, both shear wave velocity and small strain shear modulus is shown to be a function of void ratio. Different void ratio

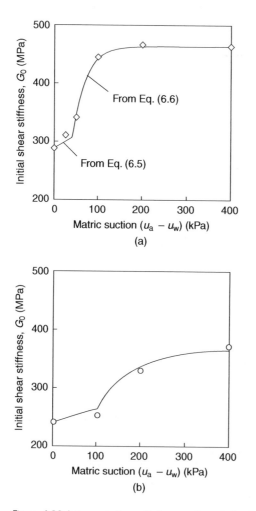

Figure 6.23 Interpretation of the results of the RC tests performed on (a) the optimum compacted material and (b) the wet of optimum compacted material at a mean stress of 400 kPa (after Mancuso *et al.*, 2002).

functions are proposed to account for the influence of void ratio on shear wave velocity and shear modulus in saturated soils (Hardin and Drnevich, 1972; Iwasaki et al., 1978; Lo Presti, 1989). For shear wave velocity, $v_{s(ij)}$, corresponding to shear wave propagating in dry and saturated soils with directions of wave propagation, i, and particle motion, j, polarized along the principal stress directions ij can be expressed as

$$v_{s(ij)} = C_{ij} \, f(e) \left(\frac{\sigma_i - u_w}{p_r} \times \frac{\sigma_j - u_w}{p_r} \right)^{n/2} \tag{6.6}$$

where C_{ij} is a material constant (with dimension m/s) reflecting soil fabric, $f(e)$ a void ratio function relating the dependency of shear wave velocity to void ratio, p_r a reference stress, $\sigma_i - u_w$ and $\sigma_j - u_w$ are the principal effective stresses in the plane in which $v_{s(ij)}$ is measured, and n is an empirical exponent.

With the adoption of the two stress state variables, $\sigma - u_a$, and $u_a - u_w$, for unsaturated soils, it is intuitive reasonable to assume the following velocity expression for shear waves travelling in unsaturated soils:

$$v_{s(ij)} = C_{ij} \, f(e) \left[\frac{(\sigma_i - u_a)}{p_r} \times \frac{(\sigma_j - u_a)}{p_r} \right]^{n/2} \left[1 + \frac{(u_a - u_w)}{p_r} \right]^{bij} \tag{6.7}$$

where bij is an empirical exponent of matric suction, reflecting the influence of matric suction on shear wave propagating in the polarization plane, ij. When soil is saturated (i.e. $u_a - u_w = 0$), Equation (6.7) becomes Equation (6.6). This equation allows a smooth transition between an unsaturated soil and a saturated soil. Assuming that the reference pressure, p_r, is 1 kPa, and the void ratio function is in the form of an exponential function, e^{aij}, where aij is an empirical void ratio exponent, Equation (6.7) can be rewritten as

$$\frac{v_{s(ij)}}{f(e)} = \frac{v_{s(ij)}}{e^{aij}} = C_{ij} \left[(\sigma_i - u_a) \times (\sigma_j - u_a) \right]^{n/2} \left[1 + (u_a - u_w) \right]^{bij} \tag{6.8}$$

Assuming that strains induced by bender elements in an elastic soil are very small (Dyvik and Madshus, 1985) and by knowing the current tip-to-tip distance, L_i, of transmitting and receiving bender elements (Viggiani and Atkinson, 1995), velocity of shear wave can be measured and calculated by:

$$v_{s(ij)} = \frac{L_i}{t_{ij}} \tag{6.9}$$

where t_{ij} is measured travelling time of shear wave propagating in the polarization plane, ij, over the distance L_i. Hence, the elastic shear modulus at very small strain, $G_{0(ij)}$, may be determined from:

$$G_{0(ij)} = \rho v_{s(ij)}^2 \tag{6.10}$$

where ρ is the bulk density of soil, which can be expressed in terms of void ratio, e, and degree of saturation, S_r, as follows:

$$\rho = \left(\frac{G_s + S_r e}{1 + e}\right) \rho_w \tag{6.11}$$

where G_s is specific gravity of soil and ρ_w density of water. By measuring the change of water content and volume of a soil specimen, soil density can be obtained using Equation (6.11) and hence $G_{0(ij)}$ can be determined.

Combining Equations (6.7), (6.10) and (6.11), the elastic shear modulus, $G_{0(ij)}$, can be expressed as

$$G_{0(ij)} = S_{ij}\, F(e) \left[\frac{(\sigma_i - u_a)}{p_r} \times \frac{(\sigma_j - u_a)}{p_r}\right]^n \left[1 + \frac{(u_a - u_w)}{p_r}\right]^{2bij} \tag{6.12}$$

where

$$F(e) = \frac{[f(e)]^2 (G_s + S_r e)}{1 + e} \quad \text{and} \quad S_{ij} = C_{ij}^2 \rho_w \tag{6.13}$$

$F(e)$ may be considered as a function of void ratio relating shear modulus to void ratio, and S_{ij} is a material constant reflecting soil fabric.

If the degree of anisotropy of unsaturated soils is expressed in terms of a shear stiffness ratio between shear modulus in the horizontal plane ($G_{0(hh)}$) to that in the vertical plane ($G_{0(hv)}$) in a triaxial test, by making use of the expression in Equation (6.12), the shear stiffness ratio can be derived as follows:

$$\frac{G_{0(hh)}}{G_{0(hv)}} = \left(\frac{v_{s(hh)}}{v_{s(hv)}}\right)^2 = \frac{S_{hh}}{S_{hv}} \left[\frac{(\sigma_h - u_a)}{(\sigma_v - u_a)}\right]^n \left[1 + \frac{(u_a - u_w)}{p_r}\right]^{2(bhh - bhv)} \tag{6.14}$$

where $v_{s(hh)}$ is velocity corresponding to a shear wave propagating horizontally with horizontal polarization, $v_{s(hv)}$ velocity corresponding to a shear wave propagating horizontally with vertical polarization, bhh and bhv are empirical exponents of matric suction reflecting the influence of matric suction on stiffness in the two directions, hh and hv, respectively. It can be seen from Equation (6.14) that the ratio of shear modulus is equal to the square of the ratio of shear wave velocity, ($v_{s(hh)}/v_{s(hv)}$), and hence the degree of stiffness anisotropy of unsaturated soils depends on soil fabric, S_{hh}/S_{hv}, ratio of net normal stresses in the horizontal and vertical directions, $(\sigma_h - u_a)/(\sigma_v - u_a)$, and matric suction, $u_a - u_w$. The influence of matric suction on the degree of stiffness anisotropy is governed by the difference between values of bhh and bhv. For saturated soils (i.e. $u_a - u_w = 0$) with isotropic fabric (i.e. $S_{hh}/S_{hv} = 1$), the degree of stiffness anisotropy in Equation (6.14) solely depends on the stress ratio, $(\sigma_h - u_a)/(\sigma_v - u_a)$ that is commonly recognized to cause stress-induced anisotropy.

On the other hand, under isotropic stress conditions (i.e. $(\sigma_h - u_a)/(\sigma_v - u_a) = 1$), Equation (6.14) can be simplified as

$$\frac{G_{0(hh)}}{G_{0(hv)}} = \left(\frac{v_{s(hh)}}{v_{s(hv)}}\right)^2 = \frac{S_{hh}}{S_{hv}} \left[1 + \frac{(u_a - u_w)}{p_r}\right]^{2(bhh - bhv)} \tag{6.15}$$

If soil is saturated, [i.e. $(u_a - u_w) = 0$], the ratio of shear modulus reveals inherent stiffness anisotropy of soil fabric (i.e. S_{hh}/S_{hv}) only. This suggests that the degree of stiffness anisotropy expressed as the ratio of the shear modulus is independent of net normal stress under isotropic conditions and it depends on both inherent fabric anisotropy and matric suction if the difference between bhh and bhv is not equal to zero. The term $[1 + (u_a - u_w)^{(bhh - bhv)}]$ in Equation (6.15) will not be equal to zero and it is governed by soil suction. This leads to so-called suction-induced anisotropy.

Testing apparatus and measuring devices

Development of a triaxial apparatus for testing unsaturated soils

For testing unsaturated soils, the principle of the axis-translation technique (Hilf, 1956) was employed to control matric suction of a soil specimen to prevent cavitation. This technique involves a translation of the reference atmospheric air pressure to a higher desirable value to preventing gauge water pressure in soil pores dropping below zero. This requires the control of u_a and u_w in a soil specimen. An existing computer-controlled triaxial apparatus originally for testing saturated soils (Li et al., 1988) was thus modified. Figure 6.24 shows a schematic diagram illustrating the modified triaxial apparatus for controlling air and water pressures independently at the top and the bottom of a soil specimen, respectively. The actual entire assembly of the triaxial apparatus together with measuring devices is shown in Figure 6.25. In this triaxial cell, air pressure is controlled through a coarse low air-entry value corundum disk placed on the top of a soil specimen, whereas water pressure is controlled through a saturated high air-entry value (3 bars) ceramic disk sealed to the pedestal of the apparatus. The use of the high air-entry ceramic disk allows the passage of water but prevents the flow of free air from the specimen to the water control and drainage system underneath it in the short term. However, since the permeability of an unsaturated soil specimen is generally very low, long duration of testing in terms of weeks and months is often encountered. It is therefore very important to design a drainage system to remove any diffused air through the high air-entry value ceramic disk.

Figure 6.26a shows a purposely designed and built spiral-shaped base pedestal for accommodating a 76 mm diameter and 150 mm specimen. This

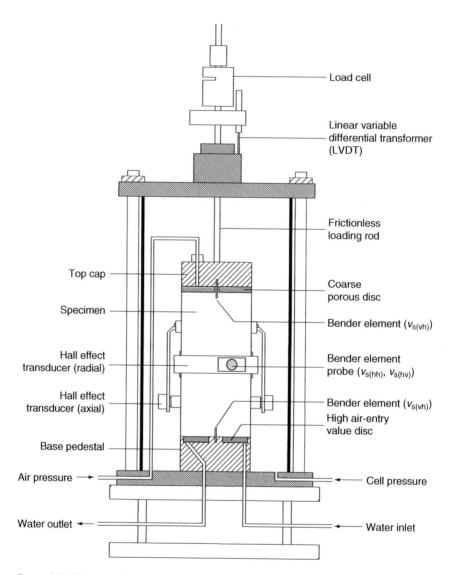

Figure 6.24 Schematic diagram of modified triaxial apparatus for testing unsaturated soils (after Ng and Yung, 2007).

pedestal was designed for circulating de-aired water to flush and remove any diffused air into the water drainage system and to house a bender element for generating shear waves during a test. Diffused air was frequently removed by allowing water circulation in one direction (from the right to the left as

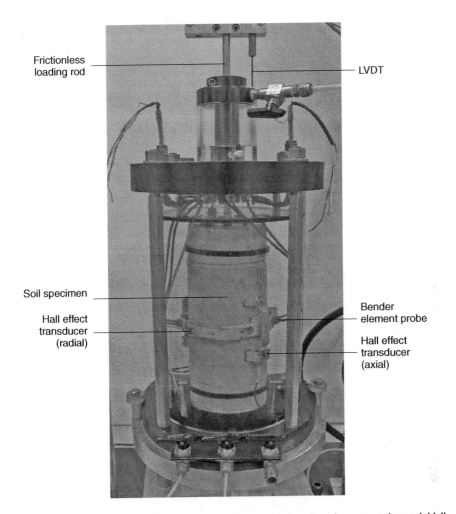

Figure 6.25 Modified triaxial apparatus equipped with bender element probe and Hall effect transducers (after Ng and Yung, 2007).

indicated by arrows in Figure 6.24) along the drainage line once every 24 h throughout a test.

Since both a high air-entry ceramic disk and a bender element were required to be embedded in the pedestal, a 16 mm diameter of recess in the centre was designed to house the bender element (see Figure 6.26a and b). This allows the installation of the bender element at the centre of the pedestal and flow of water underneath the disk. A pre-drilled high air-entry value ceramic disk was placed in the pedestal and sealed on the inner and outer circumferences of the disk with epoxy resin (see Figure 6.26b).

Arrangement of bender elements and measurement of shear wave velocity

In order to measure velocity of shear waves propagating in two planes with different polarizations and hence to determine anisotropic shear moduli of a soil specimen, a pair of bender elements was fixed in the two end platens and two pairs of bender elements were mounted at the mid-height of a specimen. The arrangement of the bender elements are shown in Figure 6.27 schematically. The vertically transmitted shear wave with horizontal polarization $(v_{s(vh)})$ is generated and received by a pair of bender elements embedded in the end platens of the triaxial cell. The mounting of these bender elements is similar to that described by Dyvik and Madshus (1985). The bender element was cut to a length of 12 mm and a width of 8 mm, with half of its length embedded in the base platen and the remaining half protruding

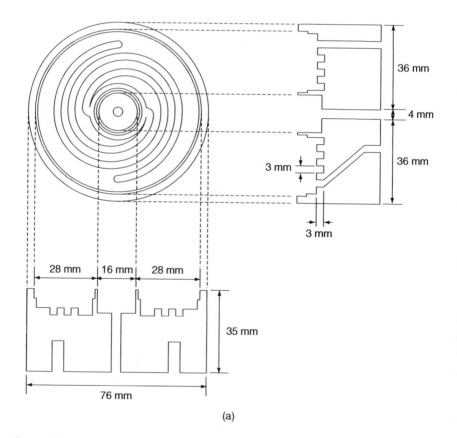

(a)

Figure 6.26 (a) Modified base pedestal; (b) details of bender element embedded in the modified base pedestal (after Ng and Yung, 2007).

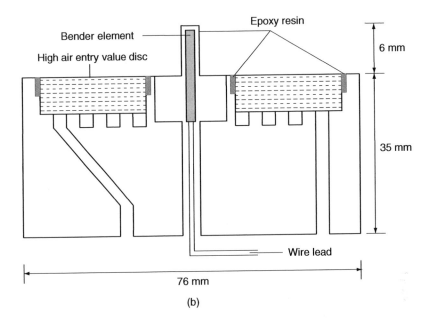

High air entry value disc

Bender element

Epoxy resin

6 mm

35 mm

Wire lead

76 mm

(b)

Figure 6.26 (Continued).

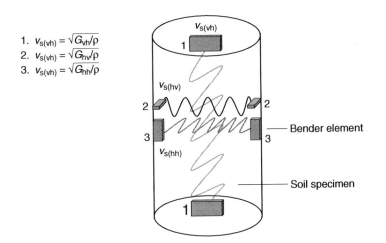

1. $v_{s(vh)} = \sqrt{G_{vh}/\rho}$
2. $v_{s(vh)} = \sqrt{G_{hv}/\rho}$
3. $v_{s(vh)} = \sqrt{G_{hh}/\rho}$

$v_{s(vh)}$

$v_{s(hv)}$

$v_{s(hh)}$

Bender element

Soil specimen

Figure 6.27 Schematic diagram showing the arrangement of bender elements (after Ng and Yung, 2007).

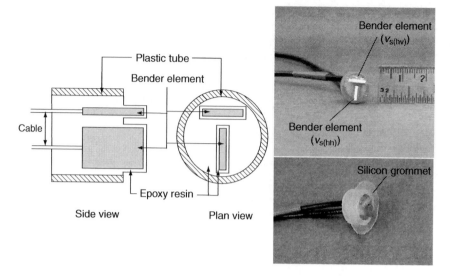

Figure 6.28 Details of the bender element probe (after Ng *et al.*, 2004b).

into a soil specimen. It is coated with epoxy resin for insulation (refer to Figure 6.26b).

The horizontally transmitted shear waves, $v_{s(hv)}$ and $v_{s(hh)}$, are generated and received by a pair of bender element probes (see Figure 6.28). Each bender element probe consists of two pieces of bender elements with a length 9 mm and a width 6 mm potted orthogonally into a short length of plastic tube. The bender elements protrude about 3 mm into a soil specimen. The probe is fitted to the mid-height of the soil specimen by means of a tailor-made silicon grommet (as shown Figure 6.28) holding the probe in position and preventing the specimen from leakage problems. The cut slot in the caliper of a Hall effect radial belt allows the cables of the bender elements to pass through (refer to Figure 6.25). More details of the bender element probe and its use for determining the anisotropic stiffness of a saturated soil are given by Ng *et al.* (2004b).

Determination of shear wave velocity and shear modulus using bender elements

As shown schematically in Figure 6.27, three pairs of bender elements are mounted onto a soil specimen to generate shear waves and to measure the velocity of shear waves propagating through the soil specimen in different directions. Excited by an external voltage, the transmitting bender element deflects and acts as an energy source to release energy into a soil specimen to generate shear stresses propagating through the soil. On the opposite end of the soil specimen, shear waves are received by a receiver element, which is aligned to detect the transverse motion of the shear waves. The transmitted and received signals of shear waves propagating in a soil specimen in two different directions are captured by an oscilloscope. For each measurement during a test, a single sinusoidal pulse with a frequency of 4–10 kHz was used as the input wave signal. The range of frequency selected was intended to obtain as clear signal as possible and to minimize the near field effect (Sanchez-Salinero *et al.*, 1986). The arriving time of each wave was determined by measuring the peak-to-peak time distance as shown in the oscilloscope, as suggested by Viggiani and Atkinson (1995).

In addition to conventional external measurements of axial strain using LVDT, the triaxial apparatus is equipped with three Hall effect transducers (Clayton *et al.*, 1989) for measuring both local axial and radial displacements. Measurements of these transducers are used to determine volume change of a soil specimen throughout a test. As a result, the current tip-to-tip travelling distance [refer to Equation (6.8)] and the current density of a soil specimen [refer to Equation (6.9)] can be accurately determined at each stage of the test.

Soil types and test specimen preparation

Soil types

The material tested was a completely decomposed tuff (CDT) extracted from a deep excavation site in Fanling, Hong Kong. The material can be described as clayey silt (ML) according to the Unified Soil Classification System.

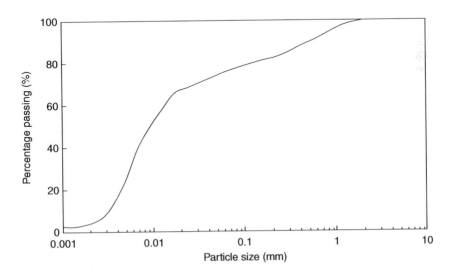

Figure 6.29 Particle size distribution of completely decomposed tuff (CDT) (after Ng and Yung, 2007).

Table 6.4 Index properties of completely decomposed tuff (CDT) (after Ng and Yung, 2007)

Index test	Completely decomposed tuff
Standard compaction tests	
Maximum dry density (kg/m³)	1760
Optimum water content (%)	16.3
Grain size distribution	
Percentage of sand (%)	24
Percentage of silt (%)	72
Percentage of clay (%)	4
D_{10} (mm)	0.003
D_{30} (mm)	0.006
D_{60} (mm)	0.015
Coefficient of uniformity (D_{60}/D_{10})	4.55
Coefficient of curvature $[(D_{30})^2/(D_{60} \times D_{10})]$	0.61
Specific gravity	2.73
Atterberg limits (grain size <425 μm)	
Liquid limit (%)	43
Plastic limit (%)	29
Plasticity index (%)	14

Figure 6.29 shows the particle size distribution of the CDT determined by sieve and hydrometer analyses (British Standards Institution, 1990). The material was yellowish-brown, slightly plastic, with very small percentage of fine and coarse sand. The liquid limit and plasticity index of the fines portion (finer than 425 μm) were 43 and 14 per cent, respectively. The physical properties of the material are summarized in Table 6.4.

Figure 6.30 shows a drying SWCC of a recompacted CDT specimen determined by a pressure plate extractor. The measured values are best-fitted by the method proposed by Fredlund and Xing (1994) and the fitted curve is also shown in Figure 6.30 for comparisons. It can be seen that the air-entry value of the recompacted CDT is about 50 kPa, determined by the intersection point between the straight sloping line and the saturation ordinate (Fredlund and Rahardjo, 1993). It should be noted that an air-entry value of a soil is referred as the matric suction that must be exceeded before air recedes into soil pores.

Specimen preparation

Each triaxial specimen of 76 mm in diameter and 152 mm in height was prepared by moist tamping. Initially the CDT sample was oven-dried to a temperature of 105 °C for 24 h and cooled inside a desiccator. De-aired water was added and mixed thoroughly with the dried soil sample at the optimum water content of 16.3 per cent inside a container. Thereafter, the well-mixed

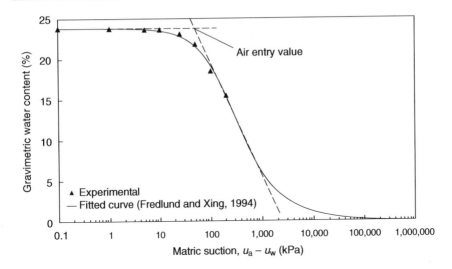

Figure 6.30 Soil–water characteristic curve of recompacted specimens of completely decomposed tuff (CDT) (after Ng and Yung, 2007).

wet CDT sample was kept inside an impervious container and stored for 24 h in a humidity- and temperature-controlled room for moisture equalization. Each specimen was prepared by ten layers of compaction directly onto the base pedestal of the triaxial cell in order to avoid disturbance caused by transportation and to improve contacts between the specimen and the high air-entry value ceramic disk. To prevent excessive densification of the lowest layers from the compaction of the uppermost layers of a soil specimen, the undercompaction procedure proposed by Ladd (1978) was adopted. Each specimen was compacted with 4 per cent denser than its desired dry density for the top three layers and 4 per cent less dense for the bottom three layers. The other layers were compacted at the desired dry density and the maximum difference in the dry densities found in any compacted specimen was generally less than 1 per cent. After compaction at the optimum water content, measured matric suction was about 54 kPa, which was slightly higher than the air-entry value determined from the drying SWCC shown in Figure 6.30.

Testing programme and procedures

In order to determine effects of soil fabric due to compaction and suction-induced stiffness anisotropy (refer to Equation 6.15), a series of isotropic compression tests were conducted on recompacted CDT specimens. Figure 8 shows the initial preparation paths (dotted lines) and also test paths (solid lines) for four suction-controlled isotropic compression test series at various constant matric suctions ranging from 0 to 200 kPa (i.e. i–0, i–50, i–100 and

Table 6.5 Summary of testing conditions for isotopic compression tests on completely decomposed tuff (CDT) (after Ng and Yung, 2007)

Specimen identity	Suction applied (kPa)	Stress range, $p-u_a$ (kPa)	Equali-sation time (days)	Initial water content (%)	Initial dry density, ρ_d (kg/m³)	Relative compac-tion (%)	Void ratio e_0	e_r	e_f
i-0	0	110–400	5	16.1	1672	95.0	0.632	0.574	0.483
i-50	50	110–500	4	16.6	1649	93.7	0.656	0.604	0.535
i-100	100	110–500	6	16.3	1679	95.4	0.626	0.579	0.522
i-200	200	110–500	7	16.4	1677	95.3	0.628	0.585	0.538

Notes
e_0: Initial void ratio.
e_r: Void ratio at net mean stress $= 110\,kPa$.
e_f: Void ratio at the maximum net mean stress.

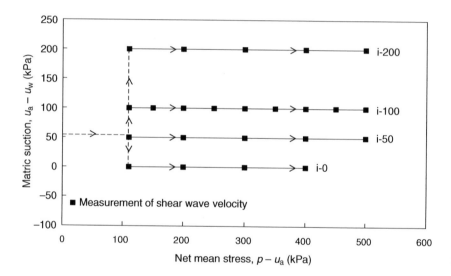

Figure 6.31 Stress paths of the triaxial tests conducted on recompacted specimens of completely decomposed tuff (CDT) (after Ng and Yung, 2007).

i–200). It should be noted that the test series was conducted at zero suction corresponding to a test on a saturated soil specimen. In order to reveal the greatest possible effects of suction on soil stiffness, the selection of suction range was based on the consideration of the measured air-entry value and the shape of the SWCC shown in Figure 6.30. Details of testing conditions are summarized in Table 6.5. As illustrated in Figure 6.31, measurements of shear wave velocity (and hence shear stiffness) were carried out at some pre-selected stress states as each isotropic loading was proceeded.

Each isotropic compression test consisted of two stages: equalization and compression. The equalization stage was to make sure that water moisture reached its equilibrium state throughout a soil specimen, i.e. pore water and pore air pressure were equalized inside the voids of a specimen at each pre-selected stress state. An equalization stage was considered to be completed when the rate of water content changes was less than 0.04 per cent per day (Fredlund and Rahardjo, 1993), which was corresponding to a water flow of about 0.5 cm³ per day. Each stage of equalization generally required 4–7 days to complete (see Table 6.5).

Once a specimen was equalized at its specified initial suction, it was then isotropically compressed to a required net mean stress, $(p - u_a)$, at a constant suction, where p is total mean stress. While maintaining the pore water and pore air pressures constant, cell pressure was ramped at 3 kPa per hour to a required target value. This rate of compression was selected to minimize any excess pore water pressure generated in a soil specimen. This compression rate was reported (Ng and Chiu, 2001) to be slow enough to ensure fully drained conditions by monitoring both changes of water flow and volume change of a soil specimen at the end of each compression stage.

Experimental results

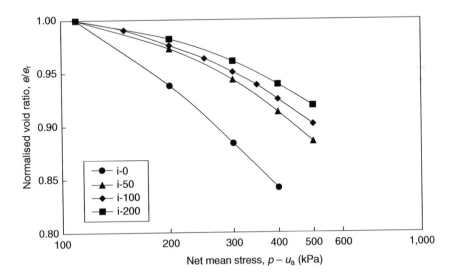

Figure 6.32 Variation of normalized void ratio, e/e_r, with net mean stress (after Ng and Yung, 2007).

Effects of change of soil suction and net mean stress on void ratio

By measuring local axial and radial strains using the Hall effect transducers, volume changes of a soil specimen and thus void ratio were evaluated during the suction-controlled isotropic compression tests. Since the void ratio, e_r, of each specimen at the start of compression at 110 kPa is somewhat different (see Table 6.5), measured void ratio along each compression path is normalized by its corresponding e_r and is plotted against the logarithm of net mean stress $(p - u_a)$ in Figure 6.32 for comparisons. As expected, the normalized void ratio, e/e_r, decreased with an increase in $p - u_a$ at all suctions but at different rates. No distinct yield stress can be identified, especially at the tests with higher suctions. For a given change of net mean stresses, soil specimen compressed at a higher constant suction showed a smaller change in normalized void ratios, i.e. smaller gradient. This demonstrates that matric suction stiffened the soil skeleton of the unsaturated soil during the isotropic compression tests for the suction ranged considered. Similar results were reported by Cabarkapa *et al.* (1999) for their compression tests at different constant suctions on quartz silt. As shown in Equation (6.7), void ratio is related to shear wave velocity. By determining void ratio, the void ratio function, $f(e)$, can be obtained with known values of matric suction and net normal stress at different states.

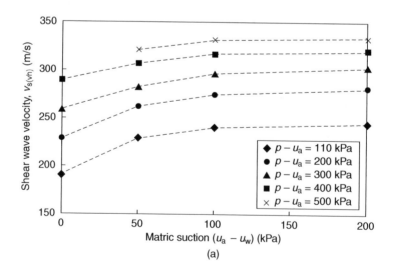

Figure 6.33 Shear wave velocity, (a) $v_{s(vh)}$, (b) $v_{s(hv)}$, (c) $v_{s(hh)}$, plotted against matric suction (after Ng and Yung, 2007).

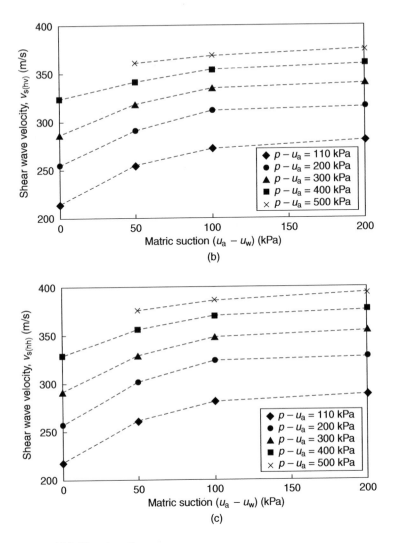

Figure 6.33 (Continued).

Relationships between shear wave velocities $\left(v_{s(vh)}, v_{s(hv)}, v_{s(hh)}\right)$ and matric suction

As shown in Figure 6.31, measurements of shear wave velocity were carried out during isotropic compression at constant suctions. Relationships between shear wave velocity and matric suction can be obtained by plotting measured shear wave velocities in three different polarization planes against matric suction at different net mean stress (see Figure 6.33a–c). In Figure 6.33, it can be seen that measured shear wave velocities, $v_{s(vh)}$, $v_{s(hv)}$

and $v_{s(hh)}$, increased with an increase in matric suction in a non-linear fashion. Similar results have been reported by Cabarkapa *et al.* (1999) and Mancuso *et al.* (2000). For any given value of $p - u_a$, as suction increased from 0 to 50 kPa, which was equal to the estimated air-entry value of the soil (see Figure 6.30), measured shear wave velocities in all polarization planes increased most significantly. Beyond the suction of 50 kPa, the measured shear wave velocities were continuously to increase but at a reduced rate up to suction of 200 kPa, i.e. the gradients of the curves decreased with an increase in matric suction. These observations were similar to those of Mancuso *et al.* (2000, 2002). The observed significant increase in shear wave velocity for suction up to the air-entry value (50 kPa) was because each specimen remained essentially saturated and bulk-water effects dominated soil stiffness (Mancuso *et al.*, 2002). Any increase in suction was practically equivalent to an increase in mean effective stress resulting in a much stiffer soil. On the other hand, as suction was increased beyond the air-entry value, each soil specimen started to desaturate and resulted in the formation of air–water meniscus (or contractile skin) at contact points of soil particles. Although meniscus water caused an increase in normal force holding the soil particles together, leading to a stiffer and hence a higher shear wave velocity, this beneficial effects on shear wave velocity did not increase indefinitely (Mancuso *et al.*, 2002) but they tended towards a limiting value, due to the progressive reduction in the meniscus radius as suction increased beyond 100 kPa as shown in Figure 6.33.

Not only does a change of shear wave velocity depend on a change in matric suction but it also depends on net mean stress. At net mean stress equal to 110 kPa, the measured shear wave velocities, $v_{s(vh)}$, $v_{s(hv)}$ and $v_{s(hh)}$ increased by 28.5, 31.0 and 33.8 per cent as suction increased from 0 to 200 kPa, respectively. On the other hand, at net mean stress equal to 400 kPa, the measured shear wave velocities, $v_{s(vh)}$, $v_{s(hv)}$ and $v_{s(hh)}$ increased only by 10.8, 12.2 and 15.0 per cent respectively for the same suction change. This clearly indicates that unsaturated soil behaviour/property (i.e. stiffness) depends on two stress state variables, matric suction and net mean stress. As the net mean stress increased, the influence of a given suction change on shear wave velocity became less effective since the soil was stiffer at higher stress.

For a given suction and net mean stress, it can be seen from Figure 6.33 that the measured $v_{s(vh)}$ was generally lower than the measured $v_{s(hv)}$ by about 20 per cent. Similar discrepancies were reported by Pennington *et al.* (2001) who suggested three possible reasons to explain the discrepancies. First, a higher transmitting frequency was normally used with the bender element probe which might lead to a higher $v_{s(hv)}$. Second, the bender element embedded in the base pedestal probably deflected with a larger magnitude than the horizontally mounted bender elements inside the probe and hence resulted in a lower $v_{s(vh)}$. Third, the discrepancies might be caused by end effects on stiffness measurements (Germaine and Ladd, 1988; Lacasse and Berre, 1988). It is generally recognized that a transmitting bender element does not only generate pure shear waves but it also produces compression waves. As compression waves travel faster than shear waves, some compression waves including those reflected at lateral boundaries may interfere with shear waves at a receiving bender element and distort the signal when it receives (Brignoli et al., 1996). Hence, the influence of compression waves on received signals depends on the sample geometry which alters the path length of the waves (Rio and Greening, 2003).

Influence of suction on the degree of fabric anisotropy

To eliminate the differences in frequency and boundary effects, $v_{s(hh)}$ and $v_{s(hv)}$ were chosen for comparisons since they were generated by the same bender element probe and they had the same travelling length and boundary conditions (Pennington et al., 2001; Ng et al., 2004). Figure 6.34a and b show the ratios, $v_{s(hh)}/v_{s(hv)}$ and $(v_{s(hh)}/v_{s(hv)})^2$, plotted against matric suction respectively. It should be noted that the ratio, $(v_{s(hh)}/v_{s(hv)})^2$ is equal to the shear stiffness ratio, $G_{0(hh)}/G_{0(hv)}$, as illustrated by Equation (6.13). Under the isotropic stress conditions, the velocity of the horizontally polarized shear wave $v_{s(hh)}$ was higher than that of vertically polarized shear wave $v_{s(hv)}$ by about 1 to 2 per cent at zero suction as shown in Figure 6.34a. The higher the applied net mean stress, the larger the ratio. Correspondingly, the deduced shear modulus $G_{0(hh)}$, was about 2.5 to 4 per cent higher

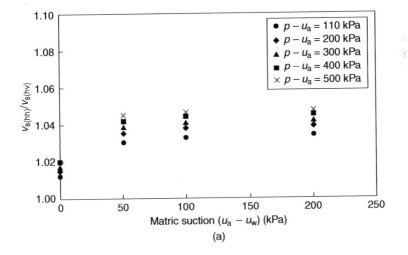

(a)

Figure 6.34 Ratio, (a) $v_{s(hh)}/v_{s(hv)}$ and (b) $(v_{s(hh)}/v_{s(hv)})^2$ or $G_{0(hh)}/G_{0(hv)}$, plotted against matric suction (after Ng and Yung, 2007).

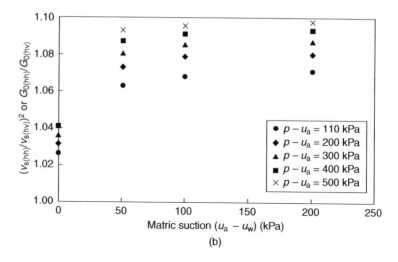

Figure 6.34 (Continued).

than $G_{0(hv)}$ when matric suction was zero. According to Equation (6.14), it is suggested that this observed stiffness anisotropy was attributed to the inherent (fabric) anisotropy (i.e. S_{hh}/S_{hv}) of the tested material during specimen preparation. The higher the net mean stress applied, the greater the anisotropic stiffness induced.

As suction increased to 50 kPa, the ratio of $v_{s(hh)}/v_{s(hv)}$ and $(v_{s(hh)}/v_{s(hv)})^2$ increased about 3.5 and 7.5 per cent on average, respectively. There was no significant increase in the velocity and stiffness ratios when suction was 100 kPa or higher. This observed suction-induced increase in $(v_{s(hh)}/v_{s(hv)})^2$, which is equal to $G_{0(hh)}/G_{0(hv)}$, is consistent with Equation (6.14), which suggests that matric suction will induce stiffness anisotropy (i.e. anisotropic shear modulus) if $b$hh is not equal to $b$hv.

Influence of net mean stress on the degree of anisotropy

Figure 6.35a and b show the measured ratios, $v_{s(hh)}/v_{s(hv)}$ and $(v_{s(hh)}/v_{s(hv)})^2$, plotted against net mean stress, respectively. The ratios, $v_{s(hh)}/v_{s(hv)}$ and $(v_{s(hh)}/v_{s(hv)})^2$, appeared to increase gently with an increase in net mean stress at almost the same rate at different suctions (i.e. the rate of increase is independent of suction). There was about 1–3 per cent increase in the stiffness ratio, $G_{0(hh)}/G_{0(hv)}$, which is equal to $(v_{s(hh)}/v_{s(hv)})^2$, when net mean stress increased from 110 to 500 kPa. Although it is generally recognized that shear stiffness will increase with an increase in mean effective stress for saturated soils, it is somewhat surprised that the stiffness ratio also

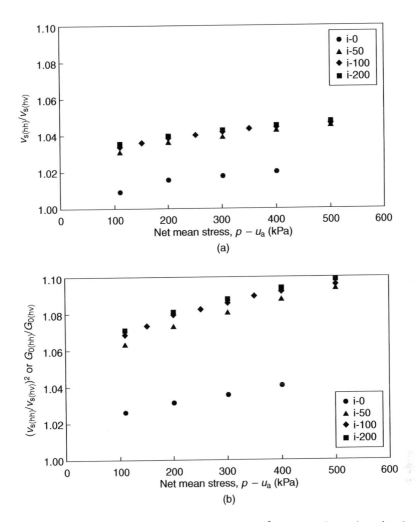

Figure 6.35 Ratio, (a) $v_{s(hh)}/v_{s(hv)}$ and (b) $(v_{s(hh)}/v_{s(hv)})^2$ or $G_{0(hh)}/G_{0(hv)}$, plotted against net mean stress (after Ng and Yung, 2007).

increase with net mean stress. This observed behaviour may be attributed to the coupling effects between hydraulic and mechanical characteristics of unsaturated soils (Wheeler *et al.*, 2003).

Verification of proposed equations

As assumed by Equation (6.7), shear wave velocity depends on void ratio, net normal stress and matric suction, the influence of void ratio on shear wave velocity may be eliminated by plotting normalized measured shear wave velocities in three polarization planes [i.e. $v_{s(vh)}/f(e)$, $v_{s(hv)}/f(e)$ and $v_{s(hh)}/f(e)$], against the product of the applied net normal stresses in the plane of shear, $\sigma_i \times \sigma_j$, in a double logarithmic scale as shown in Figure 6.36. It can be seen that the shear wave velocities increase with an increase in net normal stresses for saturated (i.e. i–0 series) and unsaturated conditions (i–50, i–100 and i–200 series), in an almost linear fashion with the product of net normal stresses. These observed results are similar to those found on saturated soils (Jamiolkowski et al., 1995; Stokoe et al., 1995; Bellotti et al., 1996; Rampello et al., 1997; Ng et al., 2004b).

By assuming $n = 0.075$ in Equation (6.7), best-fitted curves to the experimental data are also shown in Figure 6.36 for comparisons. Fitted parameters and correlation coefficient, R, for each polarization plane are listed in Tables 6.6 and 6.7, respectively. It can be seen in the figure that the fitted curves show a linear relationship between the normalized velocities and the product of stresses in the double logarithmic scale. The fitted curves appear to be fairly consistent with the measured shear wave velocities. This is reflected by the values of R^2 (where R is the correlation coefficient) for the fitted values as listed in Table 6.7. The average value of R^2 for the fitted curves is greater than 0.96, suggesting very strong correlations between the fitted curves and the measured data.

It should be noted that the parameters bij for $v_{s(hv)}$ and $v_{s(hh)}$ are different (see Table 6.6). According to Equation (6.14), the difference between bhv and bhh governs the influence of matric suction on the degree of anisotropy (i.e. suction-induced anisotropy). This follows that the suggested equation is capable of revealing the influence of suction on the anisotropy of shear wave velocity and hence shear modulus.

Summary

With the application of the axis-translation technique, a computer-controlled triaxial apparatus equipped with three pairs of bender elements and Hall effect transducers was developed to measure velocity of shear waves, $v_{s(vh)}$, $v_{s(hv)}$ and $v_{s(hh)}$, propagating in two directions with different polarizations in an unsaturated completely decomposed tuff (CDT). It was found that during the suction-controlled isotropic compression tests, the three shear wave velocities increased with an increase in suction for a given stress state in a non-linear fashion with a reduction in the rate of increase when the applied suction was higher than the air-entry value of the soil.

By using the newly derived theoretical equations in this chapter, it was possible to identify that there was an inherent stiffness anisotropy of the soil with measured $G_{0(hh)}$ at about 2.5–4 per cent higher than $G_{0(hv)}$, due to soil fabric formed during the sample preparation by moist tamping. Moreover, there was a further 7.5 per cent suction-induced anisotropy in terms of shear stiffness ratio, $G_{0(hh)}/G_{0(hv)}$, when suction was increased from 0 kPa to its air-entry value of 50 kPa. No significant suction-induced stiffness anisotropy was observed at suction higher than the air-entry value. Although it was intended to eliminate stress-induced anisotropy by carrying isotropic tests, a small increase of 1–3 per cent in the shear stiffness ratio, $G_{0(hh)}/G_{0(hv)}$, was observed when the net mean stress increased from 110 to 500 kPa. This observed behaviour may be attributed to the coupling effects between hydraulic and mechanical characteristics of the unsaturated soil.

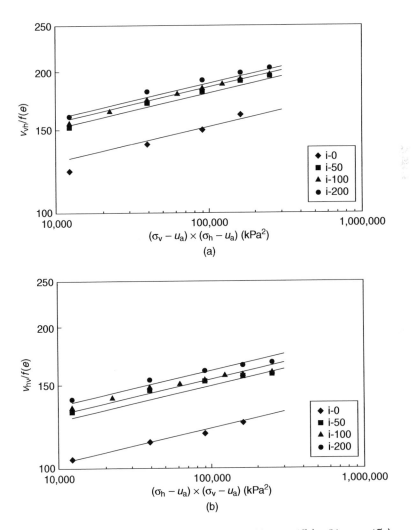

Figure 6.36 Normalized shear wave velocity, (a) $v_{s(vh)}/f(e)$, (b) $v_{s(hv)}/f(e)$ and (c) $v_{s(hh)}/f(e)$ versus product of stresses along the polarization plane (after Ng and Yung, 2007).

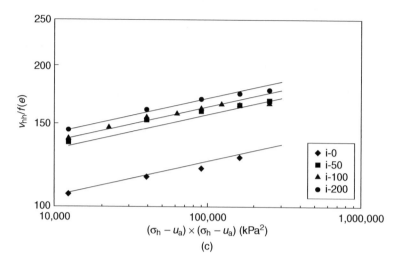

Figure 6.36 (Continued).

Table 6.6 Fitted parameters for shear wave velocities propagating in different polarization planes(after Ng and Yung, 2007)

Shear wave, $\nu_{s(ij)}$	C_{ij}	a_{ij}	b_{ij}
$\nu_{s(hh)}$	52.4	−1.3	0.058
$\nu_{s(h\nu)}$	51.7	−1.3	0.053
$\nu_{s(\nu h)}$	64.2	−0.8	0.042

Note
$\nu = 0.075$.

Table 6.7 Values of R^2 for shear waves at different suctions(after Ng and Yung, 2007)

Matric suction (kPa)	R^2		
	$\nu_{s(hh)}$	$\nu_{s(h\nu)}$	$\nu_{s(\nu h)}$
0	0.99	0.99	0.96
50	0.97	0.95	0.98
100	0.95	0.94	1.00
200	0.98	0.97	0.97

Part 3

State-dependent elasto-plastic modelling of unsaturated soil

Part 3 Frontispiece

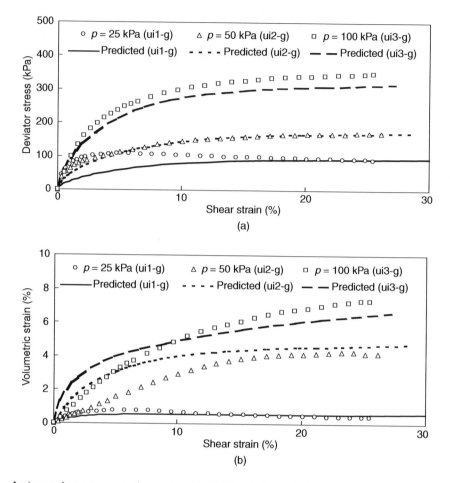

(a)

(b)

An image that represents the content of this Part 3, in particular how constitutive modelling can predict the engineering behaviour of an unsaturated soil, specifically: measured and predicted (a) stress–strain relationships; volumetric strain–shear strain relationships and (c) variation of suction with shear strain, for constant water content tests on unsaturated specimens of gravely sand at initial suction of 40 kPa (Chiu and Ng, 2003).

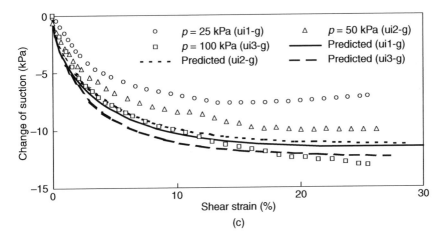

(c)

(Continued).

Chapter synopsis

Chapter 7: State-dependent elasto-plastic critical state-based constitutive model

Constitutive modelling is at the heart of modern soil mechanics because it enables numerical simulations of various boundary value problems when they are implemented in finite element or finite difference programs. This is essential not only for carrying out parametric studies to understand what is key to the engineering behaviour of soils but also as an aid to engineering design.

Here a relatively simple state-dependent elasto-plastic model for both saturated and unsaturated soils is developed. The model, which is formulated under the critical state framework extended for unsaturated soils, is based on two independent stress state variables: net normal stress and matric suction (Wheeler and Sivakumar, 1995). A modified state-dependent dilatancy is introduced to account for the effects of internal state, stress level and soil suction. Comparisons between model predictions and experimental results show the model is able to capture some key aspects of engineering behaviour of saturated and unsaturated soil behaviour at various stress paths.

Now geotechnical engineers can carry out 'what-if' parametric studies and investigate design proposals using this advanced soil model when it is implemented into a finite element or finite difference program.

A state-dependent elasto-plastic critical state-based constitutive model

Sources

This chapter is made up of verbatim extracts from the following sources for which copyright permissions have been obtained as listed in the Acknowledgements.

- Chiu, C.F. and Ng, C.W.W. (2003). A state-dependent elasto-plastic model for saturated and unsaturated soils. *Géotechnique*, 53(9), 809–829.
- Ng, C.W.W and Chiu, C.F. (2003). Laboratory study of loose saturated and unsaturated decomposed granitic soil. *J. Geotech. Geoenviron. Engrg. ASCE*, 129(6), 550–559.

Introduction

It is well known and understood that the mechanical behaviour of saturated soils can be interpreted and explained by a single stress state variable, called effective stress (Terzaghi, 1936). However, there are difficulties in applying the same approach to unsaturated soils, in particular when the effective stress fails to describe the behaviour of swelling and collapse on wetting unsaturated soils properly (Jennings and Burland, 1962). Instead, two independent stress state variables are used to describe the mechanical behaviour of unsaturated soils (Fredlund and Morgenstern, 1977). The most common independent stress state variables are net normal stress $(\sigma - u_a)$ and matric suction $(u_a - u_w)$, where σ is the total normal stress, u_a the pore air pressure and u_w the pore water pressure.

During the past 15 years, a number of constitutive models for unsaturated soils have been proposed (Alonso *et al.*, 1990; Thomas and He, 1994; Wheeler and Sivakumar, 1995; Cui and Delage, 1996). Most of these models are developed under the framework of two independent stress state variables and under an extended critical state concept for unsaturated soils. Different assumptions and constitutive relationships are used in these models to account for the behaviour of different types of unsaturated soils.

It is well understood that a correct description of dilatancy is essential for an accurate modelling of the stress–strain behaviour of soils. The traditional dilatancy formulation depends only on the stress level (or stress ratio) for modelling saturated soils. In more recent models, the effects of suction on dilatancy are included for unsaturated soils (Cui and Delage, 1996). Many experimental evidence from saturated sands (Bolton, 1986; Ishihara, 1993) and from saturated and unsaturated weathered soils (Chiu, 2001) reveal that dilatancy is also related to the internal state of soils such as void ratio. As a result, several state-dependent dilatancy relationships have been proposed to take into account the effects of state and stress level (Bolton, 1986; Ishihara, 1993; Li and Dafalias, 2000) for saturated soils. However, the dilatancy relationships used in most of the existing constitutive models for unsaturated soils do not take into account the effects of internal state.

Here a relatively simple state-dependent elasto-plastic model for both saturated and unsaturated soils is developed. The model, which is formulated under the extended critical state framework for unsaturated soils, is based on two independent stress state variables: net normal stress and matric suction. (Wheeler and Sivakumar, 1995) A modified state-dependent dilatancy is introduced to account for the effects of internal state, stress level and soil suction. In this chapter, details of the model formulation and determinations of model parameters are described and reported. By using the new model, numerical simulations of triaxial tests on both saturated and unsaturated coarse-grained decomposed granite (DG) and fine-grained decomposed volcanic (DV) weathered soils have been carried out. The simulated triaxial tests included undrained shear tests on saturated specimens, constant water content tests and wetting tests under constant deviator stress on unsaturated specimens. It should be noted that this model is not meant to capture *all* aspects of unsaturated soil behaviour.

Mathematical formulations

Basic assumptions

The proposed model is formulated on the framework of two independent stress state variables, which are the net normal stress $(\sigma - u_a)$ and matric suction $(u_a - u_w)$. The model is developed for the compressive 'triaxial space'. Effects of osmotic suction are not considered in the proposed model specifically. The model is defined in terms of four state variables: net mean stress (p), deviator stress (q), matric suction (s) and specific volume (v) as follows:

$$p = \frac{\sigma_1 + 2\sigma_3}{3} - u_a \tag{7.1}$$

$$q = \sigma_1 - \sigma_3 \tag{7.2}$$

$$s = u_a - u_w \tag{7.3}$$

$$v = 1 + e \tag{7.4}$$

where σ_1 is the major principal stress, σ_3 the minor principal stress and e the void ratio. The degree of saturation (S_r) or water content (w) is not chosen as an additional state variable in this study. Instead, the relationship between w, p and s proposed by Wheeler (1996) is used to predict the change of suction at either undrained or constant water content conditions.

The work conjugate strain rates for p and q are the volumetric strain increment $(d\varepsilon_v)$ and shear strain increment $(d\varepsilon_q)$, respectively, and are defined by:

$$d\varepsilon_v = d\varepsilon_1 + 2d\varepsilon_3 \tag{7.5}$$

$$d\varepsilon_q = \frac{2}{3}(d\varepsilon_1 - d\varepsilon_3) \tag{7.6}$$

where $d\varepsilon_1$ and $d\varepsilon_3$ are the principal strain increments.

Elasto-plasticity

The model is developed on the classical elasto-plastic framework. It takes into account the possible occurrence of irrecoverable plastic strains. The total strain increment $(d\varepsilon)$ is the sum of elastic strain increment $(d\varepsilon^e)$ and plastic strain increment $(d\varepsilon^p)$. The model assumes elastic behaviour if the soil remains inside a yield surface, and plastic strains will commence once the yield surface is reached. It is assumed that the irreversible plastic response is due to two different mechanisms: shearing and compression. Each mechanism produces both the plastic volumetric strain and plastic shear strain. As a result, $d\varepsilon_v$ and $d\varepsilon_q$ are given by

$$d\varepsilon_v = d\varepsilon_v^e + d\varepsilon_v^p{}_{(s)} + d\varepsilon_v^p{}_{(c)} \tag{7.7}$$

$$d\varepsilon_q = d\varepsilon_q^e + d\varepsilon_q^p{}_{(s)} + d\varepsilon_q^p{}_{(c)} \tag{7.8}$$

where $d\varepsilon_v^e$, $d\varepsilon_v^p{}_{(s)}$, $d\varepsilon_v^p{}_{(c)}$ are the elastic volumetric strain increment, plastic volumetric strain increment due to shearing, plastic volumetric strain increment due to compression, respectively, and $d\varepsilon_q^e$, $d\varepsilon_q^p{}_{(s)}$, $d\varepsilon_q^p{}_{(c)}$ are the elastic shear strain increment, plastic shear strain increment due to shearing, plastic shear strain increment due to compression, respectively.

Elastic strains

A yield surface marks the boundary of the region of elastically attainable states of stress. Changes of stress within the yield surface are accompanied by the purely elastic and recoverable deformations. It is assumed that the soil is isotropic and elastic within the yield surface. Regarding soil stiffness, it is postulated that elastic compressibility is influenced by p, s and S_r. The following elastic volumetric strain increment is proposed

$$d\varepsilon_v^e = \frac{\kappa \cdot d\left[p + sh(S_r)\right]}{v_0\left[p + sh(S_r)\right]} = \frac{\kappa\left[dp + h(S_r)ds + s\frac{dh(S_r)}{dS_r}dS_r\right]}{v_0\left[p + s\,h(S_r)\right]} \tag{7.9}$$

where dp is the increment of net mean stress, ds the increment of suction, dS_r the increment of degree of saturation, κ the elastic stiffness parameter with respect to a change of net mean stress at zero suction, v_0 the initial specific volume of soil and $dh(S_r)/dS_r$ the rate of change of $h(S_r)$ with respect to S_r. The function $h(S_r)$ is a correction factor for suction to take into account the effect of S_r and can be related to a soil–water characteristic curve (SWCC), as shown in Figure 7.1. Two distinct points can be identified: air-entry value (AEV) and residual saturation (RS). AEV is the value of suction where the air commences to enter into a saturated specimen. RS is the saturation level at which the effectiveness of suction to cause further drainage of the liquid phase begins to diminish and further removal of water requires vapour migration (Barbour, 1998). In the proposed model, $h(S_r)$ is chosen as follows and shown in Figure 7.2:

$$h(S_r) = 0 \quad \text{for } 0 \le S_r \le S_{ro} \tag{7.10}$$

$$h(S_r) = \frac{S_r - S_{ro}}{1 - S_{ro}} \quad \text{for } S_r \ge S_{ro} \tag{7.11}$$

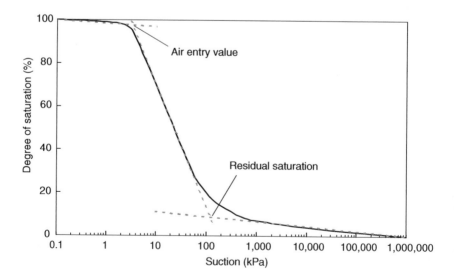

Figure 7.1 A schematic soil–water characteristic curve (after Chiu and Ng, 2003).

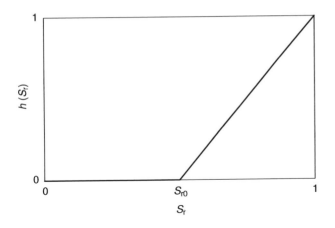

Figure 7.2 Variation of $h(S_r)$ with S_r (after Chiu and Ng, 2003).

where S_{ro} is a critical degree of saturation. For degree of saturation equal to and smaller than S_{ro}, the suction has no effect on the elastic compressibility of the soils. For degrees of saturation greater than S_{ro}, a linear variation between $h(S_r)$ and S_r is assumed. As a first approximation, S_{ro} may be taken as the RS from a SWCC (Figure 7.1). An elastic modulus K is defined as follows:

$$K = \frac{v_0 [p + s \, h(S_r)]}{\kappa} \qquad (7.12)$$

By substituting K into Equation (7.9), the elastic volumetric strain increment can be simplified as follows:

$$d\varepsilon_v^e = \frac{1}{K} d [p + s \, h(S_r)] \qquad (7.13)$$

The elastic shear strain increment is written as

$$d\varepsilon_q^e = \frac{1}{3G} dq \qquad (7.14)$$

where G is the shear modulus and can be deduced from K for an assumed value of Poisson's ratio (v),

$$G = \frac{3(1 - 2v)}{2(1 + v)} K \qquad (7.15)$$

Plastic strains

The previous section has considered a change of stress which lies within current yield surface. Consider now a change of stress, which causes the soil to yield. In the model, the yielding process is governed by both the shearing and compression mechanisms. The plastic volumetric and shear strain increments due to these two mechanisms are expressed as:

$$d\varepsilon_{v\ (s)}^{P} = \Lambda_{(s)}d_{(s)} \tag{7.16}$$

$$d\varepsilon_{v\ (c)}^{P} = \Lambda_{(c)}d_{(c)} \tag{7.17}$$

$$d\varepsilon_{q\ (s)}^{P} = \Lambda_{(s)} \tag{7.18}$$

$$d\varepsilon_{q\ (c)}^{P} = \Lambda_{(c)} \tag{7.19}$$

where $d_{(s)}$ and $d_{(c)}$ are the dilatancy for shearing and compression mechanisms, respectively; $\Lambda_{(s)}$ and $\Lambda_{(c)}$ are the non-negative loading indices for shearing and compression mechanisms, respectively. The dilatancy and loading indices are presented later.

Yield functions

There are two groups of yield surfaces in the model. Figure 7.3(a) shows schematically these yield surfaces in the stress space of $\{q : p : s\}$. A section of the yield surfaces at constant suction is plotted in Figure 7.3(b). The

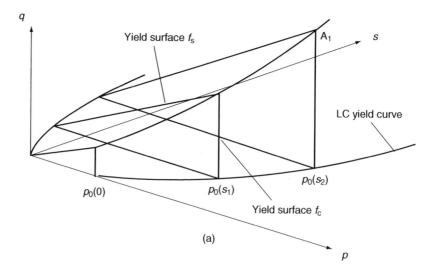

Figure 7.3 Yield surfaces: (a) in q–p–s space; (b) in q–p space at constant suction (after Chiu and Ng, 2003).

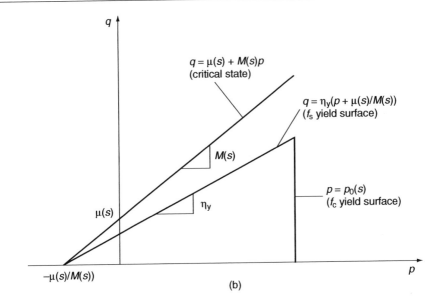

Figure 7.3 (Continued).

first group of yield surfaces (f_s) represents the shearing mechanism, which describes the plastic deformation caused by a change of stress ratio, and takes the following form:

$$f_s = q - \eta_y \left(p + \frac{\mu(s)}{M(s)} \right) \tag{7.20}$$

where η_y is the stress ratio at yielding, $M(s)$ and $\mu(s)$ are the gradient and intercept of the critical state line in the $p : q$ stress plane at a given suction, s. Implicitly, η_y has the characteristic of a hardening parameter and the corresponding hardening rule will be presented later.

The second group of yield surfaces (f_c) represents the compression mechanism, which is a vertical cut-off surface, and is defined as:

$$f_c = p - p_0(s) \tag{7.21}$$

where $p_0(s)$ is the isotropic yield stress at a given suction, s. The isotropic yield stress is a function of suction (s) and total plastic volumetric strain increment $(d\varepsilon_v{}^p)$. The relationship between $p_0(s)$ and s is described by a Loading–Collapse (LC) yield curve (Alonso et al., 1990) and is discussed in the next section. In the model, the two yield surfaces are partially coupled. f_c is influenced by the evolution of f_s, but not vice versa. That is, a constant p test causes the expansion of f_c [or an increase of $p_0(s)$], but an isotropic compression test does not affect f_s (or no change of η_y).

Loading-collapse yield curve

The intersection of the yield surface (f_c) with the $q = 0$ plane defines the LC yield curve [refer to Figure 7.3(a)]. Once the LC yield curve is reached, a further increase of p or a reduction of s leads to an expansion of the yield curve and an irrecoverable plastic compression. Any stress state on the LC yield curve (e.g. C_0 in Figure 7.4) can be related to an isotropic normal compression line of the corresponding suction in the $v : \ln p$ plane. Wheeler and Sivakumar (1995) suggested that the isotropic normal compression lines at various values of suction for unsaturated soils take the following form:

$$v = N(s) - \lambda(s) \ln \left(\frac{p_0(s)}{p_{at}} \right) \tag{7.22}$$

where $\lambda(s)$ is the gradient of the isotropic normal compression line and $N(s)$ the specific volume at the atmospheric pressure, p_{at}. Both $\lambda(s)$ and $N(s)$ vary with suction.

By considering an elastic stress path $A_0 B_0 C_0$ where stress states at A_0 and C_0 lie on the same LC yield curve at different values of suction (refer to Figure 7.4), the LC yield curve can be defined by the isotropic normal

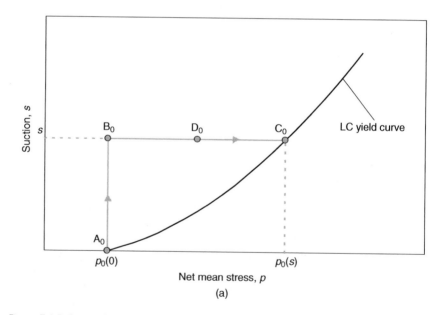

Figure 7.4 Relating the LC yield curve to the isotropic normal compression lines and unloading–reloading line at different values of suction (after Wheeler and Sivakumar, 1995; Chiu and Ng, 2003).

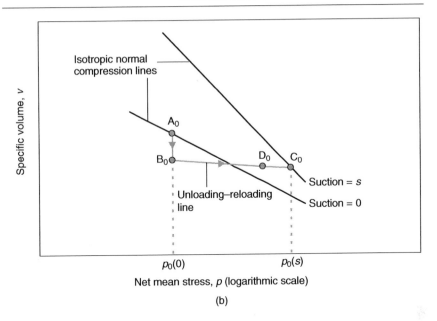

Figure 7.4 (Continued).

compression lines and unloading–reloading line at different values of suction as follows:

$$\lambda(s)\ln\left(\frac{p_0(s)}{p_{at}}\right) - \kappa\ln\left(\frac{p_0(s) + s\ h(S_r)}{p_{at}}\right) = [\lambda(0) - \kappa]\ln\left(\frac{p_0(0)}{p_{at}}\right) + N(s) - N(0)$$

(7.23)

where $\lambda(0)$ and $N(0)$ are the values of $\lambda(s)$ and $N(s)$ at zero suction, respectively, $p_0(0)$ is the isotropic yield stress at zero suction. With different values of $p_0(0)$, Equation (7.23) defines a family of yield curves in the $v : \ln p$ plane and $p_0(0)$ can be viewed as a hardening parameter. The hardening rules are presented later.

Flow rules

Different stress–dilatancy relationships have been proposed in the literature for saturated soils. Rowe (1962) postulated that dilatancy is related only to the effective stress ratio and angle of soil friction. However, experimental evidence (Cubrinovski and Ishihara, 1998; Chiu, 2001) reveals that the dilatancy of saturated soils depends on the internal state, e.g. void ratio, as well as stress level (or stress ratio). Recently, Li and Dafalias (2000) suggested the following state-dependent dilatancy for saturated soils (i.e. at zero suction),

$$\frac{d\varepsilon_v^P}{d\varepsilon_q^P} = d_1 \left(e^{m\psi} - \frac{\eta'}{M} \right) \tag{7.24}$$

where ψ is the state parameter, M the gradient of the critical state line (CSL) in the plane of effective stress (p') and deviator stress (q), η' the current effective stress ratio (q/p'), and, d_1 and m are material parameters. For saturated soils, ψ is defined as the difference between the current void ratio $[e(0)]$ and critical void ratio $[e_c(0)]$ corresponding to the current effective mean stress in the $e : lnp'$ plane (Been and Jefferies, 1985). If the current state lies above the CSL, ψ is positive and vice versa (refer to Figure 7.5). Equation (7.24) relates dilatancy to both material state and stress ratio. For unsaturated soils, dilatancy also depends on suction (Cui and Delage, 1996; Chiu, 2001). Hence, a modified Equation (7.24) for unsaturated soils is proposed as follows:

$$d_{(s)} = \frac{d\varepsilon_v^P{}_{(s)}}{d\varepsilon_q^P{}_{(s)}} = d_1(s) \left(e^{m(s)\psi(s)} - \frac{\eta}{M(s)} \right) \tag{7.25}$$

where η is the current stress ratio defined by $q/(p+\mu(s)/M(s))$, $M(s)$ the gradient of the critical state line in the $p : q$ stress plane, $\psi(s)$ the state parameter expressed in the space of $\{e : \ln p : s\}$, the two material parameters, $d_1(s)$ and $m(s)$, are no longer constants but vary with suction. For unsaturated soils,

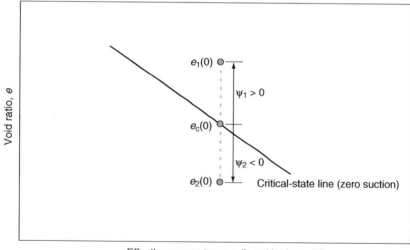

Figure 7.5 State parameters in $e - \ln p'$ plane (after Chiu and Ng, 2003).

$\psi(s)$ is defined as the difference between current void ratio $[e(s)]$ and critical void ratio $[e_c(s)]$ corresponding to the current net mean stress (p) and current suction (s) in the space of $\{e : \ln p : s\}$. The critical state of unsaturated soils is discussed later. Equation (7.25), which relates dilatancy to stress ratio, material state and suction, is used to evaluate the direction of plastic strain increments for f_s.

For evaluating the direction of plastic strain increments related to the yield surface f_c, the following flow rule is postulated,

$$d_{(c)} = \frac{d\varepsilon_v^{P}{}_{(c)}}{d\varepsilon_q^{P}{}_{(c)}} = [\lambda(s) - \kappa] d_2(s) \frac{M(s)}{\eta} \tag{7.26}$$

where $d_2(s)$ is a material parameter, which is a function of suction. Equation (7.26) shows that $d_{(c)}$ equals to infinity when η equals to zero, which yields zero plastic shear strain at isotropic compression.

Hardening rules

The loading indices $\Lambda_{(s)}$ and $\Lambda_{(c)}$ can be obtained through the plastic consistency conditions and are presented by Chiu and Ng (2003) in their Appendix 1. Volumetric hardening is assumed for both groups of yield surfaces. For f_s, plastic modulus $K_{p(s)}$ is proposed as follows,

$$K_{p(s)} = [h_1(s) - h_2(s)e_0] G \left[\frac{M(s)}{\eta} - e^{n(s)\psi(s)} \right] \tag{7.27}$$

where e_0 is the initial void ratio, $h_1(s)$, $h_2(s)$ and $n(s)$ are material parameters. $K_{p(s)}$ takes a similar form of the plastic modulus proposed by Li and Dafalias (2000) for saturated sands. Similar to dilatancy given in Equation (7.25), Equation (7.27) also relates plastic modulus to η, material state and suction. It is obvious that η is equal to the stress ratio at yielding (η_y) during plastic loading. In addition, $h_1(s)$ and $h_2(s)$ are the scaling factors, which take into account the effect of density. For f_c, the following plastic modulus $K_{p(c)}$ is proposed,

$$K_{p(c)} = v_o p_o(s) d_2(s) \frac{M(s)}{\eta} \tag{7.28}$$

It can be seen from Equation (7.28) that $K_{p(c)}$ equals to infinity when η is equal to zero, which yields a zero plastic shear strain increment at isotropic compression.

Critical state

It is assumed that an unsaturated soil will ultimately attain a critical state under continuous shearing. Wheeler and Sivakumar (1995) proposed to represent the critical states at various values of suction by the critical state lines defined by

$$q = M(s)p + \mu(s) \tag{7.29}$$

$$v = \Gamma(s) - \omega(s) \ln \left(\frac{p_0(s)}{p_{at}} \right) \tag{7.30}$$

where $M(s)$ and $\mu(s)$ are the gradient and intercept of the critical state line in the $p : q$ stress plane, respectively; $\omega(s)$ and $\Gamma(s)$ are the gradient and intercept of the critical state line in the $v : \ln p$ plane, respectively. It is assumed in the model that $M(s)$, $\mu(s)$, $\Gamma(s)$ and $\omega(s)$ all vary with suction.

Elasto-plastic variation of specific water content

In order to predict the variation of suction under either undrained or constant water content conditions, the elasto-plastic variation of specific water content (v_w, defined as $1 + S_r e$) with p and s proposed by Wheeler (1996) is used. He defined the air void ratio (e_a) as the volume of air within an element of soil containing unit volume of solids and further postulated that e_a is a function of suction and total volume of macrovoid (inter-aggregate pore) under isothermal condition. The total volume of macrovoid is a function of plastic volumetric strain, which is related to hardening parameter $p_0(0)$. As a result, e_a can be postulated to be a function of s and $p_0(0)$ and shown as follows:

$$e_a = A(s) - \alpha(s) \ln \left(\frac{p_0(0)}{p_{at}} \right) \tag{7.31}$$

$$ds = \frac{\dfrac{[\lambda(0) - \alpha(s) - \kappa]}{p_0(0)} dp_0(0) + \dfrac{\kappa}{p + sh(S_r)} dp + \kappa \dfrac{s \left[\dfrac{dh(S_r)}{dS_r} \right]}{p + sh(S_r)} dS_r}{\dfrac{d\alpha(s)}{ds} \ln \dfrac{p_0(0)}{p_{at}} - \dfrac{dA(s)}{ds} - \kappa \dfrac{h(S_r)}{p + sh(S_r)}} \tag{7.32}$$

where $A(s)$ and $\alpha(s)$ are functions of suction. Then, specific water volume (v_w) is given by subtracting e_a from v. Detailed derivation of the elastic and plastic components of the change in specific water volume is presented by Chiu and Ng (2003) in their Appendix 2. It is found that the increment of suction (ds) caused by shearing at constant water content is related to increments of p, S_r and $p_0(0)$ through the Equation 7.32.

Determination of model parameters

The parameters used in the proposed model are grouped into seven categories: elastic, water retention, isotropic compression, critical state, state-dependent dilatancy, hardening and variation of specific water content. Due to the space constraint, only the determination of the model parameters from the experimental data of the gravelly sand is described. In fact, identical determination procedures are used for sandy silt (Chiu, 2001). The calibrated values of model parameters for the gravelly sand and the sandy silt are summarized in Tables 7.1 and 7.2, respectively.

Elastic parameters

Two elastic parameters are used for the model. κ can be determined from the unloading and reloading paths of the isotropic compression tests at zero suction. κ for the gravelly sand is 0.012, while Poisson's ratio (ν) is assumed to be independent of suction and taken as 0.2 in this study.

Table 7.1 Model parameters for gravely sand (DG) (after Chiu and Ng, 2003)

Soil parameters		Suction (kPa)			
		0	*20*	*40*	*25–33*
Elastic	κ			0.012	
	ν			0.2	
Isotropic compression	$\lambda(0)$	0.115	–	–	–
	$N(0)$	1.936	–	–	–
	r_λ			1.26	
	r_N			1.02	
	β (kPa^{-1})			0.05	
Critical state	$\omega(0)$	0.110	–	–	–
	$\Gamma(0)$	1.824	–	–	–
	r_ω			1.80	
	r_Γ			1.10	
	$M(s)$			1.55	
	$\mu(s)$ (kPa)	0	–	–	7
Dilatancy	$m(s)$			5	
	$d_1(s)$	0.20	–	0.6	–
	$d_2(s)$	9.8	–	7.7	–
Hardening	$n(s)$			1	
	$h_1(s)$	0.55	–	0.24	–
	$h_2(s)$	0.05	–	0.1	–
Water retention	S_{r0}			0.47	
Specific water content	$\alpha(s)$	0.0	0.102	0.104	–
	$A(s)$	0.0	0.457	0.471	–

Table 7.2 Model parameters for sandy silt (DV) (after Chiu, 2001; Chiu and Ng, 2003)

Soil parameters		Suction (kPa)			
		0	80	150	80–150
Elastic	κ			0.012	
	ν			0.2	
Isotropic compression	$\lambda(0)$	0.09	–	–	–
	$N(0)$	1.927	–	–	–
	r_λ			2.02	
	r_N			1.18	
	β (kPa^{-1})			0.08	
Critical state	$\omega(0)$	0.087	–	–	–
	$\Gamma(0)$	1.883	–	–	–
	r_ω			2.63	
	r_Γ			1.15	
	$M(s)$			1.32	
	$\mu(s)$ (kPa)	0	–	–	33
Dilatancy	$m(s)$			3	
	$d_1(s)$	0.34	–	–	1.11
	$d_2(s)$	12.3	–	–	5.3
Hardening	$n(s)$			5	
	$h_1(s)$	0.73	–	–	0.89
	$h_2(s)$	0.33	–	–	0.47
Water retention	S_{r0}			0.3	
Specific water content	$\alpha(s)$	0.0	0.0780	0.0794	–
	$A(s)$	0.0	0.297	0.382	–

Parameters for water retention

The SWCC of the gravelly sand used in this study is not available. Therefore, the SWCC of a coarse-grained decomposed soil, which has a similar particle size distribution, as shown in Figure 7.6 is adopted. S_{r0} is estimated from the RS, which is 0.47.

Parameters for isotropic compression

The gradients [$\lambda(s)$] and intercepts [$N(s)$] of Equation (7.22) are obtained from isotropic compression tests conducted at different values of suction. Both $\lambda(s)$ and $N(s)$ vary with suction. Alonso et al. (1990) suggested the following relationship of $\lambda(s)$ with suction,

$$\lambda(s) = \lambda(0)\left\{(1 - r_\lambda)e^{-\beta s} + r_\lambda\right\} \tag{7.33}$$

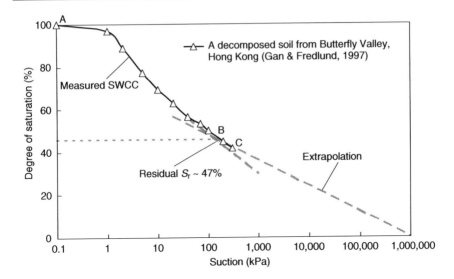

Figure 7.6 Soil–water characteristic curve of a coarse-grained decomposed soil (after Gan and Fredlund, 1997; Chiu and Ng, 2003).

where $\lambda(0)$ is the value of $\lambda(s)$ at zero suction, r_λ a parameter related to the maximum value of $\lambda(s)$ at an infinite suction and β a parameter which controls the rate of increase of $\lambda(s)$ with suction. Similarly, the following relationship of $N(s)$ with suction is assumed:

$$N(s) = N(0) \left\{ (1 - r_N) e^{-\beta s} + r_N \right\} \tag{7.34}$$

where the parameters $N(0)$ and r_N are defined similar to $\lambda(0)$ and r_λ of Equation (7.33). For simplicity, it is assumed that β is the same for both $\lambda(s)$ and $N(s)$. Equations (7.33) and (7.34) are used to fit the measured data of gravelly sand. Figure 7.7 shows the calibration of β, r_λ, r_N, $\lambda(0)$ and $N(0)$ with the measured values of $\lambda(s)$ and $N(s)$ at different suctions, and the corresponding values are summarized in Table 7.1.

Figure 7.7(a) compares the measured values of $\lambda(s)$ of gravelly sand, sandy silt, a collapsible silt (Maatouk *et al.*, 1995) and a compacted kaolin (Wheeler and Sivakumar, 1995). It can be seen that $\lambda(s)$ increases (or stiffness decreases) with suction for both gravelly sand and sandy silt (i.e. r_λ is greater than 1). As suction exceeds 80 kPa, $\lambda(s)$ approaches a steady value for sandy silt. These test results are consistent with those of the collapsible silt and the compacted kaolin for the range of suction considered. The reduction in soil stiffness as suction increases is attributed to the initial loose state of the specimens.

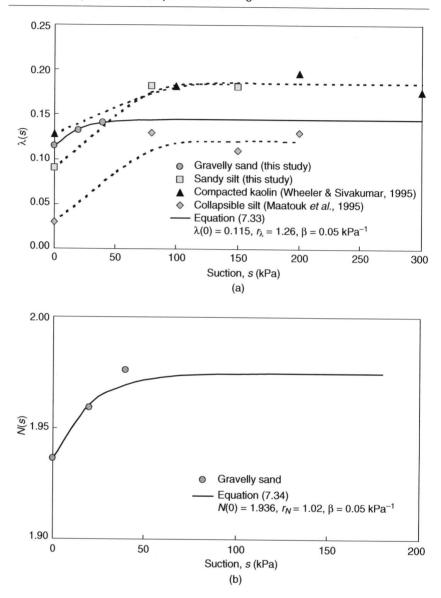

Figure 7.7 Calibrations of parameters for gravelly sand: (a) $\lambda(0)$, r_λ and β; (b) $N(0)$, r_N and β (after Chiu and Ng, 2003).

Parameters for critical state

The critical state parameters $M(s)$, $\mu(s)$, $\omega(s)$ and $\Gamma(s)$ of Equations (7.29) and (7.30) are obtained from shearing specimens to the critical state at different values of suction. Based on the test results on gravelly sand (Chiu,

2001), $M(s)$ can be taken as a constant (i.e. $M(s) = M(0) = 1.55$). On the other hand, $\mu(s)$ increases non-linearly with suction, and the measured values are summarized in Table 7.1. A bi-linear relationship between $\mu(s)$ and s is used. $\mu(s)$ increases linearly with s until $s = 25\,\text{kPa}$. For s greater than $25\,\text{kPa}$, $\mu(s)$ remains constant. Regarding $\omega(s)$ and $\Gamma(s)$, the following relationships are assumed:

$$\omega(s) = \omega(0)\left\{(1 - r_\omega)e^{-\beta s} + r_\omega\right\} \tag{7.35}$$

$$\Gamma(s) = \Gamma(0)\left\{(1 - r_\Gamma)e^{-\beta s} + r_\Gamma\right\} \tag{7.36}$$

where parameters $\omega(0)$, r_ω, $\Gamma(0)$ and r_Γ are defined similar to $\lambda(0)$, r_λ, $N(0)$ and r_N of Equations (7.33) and (7.34), respectively. For simplicity, the same value of β is used for both the isotropic normal compression and critical state lines. Then, Equations (7.35) and (7.36) are used to fit the measured data of gravelly sand. Figure 7.8 shows the calibration of the parameters r_ω, r_Γ, $\omega(0)$ and $\Gamma(0)$ with the measured values of $\omega(s)$ and $\Gamma(s)$ at different suctions, and the corresponding values are summarized in Table 7.1.

Dilatancy parameters for f_s

The dilatancy parameters $d_1(s)$ and $m(s)$ [refer to Equation (7.25)] are suction dependent. At zero suction, the best-fit values of parameters $m(0)$ and $d_1(0)$

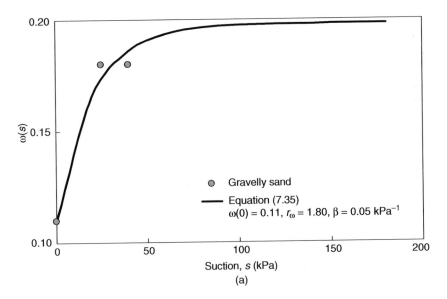

Figure 7.8 Calibrations of parameters for gravelly sand: (a) $\omega(0)$, r_ω and β; (b) $\Gamma(0)$, r_Γ and β (after Chiu and Ng, 2003).

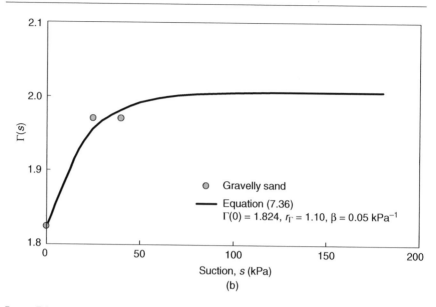

Figure 7.8 (Continued).

are determined from the undrained tests. Since there is no total volumetric strain on the saturated specimens for undrained tests, the plastic component of volumetric strain is equal to the elastic component in magnitude. The dilatancy can be rewritten as follows:

$$\frac{d\varepsilon_v{}^p}{d\varepsilon_q{}^p} \approx \frac{-d\varepsilon_v{}^e}{d\varepsilon_q} = \frac{-dp'/K}{d\varepsilon_q} \tag{7.37}$$

Based on Equation (7.37), dilatancy can be evaluated from the stress–strain curves of the undrained tests and are used to calibrate the best-fit values of $m(0)$ and $d_1(0)$. It is found that the best-fit values of $m(0)$ and $d_1(0)$ are 5 and 0.2, respectively. Figure 7.9 shows the measurements and predictions of dilatancy by Equation (7.25) against η'/M for typical undrained tests on saturated specimens of the gravelly sand.

For other values of suction, best-fit values of parameters $d_1(s)$ and $m(s)$ are determined from constant p tests, so that the measured plastic strains are induced by the shearing mechanism (f_s) only. The measured dilatancy can be evaluated from the stress-strain curves of constant p tests. Assuming $m(s)$ remains unchanged in this study (i.e. $m(s) = m(0) = 5$), it is found that the best-fit value of $d_1(s)$ at a suction of 40 kPa is 0.6. Figure 7.10 shows the measurements and predictions of dilatancy by Equation (7.25) against η/M for typical constant p tests at a suction of 40 kPa on unsaturated specimens of the gravelly sand. For suctions ranging from 0 to 40 kPa, a linear relationship between $d_1(s)$ and s is assumed.

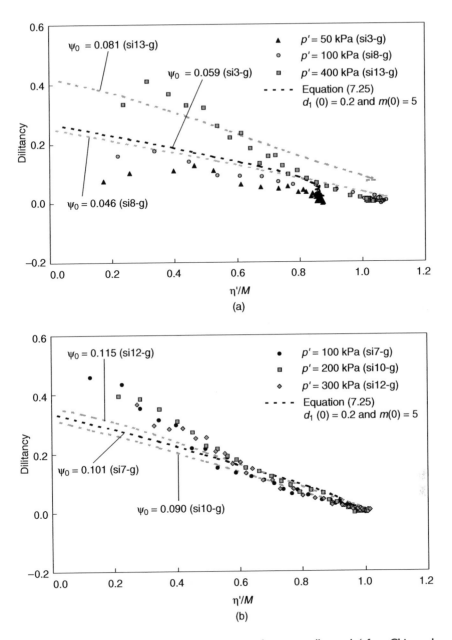

Figure 7.9 Dilatancy against η'/M at zero suction for a gravelly sand (after Chiu and Ng, 2003).

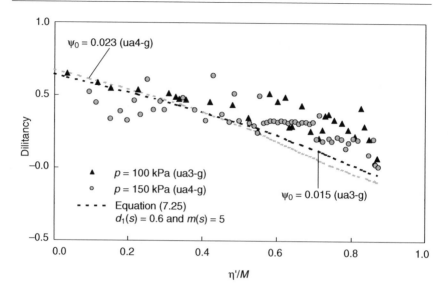

Figure 7.10 Dilatancy against η/M at suction of 40 kPa for a gravelly sand (after Chiu and Ng, 2003).

Dilatancy parameters for f_c

The parameters $d_2(s)$ of Equation (7.26) can be calibrated from constant η compression tests at different values of suction. However, no constant η test data is available in this study and the following procedures are used to estimate $d_2(s)$. At K_0 compression, there is no lateral strain (i.e. $d\varepsilon_3 = 0$), hence the left-hand side of Equation (7.26) becomes,

$$\frac{d\varepsilon_v^p{}_{(c)}}{d\varepsilon_q^p{}_{(c)}} \approx \frac{d\varepsilon_v^p}{d\varepsilon_q^p} = \frac{d\varepsilon_1}{\frac{2}{3}d\varepsilon_1} = 1.5 \tag{7.38}$$

Substituting Equation (7.38) into Equation (7.26) and rearranging into the following form:

$$d_2(s) = \frac{1.5\eta_{K_0}}{M(s)\,[\lambda(s) - \kappa]} \tag{7.39}$$

where η_{K_0} is the net stress ratio at K_0 condition. As parameters $M(s)$, $\lambda(s)$ and κ have been determined in previous sections, the only unknown of Equation (7.39) is η_{K0} or the coefficient of earth pressure at rest (K_0), which may depend on both suction and stress history of unsaturated soils. As a first approximation, it is assumed that K_0 is independent of suction and the empirical expression proposed by Jaky (1944) is used to estimate K_0 for normally compressed soil and shown as follows,

$$K_0 = 1 - \sin \phi' \tag{7.40}$$

By combining Equations (7.39) and (7.40) and assuming a constant value of K_0, the values of $d_2(s)$ are determined and summarized in Table 7.1. For suctions between 0 and 40 kPa, $d_2(s)$ decreases linearly with s.

Hardening parameters

The hardening parameters for f_s are $h_1(s)$, $h_2(s)$ and $n(s)$. At zero suction, the following stress–strain relationship is considered,

$$\frac{dq}{d\varepsilon_q} \approx \frac{dq}{d\varepsilon_q^p} = \frac{dq}{\frac{1}{K_p}(dq - \eta' \, dp')} = \frac{K_p}{1 - \eta' \frac{dp'}{dq}} \tag{7.41}$$

Substituting Equation (7.27) into Equation (7.41),

$$\frac{dq}{d\varepsilon_q} = \frac{[h_1(0) - h_2(0)e_0] \, G\left[\frac{M(0)}{\eta'} - e^{n(0)\psi(0)}\right]}{1 - \eta' \frac{dp'}{dq}} \tag{7.42}$$

Equation (7.42) is used to fit the measured stress–strain curves of the undrained tests, which only involve the change of f_s (i.e. $dp' < 0$). It is found that the best-fit values of parameters $h_1(0)$, $h_2(0)$ and $n(0)$ at zero suction are 0.55, 0.05 and 1, respectively. For other values of suction, the dp' and η' in Equation (7.42) are replaced by dp and η_y, respectively. The expression becomes

$$\frac{dq}{d\varepsilon_q} = \frac{[h_1(s) - h_2(s)e_0] \, G\left[\frac{M(s)}{\eta_y} - e^{n(s)\psi(s)}\right]}{1 - \eta_y \frac{dp}{dq}} \tag{7.43}$$

Then, Equation (7.43) is used to fit the measured stress–strain curves of the constant p tests. For simplicity, it is assumed that $n(s)$ is a constant [i.e. $n(s) = n(0)$], and $h_1(s)$ and $h_2(s)$ vary with suction. It is found that the best-fit values of parameters $h_1(s)$ and $h_2(s)$ for a suction of 40 kPa are 0.24 and 0.1, respectively. Similar to $d_1(s)$ and $d_2(s)$, $h_1(s)$ and $h_2(s)$ for suctions between 0 and 40 kPa are interpolated from the above best-fit values.

Parameters for specific water content

The isotropic normal compression lines for v_w is obtained by subtracting Equation (7.31) from Equation (7.22) and shown as follows:

$$v_w = N(s) - \lambda(s) \ln\left(\frac{p_0(s)}{p_{at}}\right) - A(s) + \alpha(s) \ln\left(\frac{p_0(0)}{p_{at}}\right) \qquad (7.44)$$

By substituting Equation (7.23) into Equation (7.44), Equation (7.44) becomes:

$$\begin{aligned} v_w = & N(s) - A(s) - \frac{\alpha(s)}{\lambda(0) - \kappa}[N(s) - N(0)] - \lambda(s)\left[1 - \frac{\alpha(s)}{\lambda(0) - \kappa}\right] \\ & \ln\left(\frac{p_0(s)}{p_{at}}\right) - \frac{\kappa\alpha(s)}{\lambda(0) - \kappa}\ln\left(\frac{p_0(s) + sh(S_r)}{p_{at}}\right) \end{aligned} \qquad (7.45)$$

Equation (7.45) can be considered to have the following form:

$$v_w = B_0(s) + B_1(s) \ln\left(\frac{p_0(s)}{p_{at}}\right) + B_2(s) \ln\left(\frac{p_0(s) + s\,h(S_r)}{p_{at}}\right) \qquad (7.46)$$

where:

$$B_0(s) = N(s) - A(s) - \frac{\alpha(s)}{\lambda(0) - \kappa}[N(s) - N(0)] \qquad (7.47)$$

$$B_1(s) = \lambda(s)\left[\frac{\alpha(s)}{\lambda(0) - \kappa} - 1\right] \qquad (7.48)$$

$$B_2(s) = -\frac{\kappa\alpha(s)}{\lambda(0) - \kappa} = -\kappa\left[\frac{B_1(s)}{\lambda(s)} + 1\right] \qquad (7.49)$$

$B_0(s)$, $B_1(s)$ and $B_2(s)$ are parameters used to define the isotropic normal compression lines for v_w. These three parameters can be measured from isotropic compression tests conducted at different suction values. By rearranging Equations (7.47) and (7.48), the following expressions of $\alpha(s)$ and $A(s)$ are given:

$$\alpha(s) = \frac{\lambda(s) + B_1(s)}{\lambda(s)}[\lambda(0) - \kappa] \qquad (7.50)$$

$$A(s) = N(s) - B_0(s) - \frac{\alpha(s)}{\lambda(0) - \kappa}[N(s) - N(0)] \qquad (7.51)$$

The measured values of $\alpha(s)$ and $A(s)$ at suctions of 0, 20 and 40 kPa are summarized in Table 7.2. The values at other suction are interpolated from the measured values.

Comparisons between model predictions and experimental results for decomposed granite

Undrained tests on saturated specimens of coarse-grained gravelly Sand (DG)

The model of Ng and Chiu (2003) is used to predict three undrained tests on saturated specimens of a coarse-grained gravelly sand (si6-g, si11-g and si12-g). All three specimens were isotropically compressed to similar void ratios but different confining pressures or different states before shearing (refer to Table 7.3); si6-g ($\psi = -0.089$) lies on the dense (dry) side of the

Table 7.3 Testing conditions for unsaturated specimens (Chiu and Ng, 2003)

Specimen identity[a]	Soil type	Initial specific volume	Net mean stress before shearing or wetting (kPa)	Specific volume before shearing or wetting	State parameter before shearing or wetting	Applied suction before shearing or wetting (kPa)
ui1-g[b]	Gravelly sand (DG)[d]	2.087	25[f]	2.076	−0.143	40
ui2-g[b]	Gravelly sand (DG)[d]	2.091	50[f]	2.053	−0.041	40
ui3-g[b]	Gravelly sand (DG)[d]	2.083	100[f]	1.983	0.014	40
ua1-g[c]	Gravelly sand (DG)[d]	2.085	25[f]	2.080	−0.139	40
ua2-g[c]	Gravelly sand (DG)[d]	2.087	50[g]	2.037	−0.057	40
ua3-g[c]	Gravelly sand (DG)[d]	2.096	100[g]	1.923	−0.046	40
ua4-g[c]	Gravelly sand (DG)[d]	2.093	150[g]	1.829	−0.067	40
ui1-v[b]	Sandy silt (DV)[e]	2.486	25[f]	2.462	−0.033	80
ui2-v[b]	Sandy silt (DV)[e]	2.503	50[f]	2.416	0.082	80
ui3-v[b]	Sandy silt (DV)[e]	2.526	150[f]	2.215	0.137	80

Notes

a 'u' represents unsaturated, 'i' represents isotropically compression, 'a' represents specimen subjected to a net stress ratio greater than zero, 'g' represents decomposed granitic soil (gravelly sand) and 'v' represents decomposed volcanic soil (sandy silt).

b Constant water content test.

c Wetting test at constant deviator stress.

d Specimens were compacted to an initial dry density of 75% Proctor maximum.

e Specimens were compacted to an initial dry density of 70% Proctor maximum.

f Shearing at constant water content after isotropic compression.

g A net stress ratio ($\eta = q/p$) of 1.4 was applied to the specimens before wetting.

critical state and si11-g ($\psi = 0.07$) and si12-g ($\psi = 0.115$) lie on the loose (wet) side of the critical state. The measurements and model predictions of the undrained tests are shown in Figure 7.11. It is noted that the proposed model is able to capture the key features of the loose and dense saturated specimens with one unique set of soil parameters.

Figure 7.11a shows the comparisons between the measured and predicted shear stress and strain relationships. It can be seen that the model can predict the strain-softening behaviour of the two loose specimens (si11-g and si12-g) well. For the si6-g test, despite the initial discrepancy between the prediction

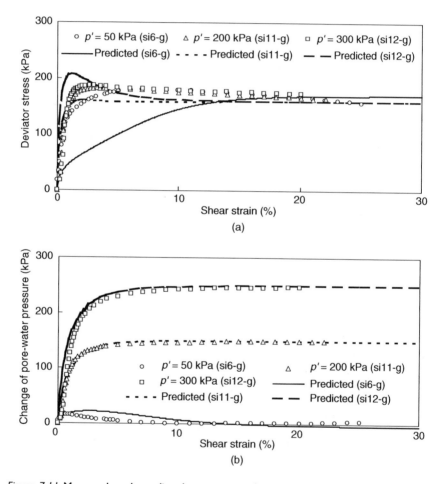

Figure 7.11 Measured and predicted parameters for gravelly sand: (a) stress–strain relationship; (b) relationship between excess pore water pressure and shear strain; (c) effective stress path for undrained tests on saturated specimens (after Chiu and Ng, 2003).

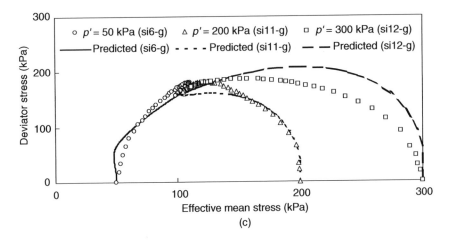

Figure 7.11 (Continued).

and measurement, the strain-hardening behaviour of the dense specimen is captured by the model well at large strains. The model gives much better predictions for the generated excessive pore water pressure (Figure 7.11b) for all three tests. Figure 7.11c shows that the predictions agree well with the three measured effective stress paths because of the incorporation of state-dependent dilatancy in the model. As the state of a specimen changes from loose (or positive ψ, e.g. si11-g and si12-g) to dense states (or negative ψ, e.g. si6-g), the proposed model predicts a change from contractive to dilative tendency accordingly.

Constant water content tests on unsaturated specimens of coarse-grained gravelly sand (DG)

Figure 7.12 compares the model predictions and measurements of three constant water content tests on unsaturated specimens of gravelly sand (ui1-g, ui2-g and ui3-g). All three specimens were isotropically compressed to different confining pressures before shearing. Two specimens, ui1-g ($\psi = -0.143$) and ui2-g ($\psi = -0.041$), lie on the dense side of the critical state and the third one, ui3-g ($\psi = 0.014$), lies on the loose side of the critical state. Similar to the saturated specimens, the model predictions, which are based on one unique set of soil parameters, are consistent with the experimental data over a range of internal state (ψ ranging from -0.143 to 0.014) qualitatively.

Figure 7.12a and b shows that the model gives reasonably good predictions for stress–strain and volumetric–shear strain relationships of the unsaturated specimens. Regarding the stress–strain relationship, the model predicts the occurrence of strain-hardening behaviour and that shear strength increases with confining pressure. Regarding volumetric strain, the model predicts volumetric contraction for the range of the confining pressure considered and the magnitude of contraction increases with confining pressure (or the amount of contraction increases with a more positive value of ψ).

Figure 7.12c shows that the model predicts monotonic reductions in suction during shearing and the amount of reduction increases slightly with confining pressure (or increases with ψ). The predicted change of suction agrees with the measurements for a confining pressure of 100 kPa (ui3-g). However, the predictions overestimate the amount of reduction for specimens compressed to confining pressures smaller than 100 kPa (ui1-g and ui2-g). This discrepancy may be due to the different forms of pore water

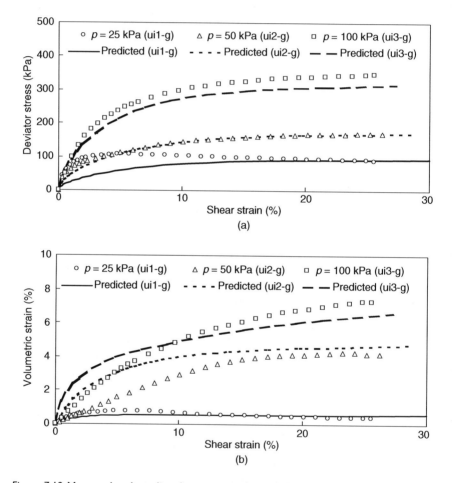

Figure 7.12 Measured and predicted parameters for gravelly sand: (a) stress–strain relationship; (b) volumetric strain–shear strain relationship; (c) variation of suction with shear strain for constant water content tests on unsaturated specimens at initial $s = 40\,$kPa (after Chiu and Ng, 2003).

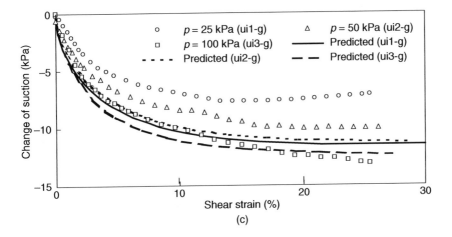

Figure 7.12 (Continued).

existed in the microstructure of the compacted soil, which contains above 20 per cent of clay and silt (see Ng and Chiu, 2003). Croney *et al.* (1958) suggested that clay specimens compacted on the dry side of optimum may exist as saturated packets (or aggregates), which contain small and water-filled (or nearly saturated) intra-aggregate voids, separated by large inter-aggregate air voids. Then, the pore water may exist in two different forms: the bulk water within saturated packets of clay and meniscus water between large inter-aggregate air voids (Gens, 1995; Delage and Graham, 1995; Wheeler and Karube, 1995). For a loosely compacted unsaturated specimen compressed to a low confining pressure (such as ui1-g and ui2-g), the inter-aggregate air void created by compaction may be preserved in comparison to other specimen subjected to a higher confining pressure. Hence, a better prediction may be obtained by considering the effects of the bulk water and meniscus water separately in deriving the variation of v_w with p, s and S_r for loosely compacted specimens compressed to a low confining pressure.

Wetting tests at constant deviator stress on unsaturated specimens of coarse-grained gravelly sand (DG)

Figure 7.13 compares the model predictions and the measurements of four wetting tests carried out at constant deviator stress on unsaturated specimens of the gravelly sand (ua1-g, ua2-g, ua3-g and ua4-g). All four specimens were anisotropically compressed to the same stress ratio but different confining pressures before wetting. All four specimens lie on the dense side of the critical state with ψ ranging from -0.139 to -0.046, and ua1-g lies the furthest away from the critical state line ($\psi = -0.139$) among the four specimens.

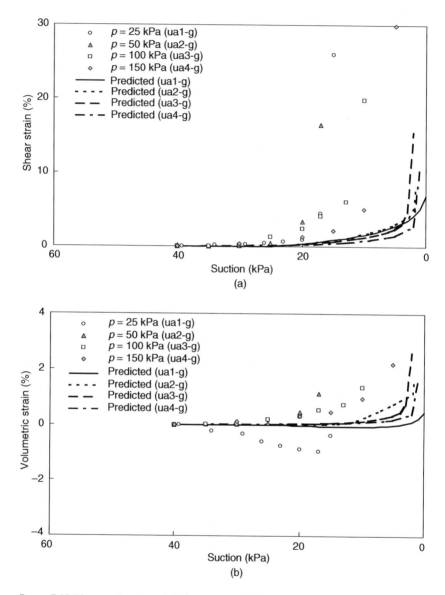

Figure 7.13 Measured and predicted variations of (a) shear strain and (b) volumetric strain with suction for wetting tests at constant deviator stress on unsaturated specimens of gravelly sand (after Chiu and Ng, 2003).

As soil suction decreases from 40 to 25 kPa, only a small strain is induced. The model predictions are in good agreement with the experimental data for this suction range because only small and elastic response occurs within the yield surface. As suction reduces further to about 15 kPa (average), a sudden and substantial increase in shear and volumetric strain can be seen from the experimental data. Although the model is unable to predict precisely when the rapid increase in strain occurred, the observed sudden increase in strain is captured reasonably well by the model in all four tests. This type of soil behaviour cannot be captured by most of the existing models. This is because the applied stress ratio ($\eta = 1.4$) of the unsaturated specimens (ua1-g to ua4-g) is smaller than the gradient of critical state line ($M = 1.55$), therefore the state of the specimens always lies below the critical state line throughout the wetting tests. In most existing unsaturated elasto-plastic models, yielding on the LC yield surface during wetting at constant deviator stress would usually give relatively modest plastic shear strains (unless the stress state is brought very close to a critical state). On the other hand, the proposed model predicts much larger plastic shear strains during wetting at constant deviator stress (as observed in the experimental tests), because of yielding on the f_s surface. In addition, the hardening rule for f_s is dependent on the internal state [Equation (7.27)] and the plastic modulus can reduce substantially as the stress states of the specimens lie below the critical state line.

Comparisons between model predictions and experimental results for decomposed volcanic soil

Undrained tests on saturated specimens of fine-grained sandy silt (DV)

To further illustrate the capability of the model, it is also used to predict the test results of a fine-grained sandy silt. Two initially identical saturated specimens (si4-v and si5-v) were isotropically compressed to different confining pressures before undrained shearing. The measurements and model predictions of the undrained tests are compared in Figure 7.14. It can be seen that the model predictions are in good agreement with the experimental data for the fine-grained soil. In particular, the model captures the phase transformation of the effective stress paths for both specimens (Figure 7.14c).

Constant water content tests on unsaturated specimens of fine-grained sandy silt (DV)

Figure 7.15 compares the model predictions and measurements of three constant water content tests on unsaturated specimens of the sandy silt (ui1-v, ui2-v and ui3-v) conducted at a suction of 80 kPa. All three specimens were isotropically compressed to different confining pressures before shearing (see Table 7.4). Test ui1-v ($\psi = -0.033$) lies on the dense side of the critical state, ui2-v ($\psi = 0.082$) and ui3-v ($\psi = 0.137$) lie on the loose side of the critical state.

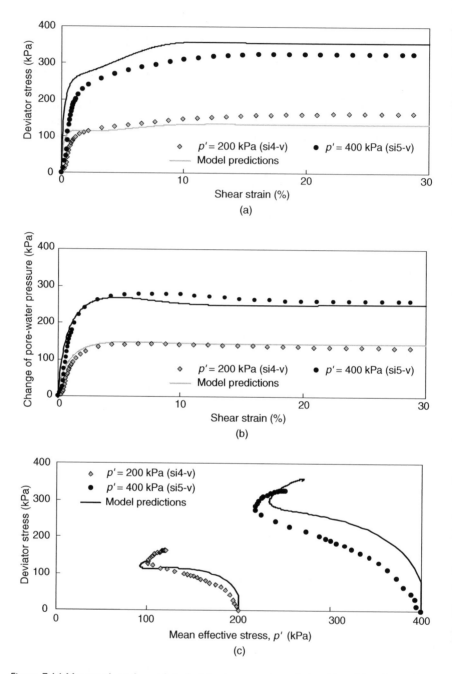

Figure 7.14 Measured and predicted (a) stress–strain relationship; (b) relationship between excess pore water pressure and shear strain; (c) effective stress path for undrained tests on saturated specimens of sandy silt (after Chiu and Ng, 2003).

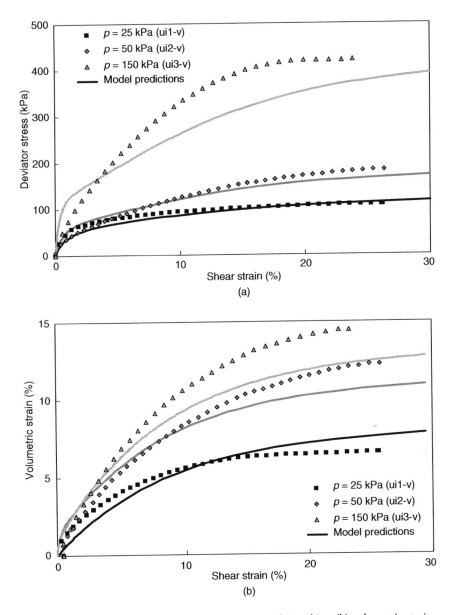

Figure 7.15 Measured and predicted (a) stress–strain relationships; (b) volumetric strain–shear strain relationships for constant water content tests on unsaturated specimens of sandy silt at initial s = 80 kPa (after Chiu and Ng, 2003).

Table 7.4 Testing conditions for undrained tests on saturated specimens (Chiu and Ng, 2003)

Specimen identity	Soil type	Initial percentage of Proctor maximum	Initial specific volume	Effective mean stress before shearing (kPa)	Specific volume before shearing	State parameter before shearing
si6-g	Gravelly sand (DG)	84	1.852	50	1.813	−0.089
si11-g	Gravelly sand (DG)	81	1.931	200	1.820	0.070
si12-g	Gravelly sand (DG)	74	2.099	300	1.820	0.115
si4-v	Sandy silt (DV)	70	2.464	200	1.882	0.059
si5-v	Sandy silt (DV)	70	2.435	400	1.797	0.035

Similar to the coarse-grained gravelly sand, the model prediction is consistent with the experimental data over a range of internal states (i.e. density), in particular the stress–strain and volumetric–shear strain relationships at confining pressures smaller than 150 kPa (i.e. ui1-v and ui2-v).

Summary

A state-dependent elasto-plastic constitutive model for both saturated and unsaturated soils has been developed to model the behaviour of soils over various ranges of densities, stress levels and soil suctions. The model is formulated under an extended critical state framework using two independent stress state variables: net normal stress and matric suction. The concept of state-dependent dilatancy is incorporated for modelling saturated as well as unsaturated soils. The effects of internal state, stress level and soil suction on dilatancy are considered in the model. Relevant testing procedures have been demonstrated to calibrate and obtain the required model parameters.

The capabilities of the model are illustrated by predicting three series of triaxial tests (undrained tests on saturated specimens, constant water content tests and wetting tests at constant deviator stress on unsaturated specimens) performed on a coarse-grained gravelly sand (DG) and a fine-grained sandy silt (DV). By using a single set of soil parameters and the incorporation of state-dependent dilatancy, the model's predictions are generally consistent with the experimental results, in particular for the effective stress paths and stress–strain relationships observed in undrained tests on saturated specimens, stress–strain and volumetric–shear strain relationships measured from constant water content tests on unsaturated specimens over various ranges of densities and stress levels, and some key features such as a sudden increase in shear and volumetric strains during wetting tests at constant deviator stress.

Field trials and numerical studies in slope engineering of unsaturated soils

Part 4 Frontispiece

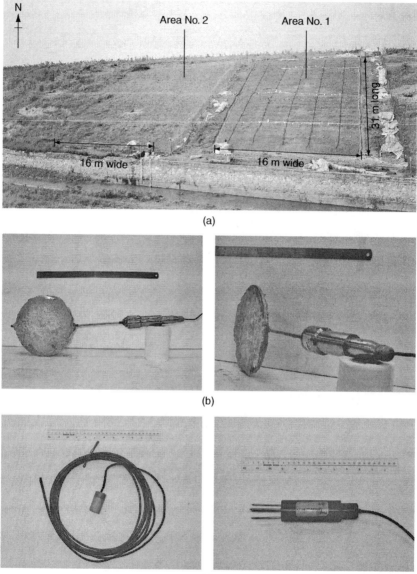

(a)

(b)

(c)

Images that represent the content of this Part 4: (a) Field trial slope; (b) Vibrating wire earth pressure cell; (c) Thermal conductivity suction sensor (left) and Theta probe for measuring volumetric water content (right); (d) Jet Fill tensiometer.

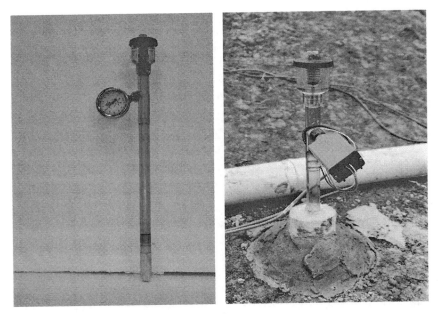

(d)

(Continued).

Chapter synopses

Chapter 8: Instrumentation and performance: A case study in slope engineering

To assist in the design of the Middle Route of the South-to-North Water Transfer Project (SNWTP) in China, two areas, each 16 m wide and 31 m long, along an 11 m high 22° cut slope in a typical medium-plastic expansive clay were extensively instrumented and their engineering performance was monitored. The 1200 km Middle Route of the SNWTP is to carry potable water from the Yangtze River region in the south to many arid and semi-arid areas in the northern regions of China, including Beijing. One of the major geotechnical problems is to design safe and economical dimensions for cut slopes that form an open channel with a trapezoidal cross-section in unsaturated expansive soils which possess high swelling and shrinkage potential.

The unique instrumentation package was designed to consider the influence of both the independent stress state variables: soil suction and net normal stress. The instrumentation package included tensiometers, thermal conductivity suction sensors, moisture probes, earth pressure cells, inclinometers, a tipping bucket rain gauge, a vee-notch flowmeter and an evaporimeter. The fundamental mechanisms of rainfall infiltration into the unsaturated expansive soil during two artificial rainfall events and the complex interaction among changes of soil suction (or water content), in situ stress state and soil deformation leading to slope failure were studied. Not only are the monitoring data useful for engineering designs of slopes in expansive unsaturated clays, but they are also invaluable for calibrating constitutive models and numerical simulation procedures. This case study is thus an invaluable resource both for the practising civil engineer as well as the numerical modeller.

Chapter 9: Engineering applications for slope stability

Limit equilibrium methods of slope stability analysis in unsaturated soils using the extended Mohr–Coulomb failure criterion are introduced. A three-dimensional numerical parametric study of rainfall infiltration into an unsaturated soil slope is described and the effect of various rainfall patterns on groundwater is found. The critical influence of antecedent rainfall on pore water pressure distributions and hence on slope instability is revealed. Other numerical parametric studies show how slope stability is influenced by state-dependent soil–water characteristic curves (SDSWCCs) undergoing wetting and drying at different stress conditions. The critical influence of damming ground water flow on slope stability is also modelled. A potential non-conservative design method that uses a conventional SWCC is described. Other factors affecting slope stability are investigated including the influence of re-compaction, surface cover and impeding layers. The role of soil nails in a potentially static liquefiable loose fill slope is investigated using a strain-softening soil model. The theory of a new conjunctive surface–subsurface modelling of rainfall infiltration is introduced and explained. This new model is capable of simulating rainfall infiltration and surface ground flow simultaneously and iteratively, without knowing the coefficient of rainfall infiltration in advance. The importance of the effects of conjunctive modelling on slope stability is reviewed. Based on the results of the numerical parametric studies presented in this chapter, design engineers will improve their understanding of rainfall infiltration mechanisms and transient flows in unsaturated soil slopes and hence be able to predict slope stability with greater confidence.

Chapter 8

Instrumentation and performance: A case study in slope engineering

Source

This chapter is made up of verbatim extracts from the following sources for which copyright permissions have been obtained as listed in the Acknowledgements.

- Ng, C.W.W., Zhan, L.T., Bao, C.G., Fredlund, D.G. and Gong, B.W. (2003). Performance of an unsaturated expansive soil slope subjected to artificial rainfall infiltration. *Géotechnique*, 53, 2, 143–157.

The South-to-North Water Transfer Project, China (Ng et al., 2003)

Introduction

Expansive soils can be found on almost all continents. The destructive effects caused by this type of soil have been reported in many countries around the world, including the United States, Australia, South Africa, India, Canada, China and Israel (Nelson and Miller, 1992; Steinberg, 1998). According to Steinberg, the annual loss due to the damage caused by expansive soils is US$10 billion and up to ¥100 million in the United States and China, respectively. Clearly, there is an urgent need to improve our understanding of the fundamental behaviour of expansive soils and design methodology for civil engineering structures constructed on these soils.

A major infrastructure project, the South-to-North Water Transfer Project (SNWTP), has been proposed to carry potable water from the Yangtze River region in the south to many arid and semi-arid areas in the northern regions of China, including Beijing. The proposed 1200 km 'middle route' is likely to be an open channel with a trapezoidal cross-section formed by cut slopes and fills. At least 180 km of the proposed excavated canal will pass through areas of unsaturated expansive soils. One of the major geotechnical problems is to design safe and economical dimensions for the cut slopes in the unsaturated expansive soils, which possess high swelling and shrinkage potential (Bao and Ng, 2000).

Although field studies of the effects of rainfall infiltration on slope stability have been carried out on residual soil slopes by some researchers (Lim *et al.*, 1996), the fundamental mechanisms of rainfall infiltration into unsaturated expansive soils during wet seasons, and the complex interaction among changes of soil suction (or water content), in situ stress state and soil deformation leading to slope failure are not fully understood. To improve our understanding of the fundamental mechanisms of rain-induced retrogressive landslides in unsaturated expansive soils, an 11 m high cut slope in a typical medium-plastic expansive clay in Zaoyang, close to the 'middle route' of SNWTP in Hubei, China, was selected for a comprehensive, well-instrumented field study of rainfall infiltration. The instrumentation includes tensiometers, thermal conductivity suction sensors, moisture probes, earth pressure cells, inclinometers, a tipping-bucket rain gauge, a vee-notch flowmeter and an evaporimeter. Calibrations of these instruments are described and discussed by Zhan *et al.* (2006).

The test site

The test site is located on the intake canal of the Dagangpo second-level pumping station in Zaoyang, Hubei, China (Figure 8.1). It is about 230 km northwest of Wuhan and about 100 km south of the intake canal for the SNW TP in Nanyang, Henan, China.

The site is in a semi-arid area with an average annual rainfall of about 800 mm, and 70 per cent of the annual rainfall is distributed from May to September.

The intake canal at the test site was excavated in 1970 with an average excavation depth of 13 m. The slope angle following excavation was 22°.

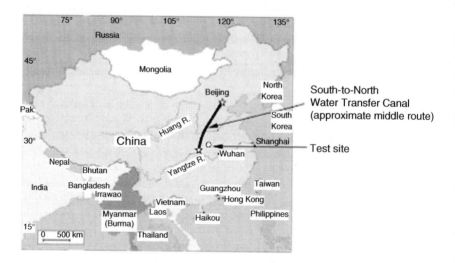

Figure 8.1 Site location for field trial (after Ng *et al.*, 2003).

Several years after construction, a number of slope failures have taken place in succession, and parts of a masonry retaining wall have been seriously deflected or destroyed. Most of the mass movement has occurred during wet seasons, and the slip surfaces are in the order of 2 m deep.

The test site areas were selected on a cut slope on the northern side of the canal (Figure 8.2). The area has a uniform slope angle of $22°$ and a uniform slope height of 11 m (measured from the top of the retaining wall). There was a 1 m wide berm at the mid-height of the slope. The slope surface was well grassed, but no trees were present. The area has a significant depth of typical unsaturated expansive soil. The ground level at the toe of the slope is approximately +96 m OD (ordnance datum). About 5 m away from the slope toe there is a 3 m high masonry retaining wall. The depth of the canal below the slope toe is about 3–5 m. Just to the west of the selected testing area there are a number of typical shallow slips and retrogressive slope failures. In this chapter, only the instruments and monitored results for the bare slope are described and reported. Details of the monitoring in the grassed area and comparisons between the results for the bare and grassed areas are given by Zhan *et al.* (2007) and Ng and Zhan (2007), respectively.

Soil profile and properties

Three groups of boreholes were drilled to a layer with hard and coarse calcareous concretions around the monitored area on the bare slope (Figure 8.3). Each group comprised two boreholes spaced 1 m apart. One borehole was

Figure 8.2 Overall view of the selected cut slope (after Ng et al., 2003).

Figure 8.3 Instrumented bare slope and layout of boreholes (after Ng et al., 2003).

for sampling and standard penetration tests (SPTs), the other was for dilatometer tests (DMT). Two boreholes (BH7 and BH8) were drilled in the monitored area and used for the installation of inclinometers after sampling and SPTs. The soil profiles and geotechnical parameters obtained from the boreholes around mid-slope (BH5 and BH6) are shown in Figure 8.4.

As shown in Figure 8.4, the predominant stratum in the slope was a yellow-brown stiff fissured clay. The clay layer was sometimes inter layered with thin layers of grey clay or iron concretions. The yellow-brown clay contained about 15 per cent hard and coarse calcareous concretions (particle size generally from 30 to 50 mm). X-ray diffraction analyses indicated that the predominant clay minerals are illite (31–35 per cent) and montmorillonite (16–22 per cent), with a small percentage of kaolinite (8 per cent) (Liu, 1997). The natural water content was generally slightly larger than the plastic limit ($w_p = 19:5$ per cent, $I_p = 30$) with the exception of a relatively low water content within the top 1 m. The dry density profile down to 2 m indicated that a relatively dense soil layer was present at a depth of about 1–5 m.

The clay was typically overconsolidated because of desiccation, as indicated by the total stress K_0 value obtained from dilatometer tests, and by its swelling pressure, which ranged from 30 to 200 kPa. The clay exhibited significant swelling and shrinkage characteristics upon wetting and drying, and an abundance of cracks and fissures were observed in the field. The width of some cracks was as large as 10 mm on the slope surface, and could be observed to extend as deep as 1 m.

The soil–water characteristics of the expansive clay near the test site were investigated by Wang (2000). Figure 8.5 shows some soil–water characteristic curves (SWCCs) measured on two undisturbed specimens with a dry density of 1.40 Mg/m^3, which is smaller than the measured average dry

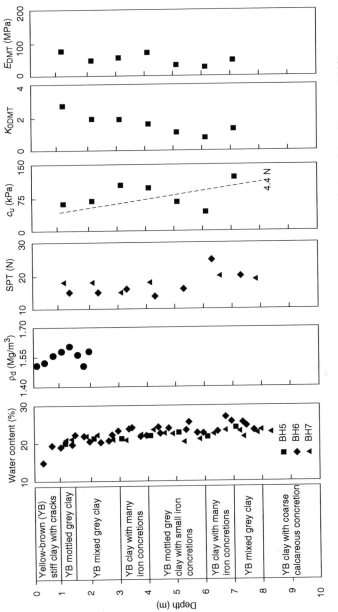

Figure 8.4 Soil profiles and geotechnical parameters from the boreholes located at mid-slope (after Ng et al., 2003).

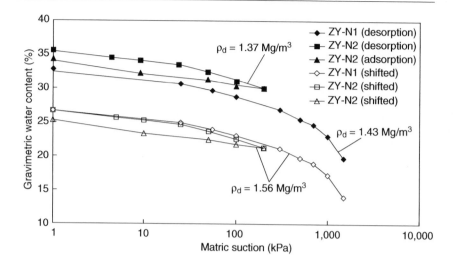

Figure 8.5 Soil–water characteristic curves for the soil from Zaoyang (after Ng et al., 2003).

density for the soils in the monitoring area (1.56 Mg/m³). Based on a study of the effects of soil density on SWCCs of a sandy silt (Ng and Pang, 2000), the SWCCs for the expansive soil with a density of 1.56 Mg/m³ may be approximately deduced and constructed by shifting the measured SWCCs at 1.40 Mg/m³, as shown in Figure 8.5. The shape of the SWCCs for the expansive soil is relatively flat, which indicates that the expansive soil possesses a high water-retention capability. The air-entry value of the soil is about 30 kPa. The drying and wetting hysteresis between the desorption and adsorption curves is relatively insignificant.

Field instrumentation programme (bare area only)

An area 16 m wide by 31 m long with a cleared surface (Figure 8.3) was selected for instrumentation and artificial rainfall simulation tests. The instruments included jet fill tensiometers, thermal conductivity suction sensors (Fredlund *et al.*, 2000), ThetaProbes for determining water content (Delta-T Devices Ltd, Cambridge, UK), vibrating-wire earth pressure cells, inclinometers, a tipping-bucket rain gauge, a vee-notch flowmeter and an evaporimeter. The layout and locations of the instruments are shown in Figures 8.6 and 8.7, and the details for each instrument are summarized in Table 8.1.

Monitoring soil suction and water content

As shown in Figures 8.6 and 8.7, there were three rows of instrumentation for soil suction and water content monitoring: R1 at the upper part, R2

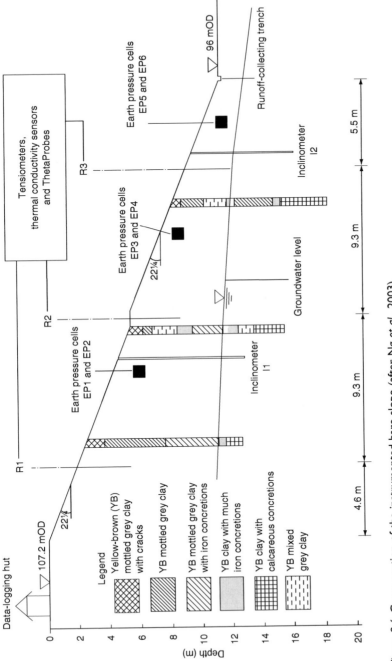

Figure 8.6 Cross-section of the instrumented bare slope (after Ng et al., 2003).

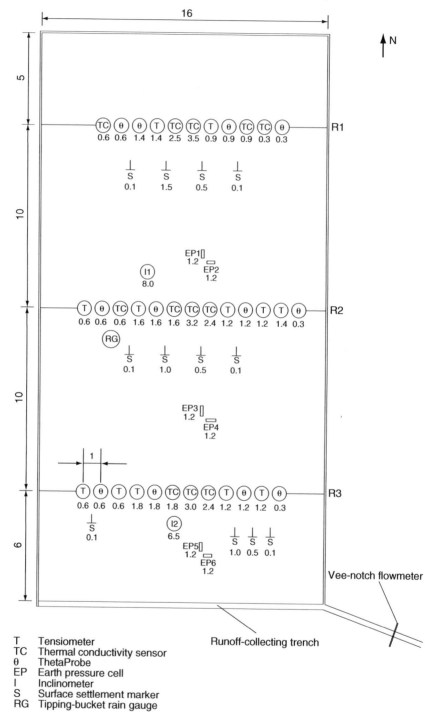

T	Tensiometer
TC	Thermal conductivity sensor
θ	ThetaProbe
EP	Earth pressure cell
I	Inclinometer
S	Surface settlement marker
RG	Tipping-bucket rain gauge

Figure 8.7 Layout of instruments in the bare area. All dimensions are in metres; numbers denote depths (after Ng *et al.*, 2003).

Table 8.1 Summary of instruments in the bare area of the trial slope (after Ng et al., 2003)

Measurement	Type of instrument	Quantity	Measuring range	Source/references
Soil suction	Jet fill tensiometer (2527A)	12	Matric suction less than 90 kPa	Soilmoisture Equipment Corporation, Santa Barbara, USA
	Thermal conductivity sensor	12	Matric suction 20–1,500 kPa	Fredlund et al. (2000)
Water content	ThetaProbe (ML2x/w)	12	Volumetric water content, 0–50%	Delta-T Devices Ltd, Cambridge, UK
Horizontal stress	Vibrating-wire earth pressure cell (EPS-36-S)	6	0–1,000 kPa	Encardio-Rite Electronics Pvte Ltd, Lucknow, India
Rainfall intensity	Tipping-bucket rain gauge (ARC 100)	2	0.2 mm/tip	Environmental Measurements Ltd, UK
Run off	Vee-notch flowmeter	1	N/A	–
Horizontal displacement	Inclinometer	2	±12°	Nil
Heave/ settlement	Movement point	12	N/A	Nil
Potential evaporation rate	Evaporimeter	1	N/A	Nil

at the middle part and R3 at the lower part of the slope. In each row, there were seven to nine suction sensors (i.e. jet fill tensiometers or thermal conductivity sensors; Fredlund et al., 2000) and four ThetaProbes, which were used for measuring volumetric water content (VWC) indirectly. These probes use the standing-wave technique to measure the apparent dielectric constant of a soil, which is then correlated with the volumetric water content in the soil. The sensors at each row were spaced 1 m apart. Most of the sensors were embedded within a depth of 2 m. For each depth, there were generally two soil suction sensors and one ThetaProbe.

Disturbed samples were taken slightly below the three rows of instrumentation using a small-diameter auger. This was done every day for determination of the gravimetric water content (GWC) profiles during the rainfall simulation period. All the auger holes were backfilled immediately after sampling.

Monitoring horizontal total stresses

Three pairs of earth pressure cells were installed for monitoring horizontal total stress in two orthogonal directions. As mentioned before, a mid-slope berm divided the slope into two parts. Two pairs of earth pressure cells were embedded 2.5 m above the toe in the upper half and the lower half of the slope, respectively. The other pair of earth pressure cells was located midway between the former two pairs. For each pair, one earth pressure cell was installed to measure the horizontal stress in the north–south direction (i.e. the inclination direction of the slope); the other was placed in the east – west direction (i.e. parallel to the longitudinal direction of the canal). All six earth pressure cells were installed vertically at a depth of 1.2 m. The installation procedure proposed by Brackley and Sanders (1992) was adopted to minimize soil disturbance due to the excavation of a slot for a pressure cell.

Monitoring horizontal movements and surface heave

Two inclinometers were installed in two orthogonal directions: one (I1) near the toe of the upper portion of the slope and the other (I2) near the toe of the lower portion of the slope. The inclinometers were bottomed at depths of 8.0 and 6.5 m, respectively (i.e. down to the hard layer with coarse and hard calcareous concretions).

In order to measure the swelling of the unsaturated expansive soil as a result of rainwater infiltration, three rows of movement points were set up near the three main rows of instrumentation (R1, R2 and R3), respectively. The movement points were constructed with concrete blocks. Each row has four movement points founded at depths of 0.1, 0.5 and 1.0 m, respectively. Two levelling datum points were constructed 20 m outside the artificial rainfall area and founded at a depth of 3 m. These two datum points were frequently monitored and checked using a city grid datum located over 100 m away from the site. The monitoring and checking confirmed that the two datum points were stable and were not affected by the artificial rainfall.

Monitoring rainfall intensity, run-off and evaporation

A tipping-bucket rain gauge was installed to record the intensity and duration of rainfall. Flowmeters were installed in the main water-supply line of a sprinkler system to record the total amount of water sprinkled onto the slope within a given time interval. A water collection channel was constructed along the toe of the slope to measure surface run-off using an automatic vee-notch flowmeter system installed at the end of the channel. An evaporimeter was installed at the middle of the slope outside the monitoring area to measure potential evaporation.

Apart from the two inclinometers and the movement points, all other instruments were connected to the computerized data acquisition system housed at the top of the slope.

Artificial rainfall simulations for the bare area

Rainfall was artificially produced using a specially designed sprinkler system. This was done to accelerate the field test programme. The sprinkler system comprised a pump, a main water-supply pipe, five branches and 35 sprinkler heads. The system could produce three levels of rainfall intensity (3, 6 and 9 mm/h).

The site was fairly dry from November 2000 to April 2001, with a total rainfall of only 60 mm. In May, when the wet season generally begins, there was only about 40 mm of rainfall. From June to 18 August, prior to the start of the rainfall simulation tests, the monitored area was protected against rainfall infiltration with a plastic membrane. The rainfall simulation tests therefore started from a relatively dry soil condition.

Figure 8.8 shows the two simulated rainfall events during the one-month monitoring period, from 13 August to 12 September 2001. Two rainfall events were simulated. The first lasted for 7 days, from the morning of 18 August to the morning of 25 August 2001, with an average daily rainfall of 62 mm. The second simulated rainfall was from the morning of 8 September to the afternoon of 10 September 2001. During both rainfall periods, in the morning of each day, the artificial rainfall was stopped for 2–3 h to allow the measurement of horizontal displacements and soil swelling, as well as to auger disturbed specimens for the determination of GWC profiles. Apart from this regular stoppage, the artificial rainfall intensity was maintained constant at 3 mm/h.

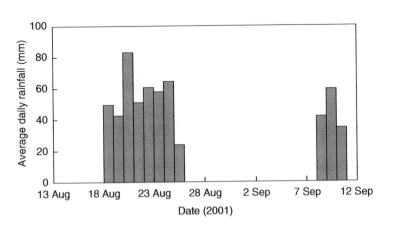

Figure 8.8 Intensity of rainfall events for the bare area during the trial period (after Ng et al., 2003).

The surface run-off from the artificial rainfall was measured by the vee-notch flowmeter. If the amount of infiltration during the two rainfall periods is assumed to be equal to the difference between the rain intensity and surface run-off, then the percentage of infiltration can be calculated by dividing the amount of infiltration by the rain intensity. During the first $1\frac{1}{2}$ days after the beginning of the artificial rainfall, the percentage of infiltration was equal to 100 per cent i.e. no run-off. Thereafter, the percentage of infiltration decreased with rainfall duration. After 4 days of rainfall, the percentage of infiltration tended towards a steady 30 per cent. Note that the measured percentage of infiltration was corrected for evaporation. Based on the measurements made by the evaporimeter, it was found that the evaporation potential ranged from 2 to 8 mm/day, which was relatively small compared with the rainfall intensity of 60 mm/day.

Observed field performance for the bare area

Effect of simulated rainfall on pore water pressures and soil suction

RESPONSES OF SOIL SUCTION OR PORE WATER PRESSURE

Since the changes of in situ pore water pressure (PWP) or soil suction, in response to the simulated rainfalls, exhibited a similar characteristic for the three different sections (R1, R2 and R3, located at the upper, middle and lower sections of the slope, respectively), typical monitored results at R2 have been selected and are shown in Figure 8.9. The results recorded by four tensiometers and four thermal conductivity sensors are presented in terms of PWP and soil suction in Figure 8.9a and b, respectively.

Immediately prior to the first artificial rainfall on 18 August 2001, negative PWPs ranging from 18 to 62 kPa were recorded by the four tensiometers (see Figure 8.9a). As expected, the higher the elevation of the tensiometer, the larger the negative PWP. With the exception of the thermal conductivity sensor (R2-TC-0-6), which showed a high soil suction of about 250 kPa at 0.6 m below ground (see Figure 8.9b), soil suctions deduced from the remaining three sensors were generally consistent with the measurements obtained from the tensiometers.

After the first heavy artificial rainfall started on 18 August 2001, the negative PWP and soil suction measured by most sensors began to decrease only after about 2 days of rainfall (about 90 mm of rain). As shown in Figure 8.9a, there was a clear delay in PWP response to rainfall infiltration, even at a depth of 0.6 m. Within the depth of 1.5 m, the duration of delay appeared to decrease as depth increased. Based on field reconnaissance and observations in trial pits, it was found that many cracks and fissures occurred near the ground surface, and a layer of relatively impermeable material was identified at about 1.5 m below the ground surface. It is postulated that, as

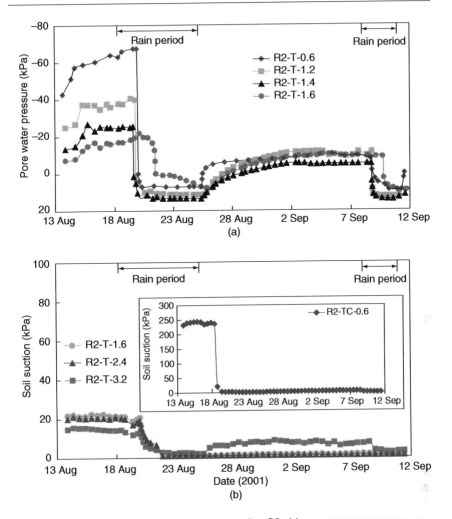

Figure 8.9 Response from suction sensors located at R2: (a) pore water pressures measured by tensiometers; (b) soil suctions measured by thermal conductivity sensors (after Ng et al., 2003).

the intact expansive clay has a relatively low water permeability, water can ingress the clay only through cracks and fissures. While rainwater flowed through the cracks and fissures initially, the tensiometers did not register any significant changes of soil suction around their tips, and this led to the initial response delay. Subsequently, when the infiltrated rainwater started to rise from the bottom of the cracks or from a perched water table formed owing to the presence of the impeding layer and seep in all directions, the lower the tensiometer, the quicker the response (i.e. the shorter the delay). Obviously, a rapid response was shown by a sharp reduction in negative PWP when water reached the locations of the tensiometers above the impeding layer. The tensiometer located below the impeding layer showed the slowest and most gradual rate of response to rainfall and the lowest magnitude of reduction of negative PWP.

After the first rainfall, the lower three tensiometers showed a gradual increase (or recovery) in negative PWP, and appeared to reach a steady state condition after 2 September, recording negative PWP from 3 to 10 kPa. The rate of recovery was very similar for the lower three tensiometers. On the other hand, the top one showed a much more rapid recovery initially. However, the final magnitudes of recovered negative PWP fell within a narrow range, and did not appear to be strongly governed by the depths of the tensiometers.

During the second artificial rainfall, the lower three tensiometers showed almost no delay in response to the rainfall. There was a change in PWP from negative to positive, although the magnitude of the change (about 10 kPa) was not very significant. On the other hand, the top tensiometer showed a 1-day delay in response to the second rainfall. However, the 'final equilibrium' PWP recorded during the second rainfall was similar to those during the first one.

The general responses of indirect soil suction measurements by thermal conductivity sensors to the two artificial rainfalls (see Figure 8.9b) were similar to those recorded by the tensiometers, except that the former showed a slower rate of response than the latter. The magnitudes of PWPs measured by the two different types of sensor were generally consistent, particularly at the depth of 1–6 m below ground. However, the inconsistency shown between the two sensors located at the depth of 0–6 m before the first rainfall may be due to the inherent limitation of tensiometers caused by cavitation at high suction.

RESPONSES OF PIEZOMETRIC LEVEL

Figure 8.10 shows the variations of piezometric level (i.e. elevation head plus PWP head) with time for sections R1, R2 and R3. The piezometric levels were calculated from the monitored results of the suction sensors installed at three different depths (0.6, 1.2 and 2.4 m) at each section. The elevation head for each sensor was calculated according to the local datum (96 m OD) located at the slope toe. The responses of piezometric levels within and below the 1 m depth are shown in Figure 8.10a and b, respectively.

Prior to the commencement of the first rainfall event, the piezometric levels at a depth of 0.6 m decreased with the elevation of the three sections (i.e. from R3 to R1: see Figure 8.10a). This suggests that there was an up-slope water flow within 1 m depth below the ground surface, resembling the capillary rise of water in an inclined column of unsaturated soil. In

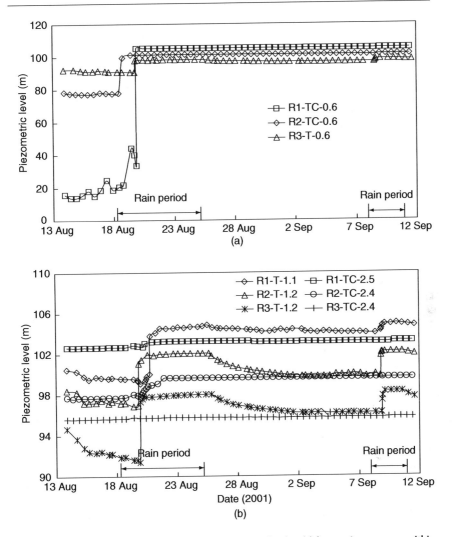

Figure 8.10 Variations of piezometric level at various depths: (a) by suction sensors within 1 m depth; (b) by suction sensors below 1 m depth (after Ng *et al.*, 2003).

contrast, at a given depth of 1 m or more below the ground surface, the piezometric levels increased with the elevation of the sections (see Figure 8.10b). This indicates a down-slope water flow below the depth of 1 m or more. The observed difference in the water flow directions above and below a depth of 1 m prior to the rainfall may be due to the presence of large numbers of opened cracks and fissures near the ground surface. These structural features enhanced the process and rate of evaporation, and hence resulted in substantial high negative PWP or soil suction near the ground surface (see Figure 8.9). By contrast, as the number of open cracks and fissures decreased significantly with depth, the influence of cracks and fissures appeared to be negligible at greater depths.

After the commencement of the first rainfall, the piezometric levels at each section increased owing to the decrease in negative PWP. At a given depth along the slope, the piezometric levels decreased with reduction in the elevations. A down-slope water flow phenomenon appeared as a result of rainfall infiltration. During the two-week non-rainfall period, the magnitude of the difference in the piezometric levels at sections R1, R2 and R3 still remained the same, even though there was a slight recovery of negative PWPs in the soil, indicating that there was a water flow in the down-slope direction after the rainfall. These results were consistent with the observed exit of groundwater near the slope toe for several days after the first rainfall period.

VARIATIONS OF IN SITU PORE WATER PRESSURE (PWP) PROFILES

Figure 8.11 shows the PWP distributions with depth. Prior to the commencement of the first rainfall period, the negative PWPs near the ground surface were substantially higher than those at greater depths, and hence the PWP profiles deviated significantly from theoretical hydrostatic conditions. The negative PWPs below a depth of 2 m were relatively low and decreased gently with an increase in depth.

After 3 days of heavy rainfall of about 180 mm, the PWPs increased significantly within the upper 2 m of soil. A positive PWP appeared at a depth of about 1–5 m below the ground at sections R2 and R3. The continued rainfall after 21 August resulted in a further increase in PWP but at a significantly reduced rate.

At the end of each of the two rainfall periods (i.e. 25 August and 10 September), the significant positive PWP was observed by tensiometers within the upper 2 m of soil, was the largest at a depth of about 1.5 m at each section. This seemed to indicate the presence of a perched groundwater table at about 1.5 m below the ground surface. The measured in situ dry density profiles demonstrated that there was a denser ($\rho_d \geq 1.60 \, \text{Mg/m}^3$) soil layer, ranging from 0.3 to 0.5 m, located at about 1.5 m (see Figure 8.4). It is believed that the dense soil layer possesses a relatively low coefficient of water permeability, and hence the infiltrated rainwater is retained above this dense layer. Owing to the impedance effect of the dense layer, the influence of rainfall on PWP below the 1.5 m depth at this site was generally insignificant. The presence of the perched groundwater table at a depth of about 1.5 m caused the development of significant positive PWPs, which led to the expansion of the initially dry expansive soil upon wetting, resulting in a reduction in the shear strength of the soil layer. This may explain

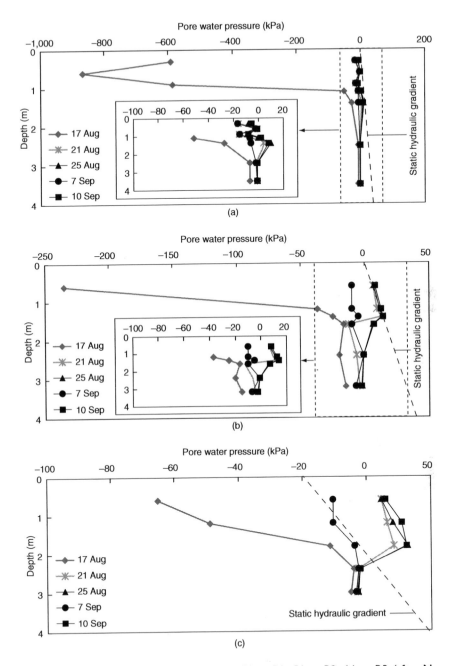

Figure 8.11 Variation of pore water pressure: (a) at R1; (b) at R2; (c) at R3 (after Ng *et al.*, 2003).

why most rain-induced landslides appear to be relatively shallow, generally found within a 2 m depth (Bao and Ng, 2000).

RESPONSE OF VOLUMETRIC WATER CONTENT (VWC)

Figure 8.12 shows the monitored results of VWC by four ThetaProbes located at the R2 section during the two artificial rainfalls. The response of VWC recorded by the ThetaProbes was generally consistent with the corresponding PWP responses shown in Figure 8.9. With the exception of the ThetaProbe located at 0–3 m below ground surface (i.e. R2-θ-0.3), there was a delay of at least 2 days in changes of VWC in response to the first artificial rainfall, which started on 18 August. The infiltration characteristics revealed by the lower three ThetaProbes were generally consistent with the pore water pressure responses shown in Figure 8.9. The rapid response of R2-θ-0.3 might be attributed to the presence of a large number of cracks and fissures near the surface. For the lower three ThetaProbes, the one at 1.2 m depth (R2-θ-1.2) responded first, followed by the one at 0.6 m depth, and finally the one at 1.6 m depth. The order of response may perhaps be explained by the presence of an impeding layer located at about 1.5 m depth, as discussed before.

After 3 days of rainfall, all measured VWCs at various depths appeared to reach a steady-state condition. About 3 days after the cessation of rain, the VWCs at different depths began to decrease progressively to another steady-state condition. The shallow probes reached new equilibrium values first, followed by the deeper ones on 2 September. After the commencement of the second rainfall on 8 September, all ThetaProbes responded quite

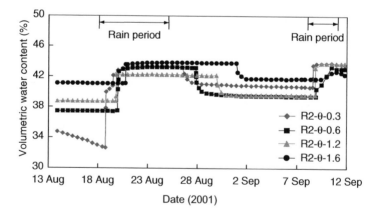

Figure 8.12 Volumetric water content changes in response to rainfall measured by ThetaProbes located at R2 (after Ng *et al.*, 2003).

rapidly, but the magnitude of the increase in VWC was generally smaller than that during the first rainfall.

VARIATIONS OF IN SITU WATER CONTENT PROFILES

Figure 8.13a–c illustrates the variations of water content profiles in response to rainfalls at sections R1, R2 and R3 respectively. In each Figure, the GWC profiles were obtained by direct sampling of the soil, and the two dashed lines in the middle denote the calculated minimum and maximum VWC profiles based on the measured GWC profiles and the measured dry density profiles. The solid lines labelled with opened symbols represent the VWC profiles measured by ThetaProbes.

Just prior to the commencement of the first rainfall event, the initial GWCs generally increased with depth within 1.5 m below ground, suggesting an upward flow of moisture via evaporation. The measured relationship of the initial GWC profiles and the measured initial negative PWP profiles (Figure 8.11) was approximately consistent with the wetting curve of the SWCCs shown in Figure 8.5. For example, the measured GWCs varied from 16.5 to 22.8 per cent within the 2 m depth at the R1 section and were reasonably consistent with the estimated range of GWC (17.5–25 per cent) deduced from the wetting curve of the SWCCs and the negative PWP profiles measured by tensiometers, as shown in Figure 8.11.

After the start of rainfall, the measured GWC profiles were generally consistent with those of the PWP responses discussed previously (Figure 8.11). A significant increase in GWC could be found within the upper 1.5 m, and the influence of the rainfall on the GWCs below 1.5 m seemed to be essentially negligible, particularly at R2. This finding supports the previous postulation that there is a low-permeability layer at mid-slope. The reduction in GWC during the two-week no-rain period appeared to be small, even in the soil layer near the ground surface.

By comparing the measured VWC profiles attained using the ThetaProbes with the calculated bounds of VWC attained using the measured GWC profiles, it can be seen that the measured VWC is, in general, significantly larger than the upper bound of VWC calculated from the GWC profiles at all three sections. Note that the lower and upper bounds of the calculated VWC, represented by the dashed lines, were obtained using the envelopes of measured GWC profiles taken on 17 August and the maximum of the two data sets recorded on 25 August and 2 September, respectively. The inconsistency between the measured and calculated VWCs may be attributed to the accuracy of the indirect measurements of VWC using the ThetaProbes. The measuring accuracy using ThetaProbes can be affected by many factors, such as variations in soil composition, dry density and cracks (Li et al., 2002). It is suggested that the measured VWC can only be interpreted as an indication of what is happening.

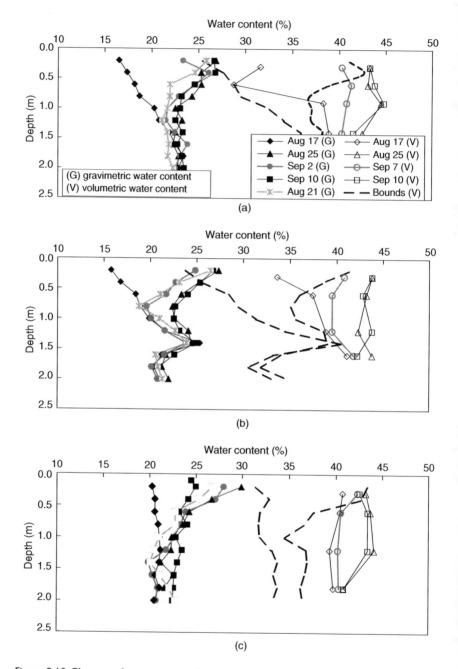

Figure 8.13 Changes of water content in response to rainfall; (a) at R1; (b) at R2; (c) at R3 (after Ng *et al.*, 2003).

RESPONSE OF HORIZONTAL TOTAL STRESSES TO CHANGES OF PORE WATER PRESSURE OR SUCTION

Figure 8.14 shows the monitored total stress ratio (σ_h/σ_v) with time from six vibrating-wire earth pressure cells (EPCs) installed in the slope. All EPCs were installed at a depth of 1.2 m, giving rise to an estimated total vertical stress (σ_v) of about 23.4 kPa, which corresponded to an average dry density of 1.56 Mg/m^3. Pressure cells EP1, EP3 and EP5 (see Figure 8.7) measured the stress changes acting in the east–west direction (i.e. perpendicular to the inclination of the slope), whereas EP2, EP4 and EP6 recorded pressures acting in the north–south direction (i.e. parallel to the inclination of the slope).

Prior to the first rainfall, the total stress ratios recorded by all the cells were lower than 0.3. An initial equilibrium stress ratio appeared to have been established for each cell shortly before the rainfall on 18 August. Two out of six cells registered a small tensile stress, probably induced as a result of soil drying. (Note: the vibrating-wire-type cells are able to record tensile stress.) During installation, the clearance between the wall of the EPC and the soil was backfilled with an epoxy resin. The thin layer of epoxy resin attached the cell securely to the soil and allowed transmission of tensile force between the cell and the soil. This installation procedure, originally proposed by Brackley and Sanders (1992), and its monitoring results demonstrated that a tensile force could be detected by the vibrating-wire cell.

After the start of the first rainfall, none of the EPCs registered any significant changes of stress for about 1½ days. The delayed response of the pressure cells was consistent with the PWP and VWC measurements, shown

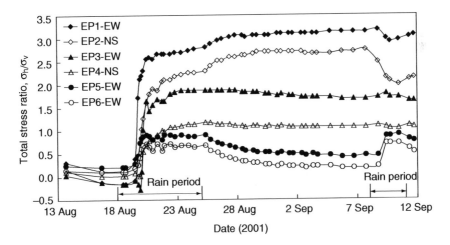

Figure 8.14 Changes of in situ total stress ratio measured by earth pressure cells (after Ng et al., 2003).

in Figures 8.9 and 8.12, respectively. Once the EPCs started to respond, the ratios (σ_h/σ_v) increased very rapidly and significantly within 1 day, and then approached a steady value during the first rainfall event. It appeared that the magnitude of increase in total horizontal stress was strongly related to the elevation of the EPCs and the initial negative PWP (see Figure 8.11). The higher the EPC's elevation, the larger the initial negative PWP present in the ground, and hence the larger the increase in σ_h/σ_v. This performance appeared to be consistent with the relationship between swelling potential of expansive soils and initial soil suction: that is, the swelling potential of an expansive soil generally increases with an increase in the initial negative PWP or suction of the soil (Fredlund and Rahardjo, 1993; Alonso, 1998). For a given pair of pressure cells located at the same elevation, the measured stress ratio in the east–west direction was always larger than that in the north–south direction. This is probably related to a higher constraint imposed as a result of sloping ground in the east–west direction as opposed to that in the north–south direction.

During the two-week no-rain period, a further increase in σ_h/σ_v was observed at EP1 and EP2. On the other hand, the EP3 and EP4 pressure cells showed a slight decrease in stress ratio throughout the no-rain period, and EP5 and EP6 recorded a larger reduction in stress ratio than EP3 and EP4. The reduction in σ_h/σ_v appeared to be due primarily to a decrease in the positive PWPs at a depth of 1–2 m during the no-rain period (see Figure 8.9). However, the continuous and gradual increase in σ_h/σ_v at EP1 and EP2 (but at a reduced rate) may be due to an ongoing 'soaking' of the soil near the location of the EPCs at R2, even after the first rainfall event.

After the start of the second rainfall event, the responses at the three pairs of EPCs were distinctly different. At EP1 and EP2, the observed σ_h/σ_v decreased rather than increased. This may be attributed to the softening of the soil after prolonged swelling during the no-rain period. For the pressure cells (EP5 and EP6) near the toe of the slope, an increase in σ_h/σ_v was recorded owing to the recovery of positive PWP during the second rainfall. The performance of EP3 and EP4 fell between the former two cases.

To make a comparison between the measured total stress ratios after the simulated rainfalls with the corresponding theoretical limiting conditions, the total stress ratios at the passive failure conditions were calculated using the *total stress* and *effective stress* approaches, respectively (Fredlund and Rahardjo, 1993). For the total stress approach, the undrained shear strength was assumed to be equal to 4.4 times the SPT N value obtained in the monitored area after the rainfalls, and the calculated total stress ratio for passive earth failure conditions was 4.4: these values were greater than the measured earth pressures except at cells EP1 and EP2, which appeared to be close to passive failure conditions.

For the effective stress approach, the calculated total passive stress ratio ranged from 2.0 to 3.1 if the saturated shear strength parameters,

$c' = 5.15\,\text{kPa}$ and $\phi' = 17°$ obtained from testing specimens with fissures and cracks (Liu, 1997), were used in the calculations. The calculated passive stress ratios appeared to be close to the measured values recorded by EP1 and EP2 after the simulated rainfalls. This seemed to suggest that the expansive soil, after the simulated rainfalls may reach passive failure along existing cracks and fissures. This finding is consistent with that observed by Brackley and Sanders (1992). This further supports the possible softening behaviour of the soil upon prolonged wetting. The high in situ stress ratios due to the swelling of expansive soils upon wetting might be one of the main reasons for the retrogressive shallow failures found near the monitored slope.

Response of ground deformations in response to the simulated rainfalls

HORIZONTAL DISPLACEMENTS DUE TO CHANGES IN SOIL SUCTION

Figure 8.15 shows the horizontal displacements of the ground in response to the simulated rainfalls. All horizontal displacements shown in the Figure were calculated by taking the rotations measured just prior to the commencement of the rainfalls as the reference datum. The calculated displacements of the inclinometers from the south to the north direction (i.e. the up-slope direction) and from the west to the east direction are defined as positive.

Figure 8.15a and b show the monitored horizontal displacements from inclinometer I1 (located just above the mid-slope) in the north–south and east–west directions, respectively. The results indicate that the ground moves towards the down-slope direction and towards the east direction. The measured horizontal displacements in the two directions illustrate similar characteristics. The horizontal displacements in the upper 1.5 m were understandably more significant than those below this depth, which looked like a 'cantilever' mode of deformation. The variations in horizontal displacement profiles were consistent with changes of PWP, which showed that the most significant changes took place at shallow depth (i.e. less than 2 m), as discussed previously in Figure 8.11.

By comparing the displacements on 19 August and 21 August, it can be seen that there was a significant increase in horizontal displacements, particularly near the ground surface, in both directions. The observed increase in displacements after 3 days of rainfall appeared to be consistent with the 2-day delayed response in pore water pressures, as discussed previously in Figure 8.9. As the rainfall continued, further changes in horizontal displacement were relatively insignificant. After the first rainfall event, a recovery of horizontal displacement (i.e. shrinkage response) was observed with respect to both directions during the two-week no-rainfall period (i.e. from 25 August to 7 September 2001), owing to the increase in soil suction or decrease in positive PWPs. The recovery of 2 mm at the ground surface

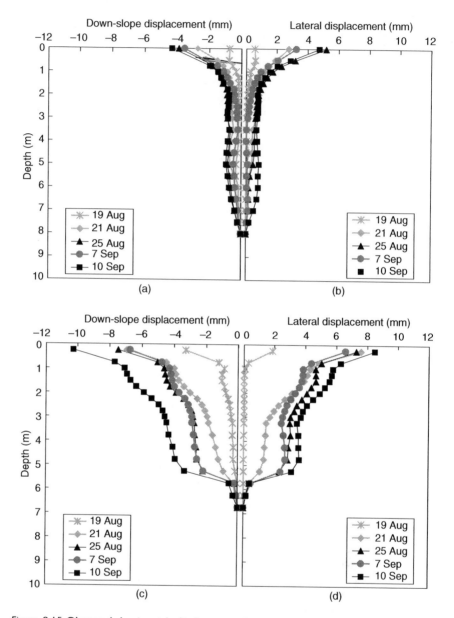

Figure 8.15 Observed horizontal displacement in response to rainfall: (a) from I1 (north–south); (b) from I1 (east–west); (c) from I2 (north–south); (d) from I2 (east–west) (after Ng *et al.*, 2003).

from the east to the west direction was far more significant than that in the up-slope direction (i.e. only about 0.2 mm). The effects of an increase in soil suction on the up-slope movements were counteracted by the influence of gravity. At the end of the second rainfall event, the observed horizontal displacements were similar to those measured at the end of the first rainfall in both directions.

Figure 8.15c and d show the monitored horizontal displacements from inclinometer I2 in the down-slope and the east direction, respectively. It can be seen that the magnitudes of displacement and deformed shapes observed were consistent in both directions, but the magnitudes were significantly larger, and the depth of influence was substantially deeper (deep-seated) than those observed at the mid-slope (i.e. at I1). It is believed that the observed larger displacements near the toe are attributed to the lower initial negative PWP (or soil suction) present at I2 than those at I1 (see Figure 8.11), resulting in a lower soil stiffness near the toe. The greater influence depth near the toe of the slope was consistent with the deeper influence of the simulated rainfalls on the GWC measured at section R3, as opposed to section R2 (see Figure 8.13).

At both I1 and I2, the consistently observed eastern movements of the ground due to rainfall infiltration might be attributed to the direction of subsurface water flow from the west to the east caused by the presence of slightly dipping geological planes. On 23 August, a seepage exit point was observed at about 15 m to the east along the lower part of the masonry wall, outside the monitoring site area. This observation seemed to support the postulation on the direction of water flow and ground movement.

SOIL SWELLING AT SHALLOW DEPTHS UPON WETTING

Figure 8.16a–c shows the measured vertical swellings in response to the simulated rainfalls near the three sections, R1, R2 and R3, respectively. It can be seen from Figure 8.16a that the top two movement points registered an upward soil movement of about 3 mm 1 day after the commencement of the first rainfall event. The rate of soil swelling was almost constant during the first 5 days of rainfall, and the soil continued to swell, but at a reduced rate, throughout the reminder of the monitoring period. On the other hand, there was no recorded swelling at the movement points embedded at both 0.5 and 1.0 m depths during the first 4 days of rainfall. Thereafter, the measured rates of swelling at the two depths were similar to those recorded at the depth of 0.1 m. As anticipated, the shallower the embedded movement point, the larger the magnitude of the measured swelling, owing to the larger changes in soil suction at shallower depth associated with rainfall infiltration (see Figure 8.11) and the cumulative swelling of the underlying soil. During the two-week no-rain period, the ground continued to swell but at a reduced rate. The continuous soil swelling at all three depths could be attributed

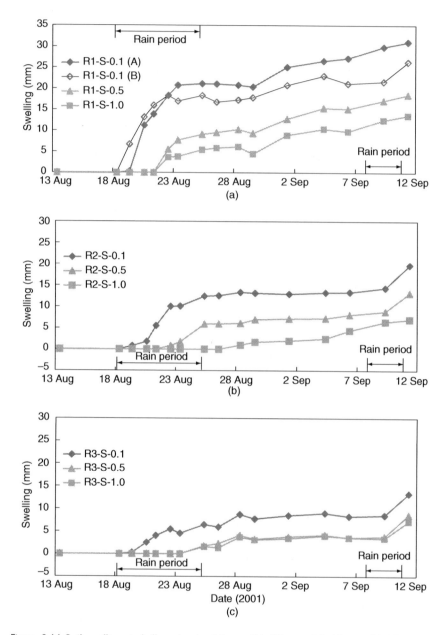

Figure 8.16 Soil swelling at shallow depths: (a) near R1; (b) near R2; (c) near R3 (after Ng *et al.*, 2003).

to the slow seepage of infiltrated water from opened cracks and fissures into the surrounding soil, leading to the secondary swelling of the expansive soil. Marked secondary swelling behaviour has been observed and reported by Chu and Mou (1973) and Sivapullaiah *et al.* (1996). Alonso (1998) has postulated that the marked secondary swelling behaviour of expansive clays is due to the slow and progressive hydration of the expansive soil microstructures. Certainly the observed characteristics of soil swelling can assist in explaining the measured increase in earth pressures at EP1 and EP2 during the no-rain period, as reported in Figure 8.15. During the second rainfall event, the rates of soil swelling increased, particularly at a depth of 0.1 m.

As shown in Figure 8.16b and c, the observed soil swelling patterns at R2 and R3 were similar to that observed at R1, except that the duration of the delayed response to rainfall was longer at a lower elevation. At a given depth, the magnitudes of soil swelling at R2 and R3 were smaller than that observed at R1. This is probably due to the smaller initial soil suction at R2 and R3 than at R1 (see Figure 8.11). Based on the measurements at the three sections, it can be generalized that the higher the initial suction, the larger the soil swelling at a given depth. Within the same section, the shallower the embedded movement point, the larger the measured soil swelling.

Summary and conclusions for the bare area (Figure 8.2)

Based on the field observations and interpreted results of the in situ measurements, the following conclusions can be drawn:

- Prior to the rainfall events, high soil suctions (or negative PWPs) were measured within the top 1 m of soil. The high initial soil suction induced an upward flux of water and moisture both in the vertical direction and along the inclined slope. On the other hand, a down-slope flux of water and moisture occurred at depths greater than 1.5 m, which may be attributed to the presence of a relatively impermeable soil layer at about 1.5 m below ground level.
- The observed responses in PWP, water content, horizontal stress and soil deformation generally showed a 1 to 2 day delay related to the initiation of the rainfall event. The combined effects of the first 3 days of rainfall (i.e. about 180 mm) on the observed responses were much more significant than the effects of the ongoing rainfall or the second rainfall event, except the down-slope deformations at the toe of the slope.
- The effect of the simulated rainfalls on the variations of PWP, water content, horizontal stress and soil deformation was generally much more significant within 2 m below ground surface than below a depth of 2 m.
- A significant perched water table was deduced at a depth of about 1.5 m below the slope surface. The perched water table is believed to

be related to the presence of a dense soil layer lacking in open cracks or fissures at that depth. The presence of the perched groundwater table caused the development of positive PWP and an expansion of the initially dry expansive soil upon wetting. This led to a reduction in the shear strength of the soil layer. This may help to explain why most rain-induced landslides occurring in similar unsaturated expansive soil slopes appear to be relatively shallow, with slip surfaces generally found within a 2 m depth.

- A significant increase in the total stress ratio (σ_h/σ_v) was observed after the simulated rainfalls, particularly in the soil layer with a high initial negative PWP or soil suction (at a high elevation in the slope). The maximum in situ total horizontal stress after the simulated rainfalls was more than three times the total vertical stress. This indicated the possibility of passive pressure failures in the softened clay after the simulated rainfalls.

- Two distinct modes of down-slope deformation were observed: a cantilever deformation within the top metre at the mid-slope and a deep-seated down-slope displacement near the toe of the slope. The observed modes of soil deformation appeared to be consistent with PWP responses and water content determinations. The deep-seated displacement was likely to be caused by the delayed subsurface water flows. An unusual but significant non-symmetric horizontal soil displacement was observed towards the east direction along the longitudinal axis of the slope at both the mid-slope and slope toe.

- A substantial soil swelling was measured in the vertical direction after the simulated rainfalls, particularly in the soil layer with a high initial negative PWP (e.g. at shallow depths and at a high elevation of the slope). The higher the initial soil suction, the larger the soil swelling. The observed soil vertical movement also revealed a marked secondary swelling characteristic in the expansive clay.

- The abundant cracks and fissures in the expansive soils played an important role in the soil–water interaction in the expansive soil slope, and greatly affected the groundwater flows and soil suction.

Chapter 9

Engineering applications for slope stability

Sources

This chapter is made up of verbatim extracts from the following sources for which copyright permissions have been obtained as listed in the Acknowledgements.

- Cheuk, C.Y., Ng, C.W.W. and Sun, H.W. (2005). Numerical experiments of soil nails in loose fill slopes subjected to rainfall infiltration effects. *Computers and Geotechnics*, Elsevier, 32, 290–303.
- Fredlund, D. G. and Rahardjo, H. (1993). *Soil mechanics for unsaturated soils*. Wiley, New York.
- Ng, C.W.W. and Pang, Y.W. (1998). Role of surface cover and impeding layer on slope stability in unsaturated soils. *Proc. 13th Southeast Asian Geotech. Conf.*, Taipei, Taiwan, ROC, pp. 135–140.
- Ng, C.W.W. and Pang, Y.W. (2000b). Influence of stress state on soil–water characteristics and slope stability. *J. Geotech. Environ. Eng.*, 26(2), 157–166.
- Ng, C.W.W. and Zhan, L.T. (2001). Fundamentals of re-compaction of unsaturated loose fill slopes. *Proc. Int. Conf. on Landslides–Causes, Impacts and Countermeasures*, Davos, Switzerland, pp. 557–564.
- Ng, C.W.W., Wang, B. and Tung, Y.K. (2001b). Three-dimensional numerical investigations of groundwater responses in an unsaturated slope subjected to various rainfall patterns. *Can. Geotech. J.*, 38, 1049–1062.
- Ng., C.W.W. and Lai, J.C.H. (2004). Effects of state-dependent soil-water characteristic and damming on slope stability. *Proc. 57th Can. Geotech. Conf.*, Géo Québec, 2004, Session 5E, 28–35.
- Tung, Y.K., Zhang, H., Ng, C.W.W. and Kwok, Y.F. (2004). Transient seepage analysis of rainfall infiltration using a new conjunctive surface-subsurface flow model. *Proc. 57th Can. Geotech. Conf.*, Géo Québec, 2004, Session 7C, 17–22.

Methods of slope stability analysis in unsaturated soils (Fredlund and Rahardjo, 1993)

Introduction

Slope stability analyses have become a common analytical tool for assessing the factor of safety of natural and man-made slopes. Two-dimensional limit equilibrium methods of slices are generally used in practice. These methods are based upon the principles of static (i.e. static equilibrium of forces and/or moments), without giving any consideration to the displacement in the soil mass. Several basic assumptions and principles used in formulating these limit equilibrium analyses are outlined prior to deriving the general factor of safety equations.

In total stress analysis of soil slopes, total stress strength parameters (c_u and ϕ_u) are often used. Pore pressures are not considered. These total stress analyses are appropriate in the short term only and not in the long term where slope stability is a minimum (Simons *et al.*, 2001). On the other hand, in an effective stress analysis, effective shear strength parameters (i.e. c' and ϕ') are generally used when soils are saturated (Simons *et al.*, 2001). The shear strength contribution from the negative pore water pressures above the groundwater table is usually ignored by setting the magnitudes to zero. The difficulties associated with the measurement of negative pore water pressures, and their incorporation into the slope stability analysis, are the primary reasons for this practice. It may be a reasonable assumption to ignore negative pore water pressures for many situations where the major portion of the slip surface is below the groundwater table. However, for situations where the groundwater table is deep or where the concern is over the possibility of a shallow failure surface, as shown in Figure 9.1, negative pore water pressures can no longer be ignored. Moreover, it is vital to consider negative pore water pressures in any forensic study assessing slope failures in unsaturated soils.

In recent years, there has developed a better understanding of the role of negative pore water pressures (or matric suctions) in increasing the shear strength of the soil. Recent developments have led to several devices which can be used to better measure the negative pore water pressures. Therefore, it is now appropriate to perform slope stability analyses which include the shear strength contribution from the negative pore water pressures. These types of analyses are an extension of conventional limit equilibrium analyses.

Several aspects of a slope stability study remain the same for soils with positive pore water pressures (e.g. saturated soils) and soils with negative pore water pressures (e.g. unsaturated soils). For example, the nature of the site investigation, the identification of the strata and the measurement of the total unit weight remain the same in both situations. On the other hand, extensions to conventional testing procedures are required with respect to

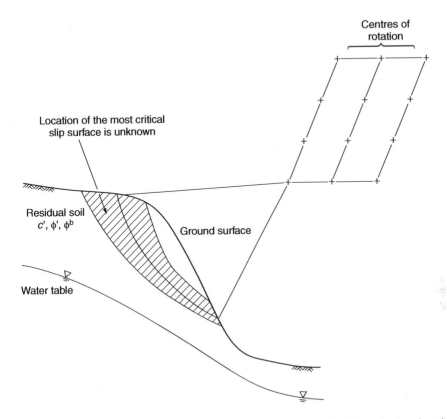

Figure 9.1 A steep natural slope with a deep groundwater table (after Fredlund and Rahardjo, 1993).

the characterization of the shear strength properties of the soil. The analytical tools used to incorporate pore water pressures and calculate the factor of safety also need to be extended.

Location of the critical slip surface

A study of the stability of a slope with negative pore water pressures involves the following steps:

- A desk study to identify any previous failures of a slope or nearby slopes,
- A survey of the elevation of the ground surface on a selected section perpendicular to the slope and to identify any unusual geological features,

- Advancement of several boreholes to identify the stratigraphy and obtain undisturbed soil samples,
- Laboratory testing of the undisturbed soil specimens to obtain suitable shear strength parameters for each stratigraphic unit (i.e. c', ϕ' and ϕ^b parameters),
- Measurement of negative pore water pressures above the groundwater table.

These steps provide the input data for performing a stability analysis. However, the location and shape of the most critical slip surface is unknown. Some combination of actuating and resisting forces along a slip surface of unknown shape and location will produce the lowest factor of safety. Of course, in the case of an already failed slope, the location of the slip surface is known.

In design, the shape of the unknown slip surface is generally assumed, while its location is determined by a trial-and-error procedure. If the shape of the slip surface is assumed to be circular, a grid of centres can be selected, and the radius varied at each centre, providing coverage of all possible conditions. When the slip surface takes on a composite shape (i.e. part circular and part linear), it is still possible to use a grid of centres and varying radii in order to search for a critical slip surface. In addition, a general shape can also be assumed using a series of straight lines to define the slip surface.

Various automatic search routines have been programmed to reduce the number of computations. Some routines start with a centre at an assumed point and seek the critical centre by moving in a zigzag manner (Wright, 1974). Others use an initially coarse grid of centres which rapidly converges to the critical centre (Fredlund, 1981).

There is probably no analysis conducted by geotechnical engineers which has received more programming attention than the limit equilibrium methods of slices used to compute a factor of safety (Fredlund, 1980). The main reasons appear to be as follows. First, the limit equilibrium method has proved to be a useful and reasonably reliable tool in assessing the stability of slopes. Its 'track record' is impressive for most cases where the shear strength properties of the soil and the pore water pressure conditions have been properly assessed (Sevaldson, 1956; Kjaernsli and Simons, 1962; Skempton and Hutchison, 1969; Chowdhury, 1980). Second, the limit equilibrium methods of slices require a limited amount of input information, but can quickly perform extensive trial-and-error searches for the critical slip surface.

General Limit Equilibrium method

The General Limit Equilibrium method (i.e. GLE) provides a general theory wherein other methods can be viewed as special cases. The elements of static used in the GLE method for deriving the factor of safety are the summation of forces in two directions and the summations of moments about a common point (Fredlund *et al.*, 1981).

These elements of static, along with the failure criteria, are insufficient to make the slope stability problem determinate (Morgenstern and Price, 1965; Spencer, 1967). Either additional elements of physics or an assumption regarding the direction or magnitude of some of the forces is required to render the problem determinate. The GLE method utilizes an assumption regarding the direction of the inter-slice forces. This approach has been widely adopted in limit equilibrium methods (Fredlund and Krahn, 1977). The various limit equilibrium slope stability methods that follow this approach have been demonstrated to be special cases of the GLE method (Fredlund *et al.*, 1981).

Calculations for the stability of a slope are performed by dividing the soil mass above the slip surface into vertical slices. The forces acting on a slice within the sliding soil mass are shown in the Figures 9.2 and 9.3 for a circular and a composite slip surface, respectively.

The forces are designated for a unit width (i.e. perpendicular direction to motion) of the slope. The variables are defined as follows:

W = the total weight of the slice of width b and height h,

N = the total normal force on the base of the slice,

S_m = the shear force mobilized on the base of each slice,

E = the horizontal inter-slice normal forces (the L and R subscripts designate the left and right sides of the slice, respectively),

X = the vertical inter-slice shear forces (the L and R subscripts designate the left and right sides of the slice, respectively),

R = the radius for a circular slip surface or the moment arm associated with the mobilized shear force, S_m, for any shape of slip surface,

f = the perpendicular offset of the normal force from the centre of rotation or from the centre of moments,

x = the horizontal distance from the centreline of each slice to the centre of rotation or to the centre of moments,

h = the vertical distance from the centre of the base of each slice to the uppermost line in the geometry (i.e. generally ground surface),

a = the perpendicular distance from the resultant external water force to the centre of rotation or to the centre of moments; the 'L' and 'R' subscripts designate the left and right sides of the slope, respectively,

A = the resultant external water forces; the L and R subscripts designate the left and right sides of the slope, respectively,

α = the angle between the tangent to the centre of the base of each slice and the horizontal; the sign convention is as follows: when the angle slopes in the same direction as the overall slope of the geometry, α is positive, and vice versa and

β = sloping angle across the base of a slice.

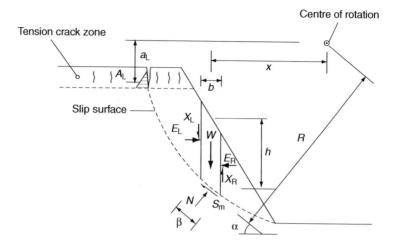

Figure 9.2 Forces acting on a slice through a sliding mass with a circular slip surface (after Fredlund and Rahardjo, 1993).

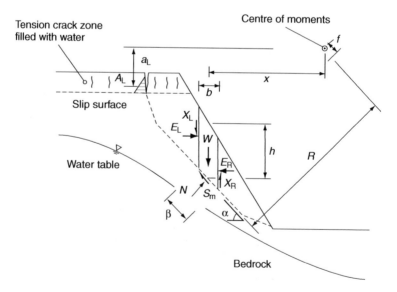

Figure 9.3 Forces acting on a slice through a sliding mass with a composite slip surface (after Fredlund and Rahardjo, 1993).

The examples shown in Figures 9.2 and 9.3 are typical of steep slopes with a deep groundwater table. The crest of the slope is highly desiccated, and there are tension cracks filled with water. The tension crack zone is assumed to have no shear strength, and the presence of water in this zone produces an external water force, A_L. As a result, the assumed slip surface in the tension crack zone is a vertical line. The depth of the tension crack is generally estimated or can be approximated analytically (Spencer, 1968, 1973). The weight of the soil in the tension crack zone acts as a surcharge on the crest of the slope. The external water force, A_L, is computed as the hydrostatic force on a vertical plane. An external water force can also be present at the toe of the slope as a result of partial submergence. This water force is designated as A_R.

Shear force mobilized equation

The mobilized shear force at the base of a slice can be written using the shear strength equation for an unsaturated soil:

$$S_m = \frac{\beta\{c' + (\sigma_n - u_a) \tan \phi' + (u_a - u_w) \tan \phi^b\}}{F} \tag{9.1}$$

where
σ_n = total stress normal to the base of a slice
F = factor of safety which is defined as the factor by which the shear strength parameters must be reduced in order to bring the soil mass into a state of limiting equilibrium along the assumed slip surface.

The factor of safety for the cohesive parameter (i.e. c') and the frictional parameters (i.e. $\tan \phi'$ and $\tan \phi^b$) are assumed to be equal for all soils involved and for all slices. The components of the mobilized shear force at the base of a slice are illustrated in Figure 9.4.

The contributions from the total stress and the negative pore water pressures are separated using the $\tan \phi'$ and $\tan \phi^b$ angles, respectively.

It is possible to consider the matric suction term as part of the cohesion of the soil. In other words, the matric suction can be visualized as increasing the cohesion of the soil. As a result, the conventional factor of safety equations does not need to be re-derived. The mobilized shear force at the base of a slice, S_m, will have the following form:

$$S_m = \frac{\beta\{c + (\sigma_n - u_a) \tan \phi'\}}{F} \tag{9.2}$$

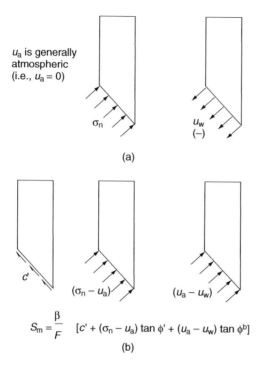

u_a is generally
atmospheric
(i.e., $u_a = 0$)

σ_n

u_w
$(-)$

(a)

c'

$(\sigma_n - u_a)$

$(u_a - u_w)$

$$S_m = \frac{\beta}{F} \; [c' + (\sigma_n - u_a)\tan\phi' + (u_a - u_w)\tan\phi^b]$$

(b)

Figure 9.4 Pressure and shearing resistance components at the base of a slice: (a) pressure components on the base of a slice; (b) contributors to shear resistance (after Fredlund and Rahardjo, 1993).

where

c = total cohesion of the soil, which has two components (i.e. $c' + (u_a - u_w)\tan\phi^b$).

This approach has the advantage that the shear strength equation retains its conventional form. It is therefore possible to utilize a computer program written for saturated soils to solve unsaturated soil problems. When this is done, the soil in the negative pore water pressure region must be subdivided into several discrete layers, with each layer having a constant cohesion. The pore air and pore water pressures must be set to zero within the computer program. This approach has the disadvantage that the cohesion is not a continuous function, and the appropriate cohesion values for each soil layer must be manually computed.

The formulations presented in the next sections are the revised derivations for the factor of safety equations that directly incorporate the shear strength contribution from the negative pore water pressures. The mobilized shear force defined using Equation (9.1) is used throughout the derivation. The effect of partial submergence at the toe of the slope, the effect of seismic loading and external line loads are not incorporated in the derivations.

It must be pointed out that the use of Equation (9.1) may overestimate the shear strength of an unsaturated soil at high suctions. This is because the shear strength of an unsaturated soil does not increase linearly with soil suction at high suction ranges. Also the value of ϕ^b is not a constant over a wide range of suctions.

Normal force equation

The normal force at the base of a slice, N, is derived by summing forces in the vertical direction (see Figure 9.2):

$$W - (X_R - X_L) - S_m \sin \alpha - N \cos \alpha = 0 \qquad (9.3)$$

Substituting Equation (9.1) into Equation (9.3) and replacing the $(\sigma_n \beta)$ term with N gives

$$W - (X_R - X_L) - \left\{ \frac{c'\beta}{F} + \frac{N \tan \phi' \beta}{F} - \frac{u_a \tan \phi' \beta}{F} + \frac{(u_a - u_w) \tan \phi^b \beta}{F} \right\}$$
$$\times \sin \alpha - N \cos \alpha = 0 \qquad (9.4)$$

or

$$N \left\{ \cos \alpha + \frac{\sin \alpha \tan \phi'}{F} \right\} = W - (X_R - X_L) - \frac{c'\beta \sin \alpha}{F} + \frac{u_a \beta \sin \alpha}{F}$$
$$\times (\tan \phi' - \tan \phi^b) + u_w \frac{\beta \sin \alpha}{F} \tan \phi^b \qquad (9.5)$$

Rearranging Equation (9.5) gives rise to the normal force equation:

$$N = \frac{W - (X_R - X_L) - \dfrac{c'\beta \sin \alpha}{F} + \dfrac{u_a \beta \sin \alpha}{F} (\tan \phi' - \tan \phi^b) + u_w \dfrac{\beta \sin \alpha}{F} \tan \phi^b}{m_\alpha}$$

$$(9.6)$$

where
$$m_\alpha = \left\{ \cos \alpha + \frac{\sin \alpha \tan \phi'}{F} \right\}$$

The factor of safety, F, in Equation (9.6) is equal to the moment equilibrium factor of safety, F_m, when solving for moment equilibrium, and is equal to the force equilibrium factor of safety, F_f, when solving for force equilibrium. In most cases, the pore air pressure, u_a, is atmospheric, and as a result, Equation (9.6) reduces to the following form:

$$N = \dfrac{W - (X_R - X_L) - \dfrac{c'\beta \sin \alpha}{F} + u_w \dfrac{\beta \sin \alpha}{F} \tan \phi^b}{m_\alpha}$$

(9.7)

If the base of the slice is located in the saturated soil, the $(\tan \phi^b)$ term in Equation (9.7) becomes equal to $(\tan \phi')$. Equation (9.7) then reverts to the conventional normal force equation used in saturated slope stability analysis. Computer coding for solving Equation (9.7) can be written such that the angle ϕ^b is used whenever the pore water pressure is negative, while the angle ϕ' is used whenever the pore water pressure is positive. The ϕ^b angle can also be considered to be equal to ϕ' at low matric suction values, while a lower ϕ^b angle is used at high matric suctions (Fredlund et al., 1987).

The vertical inter-slice shear forces, X_L and X_R, in the normal force equation can be computed using an inter-slice force function, as described later.

Factor of safety with respect to moment equilibrium

Two independent factor of safety equations can be derived: one with respect to moment equilibrium and the other with respect to horizontal force equilibrium. Moment equilibrium can be satisfied with respect to an arbitrary point above the central portion of the slip surface. For a circular slip surface, the centre of rotation is an obvious centre for moment equilibrium. The centre of moments would appear to be immaterial when both force and moment equilibrium are satisfied, as in the case for the complete GLE method. When only moment equilibrium is satisfied, the computed factor of safety varies slightly with the point selected for the summation of moments.

Consider moment equilibrium for a composite slip surface with respect to the centre of rotation of the circular portion:

$$A_L a_L + \sum W_x - \sum N_f - \sum S_m R = 0$$

(9.8)

Substituting Equation (9.1) for the S_m variable into Equation (9.8) and replacing the $(\sigma_n \beta)$ term with N yields

$$A_L a_L + \sum W_x - \sum N_f = \dfrac{\sum [c'\beta R + \{N \tan \phi' - u_a \tan \phi'\beta + (u_a - u_w) \tan \phi^b \beta\} R]}{F_m}$$

(9.9)

where F_m = factor of safety with respect to moment equilibrium. Rearranging Equation (9.9) yields

$$F_m = \dfrac{\sum \left[c'\beta R + \left\{ N - u_w \beta \dfrac{\tan \phi^b}{\tan \phi'} - u_a \beta \left(1 - \dfrac{\tan \phi^b}{\tan \phi'} \right) \right\} R \tan \phi' \right]}{A_L a_L + \sum W_x - \sum N_f}$$

(9.10)

In the case where the pore air pressure is atmospheric (i.e. $u_a = 0$), Equation (9.10) has the following form:

$$F_m = \frac{\sum\left[c'\beta R + \left\{N - u_w\beta\dfrac{\tan \phi^b}{\tan \phi'}\right\} R \tan \phi'\right]}{A_L a_L + \sum W_x - \sum N_f} \qquad (9.11)$$

When the pore water pressure is positive, the ϕ^b value can be set equal to the ϕ' value. Equation (9.10) can also be simplified for a circular slip surface as follows:

$$F_m = \frac{\sum[c'\beta R + \{N - u_w\beta - u_a\beta\} \tan \phi']R}{A_L a_L + \sum W_x - \sum N_f} \qquad (9.12)$$

For a circular slip surface, the radius, R, is constant for all slices, and the normal force, N, acts through the centre of rotation (i.e. $f = 0$).

Factor of safety with respect to force equilibrium

The factor of safety with respect to force equilibrium is derived from the summation of forces in the horizontal direction for all slices (see Figure 9.5):

$$-A_L + \sum S_m \cos \alpha - \sum N \sin \alpha = 0 \qquad (9.13)$$

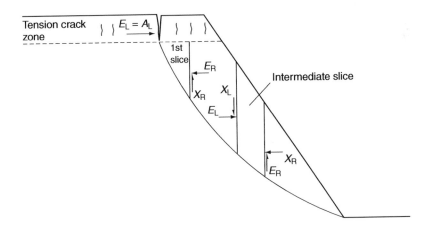

Figure 9.5 Convention for the designation of inter-slice forces (after Fredlund and Rahardjo, 1993).

The horizontal inter-slice normal forces, E_L and E_R, cancel when summed over the entire sliding mass. Substituting Equation (9.14) for the mobilized shear force, S_m, into Equation (9.13) and replacing the $\sigma_n \beta$ term with N gives

$$\frac{1}{F_f} \sum [c'\beta \cos \alpha + \{N \tan \phi' - u_a \tan \phi' \beta + (u_a - u_w) \tan \phi^b \beta\}]$$

$$= A_L + \sum N \sin \alpha \qquad (9.14)$$

where F_f = factor of safety with respect to force equilibrium.
   Rearranging Equation (9.14) yields

$$F_f = \frac{\sum \left[c'\beta R + \left\{ N - u_w \beta \frac{\tan \phi^b}{\tan \phi'} - u_a \beta \left(1 - \frac{\tan \phi^b}{\tan \phi'} \right) \right\} \tan \phi' \cos \alpha \right]}{A_L + \sum N \sin \alpha}$$

$$(9.15)$$

In the case where the pore air pressure is atmospheric (i.e. $u_a = 0$), Equation (9.15) reverts to the following form:

$$F_f = \frac{\sum \left[c'\beta R + \left\{ N - u_w \beta \frac{\tan \phi^b}{\tan \phi'} \right\} \tan \phi' \cos \alpha \right]}{A_L + \sum N \sin \alpha}$$

$$(9.16)$$

When the pore water pressure is positive, the ϕ^b value is equal to the ϕ' value. Equation (9.16) remains the same for both circular and composite slip surfaces.

Inter-slice force function

The inter-slice normal forces, E_L and E_R, are computed from the summation of horizontal forces on each slice, as follows
   Convention for the designation of the inter-slice forces

$$E_L - E_R = N \cos \alpha \tan \alpha - S_m \cos \alpha \qquad (9.17)$$

Substituting Equation (9.3) for the $(N \cos \alpha)$ term in Equation (9.17) gives the following equation:

$$E_L - E_R = \{W - (X_R - X_L)S_m \sin \alpha\} \tan \alpha - S_m \cos \alpha \qquad (9.18)$$

Rearranging Equation (9.18) gives

$$E_R = E_L + \{W - (X_R - X_L)\} \tan \alpha - S_m / \cos \alpha \qquad (9.19)$$

The inter-slice normal forces are calculated from Equation (9.19) by integrating from left to right across the slope (see Figure 9.5). The procedure is further explained in the next section. The left inter-slice normal force on the first slice is equal to any external water force which may exist, A_L, or it is set to zero when there is no water present in the tension crack zone.

The assumption is made that the inter-slice shear force, X, can be related to the inter-slice normal force, E, by a mathematical function (Morgenstern and Price, 1965):

$$X = \lambda f(x) E \qquad\qquad (9.20)$$

where
$f(x) =$ a functional relationship which describes the manner in which the magnitude of X/E varies across the slip surface and
$\lambda =$ a scaling constant which represents the percentage of the function, $f(x)$, used for solving the factor of safety equations.

Some functional relationships, $f(x)$, that can be used for slope stability analyses are shown in Figure 9.6.

Basically, any shape of function can be assumed in the analysis. However, an unrealistic assumption of the inter-slice force function can result

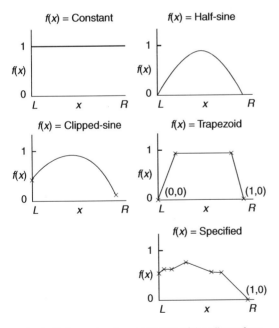

L = Left dimensional x-coordinate of the slip surface
R = Right dimensional x-coordinate of the slip surface

Figure 9.6 Various possible inter-slice force functions (after Fredlund and Rahardjo, 1993).

in convergence problems associated with solving the non-linear factor of safety equations (Ching and Fredlund, 1983). Morgenstern and Price (1967) suggested that the inter-slice force function should be related to the shear and normal stresses on vertical slices through the soil mass. Maksimovic (1979) used the finite element method and a non-linear characterization of the soil to compute stresses in a soil mass. These stresses were then used in the limit equilibrium slope stability analysis.

Other limit equilibrium methods

The GLE method can be specialized to correspond to various limit equilibrium methods. The various methods of slices can be categorized in terms of the conditions of static equilibrium satisfied and the assumption used with respect to the inter-slice forces. Table 9.1 summarizes the conditions of static equilibrium satisfied by the various methods of slices. The static equilibrium used in each of the methods of slices for computing the factor of safety is summarized in Table 9.2. Most methods use either moment equilibrium or force equilibrium in the calculation for the factor of safety. The ordinary and Simplified Bishop methods use moment equilibrium, while the Janbu Simplified, Janbu Generalized, Lowe and Karafiath and the Corps of Engineers methods use force equilibrium in computing the factor of safety. On the other hand, the Spencer and Morgenstern–Price methods satisfy both moment and force equilibrium in computing the factor of safety. In this respect, these two methods are similar in principle to the GLE method which satisfies force equilibrium and moment equilibrium in calculating the factor of safety.

Table 9.1 Elements of static equilibrium satisfied by various limit equilibrium methods (after Fredlund and Rahardjo, 1993)

Method	Force equilibrium		Moment equilibrium
	1st Direction[a] (e.g. vertical)	2nd Direction[a] (e.g. horizontal)	
Ordinary or Fellenius	Yes	No	Yes
Bishop's simplified	Yes	No	Yes
Janbu's simplified	Yes	Yes	No
Janbu's generalized	Yes	Yes	[b]
Spencer	Yes	Yes	Yes
Morgenstern–Price	Yes	Yes	Yes
Corps of Engineers	Yes	Yes	No
Lowe–Karafiath	Yes	Yes	No

Notes
[a] Any of two orthogonal directions can be selected for the summation of forces.
[b] Moment equilibrium is used to calculate interslice shear forces.

Table 9.2 Comparison of commonly used methods of slices (after Fredlund and Rahardjo, 1993)

Method	Factors of safety		Interslice force assumption
	Moment equilibrium	Force equilibrium	
Ordinary	✓		$X = 0, E = 0$
Bishop's simplified	✓		$X = 0, E \geq 0$
Janbu's simplified		✓	$X = 0, E \geq 0$
Janbu's generalised		✓	$X_R = E_R \tan \alpha_t - (E_R - E_L) t_R / b^a$
Spencer	✓	✓	$X/E = \tan \theta^b$
Morgenstern–Price	✓	✓	$X/E = \lambda f(x)$
Lowe–Karafiath		✓	$X/E =$ Average slope of ground and slip surface
Corps of Engineers		✓	$X/E =$ Average ground surface slope

Notes

a $\alpha_t =$ angle between the line of thrust across a slice and the horizontal.
   $t_R =$ vertical distance from the base of the slice to the line of thrust on the right side of the slice.
b $\theta =$ angle of the resultant interslice force from the horizontal.

The GLE method can be used to simulate the various methods of slices by using the appropriate inter-slice force assumption. The inter-slice force assumptions used for simulating the various methods are given in Table 9.2.

Three-dimensional numerical parametric study of rainfall infiltration into an unsaturated soil slope (Ng et al., 2001b)

Introduction

Rainfall-induced landslides are one of the most concerned natural disasters in many parts of the world, including Hong Kong, Japan and Brazil (Lumb, 1962, 1975; Fukuoka, 1980; Wolle and Hachich, 1989; Brand, 1995; Rahardjo et al., 1998). An example considered in this study is a cut slope located at Lai Ping Road in Shatin, Hong Kong, which has failed several times over the past 20 years due to rainfall. The most notable failure incident was in July 1997 (Sun et al., 2000a) in which a massive slope failure occurred when several days of light, intermittent rainfalls were followed by a heavy short-duration rainstorm. A three-dimensional (3D) numerical back-analysis was conducted for studying the groundwater responses of Lai Ping Road cut slope (Tung et al., 1999) using FEMWATER (Lin et al., 1997) based on the detailed hydrogeological characteristics of the failure site (Sun et al., 2000b). FEMWATER is three-dimensional finite element program for simulating subsurface flow in variably unsaturated and saturated media (Lin et al., 1997). The back-analysis of the groundwater responses was to explore the possible slope-failure mechanisms due to the rainstorm (Tung et al., 1999; Sun et al., 2000b).

In dealing with infiltration and subsequent slope stability problems, water permeability is one of the most important factors on groundwater flow in saturated and unsaturated soils. Modelling infiltration and subsurface flow ranges from simple one-dimensional (1D) Green–Ampt Model (Green and Ampt, 1911) to two-dimensional (2D) analyses (Lumb, 1962; Leach and Herbert, 1982; Lam et al., 1987; Pradel and Raad, 1993; Wilson, 1997; Rahardjo and Leong, 1997; Ng and Shi, 1998; Ng and Pang, 1998a) and field studies on rainfall infiltration (Houston and Houston, 1995; Lim et al., 1996; Zhang et al., 1997). Besides water permeability, rainfall characteristics such as rainfall duration and return periods (i.e. rainfall amount) are important factors affecting infiltration and transient seepage in unsaturated soils. Despite some recent work (Tung et al., 1999), three-dimensional analysis of transient seepage in unsaturated soils has not received adequate attention. Moreover, the role of rainfall pattern plays in three-dimensional groundwater response is rarely considered in literature.

In this section, a three-dimensional numerical parametric study on groundwater responses in an initially unsaturated soil slope subjected to various rainfall conditions is conducted. Due to the availability of abundant in situ information such as geological stratigraphy, hydrological and geotechnical data, the Lai Ping Road cut slope was chosen for this parametric study. Using FEMWATER, the finite element mesh and boundary conditions (except rainfall flux boundary) were kept the same as those used in Tung et al. (1999), in which the computed initial groundwater conditions were verified by field measurements. The main objective of this three-dimensional parametric study is to investigate the influence of typical rainfall patterns with different rainfall intensity, duration and return periods (i.e. total rainfall amount) on groundwater pressure responses. Based on the computed pore water pressure responses, an attempt is made to discuss possible causes of shallow and deep-seated slope failures in Hong Kong.

Three-dimensional subsurface flow modelling

The governing equations

The computer program, FEMWATER (Lin et al., 1997), is a three-dimensional finite element software for flow and solute transport analysis in both unsaturated and saturated media. In this study, the focus is placed on subsurface flow in the slope. Therefore, the effects of compressibility of water and soil skeleton and chemical concentration on pore water pressure and suction are not considered. Furthermore, under the assumption of isotropic flow, the governing equation for subsurface water flow in the 3D space with x, y and z coordinate system by adopted FEMWATER is

$$k_w\left(\frac{\partial^2 h}{\partial x^2}+\frac{\partial^2 h}{\partial y^2}+\frac{\partial^2 h}{\partial z^2}\right)+\frac{\partial k_w}{\partial x}\frac{\partial h}{\partial x}+\frac{\partial k_w}{\partial y}\frac{\partial h}{\partial y}+\frac{\partial k_w}{\partial z}\frac{\partial h}{\partial z}+q=F\frac{\partial h}{\partial t} \quad (9.21)$$

where F = storage coefficient; k_w = water permeability; h = hydraulic head; q = boundary flux and t = time.

The three-dimensional finite element mesh

The finite element mesh and the geological conditions employed in this study are taken from the Lai Ping Road landslide analysis (Tung *et al.*, 1999). The finite element mesh (see Figure 9.7) was constructed based on the field geometrical conditions from borehole information. The dimensions of the site are also shown in Figure 9.7. The site is idealized as three different layers of soil and rock. The top layer is colluvium with an average thickness of 20 m, the second and third are rock layers of average 30 m thick each, in which the third layer is less pervious. Details of the geological characterizations, topographical maps and the location of the landslide are described by Koor and Campbell (1998), Sun *et al.* (2000a) and Tung *et al.* (1999).

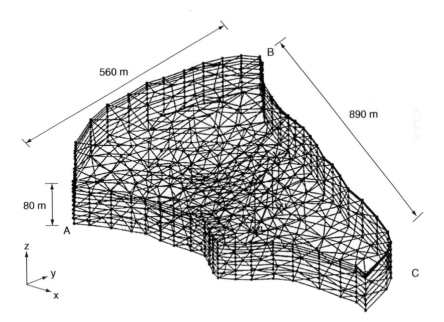

Figure 9.7 Three-dimensional finite element mesh of the Lai Ping Road site (after Tung *et al.*, 1999; Ng *et al.*, 2001b).

The grain size, shear strength, soil–water characteristic curves (SWCCs) and water permeability of the soil are reported by Sun and Campbell (1998), Ng and Pang (1998b, 2000a,b) and Sun *et al.* (2000a,b).

In this flow analysis, the SWCCs and water permeability functions adopted for various porous media are shown in Figure 9.8a and b, respectively. The saturated water permeability values for the three material

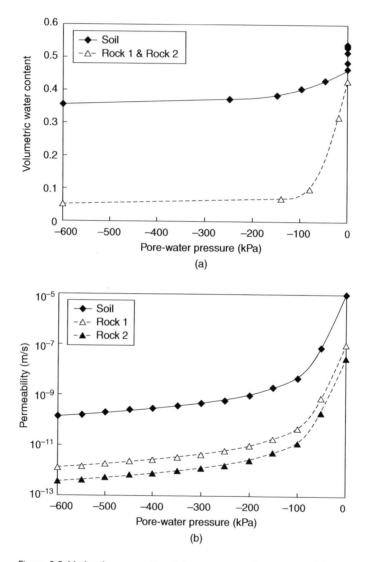

Figure 9.8 Hydraulic properties: (a) water retention curves; (b) permeability with respect to pore water pressure (after Ng *et al.*, 2001b).

layers modelled from top to bottom are 1×10^{-5}, 1×10^{-7}, 2.8×10^{-9} m/s, respectively (Tung *et al.*, 1999). For simplicity, hysteresis between drying and wetting paths (Topp and Miller, 1966; Ng *et al.*, 2000a,b; Ng and Pang, 2000b), and state-dependency of SWCCs, are not considered in the analysis.

The boundary conditions

The finite element mesh adopted for this study site has a total number of 4,005 elements with 2,500 nodes. The bottom of the mesh is assumed to be impermeable, whereas the ground surface is a flux boundary receiving rainwater infiltration. The boundary along the perimeters AB and BC in Figure 9.7 is set as a zero-flux boundary, whereas the perimeter AC is a fixed-head boundary governed by the water level in a stream along the perimeter. Any seepage face is calculated automatically by FEMWATER.

Initial steady-state conditions

All the simulations conducted in this study use the same initial groundwater condition. It is a steady state subsurface flow condition obtained from applying a very small flux of 10 mm/day to the ground surface for about 100 days and by controlling a fixed-head boundary along AC according to the hydraulic head in the stream. The hydraulic heads along the perimeters AB and BC are computed by solving the governing equation (9.21) using FEMWATER automatically. The initial flow velocity vectors over the entire mesh are shown in Figure 9.9, with OO′ being the cross-section cut across the

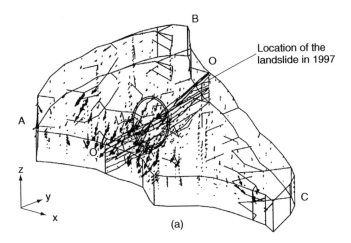

Figure 9.9 Velocity vectors in (a) the 3D mesh, and (b) in cross-section O–O′ (after Ng *et al.*, 2001b).

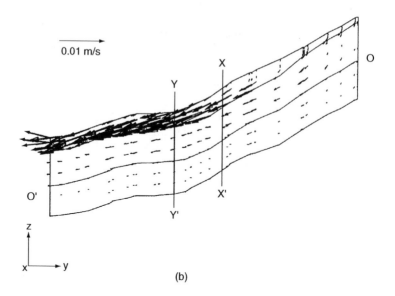

(b)

Figure 9.9 (Continued).

centre of the cut slope. Figure 9.9a shows the region with high flow velocity coincides with the landslide region, and water flow directions are generally in agreement with the directions of the failed soil movement (Tung *et al.*, 1999). For clarity, Figure 9.9b shows that the direction of the subsurface flow is generally parallel to the slope inclination angle, and the low water permeability of the two rock layers makes flow in these rock layers almost negligible, as indicated by their relatively very small velocity vectors.

The initial pore water pressure head along the cross-section OO′ is shown in Figure 9.10. The water pressure heads range from about −5 to 70 m across the depth of OO′ section. The complex pore water pressure contours at the upper slope were due to the imposed boundary flux of 10 mm/day. The two sections, XX′ and YY′, are located at the crest and the toe of the cut slope at which the initial groundwater tables at the two sections are located at the depths of 19.45 m (117.15 mPD) and 1.7 m (98.7 mPD), respectively. These two sections are selected to present simulation results because they are suitable for illustrating different groundwater responses due to an initial shallow and deep groundwater tables. Their respective initial soil suctions at the ground surface are 53 and 15 kPa, which are within the common range of measured surficial soil suction in slopes in Hong Kong. The validity of the computed initial condition was verified by comparing field monitoring data (Tung *et al.*, 1999). As shown in Figure 9.10, the measured groundwater table at borehole TT2, which is near to section XX′, was 19.5 m in May, 1998. This was consistent with the computed groundwater table of 19.45 m.

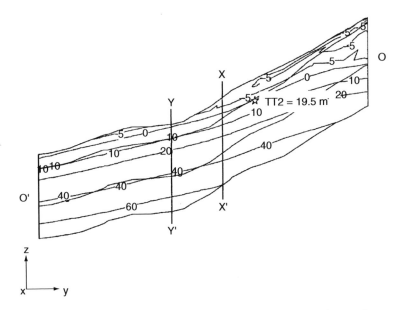

Figure 9.10 Contours of pore water pressure head for the initial groundwater conditions at cross-section O–O′. The initial water table at cross-sections X–X′ and Y–Y′ is located at 19.45 and 1.7 m, respectively, below ground surface. All values are in metres (after Ng *et al.*, 2001b).

Rainfall conditions simulated

Figure 9.11a shows six typical rainfall patterns of 24-h duration initially considered in this chapter. Dimensionless rainfall hyetographs can be easily converted to actual rainfall hyetograph by multiplying the former by the total rainfall depth. The storm patterns were extracted from the statistical analyses of hour rainfall records at 15 automatic gauges in Hong Kong (Leung, 1998). For the 24-h rainfalls, three total rainfall depths of 358, 597 and 805 mm are considered in the analysis, each corresponding to the return period of 10, 100 and 1,000 years, respectively, at the Hong Kong Observatory (Lam and Leung, 1995).

Figure 9.11b shows thee prolonged 168-h (7-day) rainfall patterns adopted for investigating the effects of different rainfall duration on groundwater responses. The input rainfall conditions considered in this investigation are summarized in Table 9.3. Since the dimensionless hyetograph divides the entire storm duration into 12 equal time intervals, then, $\Delta t = 2$ and 14 h for the 24- and 168-h storms, respectively. It should be noted that the rainfall intensity of 168-h storm is less intense than the 24-h storm for a given amount, although the total rain depth is larger for the former. With a lower intensity and longer duration, a prolonged rainstorm results in a more evenly distributed rain pattern.

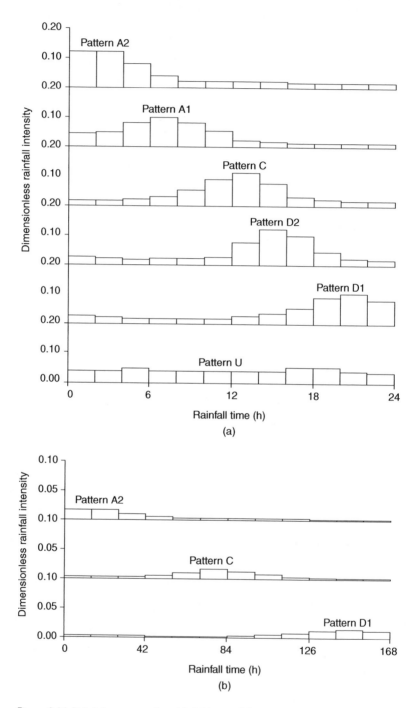

Figure 9.11 Rainfall patterns for (a) 24-h rainfall and (b) 168-h rainfall (after Ng *et al.*, 2001b).

Table 9.3 Summary of various rainfall cases simulated (after Ng et al., 2001b)

Parameter	Series			
	1	2	3	4
Return period (years)	10	100	1000	100
Duration of rainfall (h)	24	24	24	168
Total rainfall amount (mm)	358	597	805	892
Simulation time after rainfall (no rainfall) (h)	168	168	168	240

Note
Three storm patterns are considered for each series: advanced (A2), central concentrated (C) and delayed (D1).

It should be noted that the flux applied on the upper boundary in the numerical simulations is the assumed rainfall infiltration rather than the rainfall intensity. Actual infiltration pattern in the field is very difficult to measure. The potential infiltration rate is a function of time even under the constant rainfall intensity (Mein and Larson, 1973; Wilson, 1997; Hillel, 1998) – it decreases with time and finally approaches saturated water permeability. In the numerical simulation, an average value of 60 per cent of the rainfall intensity was adopted as the surface influx to the slope (Tung et al., 1999). This average value was estimated according to rainfall and surface run-off measurements at a few slopes in Hong Kong (Premchitt et al., 1992).

Preliminary study of rainfall patterns

In a series of preliminary simulations, a 10 year 24-h storm (358 mm) is used to investigate the effects of the six different rainfall patterns (refer to Figure 9.11a) on pore water pressure distributions inside the slope. It should be noted that the variations of the hydraulic head, actually reflects changes in pore water pressure head because the elevation heads are constants. Due to similarity of subsurface pore water pressure distribution, only the temporal variations of hydraulic head at the ground surface at XX' and YY' sections are shown in Figure 9.12a and b. Among the six patterns analysed, the hydraulic head distributions of patterns A2 and D1 seem to be the two extreme cases because their responses appear to envelop the other patterns. The pressure changes resemble their respective rainfall patterns. The soil is wetted most rapidly under Pattern A2 and most slowly under the rainfall of Pattern D1, as illustrated by the peak intensity occurring at the very beginning of Pattern A2 whereas most of the rainwater is distributed towards the end of Pattern D1. Consequently, these two extreme patterns (advanced

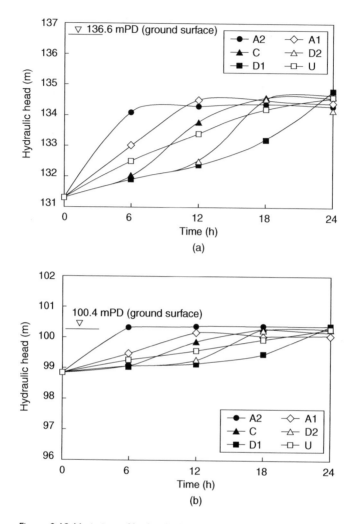

Figure 9.12 Variation of hydraulic head at the ground surface along (a) section X–X' and (b) section Y–Y', for 24-h rainfalls with a 10-year return period (after Ng *et al.*, 2001b).

type – A2 and delayed type – D1) are selected for subsequent analyses presented in this chapter. Pattern C is also selected because it behaves like a median case. Thus, the three rainfall patterns selected represent advanced, central and delayed patterns.

In order to investigate the subsurface ground water pressure responses subjected to various rainfall conditions, a total of 12 computer simulations summarized in Table 9.1 are carried out and they are grouped into four series; within each three storm patterns are considered. The first three series are for 24-h storm each corresponding to different rainfall amounts associated with return periods of 10, 100 and 1,000 years. These three series of tests aim at investigating the effects of storm pattern and rainfall amount on groundwater responses. The fourth series considers long duration (168-h) storms with a return period of 100 years. Under the same return period, the 168-h storm would have a greater rainfall amount than the 24-h storm, but they are less intense than the short-duration rainfalls. By comparing the second and the fourth series, the influence of different rainfall durations can be studied.

Typical groundwater response to rainwater infiltration

To illustrate the typical responses of pore water pressure in an initially unsaturated slope, a 24-h, 100-year rainfall with Pattern C is used. The temporal variations of vertical pore water pressure at sections XX' and YY' are shown in Figure 9.13.

At section XX', the initial soil suction around 50 kPa is nearly constant in the top 6 m. The initial groundwater table is located at the depth of 19.45 m. After raining for 12 h, the soil suction within the top 6 m of soil is reduced. The soil suction on the ground surface decreases to 10 kPa, and the soil is wetted to a depth of about 10 m at the end of storm period, that is, $t = 24$ h. It can be seen that rainwater infiltration reduces the soil suction

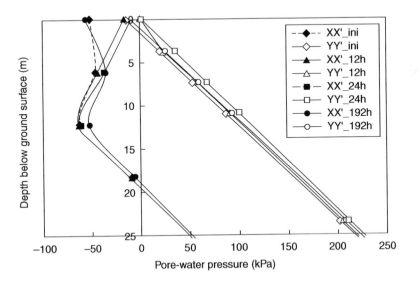

Figure 9.13 Variation of pore water pressure distribution with depth for pattern C of 24 h rainfall with a 100 year return period (after Ng et al., 2001b).

gradually, but a clear continuously moving wetting front is not observable. At $t = 192\,h$, that is, 168 h after the cessation of rainfall, the soil suction in the top 6 m below the ground surface recovers. The soil suction within the top 1 m of soil even exceeds the initial value on the ground surface. However, the pore water pressure continues to increase in deeper soils from 6 to 20 m underground during this period of no rainwater infiltration. At section XX', no visible rise in the groundwater table during the storm period. After the rain stops, however, the groundwater table rises by about 0.5 m to about 19 m at $t = 168\,h$ as water continues to flow down into the deeper soils.

From Figure 9.13, it can also be seen that the initial groundwater table is 1.7 m below the ground surface at section YY' (i.e. at the toe of the cut slope). The pore water pressure distribution along depth is nearly linear, which is quite different from that at section XX'. At $t = 12\,h$, soil suction on the ground surface decreases to zero indicating that the groundwater table rises to the ground surface. At the end of the storm ($t = 24\,h$), there is a general rise in pore water pressure along the entire soil depth. At 168 h after the cessation of the rainfall ($t = 192\,h$), the pore water pressure generally decreases but does not return to its original condition. The groundwater table is located about 1.1 m below the ground surface.

In general, the time rate of change in pore water pressure at section XX' decreases with depth, whereas the vertical rate of increase in pore water pressure at section YY' is almost constant as shown by the parallel lines. This is because that, unlike section XX', rainwater infiltration mainly affects soil suction but it has little influence on the deep groundwater level at section XX'. Infiltration at section YY' brings about a clear rise in the groundwater table and an obvious increase in pore water pressure due to its initial high groundwater table.

It can be deduced that the water permeability at section XX' is lower than that at the downstream section YY' because the soil suction at section XX' is much higher than that at section YY'. As the soil near ground surface at section XX' is wetted by infiltration, the soil suction in deeper part of the soil remains high and the water permeability decreases with depth. This gives less water flow downwards at section XX' instead more water flow laterally in the soil near the ground surface flows towards section YY'. Apart from the water flowing laterally from the upstream, the soil near section YY' also experiences relatively large amount of infiltration through ground surface due to high water permeability. As a result, the groundwater table at section YY' rises much more visibly than that at section XX'.

The above discussion generally suggests that there is an insignificant rise in groundwater table even under a 100 year rainfall. Nonetheless, the reduction of soil suction in shallow depth is rather significant. As reported by a number of researchers (Fourie, 1996; Ng and Shi, 1998), many observed shallow slope failures were not caused by a rise in groundwater table but rather could be attributed to the destruction of soil suction in the slope. Infiltration of rainfall may reduce the soil suction in the surficial soil substantially enough to trigger a shallow failure (Fourie *et al.*, 1999).

Effects of rainfall patterns on groundwater response

To study the influence of different rainfall patterns on groundwater response, 24 h rainfalls corresponding to different rainfall amounts with the three representative rainfall patterns (i.e. A2, C and D1 shown in Figure 9.11a) are used. Since the general trends on groundwater response are similar under different rainfall amounts, results corresponding to 24-h 100-year rainfall (i.e. 597 mm) under the three rainfall patterns are illustrated.

Groundwater responses to different rainfall patterns at sections XX' and YY'

Because there is no significant change in groundwater flow in the deep rock layers, time variation of hydraulic heads due to the different storm patterns at three shallow depths, namely 0, 6.1 and 18.3 m, are reported.

As shown in Figure 9.14a, the initial hydraulic heads at the three depths at section XX' are less than the elevation heads implying negative pore water pressures above the depth of 18.3 m before rainfall. The pore water pressure on the ground surface rises immediately after the rainfall starts. It rises most rapidly and significantly in response to the advanced storm pattern, followed by the central pattern and then the delayed pattern. On the ground surface, the hydraulic head within the first 12 h is the highest under the advanced pattern, whereas towards the end of the rainfall the highest hydraulic head is produced by the delay storm pattern. At the depth of 6.1 m, the pore water pressure also rises most significantly under the advanced storm pattern than the other two rainfall patterns considered at the end of the 24-h rainfall. The increase in pore water pressure induced by the advanced storm pattern is about 2 m greater than the other two rainfall patterns. No visible increase in pore water pressure can be seen at the depth of 18.3 m during the entire period of rainfall. The influence of rainfall patterns on pore water pressure is most visible on the ground surface, and such effect gradually diminishes as the depth increases.

After the cessation of the rain, the pore water pressure on the ground surface (0 m) starts to drop (see Figure 9.14a). However, the underground pore water pressure continues to rise. At depth of 6.1 m, the pore water pressures under all three storm patterns reach their peaks 42 h after the cessation of rainfall. The effect of different storm patterns on pore water pressure distribution tends to be minimal after $t = 168$ h (1 week). However, at the greater depth of 18.3 m at section XX', the pore water pressure

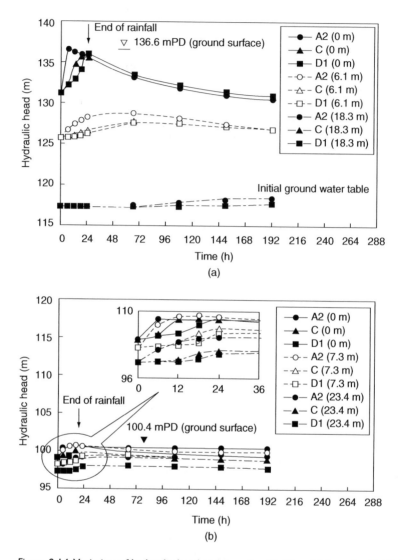

Figure 9.14 Variation of hydraulic head at (a) section X–X' and (b) section Y–Y', for 24-h rainfall with a 100-year return period (after Ng *et al.*, 2001b).

increases under all storm patterns even 168 h after the cessation of rain, particularly under the advanced storm pattern. The pore water pressure head for storm Pattern A2 rises about 2 m, which is twice the amount under storm Patterns C or D1.

Figure 9.14b illustrates the time variations of hydraulic head at section YY'. The variation of hydraulic head at section YY' clearly is much smaller as compared with that at section XX'. This suggests that the influence of rainfall patterns on hydraulic heads depend on the initial groundwater conditions. The higher the initial pore water pressure or water table, the greater the susceptibility of groundwater conditions to rainfall patterns.

At section YY' the hydraulic head under Pattern A2 reaches the ground surface at $t = 6$ h, whereas under Patterns C and D1, the hydraulic head reaches the ground surface at $t = 12$ and 24 h, respectively. This means that the groundwater table at section YY' rises to the ground surface. In other words, a seepage surface is formed at section YY' at that time which may lead to slope instability such as the slope failure at Lai Ping Road (Tung *et al.*, 1999; Sun *et al.*, 2000a). At the depth of 7.3 m, hydraulic head rises most quickly under the advanced storm pattern with the maximum rise of 2 m after 18 h. It should be pointed out that from 10 to 24 h, the hydraulic head at the depth of 7.3 m under storm Pattern A2 exceeds the ground surface elevation, indicating an upward flow from underground to the ground surface at the toe of the slope during the period and reducing slope stability. At the depth of 23.4 m, the hydraulic head curve corresponding to the advanced storm pattern is the highest among the three rainfall patterns considered. This may be due to more water infiltrating into the upstream soil under the advanced Pattern A2, as indicated by high pore water pressure at section XX', and consequently a greater amount of groundwater flows from the upstream to the toe of the slope. As a result, the pore water pressure distribution along the depth is also the highest under storm Pattern A2. This postulation can be verified by inspecting the velocity field of cross-section OO', which shows that the majority of flow vectors are downwards in the same direction as the inclination angle of the slope.

After the rain stops, the hydraulic head starts to decrease on the ground surface under both storm Patterns C and D1, but not Pattern A2 (see Figure 9.14b). For storm pattern A, the hydraulic head on the ground surface remains at 100.4 mPD, even at $t = 168$ h. This means that the groundwater table remains at the ground surface for this long period of time. It can be attributed to the great amount of groundwater that flows downstream from upstream after the rainfall. At $t = 192$ h, the difference in pore water pressure heads at the ground surface between storm Pattern A and two other patterns is about 1.5 m. A similar trend can be observed underground, with storm Pattern A consistently yielding the greatest change in pore water pressure among the three.

Considering pore water pressure changes within the slope subjected to the 24-h rainfalls, storm Pattern A has the most critical implication on slope stability at the study site, since it induces higher pore water pressure underground than the other two patterns, not only at the crest but also at the toe of the slope.

Role of hydraulic gradient (∂h/∂z)

Under one-dimensional (1D) conditions, the infiltration rate is governed by the water permeability of surficial soil and hydraulic gradient. The one-dimensional flow rate can be calculated by Darcy's Law, $v = -k_w i = -k_w \partial h / \partial z$, in which v is the infiltration rate, k_w the water permeability for a given soil suction, h the hydraulic head (total head), z the elevation (elevation head) and i the hydraulic gradient, $\partial h / \partial z$, along the z-direction. It should be noted that the infiltration rate is not a constant even under a rainfall of constant intensity. According to the model proposed by Mein and Larson (1973), the potential infiltration rate is the highest at the beginning of a rainfall and then decreases as the rainfall continues. The highest infiltration rate is primarily caused by the high hydraulic gradient at the beginning of a rainstorm, when the soil is partially saturated, i.e. when the surficial soil suction is relatively high (Wilson, 1997). As the rainfall continues, the soil near the ground surface is wetted gradually and the soil suction is reduced by infiltration, whereas the water permeability of the soil approaches saturated permeability (k_{sat}). As a result, the infiltration rate finally approaches the ultimate value, k_{sat}, when the hydraulic gradient near the ground approaches unity. If rainfall intensity is lower than k_{sat}, all rainwater can infiltrate into soil. When rainfall intensity is higher than k_{sat}, only some of the rainwater infiltrates and the excessive rainwater becomes surface run-off.

Figure 9.15a and b show the hydraulic heads along the depth at both sections XX′ and YY′ at 6 and 24 h, respectively. It can be seen from

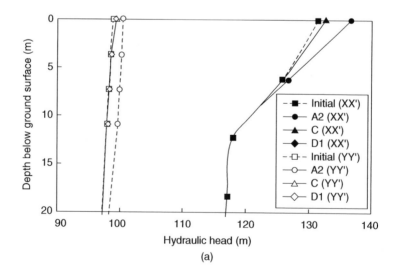

Figure 9.15 Hydraulic heads at sections X–X′ and Y–Y′ for (a) 6 h and (b) 24 h, into the 24-h rainfall with a 100-year return period (after Ng et al., 2001b).

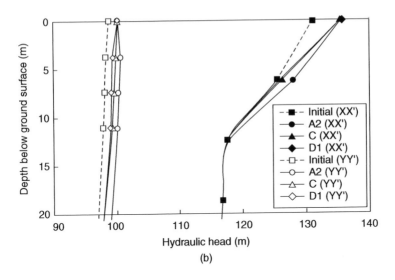

Figure 9.15 (Continued).

Figure 9.15a that the initial hydraulic gradient, $\partial h/\partial z$, in the vertical direction above 12 m underground at section XX' is greater than zero but nearly equal to zero below 12 m. At 6 h after rain started, the influence of Pattern A2 has reached as deep as 8 m, twice as deep as other two patterns (i.e. Patterns C and D1). The hydraulic gradient is clearly higher for Pattern A2. This is attributed to the high rainfall intensity at the beginning of rainfall Pattern A2. The resulting higher gradient enables the infiltrated water flow at a higher rate, which in turn saturates the soil more quickly than the other two patterns. The resulting higher water permeability leads to higher subsurface flow rate and a higher pore water pressure along the depth at section XX'.

The line of the initial hydraulic head at section YY' is nearly vertical and hence the hydraulic gradient is almost zero at this section. At this section, the groundwater table is initially at a depth of 1.7 m and the soil suction near the ground surface is low, while the water permeability is high. Although the gradient is small, the water permeability is relatively high. Therefore, there still can be significant infiltration. Moreover, the groundwater flow from upstream to downstream is significant (see Figure 9.9b). At 6 h into the rainfall, the hydraulic head along depth is increased significantly by rainfall Pattern A2, whereas the hydraulic heads of the other two patterns remain nearly unchanged. This can be ascribed to not only the higher infiltration rate at this section but also higher water flow from upstream under Pattern A2. Ultimately, it is attributed to the higher rainfall intensity of Pattern A2 during the first several hours of the rainfall.

Figure 9.15b shows the hydraulic head profiles with depth at the end of the 24-h rainfall. With the top 4 m at section YY′, the line for Pattern A2 has a slightly negative slope angle, implying an upward flow. This is consistent with Figure 9.14b in which the hydraulic head at 7.3 m below ground is higher than that at the ground surface from $t = 10$ h to $t = 24$ h. Meanwhile, it can be noticed that the slope angles of the hydraulic head lines for Pattern D1 and C are clearly positive, and the angle for Pattern D1 is slightly greater than that for Pattern C. Therefore, for these two cases, water flows in the downward direction at this section.

Among the three rainfall patterns, the higher hydraulic head and the upward water flow of Pattern A2 are due to the high infiltration rate, corresponding to the high rainfall intensity in the first several hours of this rainfall pattern. The upward flow of water is caused by water flowing from upstream. Because a certain amount of time is needed for water to flow from one place to another, the upward flow for Pattern A2 is not observed at $t = 6$ h into rainfall (Figure 9.15a). The difference in various rainfall patterns at the beginning also affects the groundwater condition later on. For Pattern D1, a significant amount of rainwater infiltration occurs at the end of the rainfall, while the pore water pressure underground is low due to its low infiltration at the beginning, resulting in a relatively larger vertical hydraulic gradient. Thus, a relatively significant downward flow occurs at section YY′ at the end of the Pattern D1 rainfall. Pattern C is a median case, so its pore water pressure distribution is right in the middle of the pore water pressure distributions of Patterns A2 and D1.

Effect of rainfall return period (or amount) on groundwater responses

For a given duration, a higher return period would correspond to a larger total rain amount and the corresponding average rainfall intensity. In this study, 24 h storm with the total rainfall depths of 398, 597 and 805 mm, each respectively corresponding to 10-, 100- and 1,000-year storms at the study site are considered to investigate the effect of rainfall amount on pore water pressure distribution in the cut slope.

The pore water pressure distributions along depth at section XX′ under storm Pattern A2 with different rainfall amounts at $t = 6$ and 192 h are shown in Figure 9.16. At $t = 6$ h, the top 10 m of soil is wetted. The soil suction on the ground surface under the 398 mm (10 year) rainfall is reduced to about 29 kPa from 53 kPa, whereas rainfall amounts of 597 mm (100 year) and 805 mm (1,000 year) rainfalls reduce it to zero on the ground surface. The pore water pressure profiles (dash lines) associated with the two higher rainfall amounts overlap. This implies surface run-off resulting from the 1000 year rainfall will be larger than that of from the 100 year rainfall.

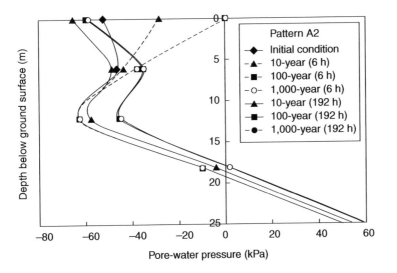

Figure 9.16 Pore water pressure distributions with depth for 24-h pattern A2 rainfall with 10-, 100- and 1,000-year return periods after 6 and 192 h (after Ng et al., 2001b).

At $t = 192$ h, there is no noticeable difference between the pore water pressure profiles corresponding to the 100 and 1000 year rainfalls. This suggests that when the 24-h rainfall amount with a return period higher than 100 year, no significant difference in ground water response can be found at the study site under the advanced rainfall pattern. Similar behaviour can also be seen at section YY′ and is not repeated here.

As for storm Pattern C, the rate of change in pore water pressure profile at section XX′ decreases as the rainfall amount increases. This can be observed that the difference in pore water pressure profiles is larger between 10 year and 100 year rainfall than between 100 year and 1000 year rainfall. The difference in rainfall amount from one return period to the next higher return period is about the same of 200 mm. For the delayed storm pattern, the trend of pore water pressure profile lies between the two other storm patterns, but closer to that of the advanced storm pattern. Details results under storm Patterns C and D1 can be found in Wang (2000).

As an increase in rainfall return period actually increases the rainfall intensity, there should be a critical rainfall return period, above which no significant increase in pore water pressure underground can be induced. For example, the value is 100 year in this study. The reason for this is that an increasing pore water pressure is due to infiltration and the infiltration rate is in turn determined by water permeability and hydraulic gradient. Water permeability in soil has its own limit, i.e. the saturated water permeability.

Although rainfall intensity is higher for a higher return period, not all of the rainwater can actually infiltrate the soil and so no significant difference in groundwater response can be induced.

Effect of rainfall patterns of prolonged rainfall on groundwater responses

In this section, comparisons of groundwater responses resulting from 24 and 168 h rainfalls of 100 year return period are made under the three different rainfall patterns. It should be noted that the rain depth of 168 h/100 year storm is 892 mm, which is about 1.5 times that of the 24 h/100 year storm of 597 mm. The average intensity of the 168 h rainfall is only about one-fifth that of the 24-h rainstorm.

The variations of hydraulic head resulting from storms of different duration at three soil depths are shown in Figures 9.17a–c. For 24 h rainfall of different patterns, the time variations of hydraulic head on the ground surface coincides well with rainfall patterns (see Figure 9.17a). The 24-h rainfalls with storm Patterns A2 and D1 result in the maximum and minimum increase in hydraulic head, respectively. However, there is no significant difference about the maximum hydraulic heads for 168 h rain under the three storm patterns. The pore water pressure of storm Pattern A2 reaches its peak at $t = 42$ h and the storm Patterns C and D1 reach their respective peaks at $t = 84$ h and $t = 168$ h. The time when the maximum pore water pressure head occurs at the ground surface coincides with the time of peak intensity of each rainfall.

Unlike the instantaneous response at the ground, there is a time lag between the peak rainfall intensity and the maximum induced hydraulic heads or groundwater pressures at 6.1 m below the ground for different patterns of 168-h rainfalls (see Figure 9.17b). For instance, the peak rainfall intensity of the Pattern A2 rainfall is at $t = 42$ h into rainfall and the maximum induced pore water pressure at the depth of 6.1 m occurs at 48 h later (i.e. 84 h into rainfall). Similarly, the peak intensity and the maximum pore water pressure response of the Pattern C rainfall occur at 84 and 126 h into rainfall, respectively. The time difference between the peak intensity of the two rainfall patterns, which is 42 h, is exactly equal to the time lag between the induced peak pore water pressures at the depth of 6.1 m for the two rainfalls.

At the depth of 18.3 m, the difference in hydraulic head or pore water pressure for various 24-h rainfalls is not significant at all (see Figure 9.17c). For the prolonged 168-h rainfalls with different rainfall patterns, the Pattern A2 rainfall appears to cause the greatest pore water pressure rise at the end of rainfall. However, the difference in maximum pore water pressure for various rainfall patterns is small and it becomes negligible at 240 h after the cessation of rainfall (i.e. at $t = 408$ h). This implies that the influence of

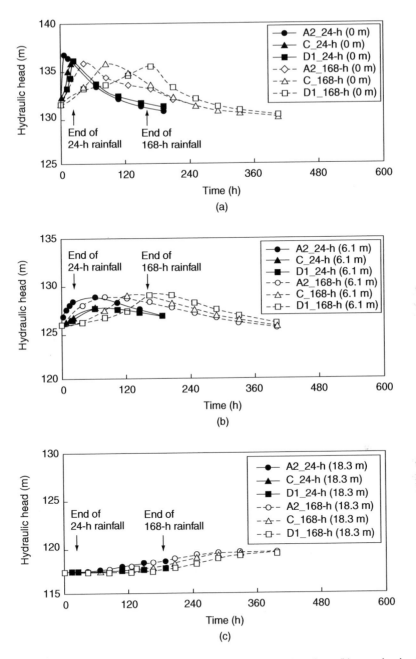

Figure 9.17 Variation of hydraulic head (a) at the ground surface, (b) at a depth of 6.1 m and (c) at a depth of 18.3 m at section X–X′ for 168-h rainfall with a 100-year return period (after Ng *et al.*, 2001b).

different rainfall patterns on pore water pressure response is not important at this depth. Comparing with the ground water responses at depth of 18.3 m under the 24-h rainfall (see Figure 8a), the 168-h prolonged rainfalls lead to a higher continuous rise of hydraulic head even after the cessation of the rainfalls. This is because the prolonged rainfall has a larger total rain depth than that of the short 24-h rainstorm. The continuous rise of hydraulic head after the cessation of the prolonged rainfalls indicates a continuous decrease in the effective stress at depth and hence potentially the decrease may induce delayed and deep-seated slope instability such as the slope failure at Lai Ping Road (Tung et al., 1999; Sun et al., 2000a).

The variation of hydraulic heads at section YY' is not presented here since the key features observed at the toe are similar to those at the crest (i.e. section XX'). Details of computed results are described by Wang (2000).

Summary and conclusions

- In this section, groundwater responses in an initially unsaturated cut slope subjected to different rainfall conditions were investigated by using a three-dimensional groundwater model. Three rainfall patterns typically found in Hong Kong with three return periods, 10, 100 and 1,000 years, were adopted for detailed study. Temporal variations of hydraulic head at different depths were studied at two selected sections, one at the crest and the other at the toe of the cut slope.

- At the crest noticeable difference in groundwater responses on the ground surface was observed under different storm patterns of the 24-h/100-year rainfall. This difference mainly derives from the first several hours of rainfall during which there is a significant difference in the infiltration rates. For the 24-h rainfalls, the influence of rainfall patterns on pore water pressure is the most significant at the ground surface but it gradually diminishes as the depth increases.

- For a given rainfall amount and duration, the advanced storm pattern is the most critical among the three patterns considered for its potential to induce the greatest increase in pore water pressure along depth in both the crest and the toe of the cut slope. The reason for this greatest increase is due to high hydraulic gradient induced by the advanced storm pattern during the first several hours of intense rainfall, which results in a higher infiltration rate than other two storm patterns.

- The groundwater responses to various rainfall patterns are different at the crest and the toe of the cut slope implying that the initial groundwater condition has a significant influence on groundwater response. At the crest, where the initial groundwater table is deeply seated, the rainwater infiltration only reduces the soil suction and the groundwater table is not affected significantly. However, at the toe where the soil is nearly saturated because of the shallow initial groundwater table, various rainfall patterns have limited influence on water permeability and the pore water pressures near the ground surface. In addition, the groundwater response at the toe is influenced by groundwater flow from the upstream of the cut slope.

- For a given rainfall pattern of 24-h intense rainstorms under a given initial hydraulic gradient, a higher the return period generally leads to a greater increase in pore water pressure. At the study site, when rainfall amount exceeds that of 100 year, the rate of increase in pore water pressure caused by rainwater infiltration decreases. Under the advance storm pattern (A2), there is no discernible difference in pore water pressures resulting from rainfall amount with return periods of 100 and 1,000 years. The main reason is that the surficial water permeability governs infiltration during the intense rainfalls and some rainwater become surface run-off when rainfall intensity far exceeds the surficial water permeability. This implies that a storm with higher intensity does not necessarily produce more adverse affect on slope stability once a critical return period is reached. This is because the infiltration is dependent on the surficial water permeability.

- Comparing with 24- and 168-h rainstorms, the latter has a lower average rainfall intensity but with greater total rainfall amount (rain depth) under a given return period. Because of the lower intensity and longer duration, the 168-h rainfalls with three different patterns induces approximately the same maximum hydraulic head on the ground surface. During the 168-h rainfalls, the difference in the maximum hydraulic heads among different storm patterns decreases significantly with depth. At 240 h after the cessation of rainfalls (i.e. at $t = 408$ h), the influence of different rainfall patterns on pore water pressure response is insignificant at any depth. By comparing with the 24-h rainfalls, however, the 168-h rainfalls lead to a continuous rise in hydraulic head at great depths and hence a gradual reduction in effective stress even after the cessation of the rainfalls. This implies that prolonged rainfalls with relatively large rainfall amount could induce slope failures at great depths (deep-seated) after the cessation of the rainfalls.

Influence of state-dependent soil–water characteristic curve and wetting and drying on slope stability (Ng and Pang, 2000b)

Introduction

Rain-induced landslides pose substantial threats to property and life and over the years have caused severe damage in many countries such as Brazil,

Italy, Japan, Malaysia, Hong Kong and mainland, China (Fukuoka, 1980; Brand, 1984; Wolle and Hachich, 1989; Malone and Pun, 1997). The physical process of rainfall infiltration into unsaturated soil slopes and the influence of infiltrated rainwater on soil suction and hence the slope instability have been investigated by researchers both in the laboratory (Fredlund and Rahardjo, 1993) and in the field (Lim et al., 1996; Rahardjo et al., 1998). Numerical simulations of rainfall infiltration have also been conducted (Anderson and Pope, 1984; Lam et al., 1987; Wilson, 1997; Ng and Shi, 1998). Similar to saturated soils, water flow through unsaturated soils is also governed by Darcy's law (Fredlund and Rahardjo, 1993). However, there are two major differences between the water flows in saturated and unsaturated soils. First, the ability of the unsaturated soils to retain water varies with soil suction has to be known. Second, the coefficient of water permeability is not a constant in unsaturated soils but it is a function of soil suction. Thus, it is essential to determine (a) the so-called soil–water characteristic curve (SWCC), which defines the relationship between the soil suction and either the water content or the degree of saturation and (b) the water permeability function that varies with soil suction for simulating transient seepage in unsaturated soil slopes. Currently, it is a common practice to derive the water permeability function from a measured saturated water permeability and a drying SWCC using the procedures established by Fredlund and Xing (1994) and Fredlund et al. (1994).

The soil–water characteristic of a soil is conventionally measured by means of a pressure plate extractor in which any vertical or confining stress is not applied and volume change of the soil specimen is assumed to be zero, whereas in the field the soil usually is subjected to certain stress. Although it is theoretically recognized that the stress state of a soil has some influence on SWCC theoretically (Fredlund and Rahardjo, 1993), few experimental results can be found in the literature. Some exceptions are perhaps the publications by Vanapalli et al. (1996, 1998, 1999), who studied the influence of the total stress state on the SWCC of a compacted fine-grained soil indirectly. The soil specimens were first loaded and then unloaded using a conventional consolidation apparatus to create a known stress history or stress state in the specimens. Subsequently, the SWCCs of the pre-loaded specimens were determined using a traditional pressure plate apparatus, in which the change of water content due to the variation of soil suction was measured under almost zero applied net normal stress $(\sigma-u_a)$. It was found that the SWCCs are significantly influenced by the stress state for specimens compacted at initial water contents dry of optimum.

Although the total net normal stress on the soil elements in an unsaturated soil slope is seldom altered, the stress state at each element is different. This may affect the soil–water characteristic of these elements, i.e. the storage capacity when subjected to various soil suctions during rainfall infiltration.

To correctly predict pore water pressure distributions and hence the slope stability of an unsaturated soil slope, it is thus essential to investigate the influence of stress state on SWCCs (or SDSWCCs). For transient flows and slope stability problems, osmotic suction is normally not very important and therefore ignored. Soil suction is generally referred to matric suction only.

Numerical simulations

In order to investigate the influence of SDSWCCs on the predictions of pore water distributions in unsaturated soil slopes and their stability, a series of finite element transient seepage and limit equilibrium analyses are carried out using SEEP/W and SLOPE/W (Geo-slope, 1998), respectively. The flow laws and governing equations for transient seepage analyses are given in Chapter 3. A typical steep unsaturated soil cut slope in Hong Kong is selected for illustrative purposes. The computed results from the transient seepage analyses are then used as input parameters for a subsequent limit equilibrium analysis of the stability of the slope.

Input parameters and analysis procedures for transient seepage analyses

As infiltration of rainwater into the soil slope is a wetting process, the measured wetting SWCC [completely decomposed volcanic (CDV)-N1] and SDSWCCs (CDV-N2 and N3 – see Chapter 3 for the measured SDSWCC) are adopted for the transient seepage analyses. In addition, a water permeability function varying with soil suction is required. For comparison, a conventional transient analysis using a drying path of the SWCC (CDV-N1) under zero net normal stress is also included.

By using the measured saturated water permeability (k_s) of the soil in a triaxial apparatus under some appropriate stress conditions (Ng and Pang, 1998a,b), these selected SWCC and SDSWCCs are fitted by a highly non-linear equation, as proposed by Fredlund and Xing (1994) for obtaining a permeability function varying with matric suction. The input parameters for predicting the permeability functions are summarized in Table 9.4. Figure 9.18 shows the permeability functions computed from the selected curves. As expected, the soil specimen loaded to a higher net normal stress has a lower permeability function. This is because the applied net normal stress led to a smaller pore size distribution inside the soil specimen.

For investigating the effects of SDSWCC on pore water pressure distributions in an unsaturated soil slope during rainstorms, a finite element mesh of a typical cut slope of height 8.6 m inclined at 55° to the horizontal in Hong Kong is created and shown in Figure 9.19. The cut slope is located

Table 9.4 Input parameters for Predicting permeability(after Ng and Yung, 2000b)

Sample identity (1)	Applied stress (kPa) (2)	a (3)	n (4)	m (5)	k_s (m/s)[a] (6)
CDV-N1 (drying)	0	3.5392	0.8726	0.2726	3.01×10^{-6}
CDV-N1 (wetting)	0	1.4636	1.0112	0.2329	3.01×10^{-6}
CDV-N2 (wetting)	40	17.8684	0.6815	0.4492	2.88×10^{-6}
CDV-N3 (wetting)	80	18.2019	0.5225	0.3301	1.17×10^{-6}

Note
a Measured by Ng and Pang (1998).

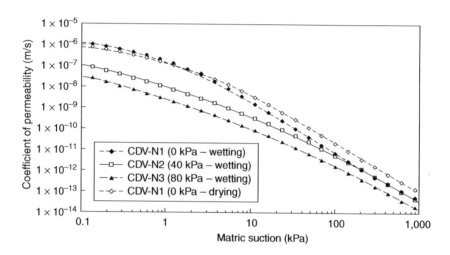

Figure 9.18 Permeability functions computed from measured soil–water characteristics (after Ng and Pang, 2000b).

on a natural hillside. The entire soil mass in the finite element mesh is idealized into three different soil layers according to their approximate stress states so that the measured SWCC and SDSWCCs and their corresponding water permeability functions can be specified. It is recognized that the current method of specifications greatly simplifies the actual complexity of the problem. However, computed results from the current simplified analyses are sufficient to reveal the important role of the SDSWCC in any transient seepage analysis.

In order to illustrate the influence of the SDSWCCs on pore water pressure distributions, two series of transient seepage analyses are conducted. In the first series of analyses, all soil layers are assumed to have the same drying SWCCs and their corresponding water permeability functions. This series is a conventional approach. In the second series of analyses, different hydraulic properties are specified in each soil layer. The wetting SWCC under 0 kPa (CDV-N1) and SDSWCCs under 40 kPa (CDV-N2) and 80 kPa (CDV-N3) applied net normal stresses and their corresponding permeability functions are specified respectively for the first, second and third soil layers as shown in Figure 9.19.

The initial groundwater conditions for each series of transient seepage analyses are established by conducting two steady-state analyses, during which a very small rainfall of intensity 0.001 mm/day is applied on the top boundary surface together with a constant hydraulic head 15 m above the Principal Datum (15 mPD or the sea level) specified on the left boundary. The bottom boundary is assumed to be impermeable and no flux is specified along the right boundary.

For the subsequent transient analyses, two rainfall patterns with an average intensity of 394 and 82 mm/day are applied on the top boundary surface in both series of analyses to simulate a short and intensive 24-h rainfall infiltration and a prolonged 7-day rainfall infiltration, respectively. The rainfall intensities adopted are based on the actual 10-year return period spanning from 1980 to 1990 (Lam and Leung, 1995). Here, it is assumed that the rate of infiltration is equal to 60 per cent of the rainfall intensity to simulate an average of 40 per cent surface run-off in Hong Kong (Tung *et al.*, 1999). The numerical simulations conducted are summarized in Table 9.5.

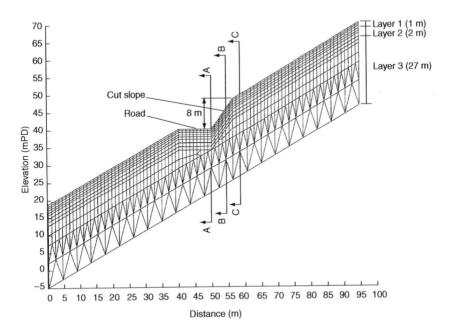

Figure 9.19 Finite element mesh for slope modelled (after Ng and Pang, 2000b).

Table 9.5 Summary of numerical simulations conducted (after Ng and Pang, 2000b)

Series number (1)	Key hydraulic parameter (2)	Type of analysis (3)	Rainfall infiltration (mm/day) (4)	Duration (5)
1	Drying SWCC (conventional)	Steady state	10.8	–
1	Drying SWCC (conventional)	Transient	236.4	24 h
1	Drying SWCC (conventional)	Transient	49.2	7 days
2	Wetting SWCC and SDSWCC	Steady state	10.8	–
2	Wetting SWCC and SDSWCC	Transient	236.4	24 h
2	Wetting SWCC and SDSWCC	Transient	49.2	7 days

Input parameters and analysis procedures for slope stability analyses

After obtaining the pore water pressure distributions from the transient seepage analyses, limit equilibrium analyses are then carried out to determine the factor of safety of the cut slope. For estimating the factor of safety using the Bishop's simplified method, some basic mechanical soil parameters are needed. In the limit equilibrium analyses, the shear strength of the soil is assumed to be governed by the extended Mohr–Coulomb failure criterion as follows:

$$\tau = c' + (\sigma_n - u_a) \tan \phi' + (u_a - u_w) \tan \phi^b \qquad (9.22)$$

The shear strength parameters include an effective cohesion c' of 2 kPa, an angle of friction ϕ' of 28° and an angle indicating the rate of increase in shear strength relative to the matric suction ϕ^b, which is equal to 14°.

Influences of SDSWCC on pore water pressure distributions

Figure 9.21a–c shows the computed distributions of pore water pressure varying with depth at sections A–A, B–B and C–C of the finite element mesh (see Figure 9.19), respectively. It is clear that there is a substantial difference between the initial pore water pressure distributions computed using the conventional drying SWCCs and the unconventional wetting SWCC and SDSWCCs in the steady-state analyses. The conventional analysis predicts a significantly higher soil suction profile than that computed by the unconventional analysis. This is because the soil in the former analysis, in comparison with the soil in the latter analysis, has a lower air-entry value and a faster rate of changing volumetric water content as values of soil suction increase (refer to Figure 9.20), and a higher water permeability function (see Figure 9.18). In other words, the soil under the applied stress has a stronger capability to retain moisture for a given soil suction due to the

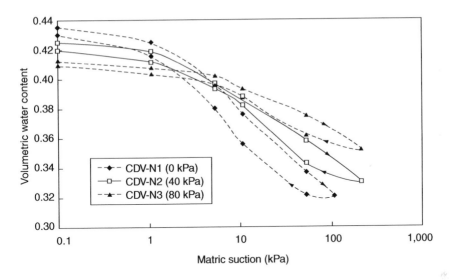

Figure 9.20 Effects of stress state on SWCCs (after Ng and Pang, 2000b).

presence of a smaller pore size distribution, as illustrated by a flatter SWCC (refer to Figure 9.20). The computed results highlight the importance of considering stress effects and drying–wetting history on SWCCs.

During the short but highly intensive rainfall (i.e. 236.4 mm/day or 2.74×10^{-6} m/s for 24 h), the pore water pressure responses at the three selected sections are similar in both the conventional and unconventional transient analyses. Only the soil suctions in the top 1–2 m depth are destroyed irrespective of the magnitude of their initial values. A relatively shallow advancing 'wetting front' (Lumb, 1975) is developed as most of the rainfall cannot infiltrate into the soil due to the relatively low water permeability to rainfall infiltration at high initial values of soil suction. On the contrary, the pore water pressure distributions predicted by the conventional and unconventional analyses are completely different during the 7-day low-intensity prolonged rainfall (49.2 mm/day or 5.69×10^{-7} m/s for 7 days), especially at sections A–A and B–B. Due to the relatively low initial values of suction present in the soil, resulting in relatively high water permeability with respect to rainfall infiltration, when the effects of stress state are considered, soil suctions at sections A–A and B–B (see Figure 9.21a and b) are totally destroyed by the advancing 'wetting front' to a depth of about 6 and 9 m from the ground surface, respectively. At section C–C (see Figure 9.21c), the advancement of the 'wetting front' is limited by the initial higher suction values than those at sections B–B and C–C. The less intensive but prolonged 7-day rainfall facilitates the advancement of 'wetting front' into the soil to some great depths and cause significant reduction in soil suction which would have some devastating effects on slope stability. For the case of the initial values of soil suction predicted using conventional drying SWCC, rainfall infiltration is hindered as a result of relatively low water permeability due to the presence of high soil suction.

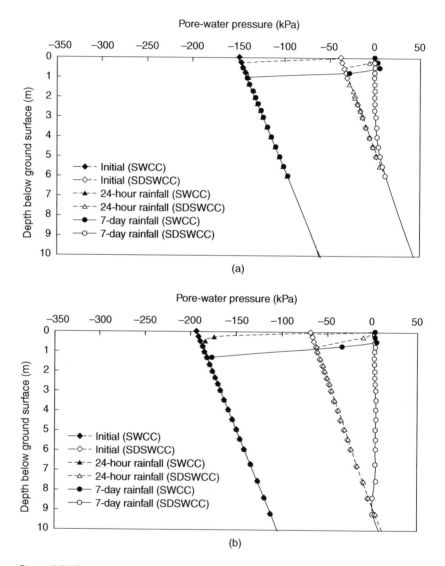

Figure 9.21 Pore water pressure distributions under various rainfall conditions along (a) section A–A; (b) section B–B; (c) section C–C (after Ng and Pang, 2000b).

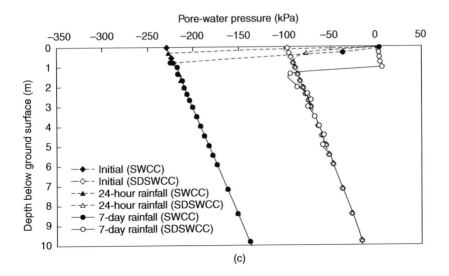

Figure 9.21 (Continued).

Influences of SDSWCC on slope stability

By using the computed pore water pressure distributions in the transient flow analyses, limit equilibrium analyses are performed on four selected non-circular slip surfaces, which pass through the toe of the slope (see Figure 9.22). It should be noted that these selected slip surfaces may not warrantee the minimum factor of safety. They were selected only for illustrating the influence of SDSWCCs on the factor of safety of some possible slips. The actual critical slip surfaces may be somewhat different.

Figure 9.23 shows the variations of factors of safety with elapsed time for the four selected slip surfaces during the 7-day rainfall. It can be seen that the limit equilibrium analyses, which adopted the pore water pressures computed by using the conventional drying SWCCs, predict substantial higher initial factors of safety at all four slip surfaces than those obtained from the analyses using the unconventional wetting SDSWCCs. This is attributed to the significant difference in the computed pore water pressure distributions (see Figure 9.21a–c) with and without considering the effects of the stress state and drying–wetting history. This implies that the traditional analyses using the conventional drying SWCCs may lead to unconservative designs.

Due to the presence of higher initial suction at shallow depths, the shallower the slip surface, the larger the initial factor of safety. With time the factor of safety decreases but at different rates. For the shallow slips (S1 and S2 – see Figure 9.22), the fall in the factors of safety is substantial when the

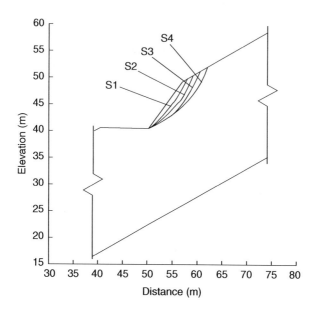

Figure 9.22 Slip surfaces considered in the slope stability analyses (after Ng and Pang, 2000b).

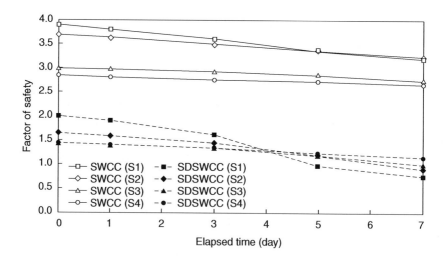

Figure 9.23 Factors of safety with respect to time from the beginning of the 7-day rainfall (after Ng and Pang, 2000b).

effects of stress state are included. At the end of rainfall, the factor of safety increases with depth, opposed to the initial safety conditions.

Conclusions

Measured SDSWCCs were adopted to derive water permeability functions for transient seepage and limit equilibrium analyses of a typical unsaturated cut slope in Hong Kong under various rainfall conditions. Based on the experimental and the simplified numerical studies, the following conclusions can be drawn:

- Numerical analyses using the measured wetting SDSWCCs and their derived water permeability functions predict a substantially higher (less negative) initial steady state pore water pressure distributions with depth than those computed by using the conventional drying SWCCs. These initial high (less negative) steady state pore water pressure distributions with depth leading to higher water permeability in the ground facilitate rainfall infiltration, which destroys soil suction to a great depth by an advancing 'wetting front'. This results in a substantially lower factor of safety during a prolonged low intensity rainfall. On the contrary, the numerical analyses suggest that only the soil suction at the top 1–2 m of soil would be destroyed under highly intensive but short-duration rainfalls, irrespective of whether the stress effects on SWCC and the drying–wetting history are considered or not.
- Based on the current experimental measurements and the simplified numerical investigations, the stress state and the drying–wetting history have a substantial influence on the soil–water characteristics of unsaturated soils. During a prolonged rainfall, analyses using wetting SDSWCCs would predict adverse pore water pressure distributions with depth and lower factors of safety than those from the conventional analyses using drying SWCCs. The wetting SDSWCCs should therefore be considered for better and safer estimations of the factor of safety for unsaturated soil slopes.

Effects of state-dependent soil–water characteristic curves and damming on slope stability (Ng and Lai, 2004)

Introduction

The objectives of this section are to study any difference in computed pore water pressure distributions using drying SWCC, wetting SWCC and wetting SDSWCC and to investigate damming effects (if any) on pore water pressure distributions and slope stability due to the presence of piles in an initially unsaturated soil slope.

Finite element mesh and numerical analysis plan

An initially unsaturated natural soil slope located in Hong Kong is chosen for this idealized parametric study. The slope inclines at approximately 26° to the horizontal. Figure 9.24 shows a dimensional finite element mesh of the slope. The 20-m thick CDV soil mass is idealized into three different soil layers according to their approximate stress states so that the measured SWCC and SDSWCCs and their corresponding water permeability functions can be specified. To study damming effects due to construction obstructions on pore water pressure distributions, two artificial hydraulic barriers such as piles are simulated in some cases. The diameter and depth of the two piles is 1 and 6 m, respectively. They are 16 m apart as shown in Figure 9.24. It should be noted that the two artificial hydraulic obstructions essentially behave as an infinite long 6 m deep impervious retaining wall in the two-dimensional seepage analyses. In other words, damming effects are very likely to be overestimated in this study, unless a three-dimensional transient seepage analysis can be carried out (Ng *et al.*, 2001b).

To investigate the influence of wetting SWCC and effects of stress states (i.e. SDSWCCs) on pore water pressure distributions, three series of transient seepage analyses are conducted. Series D consists of 'conventional' transient analyses in which the drying path of the SWCC (CDV-N1) under

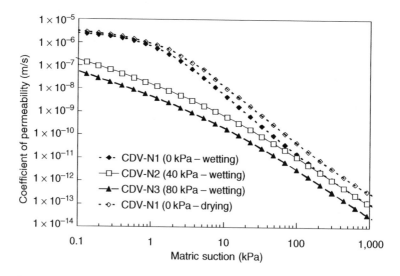

Figure 9.24 Finite element mesh and slip surface considered in the slope stability analysis (after Ng and Lai, 2004).

zero net normal stress and its corresponding water permeability function are specified for all the three soil layers in the slope. In Series W, since rainfall infiltration into the soil slope is a wetting process, all soil layers are assumed to have the same wetting SWCC (CDV-N1) under zero net normal stress and the corresponding permeability function. In Series SW, the measured wetting SWCC (CDV-N1) at zero stress and SDSWCCs (CDV-N2 and N3) are specified for soil layers 1, 2 and 3, respectively, for investigating the influence of stress-dependency of soil–water characteristics on transient seepage analyses. A summary of the three series of analyses is given in Table 9.6. For each series, there are four cases considered to study damming effect (if any) caused by the two pile obstructions. Case (a) is a reference run, in which no pile is installed in the slope. In Case (b), a pile, i.e. Pile One, is installed at the upstream only (see Figure 9.24). Similarly, Case (c) is to determine damming effect caused by Pile Two installed at downstream only. Case (d) is for studying damming effect caused by the two piles installed at both upstream and downstream, located at 16 m apart.

Numerical simulation procedures and hydraulic boundary conditions

Prior to each transient seepage analysis, a steady-state analysis is carried out to establish the initial pore water distributions in the slope. This is done by applying a uniform rainfall intensity of 0.01 mm/day (or 1.16×10^{-10} m/s) on the top of the slope surface. At the right boundary of the

Table 9.6 Summary of the numerical simulation plan (after Ng and Lai, 2004)

Series	Analysis identity	Pile one	Pile two	Soil–water characteristic
D	D-a	N	N	CDV-NI (drying)
	D-b	Y	N	
	D-c	N	Y	
	D-d	Y	Y	
W	W-a	N	N	CDV-NI (wetting)
	W-b	Y	N	
	W-c	N	Y	
	W-d	Y	Y	
SW	SW-a	N	N	CDV-NI (wetting)
	SW-b	Y	N	CDV-N2 (wetting at 40 kPa)
	SW-c	N	Y	CDV-N3 (wetting at 80 kPa)
	SW-d	Y	Y	

Notes
Average rainfall intensity of 49 mm/day are applied for 10 days.
D = Drying curve at zero stress.
W = Wetting curve at zero stress.
SW = Stress dependent wetting curves.
N = Pile is not constructed, Y = Pile is constructed.

mesh (see Figure 9.24), a constant hydraulic head of 47.6 m above the Principal Datum (or 47.6 mPD above sea level) is specified. This specified head is based on pore water pressure measured by a piezometer at the site of the selected slope. The bottom hydraulic boundary is assumed to be impermeable, and no flux boundary is specified along the left boundary of the mesh. For the cases with piles installed, the piles are modelled as an almost impermeable obstruction with a constant water permeability of 1.0×10^{-15} m/s.

During any transient seepage analysis, an average rainfall intensity of 82 mm/day is adopted. The average intensity used is based on actual measurements of rainfall intensity for a 10-year return period spanning from 1980 to 1990 in Hong Kong (Lam and Leung, 1995). To allow for surface run-off, it is assumed that the average rate of infiltration is equal to 60 per cent of the average rainfall intensity (Tung et al., 1999). In the numerical simulations, infiltration due to rainfall is thus modelled by applying a surface flux of 49 mm/day or 5.7×10^{-7} m/s (i.e. 60 per cent of 82 mm/day) across the top boundary surface of the slope for 10 days continuously in each case to simulate a prolonged rainfall event.

Slope stability analyses

After pore water pressure distributions within the slope are computed from the transient seepage analyses, factor of safety (FOS) of the slope is determined by carrying out limit equilibrium analysis using a computer software called, SLOPE/W (Geo-slope, 2002). Here computed pore water pressures from SEEP/W are used as input parameters for slope stability calculations.

Input parameters and procedures for slope stability analyses

For calculating FOS for each case, the Janbu's simplified method is used together with the extended Mohr–Coulomb failure criterion. The shear strength parameters used include an effective cohesion c' of 2 kPa, an effective angle of friction ϕ' of 28° and an angle ϕ^b of 14° which indicates the rate of an increase in shear strength with respect to matric suction. Limit equilibrium analyses are performed on a selected non-circular slip surface (see Figure 9.24). It should be noted the selected slip surface does not necessarily give the lowest FOS for each case. The selection of a slip surface is merely to compare relative FOSs computed for different cases considered as given in Table 9.6. As shown in Figure 9.24, locations 'u', 'm' and 'd' represent the intersections of the assumed slip surface and sections U–U, M–M and D–D, respectively.

Interpretation of computed results

Influences of wetting SWCC and SDSWCC on pore water pressure distributions

Figure 9.25a–c shows the computed distributions of pore water pressure with depth along section U–U of the slope without any pile obstruction (i.e. Run D-a, Run W-a and Run SW-a as summarized in Table 9.6). It is clear that there is a substantial difference among the initial pore water pressure distributions computed using the drying SWCCs (Run D-a), the wetting SWCCs (Run W-a) and the SDSWCCs (Run SW-a) in the three steady-state analyses. Run D-a predicts a much higher soil suction profile than that computed by the other two unconventional analyses (Run W-a and Run SW-a). This is because at the steady state the mobilized water permeability in the soil slope is close or equal to the applied constant infiltration rate of 0.01 mm/day (or 1.16×10^{-10} m/s) when a hydraulic gradient of unity or close to unity is set up for a soil zone near the ground surface. In other words, as shown in Figure 9.18, for the given constant infiltration rate of 1.16×10^{-10} m/s or mobilized water permeability, the corresponding soil suction calculated in Run D-a using the drying CDV-N1 permeability function is higher than that of Run W-a using the wetting CDV-N1 permeability function, which in terms is higher than that of Run SW-a using the wetting CDV-N1, N2 and N3 permeability functions.

In the transient analyses of the 10-day prolonged rainfall with the constant intensity of 49 mm/day (or 5.7×10^{-7} m/s), a wetting front is formed and it moves progressively downwards as the rainfall continues in all three runs as shown in Figure 9.25. In comparing the computed results of using a drying SWCC (i.e. Run D-a) and a wetting SWCC (i.e. Run W-a) at Day 10, the wetting front of Run W-a penetrates about 1 m deeper than that of Run D-a (Figure 9.25a and b). This is because the initial soil suction in Run W-a is lower and hence the soil is more permeable for the advancement of the wetting front. Rahardjo and Leong (1997) conducted two similar analyses to study the difference in transient seepage analyses of rainfall infiltration using a drying and a wetting SWCC. In their two analyses, however, they specified an identical initial pore water pressure distribution. They reported that the analysis using a wetting SWCC produced a shallower wetting front, which is not consistent to the results shown in Figure 9.25.

When SDSWCCs are considered, the wetting front in Run SW-a penetrates about 13 m below ground at Day 10 (see Figure 9.25c). This is substantially deeper than the computed 5-m depth in Run D-a and 6-m depth in Run W-a. For the hydraulic parameters and geometry of the slope considered in this chapter, SDSWCCs clearly facilitate the advancement of the wetting front into the soil at great depths and lead to significant reduction in soil suction. These could have some devastating adverse effects on slope stability. Therefore, neglecting the influence of stress state on SWCC in a

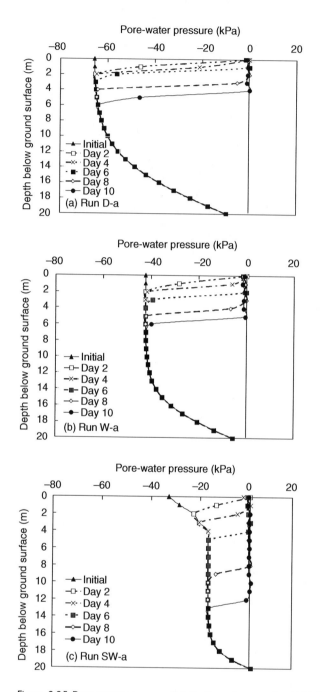

Figure 9.25 Pore water pressure distributions along section U-U: (a) Run D-a, (b) Run W-a and (c) Run SW-a (after Ng and Lai, 2004).

transient seepage analysis could result in non-conservative slope stability calculations.

Damming effect on pore water pressure distributions

Since transient seepage analyses using SDSWCCs are theoretically correct and relevant to geotechnical engineering problems, only computed results from series SW are chosen to illustrate any damming effect due to the presence of some hydraulic obstructions on pore water pressure distributions. It should be reminded that a summary of different arrangement of the two hydraulic obstructions (i.e. piles) in series SW is given in Table 9.6. Figure 9.26 shows the computed changes of hydraulic head at locations 'u', 'm' and 'd'. The depths of locations 'u', 'm' and 'd' are 5, 8 and 5 m, respectively (refer to Figure 9.24). At location 'u', due to the presence of Pile One in Run SW-b and Run SW-d, the subsurface seepage at the upstream has been dammed in these two analyses. Therefore, the initial hydraulic heads computed in Run SW-b and Run SW-d are slightly higher than those of SW-a and SW-c (Figure 9.26a). During the first five days of rainfall (Day 1 to Day 5), there is no significant change in hydraulic head in all four cases considered. This is because there is not enough time for a wetting front to reach location 'u'. As shown in Figure 9.27a, the wetting fronts of all the analyses are still above the level of location 'u' on Day 5. It should be noted that the wetting front of Run SW-b penetrates slightly deeper than that of Run SW-a (see Figure 9.27a) since seepage at the upstream face of Pile One is dammed by the pile in Run SW-b. The dammed water is forced to flow downwards around the pile tip. As the computed results of Run SW-c and Run SW-d are similar to those of Run SW-a and Run SW-b, respectively, therefore only the results from latter two runs are shown for clarity.

On Day 6, the hydraulic heads of Run SW-b and Run SW-d increase sharply, while the heads of Run SW-a and Run SW-c remain unchanged (see Figure 9.26a). This is because the wetting fronts in Run SW-b and Run SW-d seeped faster than those in Run SW-a and Run SW-c due to the damming effect, and they reached location 'u' first (Figure 9.27b). Hence, matric suction at location 'u' was dissipated (i.e. zero pore water pressure) and hence the hydraulic head at location 'u' was equal to the elevation head as shown in Figure 9.26a. In Run SW-a and Run SW-c, the wetting front has not yet reached location 'u' and so the hydraulic heads remain the same as the initial state.

From Day 7 to Day 10, the hydraulic head is almost the same for the four cases considered (see Figure 6a). This is because the wetting front in each analysis has reached and passed through location 'u' (Figure 9.27c) from Day 7 onwards.

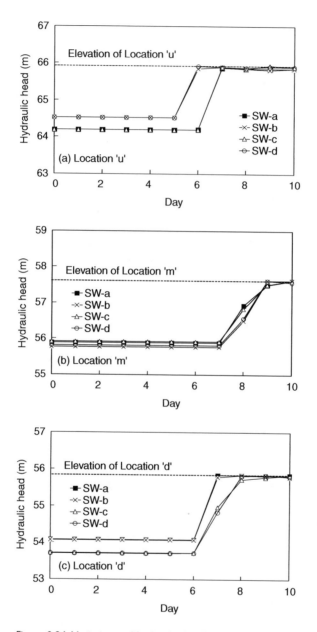

Figure 9.26 Variations of hydraulic head during the 10-day rainfall: (a) Location 'u', (b) Location 'm' and (c) Location 'd' (after Ng and Lai, 2004).

Location 'm' is situated between Pile One and Pile Two and it is 8 m away from both the piles. At this location, there is no significant difference in computed hydraulic head from the four analyses considered during the 10-day rainfall (Figure 9.26b). This implies that damming does not affect pore water pressure distribution noticeably at this location.

At location 'd', due to the presence of Pile Two which has dammed seepage from the upstream, the initial hydraulic heads of Run SW-c and Run SW-d are lower than those of Run SW-a and Run SW-b (Figure 9.26c). From Day 1 to Day 6, no major change in hydraulic head is observed. This

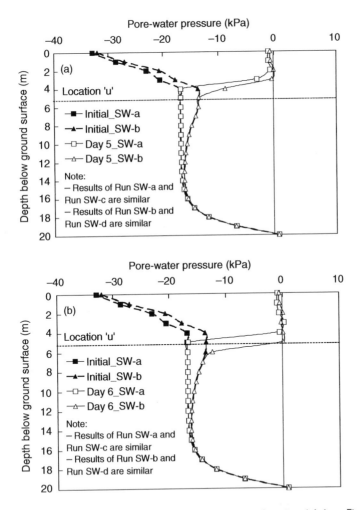

Figure 9.27 Pore water pressure distributions at location 'u' (see Figure 9.24) at (a) Day 5, (b) Day 6 and (c) Day 7 (after Ng and Lai, 2004).

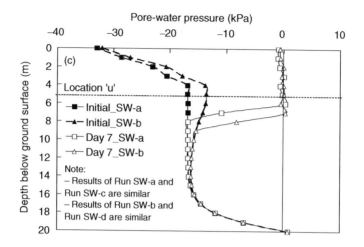

Figure 9.27 (Continued).

is because there is not enough time for wetting fronts to reach location 'd'
as shown in Figure 9.28a. The computed wetting fronts from Run SW-c and
Run SW-d are shallower than those from Run SW-a and Run SW-b due to
damming at the upstream face of Pile Two in the former two runs. On Day
7, hydraulic heads of Run SW-c and Run SW-d increase less than those of
Run SW-a and Run SW-b (Figure 9.26c). This is because the wetting fronts

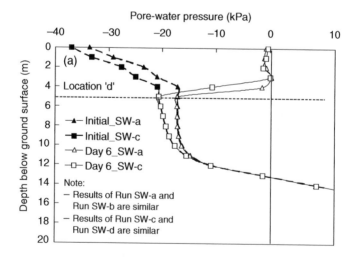

Figure 9.28 Pore water pressure distributions at location 'd' (see Figure 9.24) at
(a) Day 6, (b) Day 7 and (c) Day 8 (after Ng and Lai, 2004).

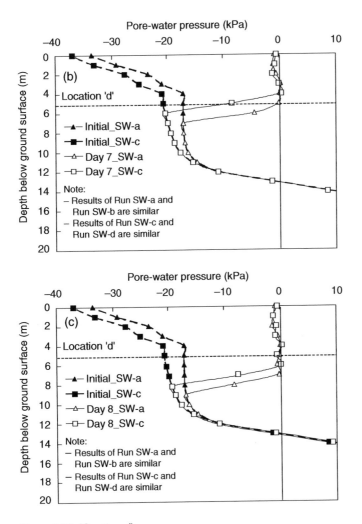

Figure 9.28 (Continued).

in Run SW-c and Run SW-d have seeped slower than those in Run SW-a and Run SW-b as a result of damming effect (refer to Figure 9.28b). From Day 8 to Day 10, there is almost no difference in hydraulic head among the four analyses (Figure 9.26c) because the wetting front in each analysis has reached and passed through location 'u' from Day 8 onwards. The hydraulic head at location 'u' is equal to the elevation head suggesting that pore water pressure is zero at this location (Figure 9.28c).

Variation of slope stability factor of safety

Influence of wetting SWCC and SDSWCC

Based on pore water pressures computed by SEEP/W, slope stability analyses of the assumed slip surface (see Figure 9.24) were carried out using SLOPE/W. Figure 9.29 shows the variations of FOSs with elapsed time for different cases during the 10-day rainfall. As expected, the computed FOS decreases as time elapses during the 10-day rainfall. It can be seen that the FOSs of the analyses using the drying SWCCs (i.e. Series D) are the highest for a given rainfall condition. On the contrary, the FOSs computed from the analyses using the wetting stress-dependent SDSWCCs (i.e. Series SW) are the lowest. This difference between the two series of analyses is attributed to the significant difference in the computed initial pore water pressure distributions (Figure 9.25a–c). This implies that an analysis using a drying SWCC would lead to non-conservative or even unsafe predictions.

Damming effect on slope stability

For the given assumed slip surface, there is no significant difference in the computed FOSs due to damming effects. This is because the pore water pressure distribution along the assumed slip surface is only slightly affected by damming. Although some damming on pore water pressure can be observed at locations 'u' and 'd', the effects are not significant (Figure 9.26a and c). At location 'm', the influence of damming is even smaller (Figure 9.26b). In fact, the extent of damming effects on pore water pressure and slope stability has already been overestimated in this study. This is because the numerical simulations are only two-dimensional. Actual three-dimensional

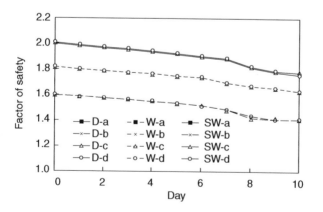

Figure 9.29 Factors of safety with respect to time for all slope analyses (after Ng and Lai, 2004).

groundwater flow around hydraulic obstacles is not permitted in the two-dimensional analyses. In other words, damming effect on FOSs would be even smaller if a three-dimensional analysis is carried out.

Conclusions

- Based on the simplified two-dimensional numerical investigations, the influence of wetting process and stress state on pore water distributions and FOS of an unsaturated soil slope have been investigated. During the 10-day prolonged rainfall, the FOS obtained from a conventional analysis using a drying SWCC is the highest, whereas an analysis using wetting stress-dependent SDSWCCs would predict the most adverse initial pore water pressure distributions with depth, and hence the lowest FOS. Thus, stress-dependency of SWCC should be considered and used in transient seepage analysis.
- For the geometry and ground conditions considered, damming of groundwater due to the presence of pile obstacles is very limited in the two-dimensional space. In other words, damming effects would be even smaller if a three-dimensional analysis is carried out.

Fundamentals of recompaction of unsaturated loose fill slopes (Ng and Zhan, 2001)

Introduction

Several thousand sizeable loose fill slopes more than 3 m high, constructed before 1977, have been identified in Hong Kong (HKIE, 1998). The majority of these loose fill slopes incline at an angle of between 30° and 45° to the horizontal, and in a loose state (Sun et al., 1998). Failures of these slopes pose a substantial threat to society and over the years have caused severe damage and disruption in Hong Kong.

Following the 1976 Sau Mau Ping landslide, the Independent Review Panel for Fill Slopes (Hong Kong Government, 1977) recommended that the 'minimum treatment' of existing loose fill slopes should consist of 'removing the loose surface soil by excavating to a vertical depth of not less than 3 m, and recompacting to an adequate Standard', and provision of 'drainage of the fill behind the recompacted surface layer at the toe of the slope'. This has become the standard practice for the upgrading of loose fill slopes in Hong Kong. However, limited transient seepage analyses have been carried out to investigate the effects of the recompaction on pore water pressure distributions in an initially unsaturated slope and hence on the slope stability.

Rainwater ingress into loose fill slopes is one of the primary contributory factors for triggering landslides (Ng and Shi, 1998; Sun *et al.*, 1998), especially in Hong Kong with abundant annual rainfall. Infiltrated water results in a reduction of soil suction and hence a reduction of soil shear strength. Moreover, the loose fill materials are generally contractive, so rainfall infiltration tends to result in the collapse of soil structure and cause flow slides or static liquefaction (Sun *et al.*, 1998). For more accurately predicting slope stability, it is vital to understand the variations of soil suction due to rainfall infiltration under various rainfall and ground conditions such as a surficial recompacted layer.

In this section, transient seepage and limit equilibrium analyses have been conducted on typical loose fill slopes with and without 3-m recompaction subjected to one day intensive rainfall. The main objective of these analyses is to investigate the fundamental effects of the recompaction on pore water pressure distributions in fill slopes and hence on their stability, assuming that no static liquefaction conditions are reached. In other words, strain-softening behaviour of loose fill material during undrained shear is not considered in this parametric analysis.

Analysis procedures

For investigating the effects of recompaction on pore water pressure distributions in fill slopes and hence on the slope stability, firstly, transient seepage analyses for saturated and unsaturated soils were carried out using a computer program called SEEP/W (Geo-Slope, 1998). After obtaining the pore water pressure distributions from the transient seepage analyses, limit equilibrium analyses were then carried out to determine the FOS of the fill slopes using SLOPE/W (Geo-Slope, 1998), in which shear strength of unsaturated soils can be represented by an extended Mohr–Coulomb failure criterion (Fredlund and Rahardjo, 1993a,b).

Here, a typical unsaturated CDV fill slope in Hong Kong is adopted. The slope is 30 m high and inclines at an angle of 32° to the horizontal. The finite element mesh for the slope is shown in Figure 9.30. The ground conditions in the slope without recompaction comprise a thick loose CDV fill layer (8 ∼ 20 m) underlain by an inclined bedrock. To simulate the effects of recompaction, a 3-m thick recompacted soil layer underneath the ground surface is modelled with different hydraulic and shear strength parameters from those of the loose fill.

A constant hydraulic head was specified along the lower vertical boundaries EF and GH according to the location of groundwater level measured at San Mao Ping (Hong Kong Government, 1977), and the upper vertical boundaries DE and AH are zero-flux boundary. The bottom boundary (GF) is assumed to be impermeable.

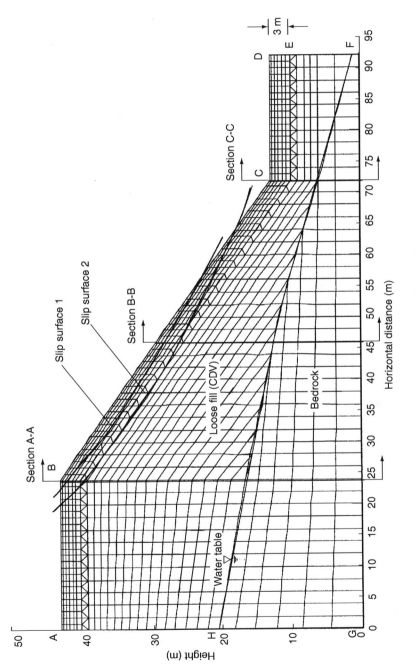

Figure 9.30 Finite element mesh for a typical loose fill slope (after Ng and Zhan, 2001).

On the ground and sloping surface (AB, BC and CD in Figure 9.30), a rainfall intensity of 394 mm/day is applied to simulate an average 24-h rainfall of a 10-year return period (Lam and Leung, 1995). It was assumed that the coefficient of infiltration is equal to 60 per cent for fill slopes in Hong Kong (Tung *et al.*, 1999). The initial groundwater conditions for transient seepage analyses are established by conducting a steady-state analysis, in which a very small rainfall infiltration rate of 1.0×10^{-9} m/s is imposed on the ground surface (from A to C, Figure 9.30).

Model parameters

To simulate transient seepage in unsaturated soils, it was essential to specify hydraulic parameters including SWCC, saturated water permeability and permeability functions for loose and recompacted fills. The SWCCs and permeability functions for the loose and recompacted CDV fills are shown in Figures 9.31 and 9.32, respectively. The saturated permeability adopted for the loose and recompacted fills are 3.0×10^{-6} and 1.0×10^{-7} m/s, respectively (Ng and Pang, 2000a).

In the limit equilibrium analyses, the shear strength of unsaturated soils is assumed to be governed by the extended Mohr–Coulomb failure criterion (see Equation 9.22). The shear strength parameters for the loose and recompacted fills are based on laboratory measurements (Leung, 1999) and GEO report (GEO, 1993), respectively. For the former, a mean value of 'collapse friction angle', 26.8° was adopted. For the latter, a mean value of effective friction angle for CDV recompacted fill in Table 8 of Geoguide 1 (GEO, 1993), of 36.5° was adopted because recompaction leads to a reduction of

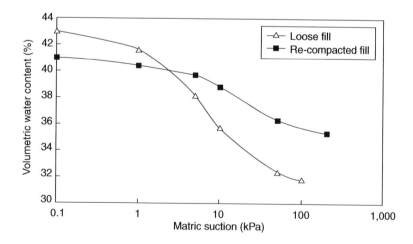

Figure 9.31 Soil–water characteristic curves for loose fill and recompacted fill (after Ng and Zhan, 2001).

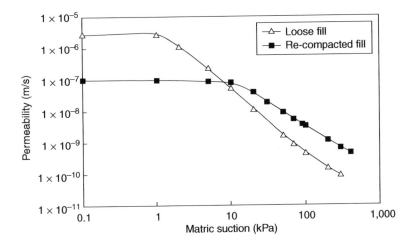

Figure 9.32 Permeability functions for the loose and recompacted soils (after Ng and Zhan, 2001).

void ratio of the soil. For both these two fills, the effective cohesion c' was taken to be 0 kPa, and the angle indicating the rate of increase in shear strength relative to the soil suction, ϕ^b, was assumed equal to half of their corresponding friction angles.

Influence of recompaction on pore water pressure distributions

Figure 9.33a–c shows the computed distributions of pore water pressure varying with depth at sections A–A (at the crest), B–B (at the middle) and C–C (at the toe) of the slope (see Figure 9.30). It can be seen that the initial pore water pressure distributions along sections A–A and B–B are similar qualitatively (refer to Figure 9.33a and b). At a depth greater than 20 and 3 m for sections A–A and B–B, respectively, there is little difference in the initial pore water pressure distributions between the loose and recompacted fill slopes. However, for shallower depths at both sections, the initial pore water in the recompacted fill slope is significantly lower than that of the corresponding loose fill slope. This can be explained by the difference in the ratio of rainfall infiltration rate to the instantaneous water permeability (q/k_w) between the loose and recompacted fill slopes. Since rainfall infiltration is a wetting process, soil is relatively dry before rainfall and it possesses a relatively high soil suction. As shown in Figure 9.32, at high soil suction (greater than 9 kPa), the hydraulic conductivity of recompacted fill is larger than that of loose fill, since the former has a higher water content. The high water permeability of recompacted fill (smaller q/k_w) leads to a low degree of saturation and hence a high initial negative pore water pressure in the recompacted fill than the loose fills for the given small initial rainfall intensity (1×10^{-9} m/s). On the other hand, at section C–C (at the slope toe), both the initial pore water pressure profiles in the loose and recompacted fill slopes are approximately at the hydrostatic conditions.

During the short but intensive rainfall (i.e. 236.4 mm/day or 2.74×10^{-6} m/s for 24 h), the corresponding pore water pressure responses along sections A–A (Figure 9.33a) and B–B (Figure 9.33b) are similar. The

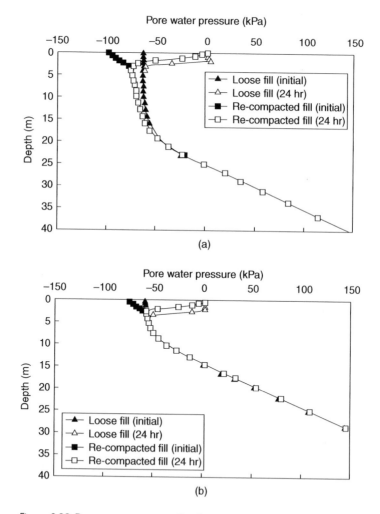

(a)

(b)

Figure 9.33 Pore water pressure distributions along (a) section A–A, (b) section B–B and (c) section C–C (after Ng and Zhan, 2001).

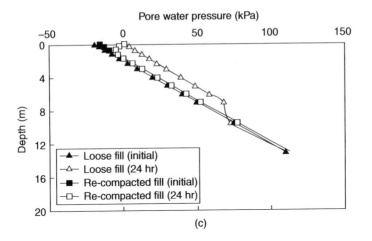

Figure 9.33 (Continued).

variation of pore water pressure due to rainfall infiltration is limited in the shallow soil layer (less than 4 m). For the loose fill slope, negative pore water pressure in the shallow soil layers (2 ~ 3 m below the ground surface) increases to zero after the 1-day rainfall, i.e. soil suctions in this soil layer are completely destroyed. For the recompacted slope, 1-day rainfall infiltration also results in an increase in the pore water pressure. However, only the negative pore water pressure at the ground surface increases to zero, and the pore water pressure in the soil layer between ground surface and the main groundwater table remains negative after the 1-day rainfall.

In Figure 9.33c, the 1-day rainfall infiltration also results in an increase in the pore water pressure at the slope toe, especially in the loose fill slope (refer to section C–C). For the loose fill slope, the main groundwater table at the slope toe rises to the ground surface and the initial soil suction is completely destroyed after the 1-day rainfall. However, for the recompacted fill slope, similar to the former two sections, the pore water pressure remains negative in the soil layer within 2 m below ground surface.

The discrepancy between the variation of pore water pressure distributions in the loose and recompacted fill slopes is attributed to the difference in the water permeability of the loose and recompacted fills at low soil suctions (Figure 9.32). Since the applied 1-day rainfall intensity $(2.74 \times 10^{-6} \, \text{m/s})$ is larger than the water permeability of the initially unsaturated fill, the surficial soil becomes saturated or nearly saturated immediately. Referring to Figure 9.32, at a low soil suction (less than 9 kPa), the water permeability of the loose fill is higher than that of recompacted fill due to its larger void ratio. This indicates more rainwater can infiltrate into the slope than that of the recompacted fill slope. However, for the recompacted fill slope, the water permeability (no more than 1.0×10^{-7} m/s) of the 3-m thick recompacted fill at low soil suctions is much lower than the applied rainfall intensity, so that relatively little rainwater can infiltrate into the recompacted fill layer, and even less can permeate into the underlain loose fill layer. Most of the rainwater runs off the slope surface. Therefore, the recompacted fill layer having low water permeability at low soil suctions acts as an impeding layer, which hinders rainfall infiltration and preserves the initial soil suction underneath.

Influence of recompaction on slope stability

After obtaining the pore water pressure distributions in the two fill slopes, two possible slip surfaces (shallow and deep-seated) were selected for limit equilibrium analyses. The first is within the 3-m thick recompacted soil layer, and the second is deeper than 3 m, i.e. passing though the loose fill layer (see Figure 9.30). It should be noted that these selected slip surfaces may not correspond to the minimum FOS. It is selected only for illustrating the influences of recompaction on the FOS of the two possible slips.

Figure 9.34 shows the variations of FOS with elapsed time (i.e. rainfall duration) for the two specified slip surfaces during the 1-day rainfall. FOS of the loose fill slope decreases with rainfall duration, especially for the shallow slip 1. On the other hand, only the FOS of the shallow slip 1 of the recompacted slope is significantly affected by rainfall infiltration. For both the shallow and deep-seated slip surfaces, FOS predicted for the recompacted fill slope is significantly larger than that for the loose fill slope during the

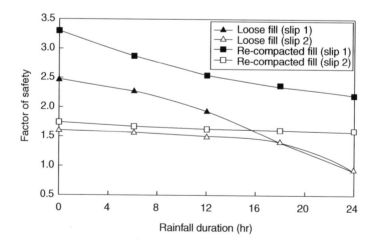

Figure 9.34 Relationships between factors of safety and rainfall duration for the loose fill and recompacted fill slopes (after Ng and Zhan, 2001).

1-day rainfall, especially for the shallow slip 1. For the case of shallow slip surface, the higher FOS of the recompacted slope is attributed to both the larger effective friction angle (ϕ') and the larger negative pore water pressure distributions than those in the loose fill slope. For the case of the deep-seated slip surface, the computed high FOS of the recompacted fill slope is mainly attributed to the high negative pore water pressure distributions present in the soil (refer to Figure 9.33a–c).

Conclusions

- Based on the transient seepage analyses, recompaction of the top 3 m of a loose fill slope leads to much higher negative pore water pressure distributions in the recompacted slope than those in the loose fill during a 1-day intensive rainfall.
- Limit equilibrium analyses predict a significantly larger FOS for the recompacted fill slope than that of the loose fill slope during the 1-day rainfall, especially for the shallow slip. Contributions of the recompaction to the slope stability are two-fold.

  o First, the recompaction leads to a reduction in void ratio and hence an increase in its shear strength.
  o Secondly, assuming that no static liquefaction will occur, the recompacted fill layer having low water permeability at low soil suctions acts as an impeding layer, which hinders rainfall infiltration, preserves the initial soil suction in the loose fill underneath, and hence improves the stability of the fill slopes.

Effects of surface cover and impeding layers on slope stability (Ng and Pang, 1998a)

Introduction

Rulon and Freeze (1985) have shown that the pore water pressure distribution in a layered slope is strongly dependent on the position of an impeding layer and it has significant effects on shear strength of soils. In their study, variations of hydraulic conductivity with respect to matric suction and transient seepage conditions were not considered.

Ng and Shi (1998) have studied the stability of unsaturated soil slopes subjected to various boundary and hydraulic conditions such as rainfall intensity, rainfall duration, boundary hydraulic head conditions and anisotropic water permeability. It is shown that the FOS is not only governed by the intensity of rainfall, initial groundwater table and the anisotropic permeability ratio, but it also depends on antecedent rainfall duration. A critical rainfall duration can be identified, at which the FOS is the lowest.

In order to maintain soil suction and minimize soil erosion, relatively impermeable surface covers are usually provided on soil slopes to improve stability. Lim *et al.* (1996) have carried out a field instrumentation program in Singapore to monitor negative pore water pressures or in situ matric suctions in a residual soil slope which is protected by different types of surface covers at various sections. It is evident that the variations of matric suction in the ground are less significant under a covered section than a bare section of the slope.

In this chapter, a section of parametric finite element transient seepage and limit equilibrium analyses of a typical steep unsaturated soil cut slope in Hong Kong are presented. The main objective of these analyses is to investigate the influence of surface covers and impeding layers on slope instability in unsaturated soils.

Finite element mesh and analysis procedures

For the parametric study presented here, a typical unsaturated cut slope in colluvium in Hong Kong is adopted. The finite element mesh for the parametric study is illustrated in Figure 9.35. It is a two-dimensional vertical cross-section through an unsaturated hillside with a steep cut slope. A constant hydraulic head is specified along the boundaries AB (at 6 mPD) and CD (at 61 mPD). The boundary BC represents bedrock and is assumed to be impermeable. Rainfall may infiltrate into the flow region along the exposed sloping surface (AD). For conducting the parametric study of transient seepage in the unsaturated soil slope, SEEP/W (Geo-Slope, 1995) is selected. The required SWCC and permeability function of this selected colluvium (Gan and Fredlund, 1997) are shown in Figure 9.36. In order to choose a reasonable rainfall intensity, one in 10-year return rainfall records collected by the Hong Kong Royal Observatory during 1980–1990 (Lam and Leung, 1995) was considered. Details of other input parameters are summarized in Table 9.7. For the transient seepage analyses, it was assumed that the rainfall intensity is equal to the rate of infiltration.

For estimating the FOS for a slope, some basic mechanical soil parameters are needed. Based on recent triaxial tests with suction measurements on the selected colluvium samples (GEO, 1994), $c' = 10$ kPa, $\phi' = 38.5°$ and $\phi^b = 22°$ are adopted for the limit equilibrium analyses. Transient pore water pressures used in the analyses are obtained directly from SEEP/W. The linear failure criterion defined by Equation (9.22) was assumed. The minimum FOS was calculated by using Janbu's simplified method.

Influence of surface cover on slope stability

In order to protect slopes from infiltration of surface water and to minimize soil erosion, a common practice in the Far East is to provide a layer of soil

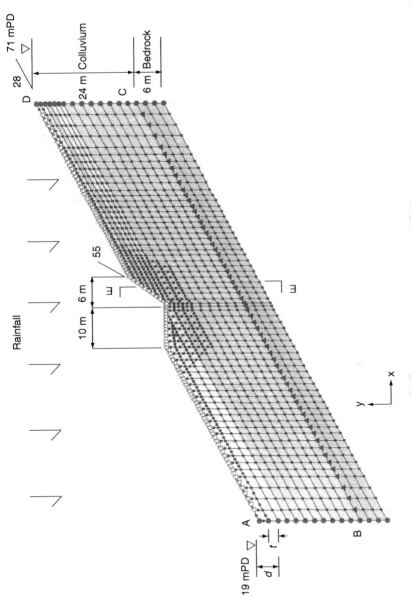

Figure 9.35 Finite element mesh for the parametric study of a typical unsaturated cut slope in colluvium in Hong Kong (after Ng and Pang, 1998a).

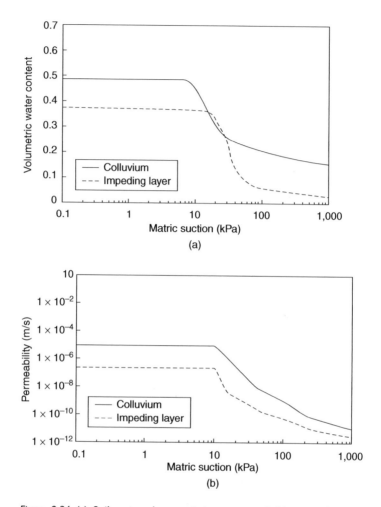

Figure 9.36 (a) Soil–water characteristic curves and (b) permeability functions for colluvium and impeding layer (after Ng and Pang, 1998a).

cement and lime plastic cover called 'chunam' on many soil slopes. If no proper maintenance is carried out, this type of surface covers will crack over a period of time in service. Under this circumstance, the effectiveness of the surface covers is in doubt and more understanding the role of the covers is needed.

Here, an impermeable surface cover on the 55° steep colluvium cut slope (see Figure 9.35) was modelled by a flux boundary with a zero percentage of applied flux (i.e. no rainwater infiltrates the soil slope). Formation and severity of cracks in colluvium were simulated by the percentage intake of

Table 9.7 Input parameters for the parametric study (after Ng and Pang, 1998a)

Parameter	Percentage of applied flux	(d/t)
c1	0%	0
c2	10%	0
c3	40%	0
c4	70%	0
c5	100%	0
i1	10%	1
i2	10%	2
i3	10%	3
i4	10%	4
i5	10%	5

the applied flux. The higher the percentage, the more severe the cracking. A 100 per cent of the applied flux is equivalent to no surface cover (i.e. 100 per cent of rainwater infiltrates the soil slope).

Figure 9.37 shows the pore water pressure distributions with depth at section E–E in the cut slope after a 5-day rainfall which has an average intensity of 120 mm/day. At one extreme, the presence of a perfect impermeable surface cover (i.e. 0 per cent of applied flux) preserves initial suction in the ground very well. As expected, no significant increase of pore water

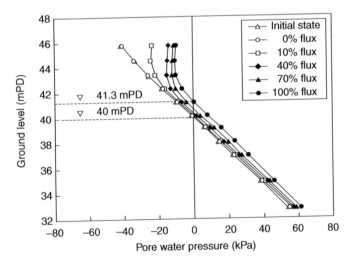

Figure 9.37 Effects of surface cover on pore water pressure distributions at section E–E (after Ng and Pang, 1998a).

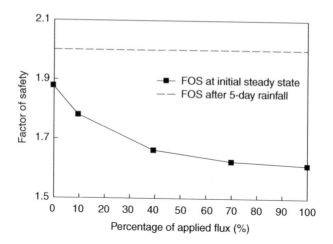

Figure 9.38 Factor of safety versus percentage of applied flux (after Ng and Pang, 1998a).

pressures and rise of the main water table can be seen after the 5-day rainfall. On the other extreme, if the surface cover becomes ineffective (i.e. 100 per cent of flux), surface water infiltrates into the soil and causes a maximum increase of pore water pressure (i.e. reduction of matric suction) of approximately 30 kPa and a rise of the main water table from 40 to 41.3 mPD. Between these two extremes, the pore water pressure responds non-linearly with the percentage of applied flux. Modelling a larger percentage of flux leads to a greater change in matric suction. Nevertheless, the most affected zone due to infiltration falls within the top few metres of the slope. This perhaps gives a theoretical explanation of why many slope failures in unsaturated residual and saprolitic soils are shallow in nature. Figure 9.38 shows the calculated FOS varying with the percentage of applied flux. The FOS decreases non-linearly as the percentage of the applied flux increases. This is consistent with the pore water pressure response shown in Figure 9.37.

Influence of impeding layer on slope stability

In many natural slopes, hydraulic properties of soils such as saturated water permeability may vary from place to place. The variations of the saturated water permeability in the ground will affect the development of pore water pressures and hence shear strength of the soil. The influence of a non-uniform water permeability system on slope stability is not fully understood. Here, a parametric study was conducted to investigate how the stability of an unsaturated slope is affected due to the presence of a soil which has a very different water permeability as compared with other soils existed in the same slope. To simplify the problem for the present study, only a single soil layer called impeding layer was assumed to lie parallel to the natural slope at varying elevations. For the purpose of the parametric study, this impeding layer had the same mechanical properties, but with very different hydraulic properties (see Figure 9.36), as the colluvium in the slope. The elevation of the impeding layer varies and an expression (d/t) is used to represent the elevation of the impeding layer for each parametric analysis, where d is the distance from the sloping ground surface to the bottom of impeding layer, and t is the thickness of impeding layer which was assumed to be 1 m in the study.

Figure 9.39 shows the contours of groundwater distributions in the uniform (i.e. colluvium only) and the layered soil slopes after the 5-day rainfall. Comparing the elevations of the main water table for the two slopes, the rise of the main water table is larger in the uniform (colluvium only) than in the layered slope. As expected, the water permeability of the impeding layer is lower than the one of the uniform slope. This impeding layer serves as an impermeable cover with respect to the colluvium located underneath it. Relatively little water can infiltrate through the layer for the given duration. Thus, a perched water table appears above

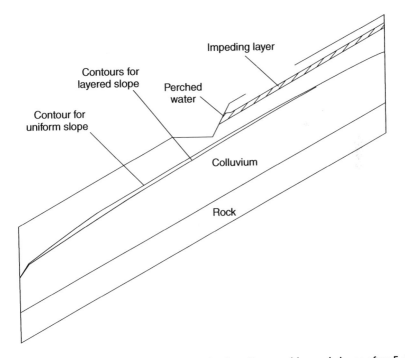

Figure 9.39 Groundwater conditions in both uniform and layered slopes after 5-day rainfall $(d/t = 3)$ (after Ng and Pang, 1998a).

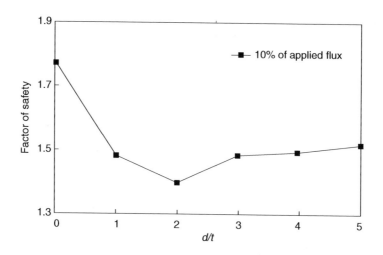

Figure 9.40 Factor of safety versus (d/t) (after Ng and Pang, 1998a).

the impeding layer as shown in the Figure 9.39. If the rainfall continues, the presence of the perched water may lead to a substantial rise of the pore water pressures above the impeding layer and leads to shallow slope failures.

Figure 9.40 shows the variations of FOS with the (d/t) ratio. Initially the FOS decreases with an increase in the (d/t) ratio until a critical value, at which the FOS is the lowest, is reached. After reaching the critical ratio, the stability of the slope improves as the (d/t) value further increases. In this particular study, the critical d/t value is equal to 2. The existence of a critical (d/t) ratio may be explained by considering two extreme cases: $d/t = 1$ and $d/t = 20$ (say). At one extreme (i.e. $d/t = 1$), this implies that the impeding layer is just located at the ground surface. Very little infiltration is expected and hence the stability of the slope is not likely to be affected by rainfall. At the other extreme (i.e. $d/t = 20$), the presence of the impeding layer at such a deep level is not likely to affect the distributions of pore water pressures at shallow depths. The stability of the slope is not likely to be affected by rainfall also. Therefore, there exists an intermediate but critical d/t ratio at which the distributions of pore water pressures and hence the stability of the slope are affected most.

Summary

- A series of finite element parametric analyses has been carried out to investigate the influence of surface cover and impeding layer on transient seepage and hence slope stability in unsaturated soils. A typical unsaturated cut slope on a hillside in Hong Kong has been selected to illustrate some key results from the parametric study.

- Based on the parametric study, it is found that the use of an impermeable surface cover on an unsaturated slope can substantially reduce the change in soil matric suction and hence improve the stability of the slope during rainfall. Once some cracks are formed in the cover, the FOS decreases non-linearly with the number and size of the cracks increase.
- When there is an impeding layer existed in an unsaturated slope, the stability of the slope is generally affected. If the layer is located at shallow depths, perched water table is likely to develop. By studying various elevations of the layer, a critical elevation can be identified at which the FOS of the slope is the lowest. For the cases considered in this chapter, the most critical d/t value is equal to 2.

Numerical experiments of soil nails in loose fill slopes subjected to rainfall infiltration effects (Cheuk et al., 2005)

Introduction

Laboratory tests have demonstrated that loose granular soils such as sands, silty sands or decomposed granitic soils may soften during undrained shearing due to structural collapse (Castro, 1969; Vaid and Chern, 1985; Pitman et al., 1994; Lade and Yamamuro, 1997; Zlatovic and Ishihara, 1997; PWCL, 1998; Yamamuro and Lade, 1998; Ng et al., 2004a). The term 'static liquefaction' derived from laboratory element tests is frequently adopted to describe soil slope failures with a long travelling distance. In fact, static liquefaction of a loose fill slope is very difficult (if not impossible) to be verified in the field since no pore pressure measurements are available at the time of failure. On the other hand, static (Take et al., 2004) and dynamic centrifuge model tests (Ng et al., 2004c) have illustrated that the generated positive pore pressures in loose decomposed granitic soil slopes are not large enough to cause complete liquefaction even at a very high ground acceleration of 0.3 g. Nevertheless, when these strain-softening soils are used as fill materials in slopes and embankments, the reduction in shear strength triggered by rainfall infiltration or rising groundwater table may lead to global instability. The high mobility of the low strength debris could cause serious destruction in urban areas. The 1976 Sau Mau Ping landslide in Hong Kong (Hong Kong Government, 1977) that killed 18 people is a typical flow failure of loose fill slopes.

Many existing loose fill slopes in Hong Kong were constructed prior to the establishment of the Geotechnical Control Office (now the Geotechnical Engineering Office) in 1977. At that time, earthwork construction in Hong Kong was not subject to any regulatory geotechnical control. Loose fill slopes were often constructed by dumping decomposed granitic or volcanic soil without proper compaction to a desired density. These slopes may be stable with the presence of suction. However, once the suction is destroyed, for example by rainfall infiltration, the loose fill material might soften and exhibit a rapid and mobile failure upon undrained shearing.

Following the recommendations set out by the Independent Review Panel for Fill Slopes (Hong Kong Government, 1977), the current practice for upgrading existing substandard loose fill slopes in Hong Kong consists of excavating the top 3 m of the loose fill and recompacting the new filling material to an adequate standard, together with the provision of a drainage blanket at the base of the compacted fill. This approach has proved successful for many existing fill slopes. However, due to the heavy machinery required for the recompaction work, this procedure may be extremely difficult and it can be hazardous during the construction stage in many heavily populated areas in Hong Kong. Moreover, slope failures during recompaction in wet weather have been reported. An alternative solution for the safe upgrading of some of the existing loose fill slopes in Hong Kong in an effective manner would be desirable.

One possible solution to this potential problem is soil nailing. Soil nails have been widely used in upgrading cut slopes that are marginally safe. The ease and speed of construction, together with the economies offered, have led to a rapid increase in the use of soil nails. However, the suitability of soil nailing in loose fill slopes has been a controversial question and has generated many technical debates concerning the failure mechanism of soil-nailed slopes. No rational conclusion can be reached due to a lack of understanding of the interaction between the strain-softening material and the soil nails (HKIE, 1998, 2003).

To investigate the possible behaviour of soil nails in loose fill slopes, numerical analyses have been conducted using the finite different program FLAC2D (Itasca, 1999). The strain-softening characteristic of loose fill material was mimicked by a user-defined constitutive model (denoted as SP model). The behaviour of loose fill slopes with and without soil nails subjected to the effects of rainfall infiltration has been assessed. The modelling procedure and findings are presented below.

Finite difference analyses

Slope geometry and soil nail arrangement

Two-dimensional plane-strain finite difference models were set up to represent a 20-m high loose fill slope with a face angle of 35° to the horizontal, and a 2-m high retaining wall at the slope toe, as shown in Figure 9.41. The

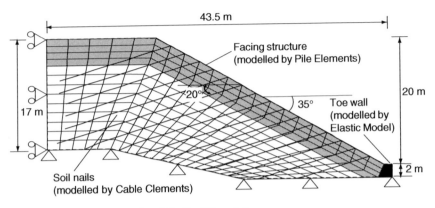

Figure 9.41 Slope geometry and finite difference mesh (after Cheuk et al., 2005).

boundary conditions of the models were taken as vertical rollers on the left boundary with full fixity at the base. Sun (1999) reviewed the data collected from about 2,000 existing fill slopes in Hong Kong and suggested that the height of these slopes ranged from 5 to 25 m with a slope angle of between 30° and 45° to the horizontal. The model employed in this study therefore represents typical fill slope geometry in Hong Kong.

The SP model was adopted in the finite difference zones at the top 3 m of the slope, in which the soil is vulnerable to strain-softening upon wetting by rainwater. The soil underneath was modelled by an elasto-plastic model with a Mohr–Coulomb yield criterion, considering that the likelihood of strain-softening due to surface infiltration at these deeper locations is relatively low. Moreover, the user-defined SP model substantially increases the computer time required in the analysis. Since the present study focuses on shallow failures, it would unnecessarily increase the analysis time if the SP model were used in all the soil zones in the slope.

For a nailed slope, 15 rows of soil nails were modelled by cable elements available in FLAC2D. The lengths of the nails varied from 9 to 25 m. The vertical spacing between each row of nails was approximately 1.5 m. The lower ends of the bottom 9 rows of nails were attached to the lower boundary of the finite difference grid to model the anchorage into the natural ground. In one of the analyses, a facing structure, modelled by pile elements in FLAC2D, was present in order to evaluate the possible beneficial effects in the mobilization of nail forces. The simulated facing is a grid structure consisting of vertical and horizontal concrete columns that connect all the nail heads together (Figure 9.42). The upper ends of the nails were connected directly to the facing structure, whilst the lower end of the facing system was attached to the toe wall.

Soil models

The SP model was developed based on the same principle as the liquefaction model developed by Gu (1992) and implemented in FLAC2D by Sun (1999) for the study of loose fill slope failures in Hong Kong. Cheuk (2001) further improved the model by allowing an elastic response of the stress path upon unloading from the collapse surface. This improvement allows a more realistic stress redistribution in a soil-nailed slope.

A simplified representation of the SP model is given in Figure 9.43a and b. The two-failure criteria at a particular void ratio, e, are plotted in q–p' plane, where q is the deviator stress and p' represents the mean effective stress. The SP model employs the collapse surface concept, which was introduced by Sladen *et al.* (1985). They observed that for a given void ratio, the peak strengths of soil specimens tested at different initial confining pressures lie on a straight line (collapse line) that passes through the critical state. This collapse line defines the stress condition for structural collapse in the q–p' plane. In the same material tested at different void ratios, collapse lines with

Figure 9.42 A typical grid facing structure (after Cheuk et al., 2005).

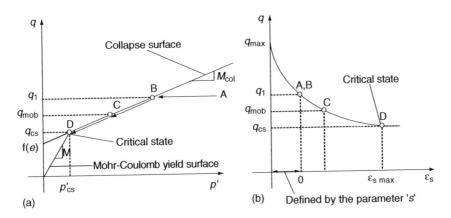

Figure 9.43 The SP model (a) q–p' plane and (b) q–ε_s plane (after Cheuk *et al.*, 2005).

the same slope were observed and these lines form the collapse surface in the three-dimensional p'–q–e space.

One of the two-failure criteria in the SP model is used to mimic the collapse surface. The Mohr–Coulomb failure criterion is used for this collapse surface, whose inclination in the q–p' plane is defined as M_{col}, which is assumed to be constant for different e. To present the three-dimensional collapse surface in the two-dimensional q–p' plane, the intercept of the collapse line at the vertical axis is defined as a function of e. For soil elements with different e values, collapse lines with different intercepts are used. This also simulates the change in the critical state shear strength of the soil element due to variations in density, as the intercept of the two-failure criteria defines the critical state shear strength. Details of this function will be discussed later.

The collapse of loose fill material is modelled assuming an undrained condition. An illustration of the modelling of strain-softening is shown in Figure 9.43a. A soil element is assumed to have an initial stress state located at Point A in Figure 9.43a. Owing to the increase in pore water pressure caused by surface infiltration, p' of the soil element decreases. Strain-softening is modelled to start under a constant e once the effective stress path intercepts the collapse surface (i.e. Point B). As the shear strain increases, the effective stress path corresponding to an increase in pore water pressure is constrained to descend along the collapse surface towards the critical state condition (p'_{cs}, q_{cs}) at Point D. When the critical state is reached, the effective stress state will not vary any further by definition.

Before the critical state is reached, the mobilized deviator stress, q_{mob}, at intermediate stress states (i.e. Point C) is determined by the mobilized shear strain, ε_s, experienced by the soil element as shown in Figure 9.43b.

A hyperbolic equation was assumed for the stress–strain relationship in the original model developed by Gu (1992). The same equation was adopted here as follows:

$$q_{mob} = q_{max} - (q_{max} - q_{cs})\sqrt{1 - \left[\frac{(\varepsilon_{s\,max} - \varepsilon_s - s)}{\varepsilon_{s\,max}}\right]^2} \qquad (9.23)$$

where

$$s = \varepsilon_{s\,max}\left[1 - \sqrt{1 - \left(\frac{q_{max} - q_1}{q_{max} - q_{cs}}\right)^2}\right]$$

The terms in Equation (9.23) are defined as shown in Figure 9.43b where q_{max} and $\varepsilon_{s\,max}$ are two input parameters that define the shape of the strain-softening curve. These parameters can be determined from laboratory undrained triaxial tests and are constant for a particular soil. The parameter s is used to define the starting point of the strain-softening curve. As shown in Figure 9.43b, this s value is actually controlled by q_1 which is the mobilized q when the effective stress path intercepts the collapse surface. In the analyses, the total shear strain, ε_s, defined as the radius of the Mohr strain circle, was calculated from the displacement vectors of a soil zone.

The other failure criterion is used to capture dilative behaviour of the soil when the stress state is located on the dry side of the critical state line. This state boundary surface below the critical; state is modelled by a Mohr–Coulomb yield criterion (Figure 9.43a) with a positive angle of dilation. The potential function is the same as the yield criterion. The details of the model parameters will be discussed later.

For soil zones that are less likely to undergo strain-softening due to surface infiltration (i.e. below 3 m) the built-in Mohr–Coulomb model in FLAC2D was used.

Model for soil nails and soil–nail interface

In this study, soil nails were modelled by cable elements (Itasca, 1999). These are one-dimensional axial elements that can yield in tension and compression, but cannot sustain a bending moment. To simulate the transfer of loading due to relative displacement between the soil and the nails, shear-coupling springs were modelled at the nodal points of a cable element. The shear behaviour of the grout annulus in a cable element during relative displacement between the reinforcement and the soil was assumed to be linear elastic and governed by the grout shear stiffness, K_{sn}. The maximum shear force that can be developed in the grout, per unit length of element, is defined by the input parameter, F_{sn}.

Considering the possibility of strength reduction around the soil–nail interface, the model was modified to allow adjustment of the frictional resistance during strain-softening. The input parameter, F_{sn}, which defines the maximum frictional resistance of the soil nails was defined as a function of the available deviator stress, q_{av}, of the confining soil element using Equation (9.24) as the soil undergoes strain-softening once its effective stress path reaches the collapse surface.

$$F_{sn} = \frac{q_{av}}{2} \times \pi \times d, \tag{9.24}$$

where d is the outside diameter of the nail–grout composite. The function defining q_{av} of the confining soil element is given by

$$q_{av} = q_{cs} + M_{col} \times \left(p' - \frac{q_{cs}}{M}\right) \quad \text{if } p' > p'_{cs} \tag{9.25a}$$

or by

$$q_{av} = M \times p' \quad \text{if } p' \le p'_{cs}. \tag{9.25b}$$

These equations represent the two-failure criteria in the q–p' plane as shown in Figure 9.43a. The M_{col} and M parameters are the inclinations of the two-failure criteria, and p'_{cs} and q_{cs} are the values of the two stress invariants at the critical state. This approach conservatively adjusts the frictional resistance of the soil–nail interface by assuming that it is the shear strength of the soil when the soil starts to soften.

Model for facing structure

The structural facing in one of the analyses was modelled by pile elements (Itasca, 1999). The interaction between the pile elements and the soil is controlled by the shear and normal coupling springs at the nodal points.

The behaviour of the shear coupling springs in the pile elements is identical to that of the cable elements described in the previous section. The stiffness of the shear coupling springs in the pile elements is denoted by K_{sf} and it has the same definition as the K_{sn} value in the cable elements. The maximum shear force, F_{sf}, that can be developed along the pile–soil interface is a function of the cohesive strength of the interface, C_{sf}, and the stress-dependent frictional resistance, ϕ_{sf}, along the interface. This is represented numerically by the equation

$$\frac{F_{sf}}{L} = C_{sf} + p' \tan \phi_{sf} \cdot P_f \tag{9.26}$$

where p' is the mean effective confining stress normal to the pile element; P_f the exposed perimeter of the pile element and L the length of the soil–nail segment.

The behaviour of the pile–soil interface in the normal direction is assumed to be linear elastic. The normal force developed during relative normal displacement between the pile nodes and the grid is governed by the stiffness of the normal coupling spring, K_{nf}. A pair of limiting normal forces is required in the model. These forces are dependent on the direction of the movement of the pile node. The maximum compressive force is defined by the input parameter C_{nf}, whereas the parameter T_{nf} controls the maximum tensile force in the normal direction.

Input parameters for the soil models

The input parameters for the SP model adopted in this study are tabulated in Table 9.8. They were obtained from published laboratory test results on Hong Kong fill materials, mainly from the investigations of the 1976 Sau Mau Ping landslides (Hong Kong Government, 1977). Some simple assumptions have been made if the parameters are not available from the laboratory test results.

Two material parameters, dry density, ρ_d, and porosity, n, of the saturated loose fill material were assumed to be $1,400\,\text{kg/m}^3$ and 0.44, respectively. The bulk modulus, K, and the shear modulus, G, were obtained based on correlations between Young's modulus, E, and SPT-N values (Chiang and Ho, 1980). A typical SPT-N value of 10 and a Poisson's ratio, v, of 0.3 were assumed.

Table 9.8 Input parameters for the SP model (after Cheuk et al., 2005)

Parameter	Value
Dry density, ρ_d	$1,400\,\text{kg/m}^3$
Porosity, n	0.44
Bulk modulus, K	$12,500\,\text{kPa}$
Shear modulus, G	$5,769\,\text{kPa}$
Effective cohesion, c'_1	0
Effective friction angle, ϕ'_1	$36.8°\ (M = 1.5)$
Angle of dilation, ψ_1	$15°$
Effective cohesion, c'_2	$f(e) = 0.25\exp(10.67 - 9.27e)$
Effective friction angle, ϕ'_2	$25°\ (M_{col} = 0.984)$
Angle of dilation, ψ_2	0
Tensile strength, T	0
Strain-softening parameter, q_{max}	$f(q_{cs}) = 4.5 \times q_{cs}$
Strain-softening parameter, $\varepsilon_{s\,max}$	0.15

In the SP model, each failure criterion requires three input parameters, including the effective cohesion, c' and the effective friction angle ϕ', as well as an angle of dilation, ψ. The parameters c'_1, ϕ'_1 and ψ_1 were adopted for the Mohr–Coulomb yield surface below the critical state line as shown in Figure 9.43a. They were deduced from the 1976 Sau Mau Ping investigation to be $c'_1 = 0$, $\phi'_1 = 36.8°$ ($M = 1.5$). The angle of dilation, ψ_1, was taken as 15°.

For the collapse surface shown in Figure 9.43a, the parameters c'_2, ϕ'_2 and ψ_2 were adopted to define it. Based on undrained triaxial test results from Hong Kong fill materials (PWCL, 1998), an average value of 25° was adopted for ϕ'_2. This is equivalent to an M_{col} of 0.984 in the $q-p'$ space. At the critical state, the angle of dilation, ψ_2, was 0°. As mentioned before, the intercept of the collapse surface at the vertical deviator stress axis, which is c'_2, is a function of e. Since c'_2 governs the critical state shear strength, the relationship between c'_2 and e can be derived from the critical state line equation, which is given by

$$q_{cs} = M \exp\left[\frac{\Gamma - e - 1}{\lambda}\right],$$ (9.27)

where M, Γ and λ are the critical state parameters. Based on measurements from the Sau Mau Ping loose fill material, the critical state parameters can be taken as

$M = 1.5$ ($\phi' = 36.8°$),
$\lambda = 0.108$ and
$\Gamma = 2.2$.

By substituting these values into Equation (9.27), the relationship between c'_2 and e can be obtained after the transformation from the $q-p'$ space to the Mohr–Coulomb $\tau - \sigma$ space. The relationship is given by

$$c'_2 = f(e) = 0.25 \exp(10.67 - 9.27e)$$ (9.28)

In the analyses, the value of e required in Equation (9.28) is determined by the stress p' of a soil zone at the end of the formation of the slope. The correlation between e and p' was obtained through laboratory tests on local fill materials (PWCL, 1998) and can be represented by the equation

$$e = 1.12 - 0.222 \log p'.$$ (9.29)

For tensile failure, the input parameter T, which governs the location of the tension cut-off line, was set to zero to ignore any tensile strength of the soil.

The parameters q_{max} and $\varepsilon_{s\ max}$ were determined from undrained triaxial tests on Hong Kong decomposed granites (PWCL, 1998). To achieve similar

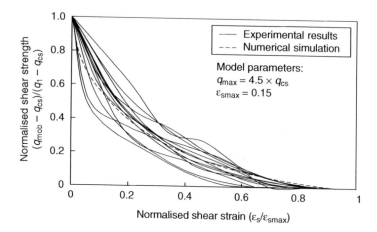

Figure 9.44 Comparison of post-peak strain-softening behaviour between experimental results and simulation by the SP model (after Cheuk *et al.*, 2005).

strain-softening pattern, the q_{max} value was set to 4.5 times the q_{cs} value and the nominal value of $\varepsilon_{s\,max} = 0.15$ was adopted in the analyses. Figure 9.44 compares the post-peak strain-softening behaviour between experimental results measured from triaxial tests and the simulation obtained by the SP model. The comparison demonstrates that the selected model parameters closely mimic the loss of shear strength, which is a key component of the SP model.

Input parameters for the soil nail system

The input parameters for the soil nails are tabulated in Table 9.9. They were determined based on a 32-mm diameter steel bar installed in each 100-mm

Table 9.9 Input parameters for the soil nails (after Cheuk *et al.*, 2005)

Parameter	Value
Area, A_n	8×10^{-4} m^2/m run
Yield force, Y_n^*	245 kN/m run
Young's modulus, E_n^*	133 GPa/m run
Shear spring stiffness, K_{sn}^*	22,000 kN/m/m run
Ultimate shear force, F_{sn}^*	$f(p')$

Note
Input values of parameters with * were divided by 1.5 to account for horizontal spacing in the 2D numerical analyses.

diameter drilled hole. The soil nails were modelled at a horizontal spacing of 1.5 m, material properties were therefore divided by 1.5 to take account of the horizontal spacing of the elements in plane-strain conditions (refer to Table 9.9). The cross-sectional area, A_n, of the cable element was taken to be 8×10^{-4} m^2/m run. The maximum tensile force that can be developed in the nails was calculated by assuming the tensile yield strength to be 460 MPa, implying a Y_n value of 245 kN/m run. The stiffness of the cable element, E_n, has the same definition as the Young's modulus and was taken to be 133 GPa/m run. The F_{sn} value was defined as a function of p' of the confining soil element (Equations 9.24 and 9.25). In order to incorporate the increase in shear stiffness of the grout, a relatively large K_{sn} of 22,000 kN/m/m run was assumed, corresponding to a shear modulus of 60,000 kN/m/m run. The stiffness of the soil–grout interface may be lower in practice, requiring larger strains for developing the nail forces. Nevertheless, parametric studies (Cheuk, 2001) have illustrated that a reduction of grout stiffness would not make significant difference to the principal observations made from the results of the present study.

The input parameters for the facing structure are tabulated in Table 9.10. The facing structure of the soil nails was assumed to consist of 300 mm wide $\times 300$ mm deep reinforced concrete columns at 1.5 m horizontal/vertical spacing. For plane-strain simplification in the numerical analyses, material properties were divided by 1.5 to allow for effects of horizontal spacing (refer to Table 9.10).

Since only the bottom face of the beams interacts directly with the soil, the exposed perimeter, P_f, was assumed to be 0.3 m/m run. The Young's modulus of the concrete beam was taken as 20 GPa/m run. The stiffness of

Table 9.10 Input parameters for the facing structure (after Cheuk et al., 2005)

Parameter	Value
Area, A_f	0.09 m^2/m run
Perimeter, P_f	0.3 m/m run
Moment of inertia, I_f	6.75×10^{-4} m^4/m run
Mass density, D_f	2,300 kg/m^3/m run
Young's modulus, E_f^*	20×10^6 kPa/m run
Shear spring stiffness, K_{sf}^*	5,769 kN/m/m run
Cohesive force for shear resistance, C_{sf}	0
Friction angle for shear resistance, ϕ_{sf}	30°
Normal spring stiffness, K_{nf}^*	12,500 kN/m/m run
Ultimate normal compressive force, C_{nf}^*	10^6 kN/m run
Ultimate normal tensile force, T_{nf}^*	0

Note
Input values of parameters with * were divided by 1.5 to account for horizontal spacing in the 2D numerical analyses.

the shear and normal springs was based on the shear and bulk modulus of the soil, which are 5769 and 12,500 kPa, respectively. The ultimate resistance of the shear coupling springs along the interface between the facing and the slope surface is governed by Equation (9.26) with $C_{sf} = 0$ and $\phi_{sf} = 30°$. This set of low $c' - \phi'$ values conservatively estimates the frictional resistance between the facing and the soil. The limiting compression force, C_{nf}, in the normal coupling spring was assumed to be very high (10^6 kN/m run), and the maximum tensile force, T_{nf}, was not input into the analysis, so it was therefore assumed to be zero.

Analysis programme and modelling procedure

Three analyses were carried out to examine the behaviour of loose fill slopes with and without soil nails subjected to undrained collapse brought about by rainfall infiltration. The first analysis established a benchmark to demonstrate the modelled behaviour of a loose fill slope without soil nails under the applied pore pressure conditions. The second case was used to study the behaviour of soil nails in loose fill slopes, but without nail heads or a facing structure at the slope surface. The third analysis was used to investigate the role of the facing structure in a nailed loose fill slope.

The modelling procedures can be divided into four stages as follows.

STAGE I – FORMATION OF THE LOOSE FILL SLOPE

The fill slope was gradually built up in layers approximately parallel to the slope surface. After the placement of each soil layer, it was assumed that static equilibrium condition was achieved. This step-by-step process simulates the initial shear stress mobilization that is likely to occur in the field during the construction of the fill slope. This construction stage was modelled in a fully drained condition. The bulk modulus of water was set to zero, so no excess pore pressure was generated at this stage. Subsequently, an initial pore water pressure distribution in the slope was assigned as shown in Figure 9.45. The distribution assumes a groundwater table located at 2 m above the in situ ground, which is the lower boundary of the grid. The soil above the groundwater table was assumed to be unsaturated. The negative pore pressures were calculated based on hydrostatic conditions. The average suction existing in the slope was limited to 50 kPa. The purpose of having suction at this stage was to ensure that the stress state in the top 3 m of soil was below the collapse surface (refer to Figure 9.43a) before wetting, as this is likely to be the situation in the field. Once the top 3 m of soil has been wetted by the effects of rainfall infiltration, full saturation was assumed. Hence, the over-estimation of soil shear strength due to suction will only affect the soil at depths deeper than 3 m.

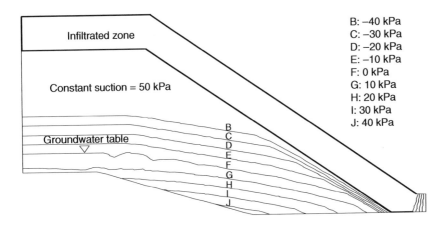

B: −40 kPa
C: −30 kPa
D: −20 kPa
E: −10 kPa
F: 0 kPa
G: 10 kPa
H: 20 kPa
I: 30 kPa
J: 40 kPa

Figure 9.45 Initial pore pressure distribution before the simulations of the effects of rainfall infiltration (after Cheuk *et al.*, 2005).

STAGE 2 – INSTALLATION OF SOIL NAILS IN LOOSE FILL SLOPES

The installation of soil nails was modelled by specifying the geometries and the structural properties of the cable elements in the loose fill slope models. In the third analysis, the properties of the pile elements which modelled the facing structure were also specified at this stage.

STAGE 3 – SIMULATION OF THE EFFECTS OF RAINFALL INFILTRATION

The actual process of rainfall infiltration was not modelled in the numerical analysis. What was simulated was the effects of stress changes in the soil caused by surface infiltration after heavy rainstorms. In order to mimic the effects, the suction profile in the top 3 m was replaced by a triangular pore water pressure distribution with 0 at the slope surface and 10 kPa at the bottom of the 3-m wetted zone. For soil at depths below 3 m, the pore water pressure (or suction) remained unchanged, as shown in Figure 9.45. This simulates the existence of a relatively impermeable layer at a depth of 3 m, while a perched water table at 3 m depth is allowed to build up. The depth of the wetted zone was selected based on the observation that the failure zone in the 1976 Sau Mau Ping landslide was generally confined to the top 3 m of the slope (Hong Kong Government, 1977). Trial runs were conducted and the results show that values smaller than 10 kPa would not trigger global failure. Consequently, a value of 10 kPa was selected for the pore pressure at the bottom of the wetted zone.

Take *et al.* (2004) conducted static centrifuge model tests to investigate the failure mechanisms of layered fill slopes formed by decomposed granitic soils. Failures were triggered by building up localized transient pore water pressures in a 1-m coarse granitic soil layer which is overlaid by a 2-m fine granitic soil to mimic possible heterogeneity formed during the construction of the fill slope. The results showed that a pore water pressure of 16 kPa was required to trigger global instability. This has a similar magnitude compared to the loading conditions employed in the present study.

STAGE 4 – EVALUATION OF GLOBAL STABILITY

The global stability of the slope was then evaluated by carrying out dynamic equilibrium calculations using FLAC2D assuming an undrained condition. Dynamic instead of static analysis was chosen because it can achieve a numerically stable solution even when the problem is not statically stable. The calculation is based on a finite difference scheme to solve the full equations of motion, using a lumped mass derived from the real density of the surrounding zones (Itasca, 1999). With this approach, the development of large strains and displacements before failure can be examined. Moreover, the dynamic formulation has an advantage in modelling physical processes, in which time is physically meaningful. As such, development of any possible failure mechanism can be explained in a manner that includes the time required for the failure to occur.

In each case, dynamic analysis was carried out for 60 prototype seconds if no numerical problem was encountered. Each time step was taken to be 10^{-5} s. The finite difference grid was updated automatically as grid displacements were calculated.

Computed results

Initial stress distributions in the loose fill slope

The distributions of p' and q at the end of the construction stage are shown in Figure 9.46. It can be seen that, due to the presence of the groundwater table (refer to Figure 9.45), which reduces the effective stress of soil underneath it, the contour lines of p' are not parallel to the slope surface at depths below 3 m. For the top several meters, where suction is assumed to be constant ($u = -50$ kPa), the inclination of the contour lines is approximately equal to the slope angle. It is important to note that the p' of the soil near the toe was generally higher than that in the soil near the crest at the same depth. This was caused by the higher lateral stress exerted on the soil near the toe. The maximum difference was about 25 kPa when a soil layer at the same depth was compared.

For q, which is not affected by the pore water pressure, the contour lines are approximately parallel. Nevertheless, they are not parallel to the slope surface. The mobilized in situ shear stress near the toe was generally higher

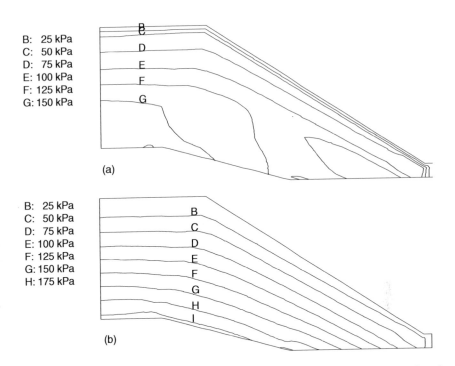

B: 25 kPa
C: 50 kPa
D: 75 kPa
E: 100 kPa
F: 125 kPa
G: 150 kPa

(a)

B: 25 kPa
C: 50 kPa
D: 75 kPa
E: 100 kPa
F: 125 kPa
G: 150 kPa
H: 175 kPa

(b)

Figure 9.46 Stress distributions at the end of formation: (a) mean effective stress, p', and (b) deviator stress, q (after Cheuk et al., 2005).

than the shear stress near the crest. Although the maximum difference of q in this case was also about 25 kPa at the same depth, the effect was more significant than in the case of p', because it represents a higher percentage difference, especially near the slope surface.

Behaviour of un-reinforced loose fill slopes

STRESS DISTRIBUTION SUBJECTED TO THE EFFECTS OF SURFACE INFILTRATION

Figure 9.47 shows the distributions of p' and q in the loose fill slope immediately after the simulation of the effects of surface infiltration. Due to an increase in pore water pressure, the p' of the top 3 m of soil decreased. Since a hydrostatic condition was assumed when the positive pore water pressure built up, the change in p' in the soil at the bottom of the 3-m wetted zone was higher than that in the soil near the slope surface. The distribution of q in the loose fill slope was found to be unchanged, because q is not affected by pore water pressure.

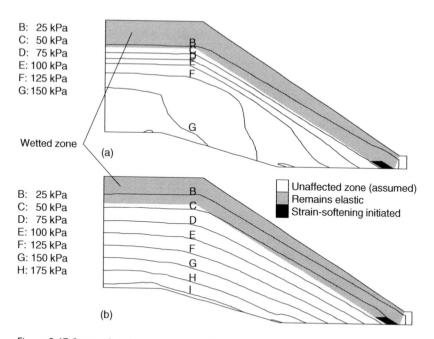

B: 25 kPa
C: 50 kPa
D: 75 kPa
E: 100 kPa
F: 125 kPa
G: 150 kPa

Wetted zone

(a)

B: 25 kPa
C: 50 kPa
D: 75 kPa
E: 100 kPa
F: 125 kPa
G: 150 kPa
H: 175 kPa

☐ Unaffected zone (assumed)
▨ Remains elastic
■ Strain-softening initiated

(b)

Figure 9.47 Stress distributions to the effects of surface infiltration: (a) mean effective stress, p', and (b) deviator stress, q (after Cheuk et al., 2005).

Figure 9.47 also shows that only one zone has initiated strain-softening after the increase in pore water pressure. In the SP model, strain-softening was assumed to have initiated when the effective stress path of a soil zone intercepts the collapse surface. This localized softening zone initiated at the toe where in situ shear stress was higher (Figure 9.45).

DEVELOPMENT OF THE FAILURE MECHANISM

The behaviour of the un-reinforced loose fill slope was monitored for only 22 s of prototype time before the grid was distorted to a state at which further calculations could not be performed. To demonstrate the development of the failure mechanism in the numerical model, the behaviour of the loose fill slope model before failure was examined. It was found that the development of the entire failure mechanism could be divided into two stages.

As shown in Figure 9.48, the first stage is the propagation of the strain-softening zone caused by the effects of surface infiltration. As mentioned before, the localized softening zone first appeared near the toe of the slope (refer to Figure 9.47). Due to the failure of this zone, additional shear stress

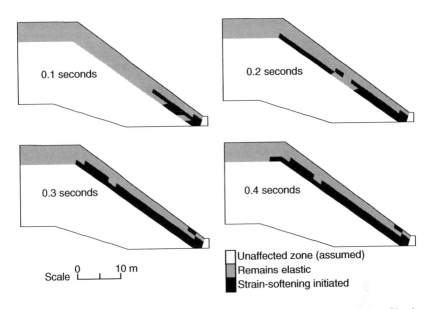

Figure 9.48 Development of progressive failure in the loose fill slope (after Cheuk *et al.*, 2005).

was mobilized at other locations in the slope due to stress redistribution. This enlarged the extent of the softening zone. At 0.3 s of prototype time, about half the soil zones in the top 3 m had begun to soften. Owing to the relatively higher in situ shear stress (refer to Figure 9.46), the softening zone was concentrated at the bottom of the 3-m wetted zone. Although strain-softening was initiated in a large number of the soil zones in the 3-m wetted zone, the associated deformation in this first stage was relatively small. The maximum total displacement in the slope at 0.4 s of prototype time was only 0.02 m, and the maximum shear strain was about 0.6 per cent. Moreover, the rate of propagation of the softening zone began to decrease at 0.3 s. The number of softening zones at 0.4 s was similar to that at 0.3 s. This indicates that the demand for shear stress was reduced due to stress redistribution.

The second stage of failure involved straining of the failed soil zones due to their inherent strain-softening characteristics. As the shear strength of the 'failed' soil zones decreased, the demand for shear stress increased again. This led to further propagation of the softening zone. Because of the continuous reduction in shear strength of the failed soil, an unbalanced force in the slope built up, and excessive displacement was therefore induced. When the shear deformation of the soil zone was large enough, calculations could no longer be performed. This occurred at 22 s in the numerical model.

DEFORMATION AND FAILURE MECHANISM

The behaviour of the un-reinforced loose fill slope at failure is summarized in Figure 9.49a–c. It can be seen from Figure 9.49a that excessive deformation was triggered in the top 3 m of the slope, and the maximum calculated displacement at this moment was over 5 m. The failure mechanism involves a well-developed shear band at shallow depths. The mobilized shear strain (Figure 9.49b) along the concentrated shear zone was generally greater than 50 per cent, indicating extremely large shear deformation. Figure 9.49c shows the extent of strain-softening at 22 s. A large proportion of the soil zones within the top 3 m have triggered strain-softening. The development of the softening zone is caused by two mechanisms as explained in the previous section.

The deformation mechanism depicted in Figure 9.49a–c resembles the observations obtained from 1976 Sau Mau Ping landslide. The report (Hong Kong Government, 1977) described the failure as an almost instantaneous conversion of the slope into a mud avalanche with considerable destructive energy. This is consistent with the very large displacements calculated from the analysis.

The results of this analysis are also consistent with the observations reported by Yoshitake and Onitsuka (1999). They used a centrifuge to study the failure mechanism of slopes filled with statically compacted decomposed granite subject to rainfall. They observed that local failure occurred initially at the toe, gradually reached the top of the slope and finally resulted in global failure of the slope.

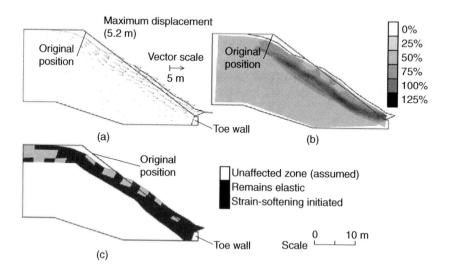

Figure 9.49 Loose fill slope at 22 s: (a) displacement vectors, (b) strain distribution and (c) extent of strain-softening (after Cheuk *et al.*, 2005).

Behaviour of nailed loose fill slopes with and without facing structure

DEFORMATION PATTERN AT 22 S PROTOTYPE TIME

To compare with the results from the un-reinforced loose fill slope, sim-ulations with the two nailed loose fill slopes (with and without a facing structure) at 22 s were examined. Figure 9.50a and b show the calculated displacements in the nailed loose fill slopes at 22 s. In both cases, the defor-mation pattern is similar to that of the loose fill slope without nails, which involves mainly deformations at shallow depths. This implies that deep-seated movement was not mobilized by the soil nails which can transfer the loading at shallow depths into deeper ground. This may be partly due to the anchorage of the soil nails in the natural ground, which prohibits the development of deep-seated failure. Suction, which remains in the deeper soil zones and increases the shear strength of the soil, also contributes to this observation.

When compared with Figure 9.49a, it can be seen in Figure 9.50a that when nails are installed in a loose fill slope without a facing structure, the maximum displacement decreased from 5.2 to 0.4 m at 22 s. As shown in Figure 9.50b, the maximum displacement decreased further to 0.1 m when the nails were connected to a facing structure at the slope surface. Apart from displacement, the maximum shear strain in the nailed slope also reduced from 100 to 10 per cent at 22 s when soil nails were installed in the slope.

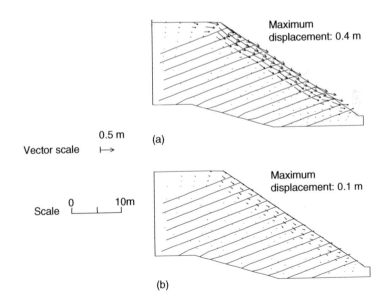

Figure 9.50 Displacement vectors in the nailed loose fill slopes at 22 s: (a) without facing and (b) with facing (after Cheuk *et al.*, 2005).

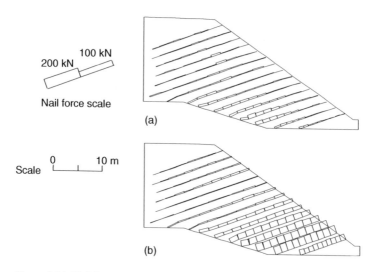

Figure 9.51 Nail forces mobilized in the nailed loose fill slopes at 22 s: (a) without facing and (b) with facing (after Cheuk et al., 2005).

For the nailed slope with a structural facing, the mobilized shear strain was in the range of 1 to 2 per cent at 22 s.

The improvement in stability observed in the nailed slopes was caused by the mobilized nail forces as shown in Figure 9.51a and b. Without a facing structure, the maximum mobilized nail force was about 100 kN (Figure 9.51a). However, it is important to be aware that the mobilized nail forces near the slope surface were relatively small, which can be attributed to the low confining stresses at shallow depths. With the facing (see Figure 9.51b), the maximum nail force increased to about 200 kN. Moreover, the mobilized nail forces were almost evenly distributed throughout the entire length of the soil nails. This suggests that the tensile forces in the soil nails were mobilized by the movement of the facing instead of by relative movement between the soil and the soil nails. This also highlights the importance of the anchorage of the nails into the natural ground, which is a key assumption in the present study. The mobilization of nail forces could only rely on the friction resistance along the nails, which can be very low in strain-softening materials, if anchorage were not present.

DEFORMATION PATTERN AT 60 S PROTOTYPE TIME

Figure 9.52a and b show the displacement vectors in the nailed loose fill slopes at 60 s. It can be seen from Figure 9.52a that, when the slope was upgraded by soil nails only, the maximum displacement could reach 6 m at 60 s prototype time. The corresponding shear strain was over 100 per cent.

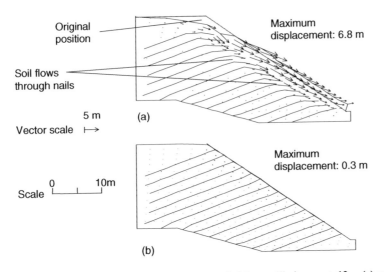

Original position

Maximum displacement: 6.8 m

Soil flows through nails

5 m   (a)

Vector scale ⊢→

0   10m
Scale ⊢__⊢__⊢

Maximum displacement: 0.3 m

(b)

Figure 9.52 Displacement vectors in the nailed loose fill slopes at 60 s: (a) without facing and (b) with facing (after Cheuk *et al.*, 2005).

A well-defined failure mechanism, which involved the development of a shallow flow slide, was observed. It is of interest to point out that extremely large relative displacement was observed between the failure mass and the upper part of the soil nails. This was caused by the low frictional resistance at the soil–nail interface at shallow depths due to the low confining stresses. Moreover, strain-softening of the surrounding soil further reduced the frictional resistance of the upper part of the soil nails. This resulted in a global failure in the nailed loose fill slope. Although the failure was delayed by the presence of soil nails, there was no reduction in the travel distance of the debris as indicated by the large displacement vectors in Figure 9.52a.

For the nailed loose fill slope reinforced with a facing structure, the maximum displacement was only 0.3 m at 60 s of prototype time, as shown in Figure 9.52b. The corresponding shear strain was about 8 per cent. Although deformation was still triggered by the effects of rainfall infiltration, the magnitude was greatly reduced. Complete static equilibrium was not achieved in the analysis at 60 s, but the numerical model was stabilizing, as indicated by the decreasing rate of deformation. Maximum incremental displacement reduced from the maximum value of 0.42 m/s at 40 s to 0.12 m/s at 60 s of prototype time after sliding failure of the loose fill began. More evidence for the stabilizing trend will be discussed when the development of the nail forces and structural forces in the slope are presented in later sections.

The tensile forces mobilized in the two nailed loose fill slopes with and without a facing at 60 s are shown in Figure 9.53a and b. As shown in

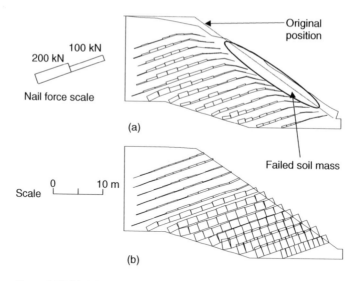

Figure 9.53 Nail forces mobilized in the nailed loose fill slopes at 60 s: (a) without facing and (b) with facing (after Cheuk *et al.*, 2005).

Figure 9.53a, the maximum induced-tensile force was about 110 kN without the facing. It should be noted that only very small ($<$ 10 kN/m) tensile forces were mobilized at shallow depths when the soil nails were not attached to any facing structure at the slope surface. This led to the flowing of the failure soil mass through the upper part of the soil nails as previously described. For the nailed slope with the facing (Figure 9.53b), the maximum tensile force developed in the soil nails was about 230 kN at 60 s of prototype time, which corresponds to a mobilization of 94 per cent of the yield force of the steel bar. Again, the uniformly distributed nail forces along the soil nails suggest that the tensile forces were induced by movement of the facing. The mobilized nail forces in the upper rows were significantly lower due to the free boundary conditions assumed at the bottom ends of these nails. This once again highlights the importance of the anchorage of the nails.

Tang and Lee (2003) and Zhang *et al.* (2006) described a series of centrifuge tests conducted on loose fill slope models reinforced by soil nails aiming at studying the behaviour of the soil nails when the models were subjected to rainfall infiltration coupled with rising groundwater table. The nails in the models penetrated deep into the slope to ensure good anchorage, while the other ends of the nails were fixed by a face plate. Although the loading conditions were different from that assumed in the present study, the results showed that the presence of the soil nails significantly reduced the deformation in the slope. These results are in good agreement with the observations obtained in the present study.

THE ROLE OF THE SOIL NAILS AND THE FACING STRUCTURE

The results of the analyses suggest that soil nails function differently depending on whether a face support structure is present at the slope surface. In the absence of a facing structure or nail heads, the soil nails bore the loading generated from the failing mass directly by frictional resistance at the soil–nail interface. However, due to the low confining stresses near the slope surface, flow-type failure around the soil nails could still occur. Although tensile forces of about 110 kN were mobilized in the soil nails (refer to Figure 9.53a), they were concentrated deeper in the slope, presumably caused by the fix boundary conditions at the ends of the nails at the lower part of the slope. The high assumed shear stiffness of the grout was also responsible for the relatively high nail forces which could be mobilized at very small soil displacements. These nail forces prevented the soil nails from pulling out of the slope. When compared to the un-reinforced loose fill slope, the soil nails undoubtedly reduced the deformation caused by strain-softening in the early stage due to the increase in the initial stiffness of the slope. However, this can only delay the development of the shallow failure, but not prevent it from happening.

When the soil nails were connected to a facing structure at the slope surface, the whole soil nail system behaved like an earth-retaining structure anchored with soil nails. When a potential flow slide was initiated by the strain-softening of the filling material, the movement of the sliding mass was restricted due to the presence of the facing. As the soil pushed the facing away from the slope, axial forces and bending moments were induced in the facing. Simultaneously, tensile forces were mobilized in the soil nails. The development of nail forces and structural forces in the facing are presented in Figure 9.54. It can be seen that the nail forces increased until 40 s of prototype time. Afterwards, the mobilized tensile forces became stable. This may indicate that the system had achieved a state of static equilibrium and the required resistance from the soil nails had been fully mobilized. Figure 9.54 also indicates that the bending moment increased until 40 s. This is consistent with the observed nail forces. The value of the axial force in the facing increased continuously until 60 s. Nonetheless, a shape change was observed at 40 s.

The development of the nail forces and structural forces in the facing indicates that the stabilizing mechanism in the nailed loose fill slope relied on the structural facing, which resisted the forces generated from the potential failure mass. These forces were eventually sustained by the pull-out resistance of the soil nails deeper in the ground through the development of tensile forces in the nails.

The pattern of the mobilized nail forces (refer to Figure 9.53b) suggests that a portion of the resistance comes from the bottom rows of the soil nails. Although this might be due to the boundary conditions assumed in the analyses, it would be beneficial from a practical point of view if soil nails

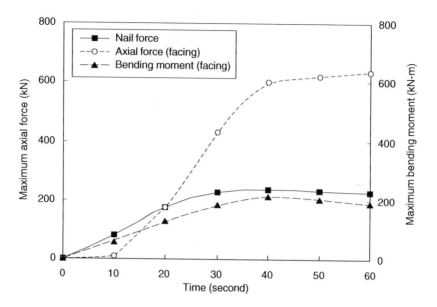

Figure 9.54 Development of nail forces and structural forces in the facing (after Cheuk *et al.*, 2005).

are only needed at the lower part of a slope, where failure is initiated as demonstrated in the analysis of an un-reinforced slope. This requires further study to substantiate.

A preliminary design guide has been published for the design and use of soil nails in loose fill slopes in Hong Kong (HKIE, 2003). The work carried out to prepare the guide has been described in subsequent publications and covers the following:

- the numerical study described above (Cheuk *et al.*, 2005),
- laboratory tests on saturated and unsaturated loose fills (Ng and Chiu, 2001, 2003; Ng *et al.*, 2004a),
- static and dynamic centrifugal model tests (Ng *et al.*, 2002b; Ng *et al.*, 2004c, 2006; Zhang *et al.*, 2006),
- pull-out tests and field trials (Tang and Lee, 2003).

Conclusions

- Three sets of finite difference analyses were conducted to explore the operating mechanism of soil nails in loose fill slopes. The characteristics of strain-softening material in the fill slopes were mimicked using a user-defined strain-softening soil model. The first analysis acted as a benchmark to illustrate the modelled behaviour of un-reinforced loose fill slopes subjected to the effects of rainfall infiltration. The numerical results indicate that shallow failure can develop in a loose fill slope when the pore water pressure in the top 3 m of fill material increases. Global failure was found to begin near the toe of the slope.

- The numerical analyses showed that when soil nails were installed in a loose fill slope without nail heads or a facing structure, global sliding failure occurred. The predicted failure mechanism was similar to that of a loose fill slope without soil nails. Although the development of global failure was delayed by the presence of soil nails, the low frictional resistance of the nails at shallow depths failed to arrest the development of a flow slide.
- The deformation of the loose fill slope was found to be greatly reduced if the soil nails were connected to a facing structure at the slope surface. The entire nail-facing structure acted as an earth-retaining structure. Global failure due to rainfall infiltration was avoided under the modelled conditions. The facing was found to play a key role in the stabilizing mechanism. As the facing was pushed by the failing soil mass, high tensile forces were mobilized in the soil nails. The applied loading generated from the potential sliding mass was then transferred to the frictional resistance between the soil and the nails deeper in the ground.

Effects of conjunctive modelling on slope stability (Tung et al., 2004)

Introduction

Landslides cause enormous economic losses and threaten public safety in many countries around the world. Among various mechanisms causing landslides, rainfall infiltration is one of the major triggering factors leading to slope failures. Infiltration is a process by which rainwater moves downwards through the earth surface to the soil underneath. As rainwater continues to infiltrate into an initially unsaturated soil slope, it would gradually reduce negative pore water pressure (suction) in the slope and weaken the slope stability. Therefore, a realistic modelling and predictive tool is essential to analyse transient seepage and pore pressure distributions in both saturated and unsaturated soil slopes. Many conventional approaches in modelling subsurface flow are to uncouple infiltration and run-off processes during a storm event. For simplicity, run-offs are assumed to be discharged instantly and the surface water depth can be thought as zero. Lam et al. (1987) is among the earliest researchers to apply finite element method to solve saturated–unsaturated seepage problem in geotechnical engineering. Ng and Shi (1998) conducted the parametric study on stability of unsaturated soil slope subjected to transient seepage using the two-dimensional finite element program SEEP/W. Furthermore, Tung et al. (1999) and Ng et al. (2001b) performed a three-dimensional numerical investigation of rainfall infiltration in an unsaturated soil slope subjected to various rainfall patterns. In most of these subsurface flow models, infiltration amount (or coefficient of infiltration) is specified in advance. In reality, variations of infiltration amount during a rainstorm event cannot be estimated easily unless interactions between rainfall, infiltration and surface run-off processes are properly considered and modelled.

The coupled process of surface–subsurface flow is initially investigated by hydraulic and hydrological engineers. Akan and Yen (1981) developed a conjunctive one-dimensional surface flow and two-dimensional subsurface flow model. Recently, Morita and Yen (2002) further presented a conjunctive two-dimensional surface flow and three-dimensional subsurface flow model. Regarding surface flow modelling in hydraulic engineering, channel slopes are rather small as compared with the slope angle of natural, fill and cut slopes encountered in geotechnical engineering. Hence, the surface flow governing equations should be modified to accommodate the steep slope condition. Kwok (2003) investigates the coupled surface–subsurface flow in steep slopes using finite difference method. While hydraulic engineers mainly concern with water balance in the flow analysis, geotechnical engineering applications are more interested in the spatiotemporal distribution and variation of pore water pressure.

In this chapter, a new conjunctive surface–subsurface flow model is developed to consider the interaction between surface and subsurface flows in transient seepage analysis of initially unsaturated soil slopes. Comparisons between computed pore water pressure distributions from uncoupled and coupled analysis will be presented and discussed. The implications of the differences in pore water distribution on slope stability will be illustrated.

Conjunctive surface–subsurface flow model

Surface flow submodel

The basic equations describing gradually varied unsteady shallow water flow are commonly known as the 'Saint Venant equations', which are developed for low inclination slopes. In order to describe the behaviour of surface flow on steep slopes, Yen (1977) modified the governing equations to accommodate the rapid varied unsteady flow condition using a new coordinate system as shown in Figure 9.55.

For one-dimension surface flow, the modified governing equations are as follows:

$$\frac{\partial h}{\partial t} + \frac{\partial (uh)}{\partial x} = r - f \tag{9.30a}$$

$$\frac{\partial (uh)}{\partial t} + \frac{\partial (u^2 h)}{\partial x} + gh \, \cos \, \theta \frac{\partial h}{\partial x} = gh(S_0 - S_f) \tag{9.30b}$$

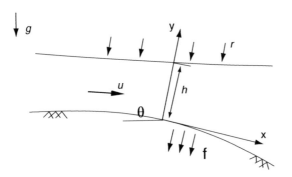

Figure 9.55 Variables used in Saint Venant equations (after Tung et al., 2004).

where the positive x-axis direction is along the ground surface downwards; the y-axis is perpendicular to the x-axis as illustrated in Figure 9.55; t is time; h is the surface water depth; u is the mean velocity along x-axis; θ is the slope angle measured from the horizontal plane; $S_0 = \sin\theta$, the bottom slope; S_f is the friction slope; r is the rainfall intensity or lateral inflow; f is the infiltration rate and g is the gravitational acceleration.

To determine the friction slope S_f in Equation 9.30, the Darcy–Weisbach equation is used by Morita and Yen (2002), i.e.

$$S_f = f_d \frac{u^2}{8gh} \tag{9.31}$$

where f_d is the frictional resistance coefficient, which is a function of Reynold's number, $R_e = (uh)/v$, with v being the kinematic viscosity, and can be determined by

$$f_d = \frac{24 + 660\left(\dfrac{r}{\sqrt[3]{gv}}\right)^{0.4}}{R_e} \tag{9.32}$$

for laminar flow with $R_e < 500$; while

$$f_d = \frac{0.223}{R_e^{0.25}} \tag{9.33}$$

for transitional flow with $500 < R_e < 30,000$; and

$$f_d = \frac{1}{4}\left[-\log\left(\frac{k_s}{12R} + \frac{1.95}{R_e^{0.9}}\right)\right]^{-2} \tag{9.34}$$

for turbulent flow with $R_e > 30,000$, where k_s is the equivalent sand grain roughness size and R is the hydraulic radius of the flow.

Subsurface flow submodel

Saturated–unsaturated subsurface flow is governed by Richard's equations. The two-dimensional Richard's equation with total head as the main variable can be expressed as:

$$\frac{\partial}{\partial x}\left(k_x \frac{\partial H_w}{\partial x}\right) + \frac{\partial}{\partial y}\left(k_y \frac{\partial H_w}{\partial y}\right) + Q = m_w \gamma_w \frac{\partial H_w}{\partial t} \tag{9.35}$$

where $m_w = \partial\theta_w/\partial u_w$ represents the slope of the SWCC; u_w is the pore water pressure; θ_w is the volumetric water content; γ_w is the unit weight of water; k_x and k_y are the hydraulic conductivities in the x- and y-directions, respectively. In the saturated zone, k_x and k_y are the saturated hydraulic conductivities in each direction, θ_w is the saturated volumetric water content, and m_w is equal to zero under the assumption of incompressibility of both water and soil structures. In the unsaturated zone, however, k_x, k_y, θ_w and m_w all vary with pore water pressure u_w. Given the boundary and initial conditions, Richards' equation can be solved numerically by the finite element method or the finite difference method.

Conjunctive surface–subsurface flow model

The conjunctive surface–subsurface flow model consists of surface flow governing equations (Equations 9.30a and b) and the subsurface flow governing equation (Equation 9.35) through for interface infiltration. Two sets of equations are solved separately and iteratively during the same time step to determine infiltration following the algorithm shown in Figure 9.56.

Here the procedure is to firstly apply the full rainfall intensity at a time step to the slope surface as a flux boundary from which the subsurface flow submodel solves for the pore water pressure distribution. If positive water pressure does not occur on the slope surface, then the slope surface is not saturated and all rainwater infiltrates into the slope. Otherwise, the slope surface is saturated and so not all rainwater can enter the slope, whence surface run-off will occur. To estimate the surface water depth, Equation (9.36) is used (Morita and Yen, 2002) as follows:

$$Y = h + r\Delta t \tag{9.36}$$

where h is the surface water depth at the previous time step; r is the rainfall intensity; and Δt is the time increment. Before run-off generation, the surface water depth h is set to be zero initially.

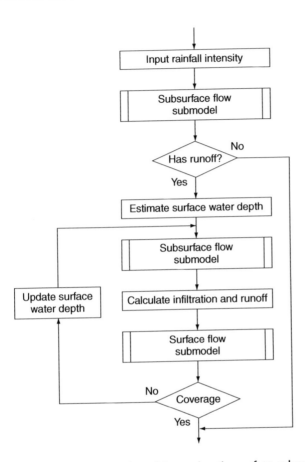

Figure 9.56 Flow chart for solving conjunctive surface–subsurface flow model (after Tung *et al.*, 2004).

Secondly, Y is used as the head boundary to calculate the subsurface flow once again. The infiltration rate through this head boundary interface can then be back-calculated. The rainfall intensity less the calculated infiltration rate yields the corresponding run-off rate.

Thirdly, by applying this estimated run-off rate, the surface flow submodel is used to calculate the corresponding surface water depth. Comparing the calculated surface water depth with the previously applied head boundary enables the model to update the head boundary according to the newly estimated surface water depth and so resolve for the subsurface flow again. The process is repeated until convergence is reached before switching to the next time step.

Numerical simulation

Introduction to the programs

CONJUNCTIVE SURFACE–SUBSURFACE FLOW ANALYSIS PROGRAM

Based on the conjunctive one-dimensional surface and two-dimensional sub-surface flow model presented above, a Fortran program was written to solve the coupled processes of rainfall, infiltration and surface run-off in steep and low inclination slopes. In the program, one-dimensional surface flow governing equations (Equation 9.30) were solved by the finite difference method. To achieve numerical stability and convergence, a four-point implicit finite difference scheme, known as the Preissmann scheme (Chaudhry, 1993), was adopted. Two-dimensional subsurface flow governing equations (Equation 9.35) were solved by the Galerkin finite element method. For numerical stability, the lumped mass matrix method was adopted. A weighted spline was used to fit the non-linear SWCC and hydraulic conductivity function. In addition, the program can automatically adjust the boundary conditions for different surface flow types, i.e. for supercritical flow, two upstream boundary conditions are given; while for subcritical flow, one upstream boundary and one downstream boundary conditions are given.

TWO-DIMENSIONAL SATURATED–UNSATURATED SEEPAGE ANALYSIS PROGRAM SEEP/W

SEEP/W is a commercial two-dimensional finite element seepage analysis program which is an uncoupled model used to solve rainfall infiltration problems (Geo-slope, 1995). It has only a subsurface flow analysis model which neglects the interactions between rainfall, infiltration and surface run-off processes.

Pore water pressure distributions

Comparisons between computed pore water pressure distributions from SEEP/W and the coupled model are presented and discussed. As shown in Figure 9.57, the length of the slope is 100 m; the slope angle is 15° and the depth of calculation zone is 11 m. Total 2,200 elements are divided for subsurface flow domain with the mesh length of each element is $\Delta z = 0.25$ m vertical and $\Delta x = 2$ m parallel to slope surface. The grid for surface flow modeling is $\Delta x = 2$m, the same as that of subsurface mesh along slope surface direction. The initial groundwater table is shown in Figure 9.57 which is obtained from a steady-state condition, where the depths of groundwater table at top and toe of the slope are fixed at 10 and 7 m, respectively. The slope surface is a flux boundary subjected to rainfall, while all other boundaries are impermeable.

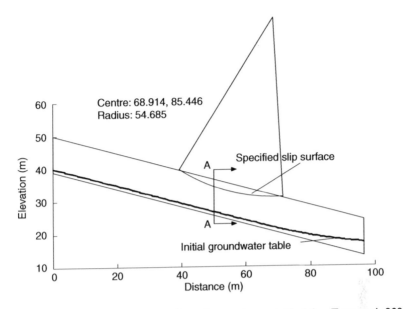

Figure 9.57 Calculation profile and initial groundwater table (after Tung *et al.*, 2004).

The slope material is homogeneous fine sand with the saturated conductivity 4.3×10^{-6} m/s (15.5mm/h), porosity 0.347 and SWCC and hydraulic conductivity function shown in Figure 9.58.

Table 9.11 lists the three cases of uniform rainfall intensity and duration used in the numerical simulation. The total rainfall applied in three cases is about 267 mm which corresponds to a 10-year return period daily rainfall at Hong Kong Observatory.

At the end of the rainfall event, the surface water depths are different with various rainfall intensities. As shown in Figure 9.59, the surface water depth increases towards the downstream direction along the slope. In addition, given the same conditions except rainfall intensity, the surface water depth is deeper, as expected, under larger rainfall intensity. In Figure 9.59, the maximum surface water depths under 22, 45 and 89 mm/h rain are about 4, 6 and 10 mm occurring at the end of storm, respectively. During rainstorms, the surface water depth on a steep slope could vary from several millimetres to several centimetres depending on the rainfall intensity and slope angle.

For a rainfall intensity of 22 mm/h, a comparison of pore water pressure distribution between coupled and uncoupled analyses at section A–A is shown in Figure 9.60a. Without considering the interaction between surface run-off and infiltration, the wetting front from the uncoupled model reaches about 1.5 m deep from the ground surface at the time rain stops. While by using coupled model, the wetting front reaches about 2.0 m in depth, an half metre deeper than that of uncoupled model.

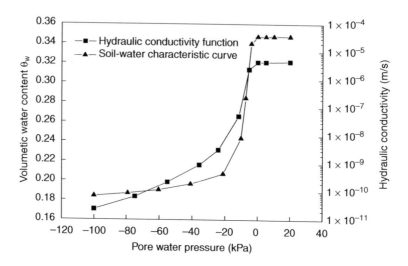

Figure 9.58 Soil–water characteristic curve and hydraulic conductivity function (after Tung et al., 2004).

Table 9.11 Rainfall characteristics of cases studied (after Tung et al., 2004)

Cases	Rainfall intensity (mm/hr)	Duration (hrs)	Rainfall volume (mm)
1	22	12	264
2	45	6	270
3	89	3	267

As shown in Figure 9.60b, the rainfall intensity of case-2 is twice as large as that of case-1. It is interesting to observe that the rainfall intensity increases, the run-off portion in total rainfall increases, whereas the corresponding portion of infiltration decreases. Under 45 mm/h rainfall intensity, the wetting front reaches about 1.0 m deep from the ground surface according to the coupled model, which is about 0.3 m deeper than the uncoupled procedure at the end of rainfall.

Now consider case-3 – a heavy rainstorm of 89 mm/h. This is two times larger than that of case-2 and four times larger than that of case-1. As shown in Figure 9.60c, most of the rainfall transforms to run-off and the remaining small portion infiltrates into the slope which affects the pore water pressure distribution only within 0.7 m in depth. Regardless of interaction between surface and subsurface flow, the wetting front reaches about 0.5 m in depth below the ground surface.

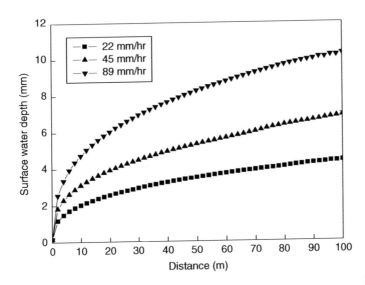

Figure 9.59 Surface water depth profiles under different rainfall intensities (after Tung et al., 2004).

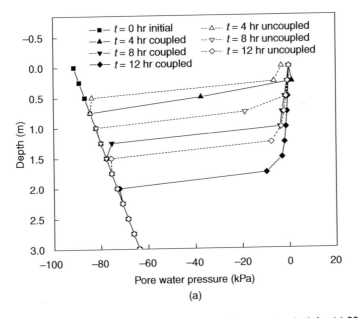

(a)

Figure 9.60 Vertical pore water pressure profile at section A–A for (a) 22-mm/h rainfall intensity, (b) 45-mm/h rainfall intensity (c) 89-mm/h rainfall intensity (after Tung et al., 2004).

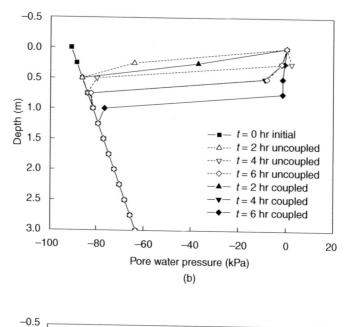

(b)

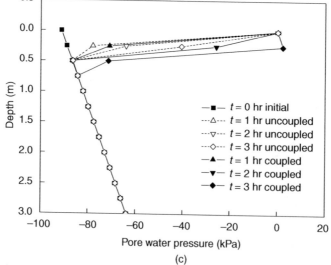

(c)

Figure 9.60 (Continued).

Slope stability analyses

As rainwater infiltrates into an unsaturated soil slope, pore water pressure increases with time and the corresponding matric suction decreases with the increase in water content. Based on the extended Mohr–Coulomb failure criterion for unsaturated soil (Fredlund et al., 1978), Equation (9.22) shows that the shear strength will decrease with loss in suction.

To illustrate the variation of FOS with pore water pressure distribution change, a single circular slip surface is specified as shown in Figure 9.57, which is about 2.5 m deep from the ground surface. The unit weight of soil is assumed to be $19.6 \, kN/m^3$, effective cohesion $c' = 0 \, kPa$, effective friction angle $\phi' = 20°$ and the friction angle related to matric suction $\phi^b = 8°$. Using the Bishop simplified method in SLOPE/W to perform the slope stability analysis, results were obtained as presented in Figure 9.61.

The slope is initially dry and the FOS of the specified slip surface is high. As the rainwater infiltrates into the slope, the FOS decreases with decreasing in matric suction. Under heavy rainstorm with intensity of 89 mm/h, and rainfall of 3 h, the wetting front only reaches 0.5–0.7 m in depth, and the FOS reduces only about 0.1. The differences in pore water pressure between uncoupled and coupled models are not significant. Correspondingly, the difference in FOS between the two results after 3-h rainfall is about 0.05. Thus, the factors of safety calculated by the uncoupled and coupled models have no significant difference either. When rainfall duration extends longer, the wetting front penetrates deeper and the difference in pore water pressure

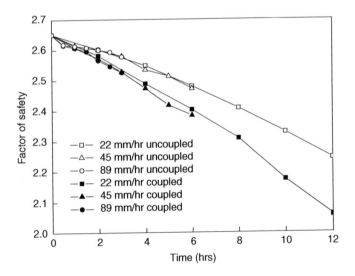

Figure 9.61 Variation of factor of safety with time for different rainfall intensities and for couple and uncoupled models (after Tung et al., 2004).

becomes more pronounced. Then, the FOS of slope drops more significantly as matric suction reduces. The differences in results between the uncoupled and coupled models become larger. After 12-h rainfall with intensity of 22 mm/h, the FOS decreases by 0.5 and the differences in FOS between the uncoupled and coupled models are about 0.2.

Conclusions

Using a new conjunctive surface–subsurface flow model, the interaction between surface and subsurface flows was used to analyse transient seepage for initially unsaturated soil slopes. Based on these studies, the following conclusions can be drawn:

- Surface water depth on a steep slope increases with the rainfall intensity, which is usually about several millimetres. Under a heavy rainstorm, surface water depth can reach several centimetres.
- Interaction between surface and subsurface flow has significant effects on infiltration process. Considering the coupled surface–subsurface flow process, more rainwater will infiltrate into the slope. Therefore, pore water pressure will change faster and the wetting front will move deeper into the soil.
- Under smaller rainfall intensity with prolonged duration, the differences in infiltration and pore water pressure between the conjunctive surface–subsurface flow model and uncoupled model were more significant.

Appendix A: definitions

Source (Fredlund and Rahardjo, 1993)

This appendix is made up of verbatim extracts from the following source for which copyright permissions have been obtained as listed in the Acknowledgements.

- Fredlund, D.G. and Rahardjo, H. (1993). *Soil Mechanics for Unsaturated Soils*. John Wiley & Sons, Inc., New York, 517p.

Definitions

Density

Density, ρ, is defined as the ratio of mass to volume. Each phase of a soil has its own density. The density of each phase can be formulated from Figure A.1. Specific volume, v_o, is generally defined as the inverse of density; therefore, specific volume is the ratio of volume to mass.

Soil particles

The density of the soil particles, ρ_s, is defined as follows:

$$\rho_s = \frac{M_s}{V_s} \tag{A.1}$$

The density of the soil particles is commonly expressed as a dimensionless variable called specific gravity, G_s. The specific gravity of the soil particles is defined as the ratio of the density of the soil particles to the density of water at a temperature of $4\,°C$ under atmospheric pressure conditions (i.e. $101.3\,kPa$).

$$G_s = \frac{\rho_s}{\rho_w} \tag{A.2}$$

The density of water at $4\,°C$ and $101.3\,kPa$ is $1,000\,kg/m^3$. Some typical values of specific gravity for several common minerals are given in Table A1.

Water phase

The density of water, ρ_w, is defined as follows:

$$\rho_w = \frac{M_w}{V_w}$$

(A.3)

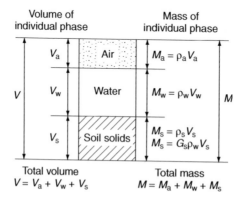

Figure A.1 Phase diagram for an unsaturated soil (after Fredlund and Rahardjo, 1993).

Table A1 Specific gravity of several minerals (from Lambe and Whitman, 1979)

Mineral	Specific gravity, G_s
Quartz	2.65
K–Feldspars	2.54–2.57
Na–Ca–Feldspars	2.62–2.76
Calcite	2.72
Dolomite	2.85
Muscovite	2.7–3.1
Biotite	2.8–3.2
Chlorite	2.6–2.9
Pyrophyflite	2.84
Serpentine	2.2–2.7
Kaolinite	2.61[a]; 2.64 $\pm$ 0.02
Halloysite ($2H_2O$)	2.55
Illite	2.84[a]; 2.60–2.86
Montmorillonite	2.74[a]; 2.75–2.78
Attapulgite	2.30

Note
a Calculated from crystal structure.

For soil mechanics problems, the variation in the density of water due to temperature differences is most significant than its variation due to an applied pressure.

Air phase

The density of air, ρ_a, is defined as follows:

$$\rho_a = \frac{M_a}{V_a} \tag{A.4}$$

The specific volume of air, v_{ao}, is

$$v_{ao} = \frac{V_a}{M_a} \tag{A.5}$$

Air behaves as a mixture of several gases. The mixture of gases is called dry air when no water vapour is present and is called moist air when water vapour is present.

Dry and moist air can be considered to behave as an 'ideal' gas under pressures and temperatures commonly encountered in geotechnical engineering. The ideal gas law can be written as follows:

$$\bar{u}_a V_a = \frac{M_a}{\varpi_a} RT \tag{A.6}$$

where
$\bar{u}_a$ = absolute air pressure (note that a bar sign indicates an absolute pressure
$\bar{u}_a = u_a + \bar{u}_{atm}$ (kPa)
u_a = gauge air pressure (kPa)
u_{atm} = atmospheric pressure (i.e. 101.3 kPa)
V_a = volume of air (m^3)
M_a = mass of air (kg)
ω_a = molecular mass of air (kg/kmol)
R = universal molar gas constant (i.e. 8.31432 J/(mol K)
T = absolute temperature (i.e. $T = t + 273.16$ K)
t = temperature (°C).

Since the right-hand side of the equation is a constant for a gas in a closed system. Under these conditions, the equation can be rewritten as Boyle's law:

$$\bar{u}_{a1} V_{a1} = \bar{u}_{a2} V_{a2} \tag{A.7}$$

Rearranging the 'ideal' gas equation gives:

$$\rho_a = \frac{\omega_a}{RT} \bar{u}_a \tag{A.8}$$

The molecular mass of air, ω_a, depends on the composition of the mixture of dry air and water vapour. The dry air has a molecular mass of 28.966 kg/kmol, and the molecular mass of the water vapour (H_2O) is 18.016 kg/kmol. The composition of air, namely, nitrogen (N_2) and oxygen (O_2) are essentially constant in the temperature. The carbon dioxide (CO_2) content in air may vary, depending on the environment, such as the rate of consumption of fossil fuels. However, the constituent of air that can vary most is water vapour. The volume per cent of water vapour in the air may range as little as 0.000002 per cent to as high as 4–5 per cent (Harrison, 1965).The molecular mass of air is affected by the change in each constituent. This consequently affects the density of air.

The concentration of water vapour in the air is commonly expressed in terms of relative humidity (RH):

$$RH = \frac{100\bar{u}_v}{\bar{u}_{vo}} \tag{A.9}$$

where
RH = relative humidity (per cent)
$\bar{u}_v$ = partial pressure of water vapour in the air (kPa)
$\bar{u}_{vo}$ = saturation pressure of water vapour at the same temperature (kPa).

The density of air decreases as the RH increases. This indicates that the moist air is lighter than the dry air.

Volume-mass relations

Porosity, n, in per cent is defined as the ratio of the volume of voids, V_v, to the total volume, V as follows:

$$n = \frac{100V_v}{V} \tag{A.10}$$

Similarly, porosity type terms can be defined with respect to each of the phases of a soil:

$$n_s = \frac{100V_s}{V} \tag{A.11}$$

$$n_w = \frac{100V_w}{V} \tag{A.12}$$

$$n_a = \frac{100V_a}{V} \qquad \text{(A.13)}$$

$$n_c = \frac{100V_c}{V} \qquad \text{(A.14)}$$

where

n_s = soil particle porosity (per cent)
n_w = water porosity (per cent)
n_a = air porosity (per cent)
n_c = contractile skin porosity (per cent).

The volume associated with the contractile skin can be assumed to be negligible or part of the water phase. The water and air porosities represent their volumetric percentages in the soil. The soil particle porosity can be visualized as the percentage of the overall volume comprised of soil particles. The sum of the porosities of all phases must equal 100%. Therefore, the following soil porosity equation can be written as

$$n_s + n = n_s + n_a + n_w = 100 \text{(per cent)} \qquad \text{(A.15)}$$

The water porosity, n_w, expressed in decimal form, is commonly referred to as the volumetric water content, θ_w, in soil science and soil physics literature. The volumetric water content notation is also used.

Void ratio, e, is defined as the ratio of the volume of voids, V_v, to the volume of soil solids, V_s as follows (Figure A.2):

$$e = \frac{V_v}{V_s} \qquad \text{(A.16)}$$

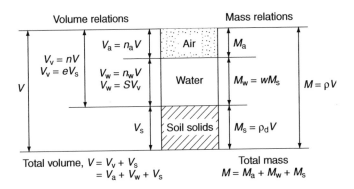

Figure A.2 Volume–mass relations (after Fredlund and Rahardjo, 1993).

The relationship between porosity and void ratio is given by the following equation:

$$n = \frac{e}{1+e} \tag{A.17}$$

The percentage of the void space which contains water is expressed as the degree of saturation, S (per cent):

$$S = \frac{100V_w}{V_v} \tag{A.18}$$

The degree of saturation, S, can be used to subdivide soils into three groups.

- Dry soils (i.e. $S = 0$ per cent): Dry soil consists of soil particles and air. No water is present.
- Saturated soils (i.e. $S = 100$ per cent): All of the voids in the soil are filled with water.
- Unsaturated soils (i.e. 0 per cent $< S < 100$ per cent): An unsaturated soil can be further subdivided, depending upon whether the air phase is continuous or occluded. This subdivision is primarily a function of the degree of saturation. An unsaturated soil with a continuous air phase generally has a degree of saturation less than approximately 80 percent (i.e. $S < 80$ per cent). Occluded air bubbles commonly occur in unsaturated soils having a degree of saturation greater than approximately 90 per cent (i.e. $S > 90$ per cent). The transition zone between continuous air phase and occluded air bubbles occurs when the degree of saturation is between approximately 80–90 per cent (i.e. 80 per cent $< S < 90$ per cent).

Gravimetric water content, w, is defined as the ratio of the mass of water, M_w, to the mass of soil solids, M_s. It is presented as a percentage [i.e. w (per cent)]:

$$w = \frac{100M_w}{M_s} \tag{A.19}$$

The volumetric water content, θ_w, is defined as the ratio of the volume of water, V_w, to the total volume of the soil, V:

$$\theta_w = \frac{V_w}{V} \tag{A.20}$$

The volumetric water content can also be expressed in terms of porosity, degree of saturation and void ratio. The volumetric water content can be written as

$$\theta_w = \frac{SV_v}{V} \tag{A.21}$$

Since V_v/V is equal to the porosity of the soil, then

$$\theta_w = Sn \tag{A.22}$$

or

$$\theta_w = \frac{Se}{1+e} \tag{A.23}$$

Notation

A_n	cross-sectional area of a cable element
A_f	cross-sectional area of a pile element
C_{sf}	cohesive strength of the shear coupling spring in a pile element
C_{nf}	maximum compressive strength of the normal coupling spring in a pile element
c'	effective cohesion
c'_1	effective cohesion of the state boundary surface on the dry side of the critical state line
c'_2	effective cohesion of the collapse surface
d	outside diameter of a cable element
D_f	mass density of a pile element
e	voids ratio
E	Young's modulus of soil
E_f	Young's modulus of a pile element
E_n	Young's modulus of a cable element
F_{sf}	maximum shear force of the shear coupling spring in a pile element
F_{sn}	maximum shear force of the shear coupling spring in a cable element
G	shear modulus of soil
I_f	moment of inertia of a pile element
K	bulk modulus of soil
K_{sn}	stiffness of the shear coupling spring in a cable element
K_{sf}	stiffness of the shear coupling spring in a pile element
K_{nf}	stiffness of the normal coupling spring in a pile
L	length of a cable element
M	stress ratio q/p' at critical state
M_{col}	inclination of the collapse surface in the $q-p'$ plane
$M(s)$	gradient of the critical state hyperline in the $q-p$ plane

$N(s)$	intercept of isotropic normal compression hyperline in plane of $v - \log p$
n	porosity of soil
p	net mean stress
p_{ref}	reference pressure 100 kPa
p'	mean effective stress $(\sigma_1' + \sigma_2' + \sigma_3')/3$
p'_{cs}	mean effective stress at critical state
q	deviator stress $(\sigma_1' - \sigma_3')$
q_{av}	available deviator stress
q_{cs}	deviator stress at critical state
q_{max}	maximum deviator stress
q_{mob}	mobilized deviator stress
q_1	mobilized deviator stress when the effective stress path intercepts the collapse surface
S_r	degree of saturation
s	matric suction $(u_a - u_w)$
T	tensile strength of soil
T_{nf}	maximum tensile force of a pile element in normal direction
u	pore pressure
u_w	pore water pressure
u_a	pore air pressure
v	specific volume
Y_n	yield force of a cable element
ε_s	total shear strain
$\varepsilon_{s\,max}$	shear strain at minimum deviator stress during strain-softening
ϕ'	effective friction angle of soil
ϕ'_1	effective friction angle of the state boundary on the dry side of the critical state line
ϕ'_2	effective friction angle of the collapse surface
ϕ_{sf}	frictional resistance of the shear coupling spring in a pile element
Γ	intercept of the critical state line in $e - \ln p'$ plane
$\Gamma(s)$	intercept of critical state hyperline in $v - \log p$ plane
λ	slope of the critical state line in $e - \ln p'$ plane
$\lambda(s)$	slope of critical state hyperline in $v - \log p$ plane
$\mu(s)$	intercept of critical state hyperline in $q - p$ plane
ν	Poisson's ratio
ρ_d	dry density of soil
σ	total normal stress
σ'	effective normal stress $\sigma - u$
σ_1	total major principal stress

σ_2 total intermediate principal stress
σ_3 total minor principal stress
τ shear stress
τ_f shear strength (shear stress at failure)
ψ angle of dilation
ψ_1 angle of dilation of the state boundary surface on the dry side of the critical state line
ψ_2 angle of dilation of the collapse surface
$\psi(s)$ slope of the critical state hyperline in $v - \log p$ plane

References

Abelev, M.Y. (1975). Compacting loess soils in the USSR. *Géotechnique*, 25(1), 79–82.

Aitchison, G.D. (1961). Relationship of moisture and effective stress functions in unsaturated soils, *Proc. Conf. on Pore Pressure and Suction in Soils*. Butterworths, London, pp. 47–52.

Akan, A.O. and Yen, B.C. (1981). Mathematical model of shallow water flow over porous media. *J. Hyd. Eng.*, ASCE, 107(4), 479–494.

Al-Homoud, A.S., Basma, A.A., Husein Malkawi, A.I. and Al-Bashabshah, M.A. (1995). Cyclic swelling behaviour of clays. *J. Geotech. Eng.*, ASCE, 121(7), 562–565.

Al-Khafaf, S. and Hanks, R.J. (1974). Evaluation of the filter paper method for estimating soil–water potential. *J. of Soil Science*, 117(4), 194–199.

Al-Mukhtar, M., Qi, Y., Alcover, J.F. and Bergaya, F. (1999). Oedometric and water-retention behaviour of highly compacted unsaturated smectites. *Can. Geotech. J.*, 36, 675–684.

Alonso, E.E. (1998). Modeling expansive soil behavior. *Proc. 2nd Int. Conf. on Unsaturated Soils*, Beijing 2, pp. 37–70.

Alonso, E.E., Gens, A. and Josa, A. (1990). A constitutive model for partially saturated soils. *Géotechnique*, 40(3), 405–430.

Anayev, V.P. and Volyanick, N.V. (1986). Engineering geologic peculiarities of construction work on loessial soils. *Proc. 5th Int. Congr. Int. Assoc. Engr. Geol.*, 2, Buenos Aires, 659–665.

Anderson, M.G. and Pope, R.G. (1984). The incorporation of soil water physics models into geotechnical studies of landslide behavior. *Proc. 4th Int. Symp. on Landslides*, 4, 349–353.

Anonymous. (1980). *Low-rise Buildings on Shrinkable Clay Soils, Part 3*. Building Research Establishment, Digest 240, Her Majesty's Stationery Office, London, p. 8.

Apfel, R.C. (1970). The role of impurity in cavitation-threshold determination. *J. Acoust. Soc. Am.*, 48, 1179–1186.

Arbhabhirama, A. and Kridakorn, C. (1968). Steady downward flow to a water table. *Water Resour. Res.*, 4, 1249–1257.

ASTM (1993). *Standard Test Method for Measurement of Collapse Potential of Soils*, D5333-92. Annual Book of ASTM Standards, 04.08, 343–345, Philadelphia, PA, ASTM.

ASTM (1998a). *Test Method for Laboratory Compaction Characteristics of Soil Using Modified Effort (56 000 ft-lbf/ft³(2700 kN-m/m³))*, D1557. Annual Book of ASTM Standards, Vol. 04.08. Soil and rock (II). Philadelphia, ASTM, 126–133.

ASTM (1998b). *Test Method for Laboratory Compaction Characteristics of Soil Using Standard Effort (12 400 ft-lbf/ft³(600 kN-m/m³))*, D698. Annual Book of ASTM Standards, Vol. 04.08. Soil and rock (II). Philadelphia, ASTM, 77–84.

Atkinson, J.H. (2000). Non-linear soil stiffness in routine design. *Géotechnique*, 50, 487–508.

Atkinson, J.H. and Sallförs, G. (1991). Experimental determination of stress-strain-time characteristics in laboratory and in-situ tests. *Proc. 10th Eur. Conf. Soil Mech. and Found. Engrg.*, Balkema, Rotterdam, Vol. 3, pp. 915–956.

Averjanov, S.F. (1950). About permeability of subsurface soils in case of incomplete saturation. *Eng. Collect.*, 7, 19–21.

Bally, R. and Oltulescu, D. (1980). Settlement of deep collapsible loessial strata under structures using controlled infiltration. *Proc. of the 6th Danube Europ. Conf. Soil Mech. & Fdn. Engrg.*, Varna, 23–26.

Bao, C.G., and Ng, C.W.W. (2000). Keynote lecture: some thoughts and studies on the prediction of slope stability in expansive soils. *Proc. 1st Asian Conf. on Unsaturated Soils*, Singapore, 15–31.

Bao, C.G., Gong, B.W, and Zhan, L.T. (1998). Keynote paper: properties of unsaturated soils and slope stability for expansive soils. *Proc., 2nd Int. Conf. on Unsat. Soils*, Vol. 2, Beijing, 71–98.

Barbour, S.L. (1987). *Osmotic flow and volume change in clay soils*. PhD Thesis, Dept. Civ. Eng., Saskatoon, SK, Canada.

Barbour, S.L. (1998). Nineteenth Canadian Geotechnical Colloquium: The soil–water characteristic curve: a historical perspective. *Can. Geotech. J.*, 35, 873–894.

Barden, L. (1965). Consolidation of compacted and unsaturated clays. *Géotechnique*, 15(3), 267–286.

Barden, L. and Pavlakis, G. (1971). Air and water permeability of compacted unsaturated cohesive soil. *J. Soil Sci.*, 22(3), 302–317.

Barden, L. and Sides, G.R. (1967). The diffusion of air through the pore water of soils. *Proc. 3rd Asian Reg. Conf. Soil Mech. Found. Eng.*, Vol. 1, pp. 135–138.

Barden, L. and Sides, G.R. (1970). Engineering behaviour and structure of compacted clay. *J. Soil Mech. Found. Div.*, ASCE, 96(SM4), 1171–1197.

Beckwith, G. (1995). Foundation design practices for collapsing soils in the Western United States. *Unsaturated Soils, Proc. 1st Int. Conf. on Unsat. Soils*, Sept. 6–8, Paris, E. E. Alonso and P. Delage, eds, Balkema Press, Rotterdam, 953–958.

Been, K. and Jefferies, M.G. (1985). A state parameter for sands. *Géotechnique*, 35, (2), 99–112.

Bell, P.G. (1992). *Engineering Properties of Soils and Rocks*. Butterworth-Heinemann Ltd, Woburn, MA, USA.

Bell, F.G. and Culshaw, M.G. (2001). Problem soils: a review from a British perspective. *Proc. Symp. on Problematic Soils*, Nottingham Trent University, Jefferson I., Murray E.J., Faragher E. and Fleming P.R., Thomas Telford eds, London, 1–35.

Bell, F.G. and Maud, R.R. (1995). Expansive clays and construction, especially of low rise structures: a view point from Natal, South Africa. *Environ. and Engrg. Geoscience*, 1, 41–59.

Bellotti, R., Jamiolkowski, M., Lo Presti, D.C.F. and O'Neill, D.A. (1996). Anisotropy of small strain stiffness in Ticino sand. *Géotechnique*, 46, (1), 115–131.

Benson, C.H. and Daniel, D.E. (1990). Influence of clods on hydraulic conductivity of compacted clay. *J. Geotech. Engrg.*, ASCE, 116(8), 1231–1248.

Benson, C.H and Gribb, M.M. (1997). Measuring unsaturated hydraulic conductivity in the laboratory and field. *Unsaturated Soil Engineering Practice, ASCE Geotech. Spec. Pub. No. 68*, ASCE New York, N.Y, 113–168.

Bishop, A.W. (1959). The principle of effective stress. Lecture delivered in Oslo, Norway, 1955; *Technisk Ukeblad*, 106(39), 859–863.

Bishop, A.W., and Blight, G.E. (1963). Some aspects of effective stress in saturated and unsaturated soils. *Géotechnique*, 13(3), 177–197.

Bishop, A.W. and Donald, I.B. (1961). The experimental study of partly saturated soil in the triaxial apparatus. *Proc. 5th Int. Conf. on Soil Mech. Found. Eng.*, Paris, Vol. 1, 13–21.

Bishop, A.W., and Wesley, L.D. (1975). A hydraulic triaxial apparatus for controlled stress path testing. *Géotechnique*, 25(4), 657–670.

Bishop, A.W., Alpan, I., Blight, G.E. and Donald, I.B. (1960). Factors controlling the shear strength of partly saturated cohesive soils. *ASCE Res. Conf. Shear Strength of Cohesive Soils*, Univ. of Colorado, Boulder, pp. 503–532.

Blight, G.E. (1965). A study of effective stresses for volume change. *Moisture equilibria and moisture changes in soils beneath covered areas*. A Symposium in Print. G.D. Aitchison ed., Butterworths, Sydney, Australia, pp. 259–269.

Blight, G.E. (1971). Flow of air through soils. *J. Soil Mech. Found. Eng. Div.*, ASCE, 97(SM4) 607–624.

Blight, G.E. (1984). Uplift forces measured in piles in expansive clay. *5th Int. Conf. on Exp. Soils*, Adelaide, pp. 240–244.

Blight, G.E., Schwartz, K., Weber, H. and Bild, B.L. (1992). Preheaving of expansive soils by flooding-Failures and Successes. *7th Int. Conf. Exp. Soils*, Dallas, Vol. 1, pp. 131–136.

Bocking, K.A., and Fredlund, D.G. (1980). Limitations of the axis translation technique. *Proc. 4th Int. Conf. on Expansive Soils*, Denver, CO, 117–135.

Boger, M. (1998). Three-dimensional analytical solutions for unsaturated seepage problem. *J. Hydrologic Eng.*, 3(3), 193–202.

Bolton, M.D. (1986). The strength and dilatancy of sands. *Géotechnique*, 36(1), 65–78.

Brackley, I.J.A. and Sanders, P.J. (1992). In situ measurement of total natural horizontal stresses in an expansive clay. *Géotechnique*, 42(2), 443–451.

Brand, E.W. (1995). Keynote paper: slope instability in tropical areas. *Proc. 7th Int. Symp. on Landslides*, 2031–2051.

Brand, E.W. (1981). Some thoughts on rain-induced slope failures. *Proc. 10th Int. Conf. on Soil Mech. and Fdn. Engrg.*, Balkema, Rotterdam, The Netherlands, 3, 373–376.

Briggs, L.J. and McLane, J.W. (1907). *The moisture equivalents of soils*. US Department of Agriculture, Bureau of Soils Bulletin No. 45.

Brignoli, E., Gotti, M. and Stokoe, K. (1996). Measurement of shear waves in laboratory specimens by means of piezo-electric transducers. *Geotechnical Testing Journal*, ASTM, 14(9), 384–397.

British Standards Institution. (1990). *British standard methods of test for soils for civil engineering purposes. Part 2.* British Standards Institution, London.

Broadbridge, P. and White, I. (1988). Constant rate rainfall infiltration: a versatile nonlinear model, 1. Analytical solution. *Water Resour. Res.* 24(1), 145–154.

Brooks, R.H. and Corey, A.T. (1964). *Hydraulic properties of porous media.* Colorado State Univ. Hydrol. Paper, No. 3, p. 27.

Brown, R.W. and Bartos, D.L. (1982). A calibration model for screen-caged Peltier thermocouple psychrometers. *U.S. Dept. Agriculture, Research Paper* INT-293.

Brown, R.W. and Collins, J.M. (1980). A screen-caged thermocouple psychrometer and calibration chamber for measurements of plant and soil water potential. *Agron. J.*, 72, 851–854.

Buckingham, E. (1907). *Studies on the Movement of Soil Moisture.* U.S. Dept. Agr.Bur. Soils Bul.38.

Burdine, N.T. (1952). Relative permeability calculations from pore size distribution data. *Trans. AIME*, 198, 71–77.

Bureau of Indian Standards (1970). *Classification and identification of soils for general engineering purposes (IS 1498).* Bureau of Indian Standards Institute, New Delhi.

Burland, J.B. (1964). Effective stresses in partly saturated soils. Discussion of: some aspects of effective stress in saturated and partially saturated soils, G.E. Blight and A. W. Bishop ed., *Géotechnique*, 14, 65–68.

Burland, J.B. (1965). Some aspects of the mechanical behavior of partly saturated soils. *Moisture equilibria and moisture changes in soils beneath covered areas.* A Symposium in Print. G.D. Aitchison ed., Butterworths, Sydney, Australia, 270–278.

Burland, J.B. (1989). Ninth Laurits Bjerrum memorial lecture: 'Small is beautiful' – the stiffness of soils at small strains. *Can. Geotech. J.*, Ottawa, 26(4), 499–516.

Cabarkapa, Z., Cuccovillo, T. and Gunn, M. (1999). Some aspects of the pre-failure behaviour of unsaturated soil. *Proc. Conf. on Pre-failure Deform. Chars of Geomaterials*, Jamiolkowski, Lancellotta and Lo Presti eds, Balkema, Rotterdam, 1, 159–165.

Casagrande, L. (1952). Electro-osmotic stabilization of soils. *Boston Soc. Civ. Eng., Contributions to Soil Mech., 1941–1953*, 285–317.

Castro, G. (1969). *Liquefaction of Sands.* PhD thesis, Harvard Soil Mechanics Series, 81, Harvard University.

Chandler, R.J. (1972). Lias Clay: weathering processes and their effect on shear strength. *Géotechnique*, 22, 403–431.

Chandler, R.J. and Gutierrez, C.I. (1986). The filter paper method of suction measurement. *Géotechnique* 36, 265–268.

Chandler, R.J., Crilly, M.S. and Montgomery-Smith, G. (1992). A low cost method of assessing clay desiccation for low-rise buildings. *Proc. Instn. Civ. Engrs.*, 92, 82–89.

Chapman, P.J., Richards, B.E. and Trevena, D.H (1975). Monitoring the growth of tension in a liquid contained in a Berthelot tube. *J. Phys. E: Sci. Instrum.*, 8, 731–735.

Chaudhry, M.H. (1993). *Open-Channel Flow.* New York, Prentice-Hall.

Chen F.H. (1988). *Foundations on Expansive Soils.* Elsevier Science Publishers, Amsterdam, 4–18.

Cheuk C.Y. (2001). *Investigations of Soil Nails in Loose Fill Slopes.* M.Phil. thesis, The Hong Kong University of Science and Technology.

Cheuk, C.Y., Ng, C.W.W. and Sun, H.W. (2001). Numerical analysis of soil nails in loose fill slopes. *Proc. 14th Southeast Asian Geotech. Conf.* Dec., Hong Kong. 1, 725–730.

Cheuk, C.Y., Ng, C.W.W. and Sun, H.W. (2005). Numerical experiments of soil nails in loose fill slopes subjected to rainfall infiltration effects. *Computers and Geotechnics*, Elsevier, 32, 290–303.

Chiang, Y.C. and Ho, Y.M. (1980). Pressuremeter method for foundation design in Hong Kong. *Proc. 6th Southeast Asian Conf. on Soil Engrg.*, Taipei; 1, 31–42.

Childs, E.C. (1969). *An Introduction to the Physical Basis of Soil Water Phenomena.* Wiley-Interscience, London, p. 493.

Childs, E.C. and Collis-George, N. (1950). The permeability of porous materials. *Proc. Royal Soc.*, 201A, 392–405.

Ching, R.K.H. and Fredlund, D.G. (1983). Some difficulties associated with the limit equilibrium method of slices. *Can. Geotech. J.*, 20(4), 661–672.

Ching, R.K.H. and Fredlund, D.G. (1984). A Small Saskatchewan copes with swelling clay problems. *Proc. 5th Int. Conf. Exp. Soils*, Adelaide, South Australia, 306–310.

Chiu, C.F. (2001). *Behavior of Unsaturated Loosely Compacted Weathered Materials.* PhD Thesis, Hong Kong University of Science and Technology, Kowloon, Hong Kong.

Chiu, C.F. and Ng, C.W.W. (2003). A state-dependent elasto-plastic model for saturated and unsaturated soils. *Géotechnique*, 53(9), 809–829.

Chiu, C.F., Ng, C.W.W., and Shen, C.K. (1998). Collapse behavior of loosely compacted virgin and non-virgin fills in Hong Kong. *Proc. 2nd Int. Conf. on Unsat. Soils*, International Academic Publishers, Beijing, 1, 25–30.

Chowdhury, R.N. (1980). A reassessment of limit equilibrium concepts in geotechnique. *Proc. Symp. Limit Equilibrium, Plasticity, and Generalized Stress Strain Applications in Geotech. Eng.*, ASCE, Annu. Conv., Hollywood, FL.

Chu, T.Y., and Mou, C.H. (1973). Volume change characteristics of expansive soils determined by controlled suction test. *Proc. 3rd Int. Con. on Expansive Soils*, Haifa, 1, pp. 177–185.

Cintra, J.C.A., Nogueira, J.B., and Filho, F.C. (1986). Shallow foundations on collapsible soils. *Proc. 5th Int. Congr. Int. Assoc. Engr. Geol.*, Buenos Aires, 2, 673–675.

Clayton, C.R.I. and Khatrush, S.A. (1986). A new device for measuring local axial strains on triaxial specimens. *Géotechnique*, 36(4), 593–597.

Clayton, C.R.I., Khatrush, S.A., Bica, A.V.D., and Siddique, A. (1989). The use of Hall effect semiconductors in Geotechnical engineering. *Geotech. Testing J.*, ASTM, 12(1), 69–76.

Clemence, S. and Finbarr, A. (1981) Design considerations for collapsible soils. *J. Geotech. Engr.*, ASCE, 107(3), 305–318.

Clevenger, W.A. (1958). Experience with loess as foundation material. *J. Soil Mech. & Fdns. Div.*, ASCE, 85, 151–180.

Coleman, J.D. (1962). Stress/strain relations for partly saturated soils. Correspondance in *Géotechnique*, 12(4), 348–350.

Corey, A.T. (1954). The interrelation between gas and oil relative permeabilities. *Producers Monthly*, 19(1), 38–41.

Corey, A.T. (1957). Measurement of water and air permeability in unsaturated soil. *Proc. Soil Sci. Amer.*, 21(1), 7–10.

Corey, A.T. (1977). *Mechanics of Heterogeneous Fluids in Porous Media.* Water Resources Publications, Fort Collins, CO, p. 259.

Corey, A.T. and Kemper, W.D. (1961). Concept of total potential in water and its limitations. *Soil Sci.*, 91(5), 299–305.

Cothren, M. (1984). Roadbed stabilization by lime slurry injection process. *5th Int. Conf. on Exp. Soils*, Adelaide. pp. 250–253.

Crilly, M.S., Schreiner, H.D. and Gourley, C.S. (1991). A simple field suction probe. *Proc. 10th Reg. Conf. for Africa on Soil Mech. & Fdn. Engrg.*, Maseru. 291–298.

Croney, D., and Coleman, J.D. (1961). Pore pressure and suction in soils. *Proc. Conf. on Pore Pressure and Suction in Soils*. Butterworths, London, pp. 31–37.

Croney, D., Coleman, J.D., and Black, W.P.M. (1958). *Movement and Distribution of Water in Relation to Highway Design and Performance*. Spec. Rep. No. 40, Highway Research Board, Washington, D.C., 40, pp. 226–252.

Croney, D., Coleman, J.D. and Bridge, P.M. (1952). *The suction of moisture held in soil and other porous materials*. Department of Scientific and Industrial Research, Road Research Technical Paper No. 24. H.M. Stationery Office, London.

Cubrinovski, M. and Ishihara, K. (1998). Modelling of sand behaviour based on state concept. *Soils Foundations*, 38(3), 115–127.

Cui, Y.J and Delage, P. (1996). Yielding and plastic behaviour of an unsaturated compacted silt. *Géotechnique*, 46(2), 291–311.

Darcy, H. (1856). *Histoire Des Foundataines Publique de Dijon*. Paris, Dalmont, pp. 590–594.

Day, R.W. (1994). Swell-shrink behaviour of expansive compacted clay. *J. Geotech. Eng.*, ASCE, 120(3), 618–623.

Delage, P., Audiguier, M., Cui, Y.-J. and Howat, M.D. (1996). Microstructure of a compacted silt. *Can. Geotech. J.*, 33, 150–158.

Delage, P. and Graham, J. (1995). The mechanical behaviour of unsaturated soils. *Unsaturated Soils: Proc. 1st Int. Conf. on Unsat. Soils*; Alonso, E.E., P. Delage, P. eds, Balkema, Rotterdam, Presses des Ponts et Chaussees, Paris, 3., pp. 1223–1256.

Denisov, H.Y. (1963). About the nature and sensitivity of quick clays. *Osnov Fudarnic Mekhanic Grant*, 5, 5–8.

Derbyshire, E. and Mellors, T.W. (1988). Geological and geotechnical characteristics of some loess and loessic soils from China and Britain: a comparison. *Eng. Geol.*, 25,135–175.

Derbyshire, E., Dijkstra, T., and Smalley, I. (1995). Genesis and properties of collapsible soils. NATO ASI Series C: *Mathematical and Physical Sciences*, Kluwer Academic Publishers, The Netherlands, 468.

Dineen, K. and Burland, J. (1995). A new approach to osmotically controlled oedometer testing. *Unsaturated Soils: Proc. 1st Int. Conf. on Unsat. Soils*; Alonso, E.E., P. Delage, P. eds, Balkema, Rotterdam, Presses des Ponts et Chaussees, Paris, 2, pp. 459–465.

Dobry, R. (1991). Low and high-strain cyclic material properties. *Proc. NSF/EPRI Workshop on Dyn. Soil Properties and Site Characteristics*, Chapter 3, CH2M HILL, Bellevue, Washington.

Donald, I.B. (1956). Shear strength measurements in unsaturated non-cohesive soils with negative pore pressures. *Proc. 2nd Aust.-New Zealand Conf. on Soil Mech. & Fdn. Engrg.*, Christchurch, New Zealand, pp. 200–205.

Dorsey, N.E. (1940). *Properties of Ordinary Water-substances*. Amer. Chemical Society. Mono. Series. New York: Reinhold. p. 673.

Dregne, H.E. (1976). *Soils of arid regions.* American Elsevier Publishing Company, Inc., New York.

Driscoll, R. (1983). The influence of vegetation on the swelling and shrinkage of clay soils in Britain. *Géotechnique,* 33, 93–105.

Dudley, J.H. (1970). Review of collapsing soils. *J. Soil Mech. & Fdns. Div.,* ASCE, 96(SM3), 925–947.

Dyvik, R. and Madshus, C. (1985). Laboratory measurement of G_{max} using bender elements. Proceedings of ASCE Annual convention: *Advances in the Art of Testing Soils under Cyclic Conditions,* Detroit, pp. 186–196.

Edlefsen, N.E. and Anderson, A.B.C. (1943). Thermodynamics of soil moisture. *Hilgardia,* 15, 31–298.

El Nimr, A.I., Fayzi, S., and Hamadto, M. (1995). Regulations to control rising ground water impacts on eastern Riyadh buildings and infrastructure. *Unsaturated Soils: Proc. 1st Int. Conf. on Unsat. Soils;* Alonso, E.E., P. Delage, P. eds, Balkema, Rotterdam, Presses des Ponts et Chaussees, Paris, 967–972.

El-Ehwany, M. and Houston, S.L. (1990). Settlement and movement in collapsible soils. *J. Geotech. Engrg. Div.,* ASCE, 116(10), 1521–1535.

Elzeftawy, A. and Cartwright, K. (1981). Evaluating the saturated and unsaturated hydraulic conductivity of soils. *Permeability and groundwater contaminant transport,* ASTM STP 746, T.F. Zimmie and C.D. Riggs, eds, ASTM, pp. 168–181.

Escario, V. (1980). Suction controlled penetration and shear tests. *Proc. 4th Int. Conf. Exp. Soils,* Denver, CO, ASCE, Vol. 2, pp. 781–797.

Esteban, V. and Saez, J. (1988). A device to measure the swelling characteristics of rock samples with control of the suction up to very high values. *ISRM Symp. on Rock Mech. & Power Plants.* Madrid, 2.

Evstatiev, D. (1995). Design and treatment of loess bases in Bulgaria Genesis and properties of collapsible soils. NATO ASI Series C: *Mathematical and Physical Sciences,* Kluwer Academic Publishers, The Netherlands, 375–382.

Fawcett, R.G. and Collis-George, N. (1967). A filter paper method for determining the moisture characteristics of soil. *Australian Journal of Experimental Agriculture and Animal Husbandry* 7, 162–167.

Feda, J. (1966). Structural stability of subsidence loess from Praha-Dejvice. *Eng. Geol.,* 1, 201–219.

Feng, M. (1999). *The effects of capillary hysteresis on the measurement of matric suction using thermal conductivity sensors.* M.Sc. Thesis. Dept. of Civ. Eng., Univ. of Saskatchewan, Saskatoon, SK.

Feng, M. and Fredlund, D.G. (2003). Calibration of thermal conductivity sensors with consideration of hysteresis. *Can. Geotech. J,* 40, 1048–1055.

Fick, A. (1855). Ueber Diffusion. *Ann. Der Phys.* Leipzig, 94, 59–86.

Fisher, R.A. (1926). On the capillary forces in an ideal soil. *J. Ag. Sci.,* 16, 492–505.

Fookes, P. and Parry R. (1994). Engineering Characteristics of Arid Soils. *Proc. of the 1st Int. Symp. on Engrg Chars of Arid Soils,* July, 1993, London, P. Fookes and R. Parry, eds, Balkema.

Fookes, P.G. and Best, R. (1969). Consolidation characteristics of some late Pleistocene periglacial metastable soils of east Kent. *Q. J. Eng. Geol.,* 2, 103–128.

Fourie, A.B. (1996). Predicting rainfall-induced slope instability. *Proc. Inst. Civ. Engrs., Geotech. Eng.,* Vol. 119, pp. 211–218.

Fourie, A.B., Rowe, D. and Blight, G.E. (1999). The effect of infiltration on the stability of the slopes of a dry ash dump. *Géotechnique*, 49(1), 1–13.

Fredlund, D.G. (1964). *Comparison of Soil Suction and One Dimensional Consolidation Characteristics on a Highly Plastic Clay*. National Research Council Canada, Division of Building Research, Ottawa, ON, Technical Report No. 245.

Fredlund, D.G. (1973). *Volume Change Behaviour of Unsaturated Soils*. PhD thesis, University of Alberta, Edmonton, AB, Canada.

Fredlund, D.G. (1975). A diffused air volume indicator for unsaturated soils. *Can. Geotech. J.*, 12(4), 533–539.

Fredlund, D.G. (1980). Use of computers for slope stability analysis. State-of-the-Art Paper, *Proc. Int. Symp. Landslides*, New Delhi, India, Balkema, Rotterdam, Vol. 2, pp. 129–138.

Fredlund, D.G. (1981). Seepage in saturated soils. Panel discussion: ground water and seepage problems. *Proc. 10th Int. Conf. Soil Mech. Found. Eng.*, Stockholm, Sweden, Balkema, Rotterdam, Vol. 4, pp. 629–641.

Fredlund, D.G. (1981). The shear strength of unsaturated soils and its relationship to slope stability problems in Hong Kong, *The Hong Kong Engrg.*, 9(4) 37–45.

Fredlund, D.G. (1987). The stress state for expansive soils. *Proc. 6th Int. Conf. on Expansive Soils*, New Delhi. Keynote address, 1–9.

Fredlund, D.G. (1996). *The Emergence of Unsaturated Soil Mechanics*. The Fourth Spencer J. Buchanan Lecture, College Station, Texas, A & M University Press, p. 39.

Fredlund, D.G. (2000). The 1999 R.M. Hardy Lecture: the implementation of unsaturated soil mechanics into geotechnical engineering. *Can. Geotech. J.*, 37: 963–986.

Fredlund, D.G. (2002). Use of soil–water characteristic curves in the implementation of unsaturated soil mechanics. *Proc. 3rd Int. Conf. on Unsat. Soils*, Recife, Brazil, Balkema, Rotterdam.

Fredlund, D.G. and Krahn, J. (1977). Comparison of slope stability methods analysis. *Can. Geotech. J.*, 14(3), 429–439.

Fredlund, D.G. and Morgenstern, N.R. (1977). Stress state variables for unsaturated soils. *J. Geotech. Engrg. Div.*, ASCE, GT5, 103, 447–466.

Fredlund, D.G. and Rahardjo, H. (1993). *Soil Mechanics for Unsaturated Soils*. John Wiley & Sons Inc. New York.

Fredlund, D.G., and Xing, A. (1994). Equations for the soil–water characteristic curve. *Can. Geotech. J.*, 31(3), 521–532.

Fredlund, D.G., Morgenstern, N.R. and Widger, A. (1978). Shear strength of unsaturated soils. *Can. Geotech. J.*, 15, 313–321.

Fredlund, D.G., Rahardjo, H. and Gan, J. (1987). Nonlinearity of strength envelope for unsaturated soils. *Proc. 6th Int. Conf. Exp. Soils*, New Delhi, India, Balkema, Rotterdam, Vol. 1, pp. 49–54.

Fredlund, D.G., Xing, A., Fredlund, M.D. and Barbour, S.L. (1996). The relationship of the unsaturated soil shear strength to the soil–water characteristic curve. *Can. Geotech. J.*, 33: 440–445.

Fredlund, D.G., Shuai, F. and Feng, M. (2000). Use of a new thermal conductivity sensor for laboratory suction measurement. *Proc. 1st Asian Conf. on Unsaturated Soils*, Singapore, 275–280.

Fredlund, D.G., Rahardjo, H. Leong, E.C. and Ng, C.W.W. (2001a). Suggestions and recommendations for the interpretation of soil–water characteristic curves. *Proc. 14th Southeast Asian Geotech. Conf.*, Hong Kong. 1, pp. 503–508.

Fredlund, D.G., Fredlund, M.D. and Zakerzadeh, N. (2001b). Predicting the permeability function for unsaturated soils. *Proc. Int. Symp. On Suction, Swelling, Permeability and Structured Clays.* IS-Shizuoka, Japan. Adachi and Fukue eds, Balkema, Rotterdam, pp. 215–221.

Freeze, R.A. and Cherry, J.A. (1979). *Groundwater.* Prentice-Hall, Englewood Cliffs, NJ, p. 604.

Fukuoka, M. (1980) Landslides associated with rainfall. *Geotech. Engrg. J.*, Southeast Asia Society of Soil Engineering, 11, 1–29.

Fung, W.T. (2001). *Experimental Study and Centrifuge Modelling of Loose Fill Slope.* M.Phil. thesis, the Hong Kong University of Science and Technology.

Fung, Y.C. (1965). *Foundations of Solid Mechanics.* Prentice-Hall, Inc., Englewood Cliffs, N.J.

Fung, Y.C. (1977). *A First Course in Continuum Mechanics.* Prentice-Hall Inc., Englewood Cliffs, NJ.

Gallen, P.M. (1985). *Measurement of soil suction using the filter paper methods: a literature review; final report.* Internal Report, Transportation and Geotech. Group, Dept. of Civ. Eng., Univ. of Saskatchewan, Saskatoon, SK, Canada, p. 32.

Gallipoli, D., Gens, A., Sharma, R. and Vaunat, J. (2003). An elasto-plastic model for unsaturated soil incorporating the effects of suction and degree of saturation on mechanical behaviour. *Géotechnique*, 53, (1), 123–135.

Gan, J.K.M. (1986). *Direct shear strength testing of unsaturated soils.* M.Sc. Thesis, Dept. Civ. Eng., Univ. Of Saskatchewan, Saskatoon, SK, Canada, p. 587.

Gan, J.K.M., and Fredlund, D.G. (1992). *Direct Shear Testing of a Hong Kong Soil Under Various Applied Matric Suctions.* Geotechnical Engineering Office, Civil Engineering Department, Hong Kong, GEO Report No. 11.

Gan J.K.M., and Fredlund, D.G. (1996). Shear strength characteristics of two saprolitic soils. *Can. Geotech. J.*, 33, 595–609.

Gan, J.K.M. and Fredlund, D.G. (1997b). *Study of the Application of Soil–Water Characteristic Curves and Permeability Functions to Slope Stability.* Final report, Geotechnical Engineering Office, Civil Engineering Department, Hong Kong.

Gan, J.K.M, Fredlund, D.G. and Rahardjo, H. (1988). Determination of the shear strength parameters of an unsaturated soil using the direct shear test. *Can. Geotech. J.*, 25(3), 500–510.

Gao, G.Y. and Wu, S.M. (1995). Settlement computation and practical confirmation of belled pier in loess. *Unsaturated Soils: Proc. 1st Int. Conf. on Unsat. Soils*; Alonso, E.E., P. Delage, P. eds, Balkema, Rotterdam, Presses des Ponts et Chaussees, Paris, 973–979.

Gardner, R. (1937). A method of measuring the capillary tension of soil moisture over a wide moisture range. *Soil Science*, 43, 227–283.

Gardner, W.R. (1958a). Laboratory studies of evaporation from soil columns in the presence of a water-table. *Soil Sci. Am.*, 85, 244.

Gardner, W.R. (1958b). Some steady state solutions of the unsaturated moisture flow equation with application to evaporation from a water-table. *Soil Sci. Am.*, 85(4), 228–232.

GCO (1998). *Guide to Rock and Soil Descriptions*. Geotechnical Control Office, Public Works Department of Hong Kong.

Geiser, F. (1999). *Comportement mecanique d'um limon non sature: etude experimental et modelisation constitutive*. PhD Thesis. Swiss Federal Institute of Technology, Lausanne, Switzerland.

Geiser, F., Laloui, L., and Vulliet, L. (2000). On the volume measurement in unsaturated triaxial test. *Unsaturated Soils for Asia*. H. Rahardjo, D.G. Toll, and E.C. Leong. A.A. Balkema ed., Rotterdam, The Netherlands, pp. 669–674.

Gens, A. (1995). Constitutive modelling: application to compacted soils. *Unsaturated Soils: Proc. 1st Int. Conf. on Unsat. Soils*, Paris, Alonso, E.E., P. Delage, P. eds, Presses des Ponts et Chaussees, Balkema, Rotterdam, 3, 1179–1200.

GEO (1993). *Geoguide 1-Guide to Retaining Wall Design*. Geotechnical Engineering Office, Civil Engineering Department, Hong Kong.

GEO (1994). *Report on the Kwun Lung Lau Landslide of 23 July 1994*, 2. GEO, Civil Engineering Department, Hong Kong Government.

Geo-slope (1995). *User's Guide for SEEP/W, Version 3*. Geo-slope International Ltd, Canada.

Geo-slope (1998). *SEEP/W (Version 4) for Finite Element Seepage Analysis and SLOPE/W for Slope Stability Analysis*. Geo-slope International, Canada.

Geo-slope (2002). *SEEP/W (Ver. 5) for Finite Element Seepage Analysis and SLOPE/W (Ver. 5) for Slope Stability Analysis*. Geo-slope International, Canada.

Germaine, J.T. and Ladd, C.C. (1988). Triaxial testing of saturated cohesive soils. *Advanced Triaxial Testing of Soil and Rock*, ASTM STP 977, Donaghe, Chaney and Silver eds, ASTM, Philadelphia, 421–459.

Gibbs, H.H. and Bara, J.P. (1962). *Predicting Surface Subsidence from Basic Soil Tests*. ASTM Special Technical Publication, 322, 231–246.

Gourley, C.S. and Schreiner, H.D. (1995). Field measurement of soil suction. *Unsaturated Soils: Proc. 1st Int. Conf. on Unsat. Soils*, Paris Alonso, E.E., P. Delage, P. eds, Presses des Ponts et Chaussees. Balkema, Rotterdam.

Gourley, C.S., Newill, D. and Schreiner, H.D. (1994). Expansive soils: TRL's research strategy. *Engineering Characteristics of Arid Soils* (Fookes, P.G. and Parry, R.H.G., ed), A.A. Balkema, Rotterdam, 247–260.

Grabowska-Olszewsla, B. (1988). Engineering geological problems of loess in Poland. *Eng. Geol.*, 25, 177–199.

Greacen, E.L., Walker, G.R and Cook, P.G. (1987). Evaluation of the filter paper method for measuring soil–water suction. *Int. Conf. on Meas. of Soil and Plant Water Status*. 137–143.

Green, W.H., and Ampt, G.A. (1911). Study on soil physics: I. flow of air and water in soils. *J. Agric. Sci.*, 4, 1–24.

Grim, R.E. (1962). *Applied Clay Mineralogy*, McGraw-Hill, New York.

Gu, W.H. (1992). *Liquefaction and Post Liquefaction Deformation Analysis*. PhD thesis, University of Alberta.

Guan, Y. and Fredlund, D.G. (1997). Use of the tensile strength of water for the direct measurement of high soil suction. *Can. Geotech. J.*, 34, 604–614.

Haines, W.B. (1930). Studies in the physical properties of soils. V. The hysteresis effect in capillary properties and the modes of moisture distribution associated therewith. *J. Agr. Sci.* 20, 97–116.

Hanafy, E.A.D.E. (1991). Swelling/shrinkage characteristic curve of desiccated expansive clays. *ASTM Geotech. Test. J.*, 14(2), 206–211.

Handbook of Chemistry and Physics (1968). 49th ed. Weast, R.C. and Selby, S.M., eds, The Chemical Rubber Co., Clevland, OH.

Handy, R.L. (1973). Collapsible loess in Iowa. *J. of Soil Science*, 37, 281–284.

Handy, R.L. (1995). A stress path model for collapsible loess. *Genesis and Properties of Collapsible Soils.* Derbyshire *et al.* eds, Kluwer Academic Publisher, 33–47.

Harrison, L.P. (1965). Fundamental concepts and definitions relating to humidity. *Humidity and moisture, fundamentals and standards.* A. Wexler and W.A. Wildhack, eds, Reinhold, New York, 3, pp. 3–69.

Harrison, B.A. and Blight, G.E. (2000). A comparison of in situ soil suction measurements. *Proc. Conf. on Unsaturated Soils for Asia*, Singapore. eds, Rahardjo, Toll and Leong, Balkema, Rotterdam, pp. 281–284.

Harvey, E.N., Barnes, D.K., McElroy, A.H., Whiteley, A.H., Pease, D.C. and Cooper, K.W. (1944). Bubble formation in animals, I. Physical factors. *Journal of Cellular and Comparative Physiology*, 24, No.1, August.

Harvey, E.N., McElroy, W.D. and Whiteley, A.H. (1947). On cavity formation in water. *J. Appl. Phys.*, 18, 162–172.

Hassler, G.L. and Brunner, E. (1945). Measurement of capillary pressures in small core samples. *Trans. Am. Inst. Mining Metall. Engrs.*, 160, 114–123.

Hayward, A.T.J. (1970). New law for liquids: don't snap, stretch. *New Scient.*, 196–199.

Henderson, S.J. and Speedy, R.J. (1980). A Berthelot-Bourdon tube method for studying water under tension. *J. Phys. E: Sci. Instrum.*, 13, 778–782.

Hilf, J.W. (1956). *An Investigation of Pore-Water Pressure in Compacted Cohesive Soils.* PhD dissertation, Tech. Memo. No. 654, U.S. Dept. of the Interior, Bureau of Reclamation, Design and Construction Div., Denver, CO, p. 654.

Hillel, D. (1982). *Introduction to Soil Physics.* Academic Press, New York, p. 364.

Hillel, D. (1998). *Introduction to Environmental Soil Physics.* Academic Press, San Diego, CA, USA, p. 364.

Hillel, D. and Mottes, J. (1966). Effect of plate impedance, wetting method and aging on soil moisture retention. *J. of Soil Science*, 102, 135–140.

Ho, D.Y.F., and Fredlund, D.G. (1982). A multistage triaxial test for unsaturated soils. *Geotech. Testing J.*, ASTM, 5, 18–25.

Ho, D.Y.F., Fredlund, D.G. and Rahardjo, H. (1992). Volume change indices during loading and unloading of an unsaturated soil. *Can. Geotech. J.*, 29, 195–207.

Holtz, W.G. and Gibbs, G.J. (1956). Engineering properties of expansive clays. *Trans. Am. Soc. Civ. Engrs*, 121, 641–677.

Holtz, W. and Hilf, J. (1961). Settlement of soil foundations due to saturation. *Proc. 5th Int. Conf. Soil Mech. Fdn. Engrg.*, Paris, DUNOD, pp. 673–679.

Hong Kong Government (1977). *Report on the Slope Failures at Sau Mau Ping, August 1976.* Hong Kong Government Printer, Hong Kong.

Hong Kong Institute of Engineers (HKIE) (1998). *Soil nails in Loose Fill–A Preliminary Study. First draft.* The Geotechnical Division of Hong Kong Institution of Engineers.

Houlsby, G.T. (1997). The work input to an unsaturated granular material. *Géotechnique*, 47(1), 193–196.

Houston, S. and EI-Ehwany, M. (1991). Sample disturbance of cemented collapsible soils. *J. Geotech. Engrg. Div.*, ASCE, 117(5), 731–752.

Houston, S. and Houston, W. (1989). State-of-the-practice mitigation measures for collapsible soil sites. *Proc. of the Fdn. Engrg. Cong.*, June 25–29, ASCE, Evanston, M., ed., pp. 161–175.

Houston, S., Houston, W. and Spadola, D. (1988). Prediction of field collapse of soils due to wetting. *J. Geotech. Engrg. Div.*, ASCE, 114(1), 40–58.

Houston, S.L. (1995). Foundations and pavements on unsaturated soils–Part one: Collapsible soils. *Unsaturated Soils: Proc. 1st Int. Conf. on Unsat. Soils*, Paris, Alonso, E.E., P. Delage, P. eds, Balkema, Rotterdam, Presses des Ponts et Chaussees, pp. 1421–1439.

Houston, S.L., Houston, W.N. and Wagner, A.M. (1994). Laboratory filter paper suction measurements. *Geotech. Testing J.*, ASTM, 17(2), 185–194.

Houston, W., Mahmoud, H., and Houston, S. (1993). A laboratory procedure for partial wetting collapse determination. *Unsaturated Soils*. Special Geotech. Publication No. 39, Houston and Wray, eds, ASCE, 54–63.

Houston, W.N., and Houston, S.L. (1995). Infiltration studies for unsaturated soils. *Unsaturated Soils: Proc. 1st Int. Conf. on Unsat. Soils*, Paris, Alonso, E.E., P. Delage, P. eds, Balkema, Rotterdam, Presses des Ponts et Chaussees, pp. 869–875.

Huang, S.Y., Barbour, S.L. and Fredlund, D.G. (1998). Development and verification of a coefficient of permeability for a deformable, unsaturated soil. *Can. Geotech. J.*, 35(3), 411–425.

Hubbert, M.K. (1940). The theory of groundwater motion. *J. Geol.*, 48, 785–944.

Hubbert, M.K. (1956). Darcy's law and the field equations of the flow of underground fluids. *Am. Inst. Min. Met. Petl. Eng. Trans.*, 207, 222–239.

Irmay, S. (1954). On the Hydraulic Conductivity of Unsaturated Soils. *Trans. Am. Geophys. Union*, 35, 463–467.

Ishihara, K. (1993). Liquefaction and flow failure during earthquakes. *Géotechnique*, 43(3), 349–416.

Itasca (1999). *FLAC – Fast Langrangian Analysis of Continua*, Version 3.4. User Manual, 1–4.

Iwasaki, T., Tatsuoka, F., and Takagi, Y. (1978). Shear moduli of sands under cyclic torsional shear loading. *Soil and Foundations*, 18(1), 39–56.

Jaky, J. (1944). The coefficient of earth pressure at rest. *J. Soc. Hungarian Archs. Engrg.*, 355–358.

Jamiolkowski, M., Lancellotta, R. and Lo Presti, D.C.F. (1995). Remarks on the stiffness at small strains of six Italian clays. *Procs. Int. Symp. on Pre-failure Deformation of Geomaterials*, S. Shibuya, T. Mitachi and S. Muira eds, Balkema 2, pp. 817–836.

Janssen, D.J. and Dempsey, B.J. (1980). *Soil-moisture properties of subgrade soils*. Presented at the 60th Annu. Transportation Res. Board Meeting, Washington, DC.

Japanese Society of Soil Mechanics and Foundation Engineering (1982). *Soil Testing Methods*. 2nd revised edition (in Japanese), Tokyo, JSSMFE.

Jardine, R.J., Gens, A., Hight, D.W. and Coop, M.R. (2004). Developments in understanding soil behaviour. Keynote paper. Advances in Geotechnical Engineering. *Proc. Skempton Memorial Conference*. Pub Thomas Telford, London, pp. 103–207.

Jefferson, I., Tye, C. and Northmore, K.J. (2001). Behaviour of silt: the engineering characteristics of loess in the UK. *Proc. Symp. on Problematic soils*, Nottingham Trent University, Thomas Telford, London, pp. 37–52.

Jennings, J. and Knight, K. (1956). Recent experiences with the consolidation test as a means of identifying conditions of heaving or collapse of foundations on partially saturated soils. *Trans. South African Inst. of Civ. Engrs.*, Aug., 255–256.

Jennings, J.E. (1960). A revised effective stress law for use in the prediction of the behaviour of unsaturated soils, pore pressure and suction in soils. *Proc. Conf. on Pore Pressure and Suction in Soils*. Butterworths, London, pp. 26–30.

Jennings, J.E. and Burland, J.B. (1962). Limitations to the use of effective stresses in partly saturated soils. *Géotechnique*, 12(2), 125–144.

Jennings, J.E. and Knight, K. (1957). The prediction of total heave from the double oedometer test. *Trans. South African Inst. of Civ. Engrs.*, 7, 285–291.

Jennings, J.E. and Knight, K. (1975). A guide to construction on or with materials exhibiting additional settlement due to collapse of grain structure. *Proc. 6th African Conf. Soil Mech. & Fdn. Engrg.*, Durban, pp. 99–105.

Jiménez-Salas, J.A. (1995). Foundations and pavements on unsaturated soils–Part two: Expansive clays. *Unsaturated Soils: Proc. 1st Int. Conf. on Unsat. Soils*, Paris, Alonso, E.E., P. Delage, P. eds, Balkema, Rotterdam, Presses des Ponts et Chaussees, pp. 1441–1464 .

Jones, W.M., Overton, G.D.N. and Trevena, D.H. (1981). Tensile strength experiments with water using a new type of Berthelot tube. *J. Phys. D: Appl. Phys.*, 14, 1283–1291.

Juca, J.F.T. and Frydman, S. (1995). Experimental techniques. *Unsaturated Soils: Proc. 1st Int. Conf. on Unsat. Soils*, Paris, Alonso, E.E., P. Delage, P. eds, Balkema, Rotterdam, Presses des Ponts et Chaussees, 3, 1257–1292.

Kasim, F.B., Fredlund, D.G., and Gan, J.K.M. (1998). Effect of steady state rainfall on long term matric suction conditions in slopes. *Proc. 2nd Int. Conf. on Unsat. Soils*, Beijing, 1, pp. 78–83.

Kassif, G., and Ben Shalom, A. (1971). Experimental relationship between swell pressure and suction. *Géotechnique*, 21(3), 245–255.

Katti, R.K. (1987). Cohesion approach to mechanics of saturated expansive soil media. *6th Int. Conf. on Exp. Soils*, New Delhi. 2, pp. 536–578.

Kaye, G.W.C. and Laby, T.H. (1973). *Tables of Physical and Chemical Constants*. 14th ed. Longman, London, p. 386.

Khanzode, R.M., Vanapalli, S.K. and Fredlund, D.G. (2002). Measurement of soil–water characteristic curves for fine-grained soils using a small-scale centrifuge. *Can. Geotech. J*, 39, 1209–1217.

Kjærnsli, B. and Simons, N.E. (1962). Stability Investigations of the north bank of the Drammen River. *Géotechnique*, 12, 147–167.

Knox, D.P., Stokoe, K.H., II and Kopperman, S.E. (1982). *Effect of State of Stress in Velocity of Low Amplitude Shear Waves Propagating along Principal Stress Directions in Dry Sand*. Report GR 82-23, University of Texas at Austin.

Komornik, A., Livneh, M., and Smucha, S. (1980). Shear strength and swelling of clays under suction. *Proc. 4th Int. Con. on Expansive Soil*, Denver, CO, 1, pp. 206–266.

Koor, N.P. and Campbell, S.D.G. (1998). *Geological Characterization of the Lai Ping Road landslide*, Geological report No. 3/98. Geotechnical Engineering Office, Hong Kong.

Koorevaar, P., Menelik, G. and Dirksen, C. (1983). *Elements of Soil Physics.* Elsevier, Amsterdam, The Netherlands, p. 228.

Krahn, J. and Fredlund, D.G. (1972). On total, matric and osmotic suction. *J. of Soil Science of America*, 114(5), 339–348.

Kunhel, R.A. and van der Gaast, S.J. (1993). Humidity-controlled diffractometry and its applications. *Advances in X-Ray Analysis*, 36, 439–449.

Kunze, RJ., Uehard, G. and Graham, K. (1968). Factors important in the calculation of hydraulic conductivity. *J. of Soil Science*, 32, 760–765.

Kwok, Y.F. (2003). *Stochastic analysis of coupled surface and subsurface flow model in steep slopes for slope stability analysis*. M.Phil Thesis. Dept. Civ. Eng., Hong Kong University of Science and Technology.

Lacasse, S. and Berre, T. (1988). Triaxial testing methods for soils. *Advanced Triaxial Testing of Soil and Rock*, ASTM STP 977, Donaghe, Chaney, and Silver, eds, ASTM, Philadelphia, pp. 264–289.

Ladd, C.C. (1960). Mechanisms of swelling in compacted clay. *Highway Res. Board Bull.*, Nat. Research Council, No. 245, 10–26.

Ladd, R.S. (1978). Preparing test specimens using undercompaction. *Geotech. Testing J.*, ASTM, 1, pp. 16–23.

Lade, P.V. and Yamamuro, J.A. (1997). Effects of nonplastic fines on static liquefaction of sands. *Can. Geotech. J.*, 34, 918–928.

Lagerwerff, J.V., Ogata, G. and Eagle, H.E. (1961). Control of osmotic pressure of culture solutions with polyethylene glycol. *Science*, 133, 1486–1487.

Lam C.C. and Leung Y.K. (1995). *Extreme Rainfall Statistics and Design Rainstorm Profiles at Selected Locations in Hong Kong*. Technical Note No. 86. Royal Observatory, Hong Kong.

Lam L., Fredlund D.G. and Barbour S.L. (1987). Transient seepage model for saturated-unsaturated soil systems: a geotechnical engineering approach. *Can. Geotech. J.*, 24, 565–580.

Lambe, T.W. (1958). The engineering behavior of compacted clay. *J. Soil Mech. Found Div.*, ASCE, 84(SM2), Paper No. 1655, 1–35.

Lambe, T.W. (1958). The structure of compacted clay. *J. Soil Mech. Found. Eng. Div.*, ASCE, 84(2), 1–34.

Lambe, T.W. (1960). A mechanistic picture of shear strength in clay. *Proc. ASCE Res. Conf. on Shear Strength of Cohesive Soils*, University of Colorado, Boulder, CO, pp. 555–580.

Lambe, T.W. and Whitman, R.V. (1979). *Soil Mechanics*. John Wiley and Sons, New York, p. 553.

Langfelder, L.J., Chen, C.F. and Justice, J.A. (1968). Air permeability of compacted cohesive soils. *J. Soil Mech. Found. Eng. Div.*, ASCE, 94, (SM4), 981–1001.

Lawton, E., Fragaszy, R., and Hardcastle, J. (1991). Stress ratio effects on collapse of compacted clayey sand. *Proc. ASCE Geotech. Engrg. Div.*, 117(5), 714–730.

Leach, B., and Herbert, R. (1982). The genesis of a numerical model for the study of the hydrology of a steep hillside in Hong Kong. *Q. J. Eng. Geol.*, 15, 243–259.

Lee, H.C. and Wray, W.K. (1992). Evaluation of soil suction instruments. *Proc. 7th Int. Conf. on Expansive Soils*, Dallas, Texas, 307–312.

Lee, I.K., and Coop, M.R. (1995). The intrinsic behavior of a decomposed granite soil. *Géotechnique*, London, 45(1), 117–130.

Leong, E.C. and Rahardjo, H. (1997). Review of soil-water characteristic curve functions. *J. Geotech. Geoenvir. Eng.*, ASCE, 123(12), 1106–1117.

Leung, K.L. (1998). *Identifying Representative Storm Profiles in Hong Kong Territory*. Final Year Project of Dept. of Civil Engineering, Hong Kong University of Science and Technology, Hong Kong.

Li, J., Smith, D.W., Fityus, S.G. and Sheng, D.C. (2002). Quantitative analysis of moisture content determination in expansive soils using neutron probes. *Proc. 3rd Int. Conf. on Unsaturated Soils*, Recife, Brazil, 1, 363–368.

Li, X.S. (2003). Technical note: effective stress in unsaturated soil: a microstructural analysis. *Géotechnique*, 53(2), 273–278.

Li, X.S. and Dafalias, Y.F. (2000). Dilatancy for cohesionless soils. *Géotechnique*, 50(4), 449–460.

Li, X.S., Chan, C.K. and Shen, C.K. (1988). An automatic triaxial testing system. *Advanced Triaxial Testing of Soil and Rock*, ASTM SPT977, 95–106.

Liakopoulos, A.C. (1965). Retention and distribution of moisture in soils after infiltration has ceased. *Bull. Int. Assoc. Sci. Hydrol.*, 10, 58–69.

Likos, W.J. and Lu, N. (2003). Automated humidity system for measuring total suction characteristics of clay. *Geotech. Testing J.*, ASTM, 26(2), 1–12.

Lim T.T., Rahardjo H., Chang M.F. and Fredlund D.G. (1996). Effect of rainfall on matric suctions in residual soil slope. *Can. Geotech. J.*, 33, 618–628.

Lin, Z. (1995). Variation in collapsibility and strength of loess with age. *Genesis and properties of collapsible soils*. NATO ASI Series C: Mathematical and Physical Sciences, 468, Kluwer Academic Publishers, The Netherlands, pp. 247–265.

Lin, H.C., Richards, D.R., Talbot, C.A., Yeh, G.T., Cheng, J.R., Cheng, H.P. and Jones, N.L. (1997). *FEMWATER: A Three-Dimensional Finite Element Computer Model for Simulation Density-Dependent Flow and Transport in Variably Saturated Media*. Technical report CHL-97-12, Coastal and Hydraulics Laboratory, Waterway Experimental Station, Mississippi.

Lin, Z.G. and Wang, S.J. (1988). Collapsibility and deformation characteristics of deep-seated loess in China. *Eng. Geol.*, 25, 271–282.

Liu, T. (1988). *Loess in China*. Springer-Verlag, Berlin.

Liu, T.H. (1997). *Problems of Expansive Soils in Engineering Construction*. Architecture and Building Press of China, Beijing (in Chinese).

Lloret, A. and Alonso, E.E. (1980). Consolidation of unsaturated soils including swelling and collapse behaviour. *Géotechnique*, 30(4), 449–477.

Lloret. A., Villar, M.V., Sanchez, M., Gens, A., Pintado, X. and Alonso, E.E. (2003). Mechanical behaviour of heavily compacted bentonite under high suction changes. *Géotechnique*, 53(1), 27–40.

Lo Presti, D.C.F. (1989). Proprietà Dinamiche dei Terreni. XIV Conferenza Geotecnica di Torino, Department of Structural Engineering, Politecnico di Torino.

Lo Presti, D.C.F., Jamiolkowski, M., Pallara, O., Cavallaro, A. and Pedroni, S. (1997). Shear modulus and damping of soils. *Géotechnique*, 47(3), 603–618.

Lu, N. and Likos, W.J. (2004). *Unsaturated Soil Mechanics*. John Wiley & Sons, New York, p. 584.

Lumb, P. (1975). Slope failures in Hong Kong. *Q. J. Engrg. Geol.*, 8, 31–65.

Lumb, P.B. (1962). Effects of rain storms on slope stability. *Proc.Symp. on Hong Kong Soils*, pp. 73–87.

Lutenegger, A. (1986). Dynamic compaction in friable loess. *J. Geotech. Engrg. Div.*, ASCE, 112(6), 663–667.

Lutenegger, A.J. and Hallberg, G.R. (1988). Stability of loess. *Eng. Geol.*, 25, 247–261.

Maatouk, A., Leroueil, S. and La Rochelle, P. (1995). Yielding and critical state of a collapsible unsaturated silty soil. *Géotechnique*, 45(3), 465–477.

Mahmoud, H. (1992). *Development of a Down-Hole Collapse Test System*. PhD dissertation, Department of Civil Engr., Arizona State University, Tempe, AZ.

Mair, R.J. (1993). Developments in geotechnical engineering research: application to tunnels and deep excavations. *Proc. Inst. Civ. Engrs.*, London, Vol. 93, pp. 27–41.

Maksimovic, M. (1979). Limit equilibrium for nonlinear failure envelope and arbitrary slip surface. *Proc. 3rd Int. Conf. Numer. Methods in Geomech.*, Aachen, Germany, pp. 769–777.

Malone, A.W. and Pun, W.K. (1997). New engineering tools for landslip risk control. *Proc. 2nd Int. Symp. on Structures and Fdns. in Civ. Engrg, Hong Kong*, Jan. 1–27. C.K. Shen, J.S. Kuang and C.W.W. Ng ed., Hong Kong University of Science and Technology.

Mancuso, C., Vassallo, R. and d'Onofrio, A. (2000). Soil behaviour in suction controlled cyclic and dynamic torsional shear tests. *Proc. Conf. on Unsaturated Soils for Asia*, Rahardjo, Toll and Leong eds, Blakema, Rotterdam, 539–544.

Mancuso, C., Vassallo, R. and d'Onofrio, A. (2002). Small strain behaviour of a silty sand in controlled-suction resonant column-torsional shear tests. *Can. Geotech. J*, 39, 22–31.

Manheim, F.T. (1966). *A hydraulic squeezer for obtaining interstitial water from consolidated and unconsolidated sediment*. U.S. Geological Survey Prof. Paper 550-C, 256–261.

Manzari, M.T. and Dafalias, Y.F. (1997). A critical state two-surface plasticity model for sands. *Géotechnique*, 47(2), 255–272.

Mao, S.Z., Cui, Y.J., and Ng, C.W.W. (2002). Slope stability analysis for a water diversion canal in China. *Proc. 3rd Int. Conf. on Unsat. Soils*, Recife, Brazil, 2, pp. 805–810.

Marinho, E.A.M., Chandler, R.J. and Crilly, M.S. (1995). Stiffness measurements on a high plasticity clay using bender elements. *Unsaturated Soils: Proc. 1st Int. Conf. on Unsat. Soils*, Paris, Alonso, E.E., P. Delage, P. eds, Balkema, Rotterdam, Presses des Ponts et Chaussees, 1, pp. 535–539.

Marinho, F.A.M. (1994). *Shrinkage Behaviour of some Plastic Soils*. PhD thesis, University of London.

Marshall, T.I. (1958). A relation between permeability and the size distribution of pores. *J. of Soil Science*, 9, 1–8.

Matyas, E.L. (1967). Air and water permeability of compacted soils. *Permeability and capillarity of soils*, ASTM STP 417, 160–175.

Maulem, Y. (1986). Hydraulic conductivity of unsaturated soils: prediction and formulas. *Method of soils analysis. Part 1. Physical and mineralogical methods*. A. Klute, ed., American Society of Agronomy, Madison, WI, 799–823.

McKee. C.R. and Bumb. A.C. (1984). The importance of unsaturated flow parameters in designing a monitoring system for hazardous wastes and environmental emergencies. *Hazardous Materials Control Research Institute National Conference*, Houston, TX, pp. 50–58.

McKeen, R.G. (1985). *Validation of procedures for pavement design on expansive soils*. Final Report, US Dept. of Transportation, Washington, DC.

McQueen, I.S. and Miller R.F. (1968). Calibration and evaluation of a wide-range gravimetric method for measuring moisture stress. *J. of Soil Science*, 106(3), 225–231.

Meigs, P. (1953). World distribution of arid and semi-arid homoclimates. In UNESCO, *Reviews of Res. on Arid Zone Hydrology, Arid Zone Res.* 203–210.

Mein, R.G. and Larson, C.L. (1973). Modelling infiltration during a steady rain. *Water Resour. Res.*, 9(2), 384–394.

Michels, W.C. (1961). *The International Dictionary of Physics and Electronics*. Second edition, Van Nostrand Co., Princeton, New Jersey.

Millington, R.J. and Quirk, J.P. (1959). Permeability of porous media. *Nature*, 183, 387–388.

Molenkamp, M.R. and Nazemi, A.H. (2003). Micromechanical considerations of unsaturated pyramidal packing. *Géotechnique*, 53(2), 195–206.

Moore, R.E. (1939). Water conduction from shallow water tables. *Hilgardia*, 12(6), 383–426.

Morgenstern, N.R. (1979). Properties of compacted soils. *Proc. 6th Pan-Am. Conf. on Soil Mech. & Fdn. Engrg.* Contribution to panel discussion Session IV, Lima, Peru, 3, pp. 349–354.

Morgenstern, N.R. and Balasubramanian, B. (1980). Effect of pore fluid on the swelling of clay-shale. *4th Int. Conf. on Expansive Soils*, Denver, Vol. 1, pp. 190–205.

Morgenstern, N.R. and Price, V.E. (1965). The Analysis of the stability of general slip surfaces. *Géotechnique*, 15, 79–93.

Morgenstern, N.R. and Price, V.E. (1967). A numerical method for solving the equation of general slip surfaces. *Br. Comput. J.*, 9, 338–393.

Morita, M. and Yen, B.C. (2002). Modelling of conjunctive two-dimensional surface-three-dimensional subsurface flows. *J. Hyd. Eng.*, ASCE, 128(2), 184–201.

Morrison, R. and Szecsody, J. (1985). Sleeve and casing lysimeters for soil pore water sampling. *J. of Soil Science*, 139, 446–451.

Mualem, Y. (1976). A new model for predicting the hydraulic conductivity of unsaturated porous media. *Water Resour. Res.*, 12, 513–522.

Mualem, Y. (1984). A modified dependent domain theory of hysteresis. *Water Resour. Res*, 12, 513–522.

Mualem, Y., and Miller, E.E. (1979). A hysteresis model based on an explicit domain-dependent function. *J. of Soil Science*, 43, 106–110.

Nelson, J.D. and Miller, D.J. (1992). *Expansive Soils: Problems and Practice in Foundation and Pavement Engineering*. New York: Wiley Interscience.

Ng, C.W.W. (2006). Stress-state Controllable Triaxial Double-Wall Volumetric Pressure Plate Extractor. US Provisional Patent Application (US60/841,616).

Ng, C.W.W. and Chen, R. (2005). Advanced suction control techniques for testing unsaturated soils (in Chinese). Keynote lecture, *2nd Nat. Conf. on Unsat. Soils*, Hangzhou, China.

Ng, C.W.W. and Chen, R. (2006). Advanced suction control techniques for testing unsaturated soils (in Chinese). *Chinese J. of Geotech. Eng.*, 28(2), 123–128.

Ng, C.W.W. and Chiu, C.F. (2001). Behaviour of a loosely compacted unsaturated volcanic soil. *J. of Geotech. and Geoenviron. Engrg.*, ASCE. 127(12), 1027–1036.

Ng, C.W.W and Chiu, C.F. (2003a). Laboratory study of loose saturated and unsaturated decomposed granitic soil. *J. of Geotech. and Geoenviron. Engrg.*, ASCE. 129(6), 550–559.

Ng, C.W.W. and Chiu, C.F. (2003b) Invited keynote paper. Constitutive modelling of unsaturated loose soil slopes subjected to rainfall infiltration. *Proc. 9th China National Conf. on Soil Mech. and Geotech. Engrg.*, Beijing. 1, pp. 187–200.

Ng., C.W.W. and Lai, J.C.H. (2004). Effects of state-dependent soil–water characteristic and damming on slope stability. *Proc. 57th Can. Geotech. Conf., Géo Québec 2004*, Session 5E, 28–35.

Ng, C.W.W. and Leung, E.H.Y. (2006). Invited paper: Small-strain stiffness of granitic and volcanic saprolites in Hong Kong. *Int. Workshop on Natural Soil 2006*, Singapore, Taylor & Francis, Vol. 4, pp. 2507–2538.

Ng, C.W.W. and Lings, M.L. (1995). Effects of modelling soil nonlinearity and wall installation on back-analysis of deep excavation in stiff clay. *J. Geotech. Engrg.*, ASCE, 121(10), 687–695.

Ng, C.W.W. and Pang, Y.W. (1998b). *Lai Ping Road Landslide Investigation-Specialist Testing of Unsaturated soils*. Tech. Rep., Geotechnical Engineering Office of the Hong Kong Special Administrative Region, Hong Kong.

Ng, C.W.W. and Pang, Y.W. (1999). Stress effects on soil-water characteristics arid pore-water pressures in unsaturated soil slopes. *Proc. 11th Asian Reg. Conf. on Soil Mech. and Geotech. Engrg.*, Korea, Vol. 1, pp. 371–374.

Ng, C.W.W. and Pang, Y.W. (2000a). Experimental investigation of soil–water characteristics of a volcanic soil. *Can. Geotech. J.*, 37(6), 1252–1264.

Ng, C.W.W. and Pang, Y.W. (2000b). Influence of stress state on soil–water characteristics and slope stability. *J. Geotech. and Geoenviron. Eng.*, ASCE. 26(2), ASCE, 157–166.

Ng, C.W.W. and Shi, Q. (1998). A numerical investigation of the stability of unsaturated soil slopes subjected to transient seepage. *Computers and Geotechnics.* 22(1), 1–28.

Ng, C.W.W. and Wang, Y. (2001). Field and laboratory measurements of small strain stiffness of decomposed granites. *Soils and Foundations*, 41(3), 57–71.

Ng, C.W.W. and Yung, S.Y. (2007). Determinations of anisotropic shear stiffness of an unsaturated decomposed soil. Provisionally accepted by *Géotechnique*.

Ng, C.W.W. and Zhan, L.T. (2001). Fundamentals of re-compaction of unsaturated loose fill slopes. *Proc. Int. Conf. on Landslides–causes, Impacts and Countermeasures*, June, Davos, Switzerland, pp. 557–564.

Ng, C.W.W. and Zhan, L.T. (2007). Comparative study of rainfall infiltration into a bare and a grassed unsaturated expansive soil slope. *Soils Found*, 47(2), 207–217.

Ng, C.W.W. and Zhou, R.Z.B. (2005). Effects of soil suction on dilatancy of an unsaturated soil. *Proc. 16th Int. Conf. Soil Mech. & Geotech. Engrg.* Osaka, Japan, 2, 559–562.

Ng, C.W.W., Bolton, M.D. and Dasari, G.R. (1995). The small strain stiffness of a carbonate stiff clay. *Soils Found.*, Tokyo, 35(4), 109–114.

Ng, C.W.W., Chiu, C.F. and Shen, C.K. (1998). Effects of wetting history on the volumetric deformations of an unsaturated loose fill. *Proc. 13th Southeast Asian Geotech. Conf.*, Taipei, Taiwan, ROC, 1, 141–146.

Ng, C.W.W., Chen, S.Y., and Pang, Y.W. (1999). Parametric study of the effects of rain infiltration on unsaturated slopes. *Rock Soil Mech.*, 20(1), 1–14 (in Chinese).

Ng, C.W.W., Pun, W.K. and Pang, R.P.L. (2000b). Small strain stiffness of natural granitic saprolite in Hong Kong. *J. Geotech. and Geoenviron. Eng.*, ASCE, 819–833.

Ng, C.W.W., Wang, B. and Tung, Y.K. (2001b). Three-dimensional numerical investigations of groundwater responses in an unsaturated slope subjected to various rainfall patterns. *Can. Geotech. J.*, 38, 1049–1062.

Ng, C.W.W., Zhan, L.T., and Cui, Y.J. (2002a). A new simple system for measuring volume changes in unsaturated soils. *Can. Geotech. J.*, 39, 757–764.

Ng, C.W.W., Zhan, L.T., Bao, C.G., Fredlund, D.G. and Gong, B.W. (2003). Performance of an unsaturated expansive soil slope subjected to artificial rainfall infiltration. *Géotechnique*, 53(2), 143–157.

Ng, C.W.W., Leung, E.H.Y. and Lau, C.K. (2004b). Inherent anisotropic stiffness of weathered geomaterial and its influence on ground deformations around deep excavations. *Can. Geotech. J.*, 41, 12–24.

Ng, C.W.W., Li, X.S., Van Laak, P.A. and Hou, D.Y.J. (2004c). Centrifuge modeling of loose fill embankment subjected to uni-axial and bi-axial earthquakes. *Soil Dynamics and Earthquake Engineering*, 24, 305–318.

Ng, C.W.W., Zhang, M., Pun, W.K., Shiu, Y.K. and Chang, G.W.K. (2006). Investigation of static liquefaction mechanisms in loose sand fill slopes. Submitted to *Géotechnique*. Under revision.

Ng, C.W.W., Cui, Y., Chen, R. and Delage, P. (2007). The axis-translation and osmotic techniques in shear testing of unsaturated soils: a comparison. *Soils and Foundations*. Vol. 47, No. 4, 678–684.

Nielsen, D.R. and Biggar, Y.W. (1961). Measuring capillary conductivity. *Soil Sci.*, 92, 192–193.

Nielsen, D.R., Jackson, R.D., Cary, J.W. and Evans, D.D. (1972). Soil water. *Am. Soc. Agron. Soil Sci.*, Madison, WI.

Nishimura, T. (2000). Direct shear properties of a compacted soil with known stress history. *Unsaturated Soils for Asia. Proc. Asian Conf. on Unsaturated Soils*, Rahardjo, H., Toll, D.G. and Leong, E.C., eds, Singapore, 557–562.

Nishimura, T. and Fredlund, D.G. (2000). Relationship between shear strength and matric suction in an unsaturated silty soil. *Unsaturated Soils for Asia. Proc. Asian Conf. on Unsaturated Soils*, Rahardjo, H., Toll, D.G. and Leong, E.C., eds, Singapore, pp. 563–568.

Noorany, I. (1992). Discussion: stress ratio effects on collapse of compacted clayey sand. Lawton, Gragaszy, and Hardcastle. *J. Geotech. Engrg. Div.*, ASCE, 188(9), 1472–1473.

Northmore, K.J., Bell, F.G. and Culshaw, M.G. (1996). The engineering properties and behaviour of the brickearth of south Essex. *Q. J. Engrg. Geol.*, 29, 147–161.

O'Neill, M.W. and Poorymoayed, A.M. (1980). Methodology for foundations on expansive clays. *J. Geotech. Engrg. Div.*, ASCE, 106, 1345–1367.

Olson, R.E. (1963). Effective stress theory of compaction. *Proc. Amer. Soc. Civil Eng.*, Vol. 89, pp. 27–44.

Oteo-Mazo, C., Saez-Aunon, J., Esteban, F. (1995). Laboratory tests and equipment with suction control. *Unsaturated Soils: Proc. 1st Int. Conf. on Unsat. Soils*, Paris, Alonso, E.E., Delage, P., eds, Balkema, Rotterdam, Presses des Ponts et Chaussees, 3, pp. 1509–1515.

Paddy, J.F. (1969). Theory of surface tension. *Surface and Colloid Science*, 1, Wiley Interscience, Toronto, Canada.

Pearsall, I.S. (1972). *Cavitation*, Mills and Boon Ltd, London.

Peck, A.J. and Rabbidge, R.M. (1969). Design and performance of an osmotic tensiometer for measuring capillary potential. *Proc. of Soil Science Society of America*, 33(2), pp. 196–202.

Pennington, D.S., Nash, D.F.T., and Lings, M.L. (1997). Anisotropy of G_0 shear stiffness in Gault Clay. *Géotechnique*, 47(3), 391–398.

Pennington, D.S., Nash, D.F.T., and Lings, M.L. (2001). Horizontally mounted bender elements for measuring anisotropic shear moduli in triaxial clay specimens, *Geotech. Testing J.*, ASTM, 24(2), 133–144.

Phene, C.J., Hoffman, G.J. and Rawlins, S.L. (1971). Measuring soil matric potential in situ by sensing heat dissipation within a porous body: theory and sensor construction. *Proc. Soil Sci. Soc. of Am.*, Vol. 35, pp. 27–32.

Philip, J.R. (1957). Theory of infiltration: I. the infiltration equation and its solution. *J. Soil Science*, 83, 345–357.

Philip, J.R. (1969). Theory of infiltration. *J. Soil Science*, 5, 215–296.

Pidgeon, J.T. (1980). The rational design of raft foundations for houses on heaving soil. *7th. Reg. Conf. for Africa on Soil Mech. and Found. Eng.*, Accra, Ghana.

Pitman, T.D., Robertson, P.K. and Sego, D.C. (1994). Influence of fines on the collapse of loose sands. *Can. Geotech. J.*, 31, 728–739.

Plesset, M.S. (1969). The tensile strength of liquids. *Cavitation State of Knowledge*, ASME, 15–25.

Poulos, S.J. (1964). *Control of Leakage in the Triaxial Test*. Soil Mech. Series No. 71, Massachusetts Institute of Technology, Cambridge, MA.

Poulovassilis, A. (1962). Hysteresis of pore water, an application of the concept of independent domains. *J. Soil Science*, 93, 405–412.

Pradel, D. and Raad, G. (1993). Effect of permeability on the surficial stability of homogeneous slopes. *J. Geotech. Engrg. Div.*, ASCE, 119(2), 315–332.

Premchitt, J., Lam, T.S.K., Shen, J.M. and Lam, H.F. (1992). *Rainstorm runoff on slopes*. GEO Report No. 12, Civil Engineering Department. Hong Kong.

Public Works Central Laboratory (PWCL) (1998). *Investigation of Fundamental Behaviour of Loose Fills Under Shear*. Soil Testing Report No. 723. Public Works Laboratories, Geotechnical Engineering Office, Hong Kong.

Pullan, A.J. (1990). The quasi linear approximation for unsaturated porous media flow. *Water Resour. Res.*, 26(6), 1219–1234.

Qian, X., Gray, D.H., and Woods, R.D. (1993). Voids and granulometry: effects on shear modulus of unsaturated sands. *J. Geotech. Eng.*, ASCE, 119(2), 295–314.

Rahardjo, H., and Leong, E.C. (1997). Soil–water characteristic curves and flux boundary problems. *Proc. Unsat. Soil Engrg. Practice*, ASCE Geotechnical Special Publication No. 68, S. L. Houston and D.G. Fredlund, eds, ASCE, New York, 88–112.

Rahardjo, H., Leong, E.C., Gasmo, J.M. and Deutscher, M.S. (1998). Rainfall-induced slope failures in Singapore: investigation and repairs. *Proc. 13th Southeast Asian Geotech. Conf.*, Taipei, Taiwan, ROC, 1, 147–152.

Ramaswamy, S.D. and Aziz, M.A. (1987). Membrane encapsulation to control swelling of subgrade for pavements. *6th Int. Conf. Exp. Soils*, New Delhi, Vol. 1, pp. 253–258.

Rampello, S., Silvestri, F. and Viggiani, G. (1995). The dependence of G_0 on stress state and history. *Proc. 1st Int. Symp. on Pre-failure Deformation of Geomaterials*, Sapporo, Japan, Vol. 2, pp. 1155–1160.

Rampello, S., Viggiani, G.M.B. and Amorosi, A. (1997). Small strain stiffness of reconstituted clay compressed along constant triaxial effective stress ratio paths. *Géotechnique*, 47(3), 475–489.

Rawlins, S.L. and Dalton, F.N. (1967). Psychometric measurement of soil–water potential without precise temperature control. *J. Soil Science*, 31, 297–301.

Richards, B.E. and Trevena, D.H. (1976). The measurement of positive and negative pressures in a liquid contained in a Berthelot tube. *J. of Physics D: Applied Physics*, 9, 1123–1126.

Richards, B.G. (1965). Measurement of the free energy of soil moisture by the psychrometric technique using thermistors. *Moisture Equilibria and Moisture Changes in Soils Beneath Covered Areas*. Butterworth, Australia, 39–46.

Richards, B.G. (1966). The significance of moisture flow and equilibria in unsaturated soils in relation to the design of engineering structures built on shallow foundations in Australia. *Symposium on Permeability and Capillarity*, American Society for Testing and Materials, Atlantic City, N.J.

Richards, L.A. (1931). Capillary conduction of liquids through porous mediums. *Physics*, NY, 1, 318–333.

Richardson, D. (1988). *Investigations of threshold effects in soil deforrnation*. PhD Thesis, City University, London.

Ridley, A.M. and Burland, J.B. (1993). A new instrument for the measurement of soil moisture suction. *Géotechnique*, 43(2), 321–324.

Ridley, A.M. and Burland J.B. (1995b). Measurement of suction in materials which swell. *Applied Mechanics Reviews*. 48(9), September.

Ridley, A.M. and Burland, J.B. (1996). A pore water pressure probe for the insitu measurement of a wide range of soil suctions. *Advances in Site Investigation Practice*. Thomas Telford, 510–520.

Ridley, A.M. and Wray, W.K. (1995). Suction measurement: A review of current theory and practices. *Unsaturated Soils: Proc. 1st Int. Conf. on Unsat. Soils*, Paris, Alonso, E.E., Delage, P. eds, Balkema, Rotterdam, Presses des Ponts et Chaussees, 1293–1322.

Rio, J. and Greening, P. (2003). Influence of sample geometry on apparent shear wave velocity measurements using bender elements. *Proc. 3rd Int. Symp. on Deformation Characteristics of Geomaterials*, Lyon.

Roesler, S.K. (1979). Anisotropic shear modulus due to stress anisotropy. *J. Geotech. Engrg Div., Proc. of the Am. Soc. Civ. Engrs.*, 105(7), 871–880.

Rollins, K. and Rogers, G.W. (1994). Mitigation measures for small structures on collapsible alluvial soils. *J. Geotech. Engrg. Div.*, ASCE, 120(9), 1533–1553.

Rollins, K., Rollins, R., Smith, T., and Beckwith, G. (1994). Identification and characterization of collapsible gravels. *J. Geotech. Engrg. Div.*, ASCE, 120(3), 528–542.

Rosenhaupt, S. and Mueller, G. (1963). Openings in masonry walls on settling supports. *J. Struct. Div.*, ASCE, 89(3), 107–131.

Rowe, P.W. (1962). The stress-dilatancy relation for static equilibrium of an assembly of particles in contact. *Proc. Royal Society of London*, Ser. A 269, pp. 500–527.

Royster, D.L. and Rowan, W.H. (1968). Highway design and construction problems associated with the loessial soils of west Tennessee. *Hwy. Research Record*, 212, 28–32.

Rubin, J., Steinhardt, R. (1963). Soil–water relation during rain infiltration: I. Theory. *J. Soil Science*, 27, 246–251.

Rulon J.J. and Freeze R.A. (1985). Multiple seepage faces on layered slopes and their implications for slope-stability analysis. *Can. Geotech. J.*, 22, 347–356.

Russam, K., and Coleman, C.D. (1961). The effect of climatic factors on subgrade moisture conditions. *Géotechnique*, 11(1), 22–28.

Russell, M.B. and Richards L.A. (1938). The determination of soil moisture energy relations by centrifugation. *Proc. Soil Sci. Soc. of Am.*, Vol. 3, pp. 65–69.

Salgado, R., Drnevich, V.P., Ashmawy, A., Grant, W.P. and Vallenas, P. (1997). Interpretation of large-strain seismic cross-hole tests. *J. Geotech. Geoenvir. Engrg*, ASCE, 123(4), 382–388.

Sanchez-Salinero, I., Roesset, J.M. and Stokoe, K.H. (1986). *Analytical Studies of Body Wave Propagation and Attenuation*. Report GR 86–15, University of Texas, Austin.

Sander, G.C. (1991). Exact solution for nonlinear, nonhystertic redistribution in vertical soil of finite depth. *Water Resour. Res.*, 27(7), 1527–1636.

Sasitharan, S., Robertson, P.K., Sego, D.C, and Morgenstern, N.R. (1993). Collapse behavior of sand. *Can. Geotech. J.*, Ottawa, 30(4), 569–577.

Satija, B.S. (1978). *Shear Behaviour of Partly Saturated Soils*. PhD thesis, Indian Institute of Technology, Delhi, India.

Schofield, R.K. (1935). The pF of water in soil. *Trans. of the 3rd International Congress on Soil Science*, 2, 37–48.

Sevaldson, R.A. (1956). The slide at Lodalen, 6 Oct., 1954. *Géotechnique*, 6, 167–182.

Shai, F., Yazdani, J., Feng, M. and Fredlund, D.G. (1998). *Supplemental report on the thermal conductivity matric suction sensor development (Year II)*. Dept. of Civ. Eng., Univ. of Saskatchewan, Saskatoon, SK.

Sheng, D., Fredlund, D.G., Gens, A. (2007). An alternative modelling approach for unsaturated soils. *Proc. 3rd Asian Conf on Unsat Soils*. Nanjing. Science Press, Beijing, 47–65.

Shi, B., Jiang, H., Liu Z. and Fang, H.Y. (2002). Engineering geological characteristics of expansive soils in China. *Eng. Geol.*, 67, 63–71.

Shuai, F., Yazdani, J., Feng, M. and Fredlund, D.G. (1998) *Supplemental Report on the Thermal Conductivity Matric Suction Sensor Development (Year II)*. Department of Civil Engineering, University of Saskatchewan, Saskatoon, Sask.

Sibley, J.W. and Williams, D.J. (1990). A new filter material for measuring soil suction. *Geotech. Testing J.*, ASTM, 13(4), 381–384.

Simons, N.E, Menzies, B.K. and Matthews, M.C. (2001). *A Short Course in Soil and Rock Slope Engineering*. Thomas Telford, London.

Simpson, B. (1992). Thirty-second Rankine lecture. Retaining structures: displacement and design. *Géotechnique*, 42(4), 541–576.

Sitharam, T.G., Sivapulliah, P.V. and Subba Rao, K.S. (1995). Shrinkage behaviour of compacted unsaturated soils. *Proc. 1st Int. Conf. on Unsat. Soils*, Paris, pp. 195–200.

Sivakumar, V. (1993). *A Critical State Framework for Unsaturated Soils*. PhD thesis, University of Sheffield, Sheffield, U.K.

Sivapullaiah, P.V., Sridharan, A. and Stalin, Y.K. (1996). Swelling behavior of soil-bentonite mixtures. *Can. Geotech. J.*, 33, 808–814.

Skempton, A.W. and Hutchinson, J. (1969). Stability of natural slopes and embankment foundations. State-of-the-Art Report, *Proc. 7th Int. Conf. Soil Mech. Found. Eng.*, Mexico City, Mexico, State-of-the-Art Vol., Balkema, Rotterdam, 291–340.

Sladen, J.A., D'Hollander, R.D. and Krahn, J. (1985). The liquefaction of sands, a collapse surface approach. *Can. Geotech. J.*, 22, 564–578.

Slatter, E.E, Jungnickel, C.A, Smith, D.W. and Allman, M.A. (2000). Investigation of suction generation in apparatus employing osmotic methods. *Proc. Asia Conf. on Unsat. Soils*, Singapore, pp. 297–302.

Soilmoisture Equipment Corp. (1985). Commercial Publications. Santa Barbara, CA.

Songyu, L., Heyuan, L., Peng, J. and Yanjun, D. (1998). Approach to cyclic swelling behaviour of compacted clays. *Proc. 2nd Int. Conf. on Unsat. Soils*, Beijing, China, Vol. 2, pp. 219–225.

Spencer, E. (1967). A method of analysis of the stability of embankments assuming parallel interslice forces. *Géotechnique*, 17, 11–26.

Spencer, E. (1968). Effect of tension on stability on embankments. *J. Soil Mech. Found. Eng. Div.*, ASCE, 94 (SM5), 1159–1173.

Spencer, E. (1973). Thrust line criterion in embankment stability analysis. *Géotechnique*, 23, 85–100.

Sposito, G. (1981). *The Thermodynamics of Soil Solutions*. London: Oxford Clarendon Press.

Srivastava, R. and Yeh, T.C.I. (1991). Analytical solutions for one dimension, transient infiltration toward the water table in homogeneous and layered soils. *Water Resour. Res.*, 27(5), 753–762.

Stannard, D.I. (1992). Tensiometers – theory, construction and use. *Geotech. Testing J.*, ASTM, 15, No.1, 48–58.

Steinberg, M. (1998). *Geomembranes and the Control of Expansive Soils in Construction*. New York: McGraw-Hill.

Stokoe, K.H., II, Hwang, S.K., and Lee, J.N.K. (1995). Effect of various parameters on the stiffness and damping of soils at small strains. *Proc. Int. Symp. on Pre-Failure Deformation Characteristics of Geomaterials*, Hokkaido 1994, Japan, S. Shibuya, T. Mitachi, and S. Miura, eds, 2, Balkema, Rotterdam, pp. 785–816.

Subba Rao, K.S. and Satyadas, G.C. (1987). Swelling potentials with cycles of swelling and partial shrinkage. *Proc. 6th Int. Conf. on Expansive Soils*, New Delhi, India, Vol. 1, pp. 137–147.

Sun, H.W. (1999). *Review of Fill Slope Failures in Hong Kong*. GEO Report No. 96, Geotechnical Engineering Office, Hong Kong.

Sun, H.W. and Campbell, S.D.G. (1998). *The Lai Ping Road Landslide of 2 July 1997*. Landslide study report (LSR 27/98). Geotechnical Engineering Office, Hong Kong.

Sun, H.W., Wong, H.N. and Ho, K.K.S. (1998). Analysis of infiltration in unsaturated ground. *Slope Engineering in Hong Kong*, 101–109.

Sun, H.W., Ho, K.K. S., Campbell, S.D.G. and Koor, N.P. (2000a). The 2 July 1997 Lai Ping Road Landslide, Hong Kong–Assessment of Landslide Mechanism. *Proc. 8th Int. Sym. on Landslides*, Cardiff, June, Vol. 3, 1423–1430.

Sun, H.W., Law, W.H.Y., Ng, C.W.W., Tung, Y.K. and Liu, J.K. (2000b). The 2 July 1997 Lai Ping Road Landslide, Hong Kong–Hydrogeological Characterisation. *Proc. 8th Int. Sym. on Landslides*, Cardiff, June, 3, pp. 1431–1436.

Swarbrick, G.E. (1995). Measurement of soil suction using the filter paper method. *Unsaturated Soils: Proc. 1st Int. Conf. on Unsat. Soils*, Paris, Alonso, E.E., Delage, P. eds, Balkema, Rotterdam, Presses des Ponts et Chaussees.

Tadepalli, R. and Fredlund, D.G. (1991). The collapse behaviour of a compacted soil during inundation. *Can. Geotech. J.*, 28, 477–488.

Take, W.A. and Bolton, M.D. (2003). Tensiometer saturation and the reliable measurement of soil suction. *Géotechnique*, 53, 159–172.

Take, W.A., Bolton, M.D., Wong, P.C.P. and Yeung, F.J. (2004). Evaluation of landslide triggering mechanisms in model fill slopes. *Proc. 8th Int. Sym. on Landslides*, Cardiff, June, 1, pp. 173–184.

Tang, W.H. and Lee, C.F. (2003). Potential use of soil nails in loose fill slope: an overview. *Proc. Int. Conf. on Slope Engrg.*, Hong Kong, China, 2003, 974–997.

Tarantino, A. and Mongiovi, L. (2000a). A study of the efficiency of semi-permeable membranes in controlling soil matrix suction using the osmotic technique. *Proceedings Unsaturated Soils for Asia*. Singapore, 303–308.

Tarantino, A. and Mongiovi, L. (2000b). Experimental investigations on the stress variables governing unsaturated soil behaviour at medium to high degrees. *Proc. Exp. Evidence and Theoretical Approaches in Unsat. Soils*. Italy, 3–19.

Tatsuoka, F. and Kohata, Y. (1995). Stiffness of hard soils and soft rocks in engineering applications. *Proc. Int. Symp. on Pre-Failure Deformation Characteristics of Geomaterials*, Hokkaido 1994, Japan, S. Shibuya, T. Mitachi and S. Miura, eds, Vol. 2, Balkema, Rotterdam, pp. 947–1063.

Taylor, S.A. and Ashcroft, G.L. (1972). *Physical Edaphology*. San Franscisco, CA. 533p.

Temperley, H.N.V. and Chambers L.L.G. (1946). The behaviour of water under hydrostatic tension: I. *Proceedings of the Physical Society*, 58, 420–436.

Terzaghi, K. (1936a). The shear resistance of saturated soils. *Proc. 1st Int.Conf. Soil Mech. & Fdn. Engrg.*, Cambridge, Mass., Harvard University, (1), 54–56.

Terzaghi, K. (1936b). Stability of slopes of natural clay. *Proc. 1st Int.Conf. Soil Mech. & Fdn. Engrg.*, Cambridge, Mass., Harvard University, 1, pp. 161–165.

Thomas, H.R. and He, Y. (1994). Analysis of coupled heat moisture and air transfer in a deformable unsaturated soil. *Géotechnique*, 44(5), 677–689.

Tinjum, J.M., Benson, C.H. and Blotz, L.R. (1997). Soil-water characteristic curves for compacted clays. *J. Geotech. and Geoenvir. Eng.*, ASCE, 123(11), 1060–1069.

Topp, G.C. and Miller, E.E. (1966). Hysteresis moisture characteristics and hydraulic conductivities for glass-bead media. *J. Soil Science*, 30, 156–162.

Trevena, D.H. (1987). *Cavitation and tension in liquids*. Bristol: Adam Hilger.

Tripathy, K.S., Subba Rao, K.S. and Fredlund, D.G. (2002). Water content – void ratio swell-shrink paths of compacted expensive soils. *Can. Geotech. J*, 39, 938–959.

Tung, Y.K., Ng, C.W.W. and Liu, J.K. (1999). *Lai Ping Road Landslide Investigation Three Dimensional Groundwater Flow Computation*. Technical Report - Part I, Geotechnical Engineering Office of the Hong Kong Special Administrative Region, Hong Kong.

Tung, Y.K., Zhang, H., Ng, C.W.W. and Kwok, Y.F. (2004). Transient seepage analysis of rainfall infiltration using a new conjunctive surface-subsurface flow model. *Proc. 57th Can. Geotech. Conf., Géo Québec 2004*, Session 7C, 17–22.

Turnbull, W. (1968). Construction problem experiences with loess soils. *Hwy. Research Record*, 212, 10–27.

Vaid, Y.P. and Chern, J.C. (1985). Cyclic and monotonic undrained response of saturated sand. *Advances in the Art of Testing Soils Under Cyclic Conditions*. Vijay Khosla ed., American Society of Civil Engineers, New York, 120–147.

van Amerongen, G.J. (1946). Permeability of different rubbers to gases and its relation to diffusivity and solubility. *J. Appl. Phys.*, 17(11), 972–985.

Van Der Merwe, D.H. (1964). The prediction of heave from the plasticity index and the percentage clay fraction. *The Civil Engineer in South Africa*, 6(6), 103–107.

van Genuchten, M.T. (1980). A closed form equation for predicting the water permeability of unsaturated soils. *J. Soil Science*, 44, 892–898.

Vanapalli, S.K., Fredlund, D.G., Pufahl, D.E. and Clifton, A.W. (1996). Model for the prediction of shear strength with respect to soil suction. *Can. Geotech. J.*, 33(3), 379–392.

Vanapalli, S.K., Siller, S.W., and Fredlund, M.D. (1998b). The meaning and relevance of residual state to unsaturated soils. *Proc., 51st Canadian Geotechnical Conf.*, Edmonton, Alta., Canada, 1–8.

Vanapalli, S.K., Pufahl, D.E. and Fredlund, D.G. (1999). The effect of soil structure and stress history on the soil–water characteristics of a compacted till. *Géotechnique*, 49(2), 143–159.

Viana Da Fonseca, A., Matos Fernandes, M. and Silva Cardoso, A. (1997). Interpretation of a footing load test on a saprolitic soil from granite. *Géotechnique*, 47(3), 633–651.

Viggiani, G. and Atkinson, J.H. (1995). Stiffness of fine-grained soil at very small strains. *Géotechnique*, 45(2), 249–265.

Vinale, F., d'Onofrio, A., Mancuso, C., Santucci de Magistris, F. and Tatsuoka, F. (1999). The prefailure behaviour of soils as construction materials. 2nd *Int. Conf. on Pre-failure Behaviour of Geomaterials*, Turin, Balkema, Rotterdam.

Walsh, K., Houston, W. and Houston, S. (1993). Evaluation of in-place wetting using soil suction measurements. *J. Geotech. Engrg. Div.*, ASCE, 119(5), 862–873.

Wan, A.W.L. and Gray, M.N. (1995). On the relations of suction, moisture content and soil structure in compacted clays. *Unsaturated Soils, Proc. 1st Int. Conf. on Unsat. Soils*, Sept. 6–8, Paris, E.E. Alonso and P. Delage, eds, Balkema Press, Rotterdam, Vol. 1, pp. 215–222.

Wan, R.G. and Guo, R.G. (1998). A simple constitutive model for granular soils: modified stress-dilatancy approach. *Comput. Geotech.*, 22(2), 109–133.

Wang, B. (2000). *Stress Effects on the Soil–Water Characteristics of Unsaturated Expansive Soils*. M.Phil. thesis. Department of Civil Engineering, Hong Kong University of Science and Technology.

Wang, Y. and Ng, C.W.W. (2005). Effects of stress paths on the small-strain stiffness of completely decomposed granite. *Can. Geotech. J.*, 42(4), 1200–1211.

Warkentin, B.P. and Bozozuk, M. (1961). Shrinkage and swelling properties of two Canadian clays. *Proc. 5th Int. Conf. on Soil Mech. and Found. Eng.*, Dunod, Paris, pp. 851–855.

Wheeler, S. and Sivakumar, V. (1992). Critical State concepts for unsaturated soils. *Proc. of the 7th Int. Conf. on Expansive Clays*, Dallas, TX, Aug. 3–5, 1, pp. 167–172.

Wheeler, S.J. (1986). *The Stress-Strain Behavior of Soils Containing Gas Bubbles.* D.Phil. thesis, Oxford University, Oxford, UK.

Wheeler, S.J. (1996). Inclusion of specific water volume within an elasto-plastic model for unsaturated soil. *Can. Geotech. J.*, 33, 42–57.

Wheeler, S.J. (2006). Keynote oral presentation: *Constitutive Modelling of Unsaturated Soils. 4th Int. Conf. on Unsaturated Soils (UNSAT06)*, Arizona, USA. Unpublished.

Wheeler, S.J. and Karube, D. (1995). State of the art Report – Constitutive modelling. *Proc. 1st Int.Conf. on Unsaturated Soils*, Paris, Balkema, Rotterdam, Vol. 3, 1323–1356.

Wheeler, S.J. and Sivakumar, V. (1995). An elasto-plastic critical state framework for unsaturated soil. *Géotechnique*, 45(1), 35–53.

Wheeler, S.J., Sharma, R.S. and Buisson, M.S.R. (2003). Coupling of hydraulic hysteresis and stress-strain behaviour in unsaturated soils. *Géotechnique*, 53, 1, 41–54.

Williams, A.A.B. (1980). Severe heaving of a block of flats near Kimberley. *7th.Reg. Conf. for Africa on Soil Mech. and Found. Eng.*, 1, pp. 301–310.

Williams, A.A.B. and Pidgeon, J.T. (1983). Evapotranspiration and heaving clays in South Africa. *Géotechnique*, 33, 141–150.

Williams, J. and Shaykewich, C.F. (1969). An evaluation of polyethylene glycol (PEG) 6000 and PEG 20000 in the osmotic control of soil–water matric potential. *Can. J. Soil Science*, 102(6), 394–398.

Wilson, G.W. (1997). Surface flux boundary modelling for unsaturated soils. *Proc., Unsaturated Soil Engineering Practice*, ASCE Geotechnical Special Publication No. 68, S.L. Houston and D. G. Fredlund, eds, ASCE, New York, pp. 38–65.

Wolle, C.M. and Hachich, W. (1989). Rain-induced landslides in south-eastern Brazil. *Proc. of 12th Int. Conf. Soil Mech. & Fdn. Engrg.*, Rio de Janeiro, Balkema, Rotterdam, 3, 1639–1644.

Wright, S.G. (1974). *SSTABI – A general computer program for slope stability analysis.* Res. Report No. GE-74-1, Dep. Of Civil Eng., Univ. of Texas at Austin.

Yamamuro, J.A. and Lade, P.V. (1997). Static liquefaction of very loose sands. *Can. Geotech. J.*, 34, 905–917.

Yamamuro, J.A. and Lade, P.V. (1998). Steady state concepts and static liquefaction of silty sands. *J. Geotech. Engrg. Div.*, ASCE, 124(9), 868–877.

Yen, B.C. (1977). Stormwater runoff on urban areas of steep slope. Environmental protection technology series. EPA-600/2, 77–168.

Yong, R.W. and Warkentin, B.P. (1975). *Soil Properties and Behaviour.* Elsevier Scientific, New York.

Yoshitake, S. and Onitsuka, K. (1999). Centrifuge model tests and stability analysis on mobilizing process of shear strength of decomposed granite soil slope. *Slope Stability Engineering*, Yagi, Yamagami and Jiang.

Youd, T.L. (1973). Factors controlling maximum and minimum densities of sand. *Evaluation of relative density and its role in geotechnical projects involving cohesionless soils.* ASTM STP523, American Society for Testing and Materials, Philadelphia, PA, 98–112.

Young, R.F. (1989). *Cavitation.* McGraw-Hill Book Company, London.

Zhan, L.T. (2003). *Field and Laboratory Study of an Unsaturated Expansive Soil Associated with Rain-Induced Slope Instability.* PhD Thesis. Hong Kong University of Science and Technology.

Zhan, L.T. and Ng, C.W.W. (2004). Analytical Analysis of Rainfall Infiltration Mechanism in Unsaturated Soils. *International Journal of Geomechanics*, ASCE. 4(4), 273–284.

Zhang, J.F., Yang, J.Z. and Wang, F.Q. (1997). Rainfall infiltration test for high slope of shiplock in Three Gorge Dam. *Yangtze River Scientific Research Institute, Technical Report*, No. 97–264 (In Chinese).

Zhan, L.T., Ng, C.W.W. and Fredlund, D.G. (2006). Instrumentation of an unsaturated expansive soil slope. *Geotech. Test. J.*, 30(2), 1–11.

Zhan, L.T., Ng, C.W.W. and Fredlund, D.G. (2007). Field study of rainfall infiltration into a grassed unsaturated expansive soil slope. *Can. Geotech. J.*, 4, 392–408.

Zhang, W. and Zhang (1995). Development of loess engineering properties research in China, *Unsaturated Soils: Proc. 1st Int. Conf. on Unsat. Soils*, Paris, Alonso, E.E., P. Delage, eds, Balkema, Rotterdam, Presses des Ponts et Chaussees.

Zhang, M., Ng, C.W.W., Take, W.A., Pun, W.K., Shill, Y.K. and Chang, G.W.K. (2006). The role and mechanism of soil nails in liquefied loose sand fill slopes. *Proc. 6th Int. Conf. Physical Modelling in Geotechnics (TC2)*, Hong Kong, Taylor & Francis. Vol. 1, pp. 391–396.

Zhou, R.Z.B., Take, A. and Ng, C.W.W. (2006). A case study in tensiometer interpretation: centrifuge modelling of unsaturated slope behaviour. *Proc. 4th Int. Conf. on Unsaturated Soils*, ASCE, Carefree, Arizona. Vol. 2, pp. 2300–2311.

Zlatovic, S. and Ishihara, K. (1997). Normalized behaviour of very loose non-plastic soils: effects of fabric. *Soils and Found.*, 37(4), 47–56.

Zur, B. (1966). Osmotic control of the matric soil–water potential: I. Soil–water system. *J. Soil Science*, 102(6), 394–398.

Author Index

Abelev, M.Y., 251
Akan, A.O., 600
Al-Homoud, A.S., 253
Al-Khafaf, S., 67
Al-Mukhtar, M., 54, 56, 57
Alonso, E.E., 5, 20, 31, 101, 227, 230, 240–1, 247–8, 327, 331, 340, 356, 406, 435, 441, 448, 494, 499
Ampt, G.A., 194, 205, 210, 516
Anayev, V.P., 251
Anderson, A.B.C., 538
Anderson, M.G., 538
Andrei, S., 250
Apfel, R.C., 27
Arbhabhirama, A., 106
Ashcroft, G.L., 9
Atkinson, J.H., 381–2, 409, 417
Aziz, M.A., 278

Balasubramanian, B., 275
Bally, R., 251
Bao, C.G., 197, 252, 473, 490
Bara, J.P., 232
Barbour, S.L., 93, 121, 130, 279, 438
Barden, L., 114–17, 121, 401, 405
Bartos, D.L., 67
Beckwith, G., 230–1, 249–50
Been, K., 366, 444
Bell, F.G., 159, 163, 227, 230–1, 233, 252–6
Bell, P.G., 159, 163
Bellotti, R., 408, 427
Ben Shalom, A., 49–50, 295
Benson, C.H., 145, 156
Berre, T., 424
Best, R., 232, 319
Biggar, Y.W.109

Bishop, A.W., 14–15, 41, 288–9, 295, 311, 394
Blight, G.E., 14, 31, 70, 87, 89–92, 113–14, 277, 278, 295
Bocking, K.A., 295
Boger, M., 195
Bolton, M.D., 83, 333, 366, 436
Bozozuk, M., 253
Brackley, I.J.A., 482, 493, 495
Brand, E.W., 336, 361, 515, 538
Briggs, L.J., 168
Brignoli, E., 424
Broadbridge, P., 195, 199
Brooks, R.H., 102–3, 105–6, 115, 197, 322–3, 325
Brown, R.W., 66–7
Buckingham, E., 32, 100, 107
Bumb, A.C., 322–4
Burdine, N.T.102, 137
Burland, J.B., 6, 14, 17, 28, 48, 50, 53, 83–4, 295–6, 310, 381, 401, 435
Brunner, E., 168

Cabarkapa, Z., 380, 394–9, 423
Campbell, S.D.G., 148, 517, 518
Casagrande, A., 95, 341
Casagrande, L., 95, 341, 347
Castro, G., 575
Chambers, L.L.G., 37
Chandler, R.J., 25–7, 30, 67, 69, 83, 257
Chaudhry, M.H., 604
Chowdhury, R. N., 504
Chen, F.H., 11
Chen, R., 31, 41, 280
Chapman, P. J., 30
Cartwright, K., 108

Chern, J.C., 575
Cherry, J. A., 99
Cheuk, C.Y., 365, 501, 575, 577–9,
    584–5, 587, 589–6, 598
Childs, E.S., 100, 101, 107
Ching, R.K.H., 67, 514
Chiu, C.F., 20, 38, 42, 44, 78, 159,
    227, 230, 245, 279, 289, 327,
    328, 331, 334–5, 338–42, 344–8,
    350, 353–5, 357–8, 360–2, 364,
    367–8, 371–6, 421, 432, 435–6,
    438–40, 442–7, 449–1, 453–4,
    457–8, 460–2, 464–5, 598
Chu, T.Y., 499
Cintra, J.C., 251
Clayton, C.R.I., 339, 382, 417
Clemence, S., 250
Clevenger, W.A., 230–1
Coleman, J.D., 14, 19, 155, 314
Collins, J.M., 66
Collis-George, N., 67, 100, 107
Coop, M.R., 341, 359, 374
Corey, A.T., 99, 102–6, 109, 111, 115,
    197, 322–3, 325–6
Cothren, M., 276
Crilly, M.S., 67
Croney, D., 155, 168, 314, 351, 461
Cubrinovski, M., 443
Cuccovillo, T., 380
Cui, Y.J., 31, 49, 50, 241, 279, 295,
    298–299, 327–8, 333–4, 435–6,
    444
Culshaw, M.G., 227–8, 230–1, 233,
    252–6

d'Onofrio, A., 380
Dafalias, Y.F., 366, 436, 443, 445
Dalton, F.N., 67
Darcy, H., 601
Daniel, D.E., 156
Day, R.W., 253, 257
de Sousa Pinto, C., 5, 28–9
Delage, P., 5, 31, 49–51, 53–4, 57,
    227, 241, 280, 295–6, 298–299,
    327–8, 333–4, 401, 405, 435–6,
    444, 461
Dempsey, B.J., 34
Denisov, H.Y., 232
Derbyshire, E., 230, 232
Dineen, K., 48, 50, 53, 295–6
Dobry, R., 382
Donald, I.B., 288, 317–18

Dorsey, N.E., 9
Dregne, H.E., 11
Drnevich, V.P., 408
Dudley, J.H., 11
Dyvik, R., 396, 409, 414

Edlefsen, N.E., 12
El-Ehwany, M., 70, 230, 235
El Nimr, A.I., 230, 240
Elzeftawy, A., 108
Esteban, V., 36, 54

Fawcett, R.G., 67
Feda, J., 232
Feng, M., 31, 71–6
Finbarr, A., 250
Fisher, R.A., 404
Fookes, P., 228, 232
Fourie. A.B., 526, 527
Fredlund, D.G., 5–14, 16, 19–23, 31,
    34–5, 38, 42, 60, 66–8, 70, 72–6,
    78, 81–6, 93–7, 99, 101–3, 105,
    107–10, 111, 116–17, 121–6,
    128, 130–3, 135–40, 143–5,
    147–8, 150, 166, 174–5, 177–9,
    182, 184–91, 196–7, 227, 229,
    245, 279–3, 285–7, 295, 298,
    300–1, 307, 309, 311, 311–15,
    317–26, 329–30, 352, 354,
    359–60, 374, 418–20, 435, 449,
    473, 478, 481, 494, 501–6, 508,
    510–11, 513–14, 538–39, 560,
    568, 609, 611–12, 615
Freeze, R.A., 99, 567
Frydman, S., 295
Fukuoka, M., 518, 538
Fung, Y.C., 13, 315
Fung, W.T., 365–6

Gallen, P. M., 67
Gallipoli, D., 19–20
Gan, J.K.M., 38, 46–7, 279, 286, 305,
    309, 318–20, 330, 352, 354,
    359–60, 374, 449, 568
Gao, G.Y., 251
Gardner, H.R., 106–7, 136
Gardner, R., 67–8, 168–9
Geiser, F., 288, 289
Gens, A., 461
Gibbs, G. J., 259
Gibbs, H.H., 232, 259
Gong, B.W., 94, 473

Gourley, C.S., 67, 257
Grabowska-Olszewsla, B., 232
Graham, J., 461
Gray, D.H., 380, 401
Gray, M.N., 401
Greacen, E.L., 69
Green, W.H., 194, 205, 210, 516
Greening, P., 424
Gribb, M.M., 145
Grim, R.E., 254
Gu, W.H., 578, 580
Guan, Y., 83
Gunn, M., 380
Guo, R.G., 366
Gutierrez, C.I., 67, 69

Hachich, W., 515, 538
Haines, W.B., 140, 142
Hallberg, G.R., 234
Hanafy, E.A.D.E., 257, 271
Handy, R.L., 11, 232
Hanks, R.J., 68
Hardin, B.O., 408
Harrison, B.A., 31, 70, 87, 89–91, 614
Harvey, E.N., 26, 28, 30, 83
Hassler, G.L., 168
Hayward, A.T.J., 26
He, Y., 435
Henderson, D.S., 30
Herbert, R., 194, 216, 516
Hilf, J.W., 36–7, 78, 166, 294–5, 339, 411
Hillel, D., 93, 109, 140, 141–2, 155, 198, 203, 207, 213, 218–19, 523
Ho, D.Y.F., 268, 300–1, 582
Holtz, W., 251, 259
Houlsby, G.T., 5, 19–20
Houston, S.L., 70, 214, 227–8, 230, 234–5, 238–40, 249–50, 516
Houston, W.N., 214, 235, 238–40, 249–50, 516

Ishihara, K., 336, 341, 343, 366, 436, 443, 575
Iwasaki, T., 408

Jaky, J., 455
Jamiolkowski, M., 380, 408, 427
Janssen, D.J., 34
Jardine, R.J., 382
Jefferies, M.G., 366, 444
Jefferson, I., 11, 227, 230

Jennings, J.E., 6, 17, 232, 239, 257, 310, 435
Jiménez-Salas, J.A., 227, 273
Jones, W.M., 30
Juca, J.F.T., 295

Karube, D., 16–19, 340, 461
Kasim, F.B., 194, 206, 209
Kassif, G., 49–50, 295
Katti, R.K., 275
Kay, G. W. C., 22, 23
Kemper, W.D., 99
Khanzode, R.M., 93, 168–9, 171–4
Khatrush, S.A., 382
Kjaernsli, B., 504
Knight, K., 232, 239, 257
Knox, D.P., 408
Kohata, Y., 381–2
Komornik, A., 49–50, 295
Koor, N.P., 517
Krahn, J., 84, 123, 505
Kridakorn, C., 106
Kunhel, R.A., 54–5
Kunze, R.J., 108, 147
Kwok, Y.F., 501, 600

Laby, T. H., 22, 23
Lacasse, S., 424
Ladd, R.S., 115, 361, 401, 419, 424
Lade, P.V., 336, 343, 575
Lagerwerff, J.V., 49, 295
Lai, J.C.H., 547, 548, 552, 554–6, 558
Lam, C.C., 521, 541, 550, 562, 5689
Lam, L., 93, 516, 538, 599
Lambe, T.W., 100, 154–5, 281, 405
Lancellotta, R., 380
Langfelder, L.J., 114–15
Larson, C.L., 214, 523, 530
Lawton, E., 238
Leach, B., 194, 217, 516
Lee, C.F., 596, 598
Lee, H.C., 70
Lee, I.K., 341, 359, 374
Leong, E.C., 31, 93, 137, 145–7, 195, 380, 516, 551
Leung, C.C., 521
Leung, K.L., 521
Leung, Y.K., 541, 550, 562, 568
Li, J., 491
Li, X.S., 17, 366, 409, 411, 436, 443, 445
Liakopoulos, A. C., 109

Likos, W.J., 55–8
Lim, T.T., 474, 516, 538, 568
Lin, H.C., 233, 515, 516
Lin, Z.G., 231
Lings, M.L., 381
Liu, T., 11, 230, 231
Liu, T.H., 476, 495
Lloret, A., 54, 101
Lo Presti, D.C.F., 380, 382, 408
Lu, N., 55, 57, 58, 201, 203
Lumb, P.B., 515–16, 543
Lutenegger, A., 234, 251

Maatouk, A., 341, 345–8, 352, 354, 449–50
McKee, C.R., 322–5
McKeen, R.G., 67
McLane, J.W., 168
McQueen, I.S., 67, 257
Madshus, C., 396, 409, 414
Mahmoud, H., 235, 250
Mair, R.J., 381
Maksimovic, M., 514
Mancuso, C., 380, 399–3, 406–7, 423–4
Manea, S., 249
Manzari, M.T., 366
Mao, S.Z., 297, 301–2, 560
Manheim, F. T., 85
Marinho, E.A.M., 5, 24–29, 83
Marshall, T.I., 108, 197
Matyas, E.L., 111, 113–14, 116–17
Maulem, Y., 137, 147
Meigs, P., 11
Mein, R.G., 214, 523, 530
Mellors, T.W., 232
Michels, W.C., 13
Miller, D.J., 276, 473
Miller, E.E., 109, 143, 519
Miller, R.F., 67, 257
Millington, R.J., 108
Molenkamp, M.R., 17
Mongiovi, L., 53, 295–6, 301
Morgenstern, N.R., 6, 15, 19, 38, 275, 295, 435, 505, 513–14
Morita, M., 600, 601, 602
Morgenstern, N .R., 6, 15, 19, 38, 275, 295, 435, 505, 513, 514
Moore, R. E., 107
Morrison, R., 207
Mou, C.H., 499
Mualem, Y., 143

Nazemi, A.H., 17
Nelson, J.D., 276, 473
Ng, C.W.W., 2, 11, 20, 31, 38–46, 47–54, 60, 83, 93–4, 123, 130, 148–52, 154, 156–7, 159–63, 166–7, 194–5, 205, 209, 214–15, 222, 227, 230, 240, 243–8, 252, 279–80, 287, 289–1, 294, 298–300, 302–3, 304, 306–8, 326–29, 331–5, 338, 340–2, 344–8, 350, 353–5, 357–9, 360, 362–3, 364, 367–8, 371–6, 380–2, 408, 412–18, 420–2, 425, 427–9, 432, 435, 438–40, 442, 444–6, 449–1, 453–4, 458, 460–2, 464–5, 473–80, 483, 485, 487, 489–90, 492–3, 496, 498, 501, 515–19, 521–2, 524–6, 528, 530, 533, 535, 538–41, 543–4, 546–8, 552, 554–6, 558–64, 566–7, 569–5, 598–9
Nielsen, D.R., 109
Nishimura, T., 329, 334–5
Northmore, K.J., 232–3

O'Neill, M.W., 256
Olson, R.E., 115
Oltulescu, D., 251
Onitsuka, K., 592
Oteo-Mazo, C., 54, 58

Paddy, J.F., 6
Pang, R.P.L., 20, 39, 40, 380
Pang, Y.W., 2, 20, 38, 40, 93, 123, 130, 148–52, 154, 156–7, 159–63, 166, 195, 209, 478, 501, 516, 518–19, 539–1, 543–4, 546, 562, 567, 570–4
Parry, R., 228
Pavlakis, G., 114–16
Pearsall, I.S., 5, 26
Peck, A.J., 53
Pennington, D.S., 408, 424–5
Phene, C.J., 72
Philip, J.R., 195, 199–200, 207
Pidgeon, J.T., 278
Pitman, T.D., 575
Plesset, M.S., 26
Poorymoayed, A.M., 256
Popescu, M.E., 253
Poulos, S. J., 120
Poulovassilis, A., 140, 143

Pradel, D., 206, 516
Premchitt, J., 523
Price, J.C., 505, 513–14
Price, V. E., 505, 513, 514
Pullan, A.J., 195, 199
Pun, W.K., 380, 538

Qian, X., 380, 382, 384–93
Quirk, J.P., 108

Raad, G., 206
Rabbidge, R.M., 53
Rahardjo, H., 5, 11, 13, 20–3, 31,
    34–5, 38, 66–8, 70, 81–6, 93–7,
    99, 101–3, 105, 107–11, 116–17,
    121, 130–3, 136–7, 144–8, 150,
    166, 174, 177–9, 182, 184–91,
    195–6, 279–3, 285–7, 307, 309,
    311–13, 330, 380, 419–20, 494,
    501–3, 506, 508, 511, 513,
    515–16, 538, 551, 560, 611,
    612, 615
Ramaswany, S.D., 277
Rampello, S., 404–5, 408, 427
Rawlins, S.L., 67
Richards, B.E., 30
Richards, B.G., 66, 314
Richards, L.A., 100, 107, 168, 199
Richardson, D., 382
Ridley, A.M., 28, 31–2, 63–7, 68–71,
    77–9, 83–4, 401
Rio, J., 424
Roesler, S.K., 408
Rogers, G.W., 250
Rollins, K., 240, 250
Rosenhaupt, S., 278
Rowan, W.H., 251
Rowe, P.W., 340, 366, 443
Royster, D.L., 251
Rubin, J., 214
Rulon, J.J., 567
Russam, K., 314
Russell, M.B., 168

Saez, J., 36, 54
Salgado, R., 382
Sällfors, G., 381–2
Sanchez-Salinero, I., 396, 417
Sander, G.C., 195, 199
Sanders, P.J., 482, 493, 495
Sasitharan, S., 336, 375
Satyadas, G.C., 253

Schofield, R.K., 77
Schreiner, H.D., 67
Semkin, V.V., 251
Sevaldson, R.A., 504
Shaykewich, C.F., 48–9, 53, 59, 297
Shen, C.K., 227
Shi, B., 11, 94, 195, 205, 214–15, 516,
    526, 538, 560, 567, 599
Shuai, F., 71–2
Sibley, J.W., 69
Sides, G.M., 117, 121, 401, 405
Simons, N.E., 502, 504
Simpson, B., 381
Sitharam, T.G, 268
Sivakumar, V., 16, 241, 288, 292, 327,
    331, 341, 345–7, 352, 354, 356,
    372, 374, 433, 435–6, 442, 446,
    449–50
Sivapullaiah, P.V., 499
Skempton, A. W., 504
Sladen, J.A., 336, 578
Slatter, E.E., 53, 296
Sokolovich, V.E., 251
Songyu, L., 253
Sousa Pinto, C., 5, 28, 29
Souza, A., 240, 251
Speedy, R.J., 30
Spencer, J., 5, 505, 507
Sposito, G., 12, 32
Srivastava, R., 195–6, 199–202, 217
Stannard, D.I., 78
Steinberg, M., 473
Steinhardt, R., 214
Stokoe, K.H., 382, 408, 427
Subba Rao, K.S., 227, 253
Sun, H.W., 148, 501, 515–18, 529,
    536, 559–60, 577–8
Swarbrick, G.E., 69
Szecsody, J., 207

Take, W.A., 30, 83, 576, 587
Tang, W.H., 596, 598
Tarantino, A., 53, 295, 296, 301
Tatsuoka, F., 381–2
Taylor, S.A., 9
Telford, Thomas, 227
Temperley, H.N.V., 37
Terzaghi, K., 5, 13–14, 435
Thomas, H.R., 435
Tinjum, J.M., 155, 157
Trevena, D.H., 27, 30

Tripathy, K.S., 227, 252, 257–62, 264–6, 269–70, 272, 274
Tung, Y.K., 501, 515–17, 519, 520, 523, 529, 536, 541, 550, 562, 599, 601, 603, 605–7, 609
Turnbull, W., 250

Vaid, Y.P., 575
van Amerongen, G. J., 119
van der Gaast, S.J., 54–5
Van Der Merwe, D.H., 255–6
Vanapalli, S.K., 93, 148, 156–7, 166, 209, 335, 405, 538
Vassallo, R., 380
Viana Da Fonseca, A., 382
Viggiani, G.M.B., 409, 417
Vinale, F., 401
Volyanick, N.V., 251

Walsh, K., 249–50
Wan, A.W.L., 401
Wan, R.G., 366, 401
Wang, B., 94, 298, 299, 476, 501, 533, 536
Wang, S.J., 233
Wang, Y., 382
Warkentin, B.P., 253
Wesley, L.D., 41, 289, 394
Wheeler, S.J., 16–20, 241, 288, 292, 327, 331, 340–1, 346–7, 352, 354, 356, 372, 374, 426, 433, 435–7, 442, 446, 449–50, 461
White, I., 66, 195, 199
Whitman, R.V., 100, 405

Williams, D.J., 69
Williams, J., 48–9, 53, 59, 277, 297
Wilson, G.W., 210, 516, 523, 530, 538
Wolle, C.M., 515, 538
Woods, R.D., 380
Wray, W.K., 31–2, 63–7, 68–71, 77–9, 84
Wright, S.G., 504
Wu, S.M., 251

Xing, A., 136–40, 279, 298, 314, 330, 418, 538–9

Yamamuro, J.A., 336, 343, 575
Yeh, T.C.I., 195–6, 199–200, 201–2, 217
Yen, B.C., 600, 601, 602
Yong, R.W., 253
Yoshitake, S., 592
Youd, T.L., 382
Young, R.F., 5, 24–5, 37
Yung, S.Y., 380, 41215, 418, 420–2, 425, 427–9

Zakerzadeh, N., 93
Zhan, L.T., 38, 41, 42, 44, 46, 60, 63, 94, 279, 330, 473, 501, 559, 561–4, 566
Zhang, H., 501
Zhang, J.F., 516
Zhang, W., 230, 240, 250
Zhou, R.Z.B., 30, 38, 45, 47, 279, 326–29, 331–4
Zlatovic, S., 575

Subject Index

adsorption, state-dependent SWCC
    and 148
    see also drying; wetting
Agsco 382
air flow, *see* flow of air
air phase 7, 21
    coefficient of permeability with
        respect to 113
    density 613
    flow of 111
    Fick's law for 111–14
    see also three-phase system
air–water interface 20, 22
    see also contractile skin
anisotropic consolidation 356
anisotropic shear stiffness of
        completely decomposed tuff
        (clayey silt) 408
    bender elements arrangement
        413–16
    shear modulus determination
        using 416
    shear wave velocity measurement
        using 416
    completely decomposed tuff (CDT)
        417–18
    experimental results 421
        net mean stress and degree of
            anisotropy 426
        proposed equations verification
            426–30
        shear wave velocities and matric
            suction, relationship between
            422–4
        soil suction and net mean stress
            effect on void ratio 421

suction and degree of fabric
        anisotropy 425–6
    shear modulus measurement 416
    shear wave velocity measurement
        413–16
    soil types 417
    test specimen preparation 417, 419
    testing
        apparatus and measuring devices
            411
        programme and procedures
            419–21
        triaxial 411–12
    theoretical considerations 408–11
    triaxial apparatus for testing
        unsaturated soils 411–12
anisotropic steady-state seepage
        176–7
anisotropically consolidated
        unsaturated specimens 355, 358
anisotropy 176
antecedent infiltration rate 215–16,
        220
    combined effect of saturated
        permeability and 216–17
    influence on flows 213–16
arid regions 10
artificial rainfall simulations (field
        performance aspects) 483
    ground deformations responses
        495
    horizontal
        displacements due to soil suction
            changes 495–7
        total stresses response 493–5
    in situ
        PWP profiles variations 488–9

water content profiles variations
  491–2
piezometric level responses 486–8
pore water pressure and
  horizontal total stresses response
    493–5
  in situ PWP profiles variations
    488–9
  in situ water content profiles
    variations 491–2
  piezometric level responses to
    486–8
  volumetric water content
    responses 490
soil suction or pore water pressure
  responses 484–6
soil swelling at shallow depths upon
  wetting 497–9
volumetric water content responses
  490
see also slope engineering
axis-translation technique 36, 58
  advantages and limitations 47
  applications 38–41
  matric suction controlling 36–7
  SDSWCC measurements 38–41
  shearing test in direct shear box
    44–7
  stress path testing in triaxial
    apparatus 41–44
  SWCC measurements 38–41
  working principle 37–8
  see also osmotic technique
axis-translation techniques and
    osmotic techniques, comparison
    between 59, 294–5, 297
  test results
    axis-translation technique 302–4
    comparison 305–7
    osmotic technique test 304
  testing
    equipment 297–8
    material and specimen preparation
      298–9
    procedures 300–2
    programme 300

bender elements arrangement 413–16
  shear modulus determination using
    416
  shear wave velocity measurement
    using 416

see also anisotropic shear stiffness of
  completely decomposed tuff
  (clayey silt)
Bishop's simplified method, 542, 609
boiling 25
  see also cavitation
BRE, see Building Research
  Establishment
bubbling pressure 104, 322
Building Research Establishment (BRE)
  255

calcium montmorillonite 254
calibration 72
  apparent volume change 293
  differential pressure
    transducer 293
  thermal conductivity sensors 71
  volume change measuring system
    293
capillary effect 17
  see also stress state variables
capillary fringe 7–8
capillary model 32
  equilibrium 32
  soil–water interface, equilibrium at
    32–3
  of water column and capillary
    height 33
  see also suction
carbonation, lime 276
Casagrande graphical method 341
cavitation 25, 27
  avoidance 28
  boiling 25
  degassing 25
  gaseous 25
  inception and nuclei 26
  nucleation aspects 27
  pore water under tension and 26
  surface tension and 24
  vaporous 25
CDG, see completely decomposed
  granite
CDT, see completely decomposed
  tuff
cementation 275, 276
centrifuge for state-dependent SWCC
  168–9
centrifuge technique
  principle 169–70
  water-retention 168

ceramic cylinders   170–4
    porous   171
    SWCC   173
chemical diffusion through
        water   121
clay minerals   253–4
clay soils
    expansive   254
    swelling in   254
clayey silt, anisotropic shear stiffness of
    bender elements arrangement
        413–16
        shear modulus determination
            using   416
        shear wave velocity measurement
            using   416
    completely decomposed tuff (CDT)
        417–18
    experimental results   421
        net mean stress and degree of
            anisotropy   426
        proposed equations verification
            426–30
        shear wave velocities and matric
            suction, relationship between
            422–4
        soil suction and net mean stress
            effect on void ratio   421
        suction and degree of fabric
            anisotropy   425–6
    shear modulus measurement   416
    shear wave velocity measurement
        413–16
    soil types   417
    suction
        degree of fabric anisotropy
            425–6
        matric suction   422–4
        void ratio, effect on   421
    test specimen preparation
        417, 419
    testing
        apparatus and measuring devices
            411
        programme and procedures
            419–21
        triaxial   411–12
    theoretical considerations   408–11
    triaxial apparatus for testing
        unsaturated soils   411–12
clayey soils swelling   253
clays, plastic   11

climate
    in situ suction measurements
        and   87
    vadose zone and   8
CNSS, see cohesive nonswelling soil
coarse-grained gravelly sand (DG)
    constant water content tests on
        459–61
    saturated   457–9
    undrained tests on   457–9
    unsaturated   459–63
    wetting tests on   461–3
    see also decomposed volcanic (DV)
        soils; state-dependent
        elasto-plastic modelling
cohesive nonswelling soil (CNSS)   275
collapse
    behaviour of
        non-virgin fill   247
        virgin fill   244–6
    loading-collapse yield curve   442–3
    wetting and
        collapse caused by wetting
            227–8
        stress state variable and   17
collapsible soils   228
    conditions leading to   230
    double oedometer test   239
    field testing   240
    foundation settlements estimation
        249
    identification   231
        collapse indicators   231
        double oedometer test   232–3
        geological reconnaissance   231
        laboratory testing   231
    mitigation measures against
        250–2
    non-virgin fills and   240
    occurrence and microstructure of
        228
    single point, multiple specimen test
        239
    single specimen test   238
    virgin fills and   240
    wetting tests   239
    wetting tests, laboratory response to
        235–7
        comparison of commonly used lab
            test methods   238
        naturally occurring collapsible
            soils (Case I)   234–5, 238

sample compacted in the
oedometer (Case III) 237–8
sample taken from existing
compacted fill (Case II)
237–8
sampling, loading and wetting
effect of 234
see also swelling
compacted expansive soils 257
see also under expansive soil
completely decomposed granite (CDG)
45, 329
see also decomposed granite (DG)
soils
completely decomposed tuff (CDT)
see also anisotropic shear stiffness of
completely decomposed tuff
(clayey silt) 417
completely decomposed volcanic
(CDV) specimen 40
see also decomposed volcanic (DV)
soils
compression 14
dependent dilatancy parameters
454, 455
isotropic 397–9, 448
shear stiffness of quartz silt
isotropic 397–9
under constant suctions 397
suctions and 397
confining pressure and void ratio
effect on
degree of saturation of sands
385–8
shear modulus ratio of sands
385–8
conjunctive modelling effects on slope
stability 599–600
conjunctive surface–subsurface flow
analysis program 604
conjunctive surface–subsurface flow
model 602–3
subsurface flow submodel 602
surface flow submodel 600–1
numerical simulation
conjunctive surface–subsurface
flow analysis program 604
two-dimensional
saturated–unsaturated
seepage analysis program
SEEP/W 604

pore water pressure distribution
604–8
slope stability analyses 609–10
consolidated conditions, shear strength
of unsaturated soils
and 281
consolidated drained (CD) test 282,
284
consolidated drained triaxial
compression test 282
consolidated saturated decomposed
volcanic soil, isotropically 339
initially unsaturated
specimens during
saturation 339–41
isotropic compression 341
shear behaviour under undrained
conditions 342–4
consolidated unsaturated decomposed
volcanic soil, isotropically
isotropic compression 345–6
shear behaviour under constant
water content 348–54
yielding behaviour 346–8
consolidated undrained (CU) test 284
consolidation
direct shear tests on unsaturated
soils and 286
soil mass 9
constant water (CW) content test 284
contact–angle effect 141
contractile skin 20–4
see also surface tension
cracking, soil mass 8
critical slip surface location 503–4
see also slope stability
critical state (elasto-plastic modelling
aspects) 446
critical state parameters 450–1
see also state-dependent
elasto-plastic modelling
cyclic swell–shrink tests 261–3
see also expansive soils

damming
factor of safety variation 558
influence on slope stability 547
finite element mesh and numerical
analysis plan 548–9
input parameters and procedures
for slope stability analyses
550

damming (*Continued*)
  numerical simulation procedures
    and hydraulic boundary
    conditions 549–50
  pore water pressure distributions
    and 553–7
Darcy's law 99
  of air phase 114
  for saturated soils 99
  for unsaturated soils 99–100
  see also Fick's law
DAVI, *see* diffused air volume
  indicator
decomposed granite (DG) soils 436
  shear stiffness of 399
    empirical equations for 405–8
    RCTS testing procedures 399
    suction and 402–8
    system to measure changes in
      water content 400
    tested material 401
    state-dependent elasto-plastic model
      predictions
    constant water content tests on
      459–61
    saturated 457–9
    undrained tests on 457–9
    unsaturated 459–63
    wetting tests on 461–3
  suction effect on elastic shear
    modulus of DG (silty sand)
    394–408
  see also decomposed volcanic (DV)
    soils; quartz silt
decomposed volcanic (DV) soils 436
  fine-grained sandy silt
    undrained tests on 463
    water content tests on 463, 465
  saturated (isotropically consolidated)
    initially unsaturated specimens
      during saturation 339–41
    isotropic compression 341
    shear behaviour under undrained
      conditions 342–4
  state-dependent elasto-plastic
    modelling and
    saturated 463
    undrained tests on 463
    unsaturated 463, 465
    water content tests on 463, 465
  undrained tests on
    fine-grained sandy silt 463

state-dependent elasto-plastic
  modelling 463
unsaturated (isotropically
  consolidated)
  isotropic compression 345–6
  shear behaviour under constant
    water content 348–54
  yielding behaviour 346–8
  water content tests on 463, 465
deep foundations 278
deformation
  expansive soils 271
  pattern at
    22 s prototype time 593
    60 s prototype time 594–6
  vertical 271
deformation mechanism
  failure and 592
  un-reinforced loose fill slopes 592
deformation responses, ground
  horizontal displacements due to soil
    suction changes 495–7
  soil swelling at shallow depths upon
    wetting 497–9
  see also slope engineering
degassing 25
degree of anisotropy of CDT
  net mean stress effect on 426
  suction effect on 425, 426
degree of saturation
  friction angle (ϕ^b) and degree of
    saturation (χ), relationship
    between 311–13
  permeability
    function and hysteresis 108
    coefficient and 100–10
  sands
    confining pressure and void ratio
      effect on 385–8
    grain shape effect on 392–3
    grain size distribution effect on
      388–92
  SWCC and 127
density
  air phase 613
  defined 611
  soil particles 611
  water phase 612
desaturated coefficient of permeability
  207–9
desaturated permeability, infiltration
  and 207–9

desaturation 9
  coefficient 207
  saturated soils (man-made activities
    and) 12
desiccation, soil mass 8
desorption 140, 142, 148
  see also drying; wetting
DG, see decomposed granite (DG) soils
differential pressure transducer (DPT)
    42, 293
diffused air volume indicator (DAVI)
    41
diffusion 117
  air diffusion through water 118–31
  chemical diffusion through water
    121
  Fick's law for 118
  flow of air 116
  flow of water 116
digital transducer interface (DTI) 41
dilatancy 47
  granite soils
    saturated 366
    unsaturated 370
  maximum 334–5
  negative 331
  parameters
    compression-dependent 454–5
    suction-dependent 451–4
  positive 331
  soil suction and 326–8
    during shear, evolution of 331–3
    maximum dilatancy 334–5
    soil type and test procedure
      329–30
    stress–dilatancy relationship
      333–4
    test results 331–3
  see also state-dependent
    elasto-plastic modelling
dilatometer tests (DMT) 476
direct shear box test 44–7, 280
  on unsaturated soils 286
    consolidation 286
    failure envelope 286
    shearing aspects 286
  suction equalization and 60
  see also triaxial tests
displacement
  down-slope 496
  due to soil suction changes 495–7
  horizontal 495–7

lateral 496
  simulated rainfall and 495–7
  see also slope engineering
distortion 14
double oedometer test, collapsible soils
    and 239
down-slope displacement 496
downward flux 8–9
  see also precipitation
drained tests 282
  see also undrained test
drier climatic regions, soil mechanics
    problems in 12
drift condition 80
drill hole lime 276
driving potential
  air phase 111
  water phase 97, 99
dry-bulb 66
drying 140, 142–3
  drying SDSWCC influence on slope
    stability 537–40, 545
    influence on pore water pressure
      distributions 542
    transient seepage analyses 539,
      540
  hysteresis and 140
  recompacted soils SWCC and 158
  slope stability aspects 537–8
    pore water pressure distributions
      542
    transient seepage analyses
      539–40
  state-dependent SWCC and 147
  SWCC and 165, 537–40, 545
  soil mass 8
  see also hysteresis; wetting
DV, see decomposed volcanic (DV)
  soils

earth pressure cells (EPCs) 493
effective stress 14, 16, 494
effective stress equation 13–14
  Bishop's 14–15
  saturated behaviour 16
  single-valued 14
  unsaturated behaviour 14, 16
  see also stress state variables
elastic body, stress state variables
  for 13

elastic parameters 447
see also elasto-plastic modelling,
  state-dependent
elastic strains 437–9
elasto-plastic modelling,
  state-dependent
  coarse-grained gravelly sand (DG)
    undrained tests on 457–9
    water content tests on 459–61
  dilatancy parameters
    compression-dependent 454–5
    suction-dependent 451–4
  fine-grained sandy silt (DV)
    undrained tests on 463
    water content tests on 463, 465
  mathematical formulations
    basic assumptions 436–7
    critical state 446
    elastic strains 437–9
    elasto-plastic variation of specific
      water content 446
    elasto-plasticity 437
    flow rules 443–5
    hardening rules 445
    loading-collapse yield curve
      442–3
    plastic strains 440
    yield functions 440–1
  model parameters determination
    critical state parameters 450–1
    dilatancy parameters 451–5
    elastic parameters 447
    hardening parameters 455
    isotropic compression parameters
      448–9
    specific water content parameters
      456
    water retention parameters 448
  specific water content 446, 456
  undrained tests on saturated
    specimens of
    coarse-grained gravelly sand (DG)
      457–9
    fine-grained sandy silt (DV) 463
  water content tests on unsaturated
    specimens of
    coarse-grained gravelly sand (DG)
      459–61
    fine-grained sandy silt (DV) 463,
      465

wetting tests on unsaturated
  specimens of coarse-grained
  gravelly sand (DG) 461–3
elasto-plasticity 437
EPCs, see earth pressure cells
equalization, suction 62
  see also suction
equilibrium
  soil–water interface 32–3
  of water column and capillary height
    33
equipotential lines 184
evaporation 8, 482
  see also slope engineering
evapotranspiration 8, 9
expansive soils 252
  clay soils 254
  compacted 257
  cyclic swell–shrink tests 261–3
  foundations design 273
  initial placement conditions 268–9
  potential swell determination
    oedometer methods 255, 257
    soil suction methods 255–6
  potential swell determination 255
  surcharge pressures 270–1
  swelling processes 253
  swell–shrink tests 253
  volumetric and vertical deformation
    271–2
  water content-void ratio
    swell–shrink paths of
      compacted 257–9, 263–8
    initial placement conditions
      influence 268–9
    soils used and testing procedures
      259–63
    surcharge pressure influence
      270–1
explosive vaporization 25
extended Mohr–Coulomb failure
  criterion 307–11
external stress 17–18
  see also stress

facing structure
  nailed loose fill slopes 593
  models 581–2
  soil nails and facing structure, role
    of 597–8
factor of safety (FOS)
  damming effect on 558

wetting SWCC and SDSWCC effect
on 558
see also safety equations
failure criterion 307
see also extended Mohr–Coulomb
failure criterion
failure envelope
direct shear tests on unsaturated
soils and 285
matric suction and 312
failure mechanism
deformation and 592
development 590–1
un-reinforced loose fill slopes
590–1
FEMWATER 515–17
see also rainfall infiltration into
unsaturated soil slope (3D
numerical parametric study)
Fick's law
for air phase 111–14
for diffusion 117
see also Darcy's law
field instrumentation (slope
engineering aspects)
artificial rainfall simulations 483–4
evaporation monitoring 482
horizontal
movements and surface heave
monitoring 482
total stresses monitoring 482
rainfall intensity monitoring 482
runoff monitoring 482
soil suction and water content
monitoring 478, 481
see also slope engineering
field performance (slope engineering
aspects)
ground deformations responses
495
horizontal
displacements due to soil suction
changes 495–7
total stresses response 493–5
in situ
PWP profiles variations 488–9
water content profiles variations
491–2
piezometric level responses 486–8
soil suction or pore water pressure
responses 484–6

soil swelling at shallow depths upon
wetting 497–9
volumetric water content responses
490
see also artificial rainfall
simulations; slope engineering
field testing, collapsible soils 240
fills
non-virgin 248
collapse behaviour 247
collapsible soils and 240
testing procedures 243–4
testing program 241–2
virgin fills
collapse behaviour 244–6
collapsible soils and 240
testing procedures 243–4
testing program 241–2
filter paper method 67
filter papers
Schleicher and Schuell No. 589
67–8
Whatman No. 42 68–9
fine-grained sandy silt (DV)
saturated 463
undrained tests on 463
unsaturated 463, 465
water content tests on 463, 465
see also coarse-grained gravelly sand
(DG); state-dependent
elasto-plastic modelling
finite difference analyses
facing structure models 581–2
input parameters for
soil models 582–4
soil nail system 584–5
slope geometry and soil nail
arrangement 576–7
soil models 578–80
input parameters for 582–4
soil–nail interface models 580–1
soil nails and soil–nail interface
models 580–1
see also soil nails in loose fill slopes
subjected to rainfall infiltration
effects
finite element mesh
damming and 548
three-dimensional 517
FLAC2D 576

flexible structures 277
  see also deep foundations; rigid
    structures
flocculation 275
flow laws 93, 121
  for air 94
    see also under flow of air
  state-dependent elasto-plastic
    modelling 443–5
  for water 94
    see also under flow of water
  see also flows
flow modelling
  conjunctive surface–subsurface
    602, 603
  subsurface 602
  subsurface three-dimensional
    3D finite element mesh 517–18
    boundary conditions 519
    governing equations 516–17
    initial steady-state conditions
      519–21
    rainfall conditions simulated
      521–3
  surface flow submodel 600
  see also flow of air; flow of water
flow of air 111
  air flow 120
    diffusion through water and flow
      laws for 94
  coefficient of permeability and
    matric suction 115
  Darcy's law 114
  diffusion aspects 117
  driving potential for 111
  Fick's law 111–14
  permeability coefficient and air
    phase 114
  steady state
    one-dimensional 188–90
    two-dimensional 191–2
  see also flow of water; flows
flow of water 94
  Darcy's law for
    saturated soils and 99
    unsaturated soils and 99–100
  diffusion aspects 117
  driving potential for water phase
    97–9
  hydraulic head gradient 97
  matric suction gradient 95–6
  osmotic suction and 99

permeability coefficient and water
    phase 100
SDSWCC 123
slope of infinite length 184
steady-state 175
  heterogeneous, isotropic
    steady-state seepage 176
  permeability coefficient variation
    176
  transient 192–4
  two-dimensional 191–2
  water content gradient 95
  water phase
    driving potential for water phase
      97–9
    permeability coefficient and 100
  see also flow modelling; flow of air;
    flows
flows
  antecedent and subsequent
    infiltration rate 213–16
  combined effect of saturated
    permeability and 216
  influence of 213–16
  infiltration in unsaturated soils
    antecedent and subsequent
      infiltration rate 213–16
    one-dimensional 199–202
    rainfall infiltration mechanism
      194–6
  hydraulic properties 196–9
  parametric studies 202
    antecedent and subsequent
      infiltration rate 213–16
    desaturated permeability 207–9,
      212
    one-dimensional vertical transient
      infiltration in unsaturated
      soils 203
    relative sensitivity of hydraulic
      parameters 211
    saturated permeability 205–6,
      212–13
    steady-state infiltration 203
    water storage capacity 209–10
  rainfall infiltration mechanism in
    unsaturated soils 194–6
  saturated permeability
    antecedent and subsequent
      infiltration rate and 216
    influence on 205–6
    parametric studies 211

soil heterogeneity   217
steady-state   174–5
transient   174
unsteady-state   175
water storage capacity and   209–10
fluid medium components,
    permeability coefficient and   101
fluid phases
    saturated soil mechanics   8
    unsaturated soil mechanics   8
flux
    surface   8
    upward   8
force equilibrium   511–12
    see also slope stability analyses
force function, inter-slice   512–14
    see also slope stability
foundation settlements, collapsible
    soils   249
foundations design, expansive swelling
    soils   273
    deep foundations   278
    flexible structure   277
    ground stabilization   275–6
    moisture changes, defence against
        277
    moisture content stabilization
        276–7
    passive stabilization   277
    rigid structure   278
    substitution of ground   274–5
foundations, deep   278
    see also structures
four-phase materials   20
four-phase unsaturated soil system   21
free water   26
friction angle (ϕ^b) and degree of
    saturation (χ), relationship
    between   311–13

gaseous cavitation   25
gauge, vacuum   80
General Limit Equilibrium (GLE)
    method   504–7
    inter-slice force function   512–14
    normal force equation   509–10
    safety equation and
        force equilibrium   511–12
        moment equilibrium   510–11
    shear force mobilized equation
        507–9
    see also slope stability

Glazier Way sand   382–3, 386
GLE, see General Limit Equilibrium
    method
gradient
    hydraulic head   97
    matric suction   95–6
    water content   95
grain shape effect on
    degree of saturation of sands
        392–3
    shear modulus ratio of sands
        392–3
grain size distribution effect on
    degree of saturation of sands
        388–92
    shear modulus ratio of sands
        388–92
granite soil, loosely saturated and
        unsaturated decomposed
    shear behaviour   363–6, 368–74
    soil type and specimen preparation
        358–61
    testing program and procedures
        361–3
    wetting behaviour   374–9
granite soils
    undrained tests   361
    wetting tests   361
granite, decomposed (silty sand)   394
    shear stiffness of   399
        empirical equations for   405–8
        RCTS testing procedures   399
        suction and   402–8
        system to measure changes in
            water content   400
        tested material   401
    state-dependent elasto-plastic model
        predictions   457–63
    suction effect on elastic shear
        modulus of   399
gravelly sand (DG) and state-dependent
        elasto-plastic modelling
    constant water content tests on
        459–61
    saturated   457–9
    undrained tests on   457–9
    unsaturated   459–63
    wetting tests on   461–3
gravimetric water content   616
    simulated rainfall and   491
    SWCC and   127–8
    see also volumetric water content

gravitational head 98
  infinite slope 184, 186
ground
  stabilization 275–6
  substitution 274
groundwater response
  rainwater infiltration 525–6
  rainfall patterns effect on
    hydraulic gradient role 530–2
    patterns at sections XX' and YY'
      527–9
  rainfall patterns of prolonged
    rainfall effect on 534–6
  rainfall return period (or amount)
    effect on 532–4
groundwater table 7
  climate changes and 8
  pore water pressure and 10
gypsum block 70–1
  see also suction

Haines jumps 142
hardening
  parameters determination 455
  rules 445
  state-dependent elasto-plastic
    modelling 446
heave monitoring 482
  see also slope engineering
heterogeneity, soil 217, 220
heterogeneous
  steady-state seepage
    anisotropic 176–7
    isotropic 175
  nucleation 27
high-suction tensiometer
  probe 83
homogeneous
  nucleation 27
  saturated soil 181
horizontal displacements due to soil
  suction changes, simulated rainfall
  and 495–7
horizontal movements
  slope engineering aspects 482
  surface heave and 482
  see also vertical deformation
horizontal total stresses
  response to PWP or suction changes
    493
  simulated rainfall and 493
  slope engineering aspects 482

humidity control techniques 54, 58
  advantages and limitations 57–8
  applications 56–7
  working principle 54–5
  see also osmotic technique
hydraulic conditions, damming
    influence on slope stability and
    549–50
hydraulic gradient
  head gradient 97, 114
  slope stability and 530
hydraulic head
  distribution 183
  flow of air and 114
  gravitational 98
  head gradient 97
  infinite slope 184
  pressure 97
  velocity 97
hydraulic head gradient 97, 114
hydraulic properties of unsaturated soils
  matric suction aspects 198
  pore size distribution 198
  pore water pressure 196–7
  SWCC 196
  water coefficient of permeability
    196–8
hysteresis
  hysteretic soil moisture characteristic
    142
  in situ suction measurements and
    88–9
  permeability function 108–10
  SWCC 165
    contact–angle effect 141
    desorption 140
    drying 140, 142–3
    main branches 142
    reversible and irreversible
      relationship 141
    scanning curves 142–3
    sorption 140–2
    wetting 140, 142–3
  thermal conductivity sensors 71–6

impeding layer 567–9, 572–4
  see also slope stability; surface cover
in situ PWP profiles variations 488–9
in situ suction measurements 87
  case study 87, 92
  climate description 87
  hysteresis effect and 90

moisture continuity and   89, 90
sites description and   87
subsoils and   87–8
in situ water content, simulated rainfall
    and   491, 492
infiltration
    antecedent infiltration rate   215–16,
        220
        combined effect of saturated
            permeability and   214
        influence on flows   213–16
    one-dimensional infiltration in
        unsaturated soils   199–202
    parametric studies   202–3
        antecedent and subsequent
            infiltration rate, combined
            effect of saturated   213–16
        desaturated permeability   207–9,
            210, 212
        relative sensitivity of hydraulic
            parameters   211–13
        saturated permeability   205–6,
            212–13
        steady-state flow   205
        transient flow   203
        water storage capacity   209–10
    steady-state   220
    see also under infiltration, rainfall
infiltration, rainfall   194–6
    antecedent and subsequent
        infiltration rate   213–16
    hydraulic properties of unsaturated
        soils   196–9
    in unsaturated soils   196–9
    soil heterogeneity and   217–19
    soil nails in loose fill slopes subjected
        to   575–6
        analysis programme and
            modelling procedure   586
        deformation mechanism   592
        deformation pattern at 22 s
            prototype time   593
        deformation pattern at 60 s
            prototype time   594–6
        facing structure models   581–2
        failure mechanism   590–2
        finite difference analyses   576–85
        global stability evaluation   588
        initial stress distributions in loose
            fill slope   588–9
        loose fill slope formation   586

        nailed loose fill slopes with and
            without facing structure
            593–8
        simulated rainfall infiltration
            effects   587
        slope geometry and soil nail
            arrangement   576–7
        soil models   578–80
        soil models, input parameters for
            582–4
        soil nail system, input parameters
            for   584–5
        soil nails and facing structure, role
            of   597–8
        soil nails and soil–nail interface
            models   580–1
        soil nails installation in loose fill
            slopes)   587
        SP models   578
        stress distribution subjected to
            surface infiltration effects
            589–90
        un-reinforced loose fill slopes
            589–92
    SWCC   195, 196
    unsaturated soil slope (3D numerical
        parametric study) and   515
        3D finite element mesh   517–18
        boundary conditions   519
        FEMWATER   515–16
        governing equations   516–17
        hydraulic gradient role   530–2
        initial steady-state conditions
            519–21
        patterns at sections XX' and YY'
            527–9
        prolonged rainfall effect   534–6
        rainfall conditions simulated
            521–3
        rainfall patterns, preliminary study
            of   523–4
        rainfall patterns effect on
            groundwater response
            527–36
        rainfall return period (or amount)
            effect   532–4
        subsurface flow modelling
            516–23
        typical groundwater response to
            infiltration   525–6
ink bottle effect   155

instrumentation, slope engineering and
    field   478–81
  artificial rainfall simulations   483–4
  evaporation monitoring   482
  horizontal
    movements and surface heave
        monitoring   482
    total stresses monitoring   482
  rainfall intensity monitoring   482
  runoff monitoring   482
  soil suction and water content
      monitoring   478, 481
inter-particle forces   17–18
  see also stress state variables
inter-slice forces   512–14
isotropic compression   397–9
  shear stiffness of quartz silt   397–9
  state-dependent elasto-plastic
      modelling   448–9
isotropic steady-state seepage   176
isotropically consolidated saturated
      decomposed volcanic soil   339
  initially unsaturated specimens
      during saturation   339–41
  isotropic compression   341
  shear behaviour under undrained
      conditions   342–4
isotropically consolidated unsaturated
      decomposed volcanic soil   355–7
  isotropic compression   345–6
  shear behaviour under constant
      water content   348–54
  yielding behaviour   346–8
Itasca   576, 580–1, 588

jet fill tensiometer   82, 88
  see also tensiometers

kaolinite   254
Kelvin's equation   12
  see also suction

lateral displacement   496
lime
  carbonation   275, 276
  drill hole   276
  pressure injected   276
  stabilization   275
liquefaction, static   336
liquid–gas interface   26
loading, collapsible soils   234

loading–collapse (LC) yield curve
    230, 442–3
loess soils   11
loose fill slopes subjected to rainfall
    infiltration effects, soil nails in
    575–6
  analysis programme and modelling
      procedure
    Stage 1 (formation of the loose fill
        slope)   586
    Stage 2 (soil nails installation in
        loose fill slopes)   587
    Stage 3 (simulated rainfall
        infiltration effects)   587
    Stage 4 (global stability
        evaluation)   588
  analysis programme and modelling
      procedure   586
  finite difference analyses
    facing structure models   581–2
    slope geometry and soil nail
        arrangement   576–7
    soil models   578–80
    soil models, input parameters for
        582–4
    soil nail system, input parameters
        for   584–5
    soil nails and soil–nail interface
        models   580–1
    SP models   578
  initial stress distributions in loose fill
      slope   588–9
  nailed loose fill slopes with and
      without facing structure
    deformation pattern at 22 s
        prototype time   593
    deformation pattern at 60 s
        prototype time   594–6
    soil nails and facing structure, role
        of   597–8
  un-reinforced loose fill slopes
    deformation mechanism   592
    failure mechanism   590–1, 592
    stress distribution subjected to
        surface infiltration effects
        589–90
loosely compacted volcanic soil   335
  isotropically consolidated DV soil
      (saturated)   335–6
  initially unsaturated specimens
      during saturation   339–41
  isotropic compression   341

shear behaviour under undrained
     conditions   342–4
isotropically consolidated DV soil
     (unsaturated)   355–7
isotropic compression   345–6
shear behaviour under constant
     water content   348–54
yielding behaviour   346–8
see also loosely decomposed granitic
     soil
loosely decomposed granitic soil
saturated   358–69
shear behaviour   363–9
unsaturated decomposed granitic soil
     358–62
shear behaviour   369–74
wetting behaviour   374–9
see also decomposed granite (DG)
     soils

man-made activities, desaturation of
     saturated soils and   12
man-made structures, soil mechanics
     and   10
manometer, mercury   79
matric suction   13, 19, 63
air coefficient of permeability and
     114
axis-translation technique and   36
controlling   48
failure envelope and   312
osmotic technique   49
gradient   95–6
gypsum block and   70–1
hydraulic properties of unsaturated
     soils   196
measuring devices   63
shear testing and   295
shear wave velocities relationship
     with anisotropic shear stiffness
     of CDT and   422–4
water coefficient of permeability and
     106–7
see also osmotic suction
maximum dilatancy   334, 335
mechanical stabilization   275
mercury manometer   79
microstructure, collapsible soils   228
Milipore MF 0.025   69
Milipore MF 0.05   69
mitigation measures, collapsible soils
     250, 251, 252

mixing in place procedures   276
mobilized shear force equation   507–9
see also slope stability
Mohr–Coulomb failure criterion,
     extended   307–11
moisture
content stabilization   276–7
expansive soils foundation design
     and   277
in situ suction measurements and
     87–9
moment   510, 511
montmorillonite   11, 254
Morgenstern–Price methods   514
mortar   382
movement, horizontal (slope
     engineering aspects)   482

nailed loose fill slopes subjected to
     rainfall infiltration effects   575–6
analysis programme and modelling
     procedure
Stage 1 (formation of the loose fill
     slope)   586
Stage 2 (soil nails installation in
     loose fill slopes)   587
Stage 3 (simulated rainfall
     infiltration effects)   587
Stage 4 (global stability
     evaluation)   588
analysis programme and modelling
     procedure   586
finite difference analyses
facing structure models   581–2
slope geometry and soil nail
     arrangement   576–7
soil models   578–80
soil models, input parameters for
     582–4
soil nail system, input parameters
     for   584–5
soil nails and soil–nail interface
     models   580–1
SP models   578
initial stress distributions in loose fill
     slope   588–9
nailed loose fill slopes with and
     without facing structure
deformation pattern at 22 s
     prototype time   593
deformation pattern at 60 s
     prototype time   594–6

nailed loose fill slopes subjected to
rainfall infiltration effects
(*Continued*)
soil nails and facing structure, role
of 597–8
un-reinforced loose fill slopes
deformation mechanism 592
failure mechanism 590–1, 592
stress distribution subjected to
surface infiltration effects
589–90
natural soil specimen 152
natural soils
and recompacted soils, comparison
between 163–4
constant volume assumption
verification 159–61
normal stress influence on 161–3
stress state influence on 159
SWCC and 154, 165
see also recompacted soils
negative dilatancy 331
non-virgin fills 248
collapse behaviour 247
collapsible soils and 240
testing procedures 243, 244
testing program 241, 242
Noorany, I. 234, 238
normal force equation 509–10
see also slope stability
normal shrinkage 268
normal stress 17
normal stress influence on
natural soils 159
recompacted soils 163
nucleation
heterogeneous 27
homogeneous 27
see also cavitation
nuclei, cavitation inception and 26
numerical analysis, damming and slope
stability 548
numerical simulation
conjunctive modelling effects on
slope stability
conjunctive surface–subsurface
flow analysis program 604
two-dimensional
saturated–unsaturated
seepage analysis program
SEEP/W 604

damming influence on slope stability
549–50
SDSWCC influence on slope stability
input parameters and analysis
procedures for stability
analyses 542
transient seepage analyses
539–41

oedometer methods
double oedometer test 232–3
swelling determination and 255,
257
oedometer test, double 232–3
one-dimensional flow 178–9
of air 188–90
formulation 179–81
see also two-dimensional flow
one-dimensional infiltration in
unsaturated soils 199–202
osmotic suction 13, 63
flow of water in soils and 99
measuring devices 63
squeezing technique 83
see also matric suction
osmotic technique 58, 296
advantages and limitations 53
applications 50–3
comparison with axis-translation 58
working principle 49–50
see also axis-translation technique
osmotic techniques and
axis-translation, comparison
between 294–5, 297
axis-translation technique test results
302–4
osmotic technique test results 304
test results comparison 305–7
testing
equipment 297–8
material and specimen preparation
298–9
procedures 300–2
programme 300
Ottawa F-125 sands 382, 383, 387–8

parametric studies, infiltration 202
antecedent and subsequent
infiltration rate, combined effect
of saturated 213–16
desaturated permeability 207–9,
213

relative sensitivity of hydraulic
parameters 211
saturated permeability 205–6,
212–13
steady-state flow 203
transient flow 203
water storage capacity 209–10
partial vapour pressure 12
see also suction
pascals 127
passive stabilization 277
PEG solution 296
see also osmotic technique
Peltier effect 66
performance, slope engineering and
field
artificial rainfall simulations
ground deformations responses
495
horizontal displacements due to
soil suction changes 495–7
horizontal total stresses response
493–5
in situ PWP profiles variations
488–9
in situ water content profiles
variations 491–2
piezometric level responses
486–8
soil suction or pore water pressure
responses 484–6
soil swelling at shallow depths
upon wetting 497–9
volumetric water content
responses 490
permeability 10
desaturated permeability influence
on infiltration 207–9
saturated permeability influence on
infiltration 205–6
volume–mass relationship and 101
permeability coefficient
infinite slope 184
steady-state water flow and
coefficient variation 175
permeability coefficient and air phase
114
degree of saturation 115
matric suction and 115
saturation variation effect on 115
see also flow of air

permeability coefficient and water
phase 100
degree of saturation 101–6
fluid and porous medium
components 101
hysteresis 108–10
matric suction and 106–7
permeability and volume–mass
relationship 101
saturation variation effect on 102
volumetric water content and 107,
108
see also flow of water
pF scale 125–6
pF unit of soil suction 126
piezometric level responses, simulated
rainfall and 486–8
see also artificial rainfall
simulations (field performance
aspects)
placement conditions, expansive soil
and 268–9
plaster of Paris 71
see also gypsum block
plastic
clays 11
strains 440
volumetric strains 17
plasticity 255, 437
see also elasto-plastic modelling,
state-dependent
plates
pressure 77–8
suction 77
pore size distribution, hydraulic
properties of unsaturated soils
198
pore water pressure (PWP)
distribution 9
conjunctive modelling effects on
slope stability 599–603
damming influence on 553–7
recompaction influence on
563–6
SDSWCC influence and 542–5
slope stability and 542–5
wetting SWCC and SDSWCC on
551–2
horizontal total stresses response to
493
hydraulic properties of unsaturated
soils 196

pore water pressure (PWP) (*Continued*)
  in situ 488–9
    PWP profiles variations 488–9
    water content profiles variations
        491–2
  infinite slope and 184–8
  profiles variations and simulated
      rainfall 488–9
  saturate soil and 8
  simulated rainfall and 484–6
    horizontal total stresses response
        493–5
    in situ PWP profiles variations
        488–9
    in situ water content profiles
        variations 491–2
    piezometric level responses to
        486–8
    PWP responses to 484–6
    volumetric water content
        responses 490
  stress state variables and 14
pore-fluid 176
pore-fluid volume distribution in
    unsaturated soil 176
porosity 614
  and void ratio 615
  water 615
porous
  blocks 70
  ceramic cylinder 170
  medium components 101
  permeability coefficient and 100
positive dilatancy 331
  see also dilatancy
pozzolanic reaction 275–6
precipitation 8
Preissmann scheme 604
pre-pressurization 30
pressure as stress state variable 13
pressure
  head 97
  injected lime 276
  plates 77–8
  transducer 80
psychrometer 64–5, 88
  thermistor/transistor 65
  thermocouple 66
  see also tensiometers; thermal
      conductivity sensors
PWP, see pore water pressure

quartz silt 394
  compression and swelling
      characteristics under constant
      suctions 397
  isotropic compression 397–9
  material tested 396–7
  suction effect on elastic shear
      modulus of 394–9
  triaxial equipment and
      instrumentation 394
  see also decomposed granite (DG)
      soils; decomposed volcanic (DV)
      soils

radius of curvature of curved water
    13
rainfall infiltration
  antecedent and subsequent
      infiltration rate 213
  hydraulic properties of unsaturated
      soils 196–9
  in unsaturated soils 194
  soil heterogeneity and 217–19
  soil nails in loose fill slopes subjected
      to 575–6
    analysis programme and
        modelling procedure 586
    deformation mechanism 592
    deformation pattern at 22s
        prototype time 593
    deformation pattern at 60s
        prototype time 594–6
    facing structure models 581–2
    failure mechanism 590–2
    finite difference analyses 576–85
    global stability evaluation 588
    initial stress distributions in loose
        fill slope 588–9
    loose fill slope formation 586
    nailed loose fill slopes with and
        without facing structure
        593–8
    simulated rainfall infiltration
        effects 587
    slope geometry and soil nail
        arrangement 576–7
    soil models 578–80
    soil models, input parameters for
        582–4
    soil nail system, input parameters
        for 584–5

soil nails and facing structure, role
    of 597–8
soil nails and soil–nail interface
    models 580–1
soil nails installation in loose fill
    slopes) 587
SP models 578
stress distribution subjected to
    surface infiltration effects
    589–90
un-reinforced loose fill slopes
    589–92
SWCC 195, 196
rainfall infiltration into unsaturated
    soil slope (3D numerical
    parametric study) 515
FEMWATER 515–16
rainfall patterns effect on
        groundwater response
    hydraulic gradient role 530–2
    patterns at sections XX' and YY'
        527–9
    prolonged rainfall effect 534–6
    rainfall return period (or amount)
        effect 532–4
rainfall patterns, preliminary study of
    523–4
subsurface flow modelling
    3D finite element mesh 517–18
    boundary conditions 519
    governing equations 516–17
    initial steady-state conditions
        519–21
    rainfall conditions simulated
        521–3
typical groundwater response to
    infiltration 525–6
rainfall intensity, slope engineering
    aspects 482
rainfall simulations, artificial (field
    performance aspects) 483
ground deformations responses
    495
horizontal
    displacements due to soil suction
        changes 495–7
    total stresses response 493–5
in situ
    PWP profiles variations 488–9
    water content profiles variations
        491–2
piezometric level responses 486–8

pore water pressure and
    horizontal total stresses response
        493–5
    in situ PWP profiles variations
        488–9
    in situ water content profiles
        variations 491–2
    piezometric level responses to
        486–8
    volumetric water content
        responses 490
soil suction or pore water pressure
    responses 484–6
soil swelling at shallow depths upon
    wetting 497–9
volumetric water content responses
    490
see also slope engineering
RCTS 399–400
see also shear stiffness
recompacted soils
and natural soils, comparison
    between 163–4
constant volume assumption
    verification 159–61
normal stress influence on 161–3
specimen 151–2
SWCC and 165
    drying and wetting influence on
        158–9
    initial dry density on influence on
        154–5
    initial water content influence on
        154–8
    ink bottle effect 155
    stress state influence on 159
recompaction (unsaturated loose fill
    slopes) 559
analysis procedures 560–2
model parameters 562–3
recompaction influence on
    pore water pressure distributions
        563–6
    slope stability 566–7
reduced stress variables 14
see also stress state variables
residual shrinkage 268
residual soils 11
resonant column (RC) tests 402
resonant column-torsional shear
    (RCTS) test 399
see also shear stiffness

rigid structures 278
runoff, slope engineering aspects 482

safety equations 510–12
  force equilibrium and 511–12
  moment equilibrium and 511
  see also factor of safety (FOS); slope
    stability
Saint Venant equations 600
sampling, collapsible soils 234
sands
  Agsco 382
  Glazier Way 382–3, 386
  mortar 382
  Ottawa F-125, 382–3, 387
  silty, *see under* silt
  see also decomposed granite (DG)
    soils; decomposed volcanic (DV)
    soils
sands (shear stiffness measurement)
  clayey silt (anisotropic shear
    stiffness)
    bender elements arrangement
      413–16
    completely decomposed tuff
      (CDT) 417–18
    matric suction and 422–4
    net mean stress and degree of
      anisotropy 426
    proposed equations verification
      426–30
    shear modulus measurement 416
    shear wave velocity and 413–16,
      422–4
    soil suction and net mean stress
      effect on void ratio 421
    soil types 417
    suction and degree of fabric
      anisotropy 425–6
    void ratio effect on suction 421
    test specimen preparation 417,
      419
    testing apparatus and measuring
      devices 411
    testing programme and procedures
      419–21
    triaxial testing 411–12
    theoretical considerations
      408–11
  initial water contents effect on
    382–5

maximum shear modulus ratio and
    optimum degree of saturation
  confining pressure and void ratio
    effect on 385–8
  grain shape effect on 392–3
  grain size distribution effect on
    388–92
quartz 394
  compression and swelling
    characteristics under constant
    suctions 397
  isotropic compression 397–9
  material tested 396–7
  suction effect on elastic shear
    modulus of 394–9
  triaxial equipment and
    instrumentation 394
suction effect on elastic shear
    modulus of
  decomposed granite (silty sand)
    399–408
  quartz silt 394–9
saprolitic soils 11
saturated permeability 212
  combined effect of antecedent and
    subsequent infiltration rate and
    216–17
  infiltration and 204–5
saturated soils
  Darcy's law for 99
  effective stress equation 16
  flow of water in 99
  homogeneous 181
  man-made activities and
    desaturation 12
  mechanics 6
    fluid phases and 8
    seepage analyses and 10
    surface flux boundary condition
      and 9
  pore water pressure and 8
  stress state variables for 13
  suction 12
  see also unsaturated soils
saturated specimens
  state-dependent elasto-plastic
    modelling
    coarse-grained gravelly sand (DG)
      457–9
    fine-grained sandy silt (DV) 463
  undrained tests on 457–9
  granite soils 362, 369

shear behaviour
   granite soils   362, 369
   volcanic soils   339, 341–2
volcanic soils
   isotropic compression   341
   isotropically consolidated   339
   under undrained conditions   342
saturated–unsaturated seepage analysis
   program 2D   604
saturation variation effect on
   permeability coefficient   102–6,
   115
scanning curves   142–3
Schleicher and Schuell No. 589   67,
   68–9
SDSWCC, *see* state-dependent
   soil–water characteristic curve
Seebeck effect   66
SEEP/W   604
seepage   93
   analyses
      saturated soil mechanics
         and   10
      transient   539–41
      unsaturated soil mechanics and
         10
   heterogeneous
      anisotropic steady-state   176–7
      isotropic steady-state   176–7
   problems   10
   see also drying; wetting
sensors, thermal conductivity   71
shear behaviour of unsaturated soils,
   see shear strength of unsaturated
   soils
shear box   41
shear force mobilized equation
   507–9
   see also slope stability
shear modulus
   bender elements and   416
   decomposed granite (silty sand)
      399–408
   elastic   394
   quartz silt   394–9
   measurement   416
   suction effect on elastic shear
      modulus of
      decomposed granite (silty sand)
         399–408
      quartz silt   394–9
   see also stress state variables

shear modulus ratio, sands
   confining pressure and void ratio
      effect on   385–8
   grain shape effect on   392, 393
   grain size distribution effect on
      388–92
shear stiffness
   decomposed granite (silty sand)   399
      empirical equations for   405–8
      RCTS testing procedures   399
      suction and   402–8
      system to measure changes in
         water content   400
      tested material   401
   quartz silt   394
      compression and swelling
         characteristics under constant
         suctions   397
      isotropic compression   397, 398,
         400
      material tested   396, 397
      triaxial equipment and
         instrumentation   394
   see also shear strength; stress state
      variables
shear stiffness measurement   380
   maximum shear modulus ratio and
      optimum degree of saturation
      confining pressure and void ratio
         effect on   385–8
      grain shape effect on   392–3
      grain size distribution effect on
         388–92
   sands, initial water contents effect
      on   382–5
   shear modulus of
      decomposed granite (silty sand)
         399–408
      quartz silt   394–9
shear stiffness measurement (completely
   decomposed tuff)   408
   bender elements arrangement
      413–16
      shear modulus determination
         using   416
      shear wave velocity measurement
         using   416
   completely decomposed tuff (CDT)
      417–18
   experimental results   421
      net mean stress and degree of
         anisotropy   426

shear stiffness measurement
(completely decomposed tuff)
(*Continued*)
proposed equations verification
426–30
shear wave velocities and matric
suction, relationship between
422–4
soil suction and net mean stress
effect on void ratio 421
suction and degree of fabric
anisotropy 425–6
shear modulus measurement 416
shear wave velocity measurement
413–16
soil types 417
test specimen preparation 417, 419
testing
apparatus and measuring devices
411
programme and procedures
419–21
triaxial 411–12
theoretical considerations 408–11
triaxial apparatus for testing
unsaturated soils 411–12
shear strength 15
anisotropically consolidated
unsaturated specimen 358
granite soil, loosely saturated and
unsaturated decomposed
shear behaviour of saturated
specimen 363–9
shear behaviour of unsaturated
specimen 369–74
soil type and specimen
preparation 358–61
testing program and procedures
361–2
wetting behaviour of unsaturated
specimen 374–9
isotropically consolidated
saturated specimen 357
unsaturated specimen 357
suction aspects 294
see also under shear strength of
unsaturated soils
shear strength of unsaturated soils
279–80
axis-translation techniques and
osmotic techniques, comparison
between 59, 294–5, 297

axis-translation technique 302–4
comparison 305–7
osmotic technique test 304
testing equipment 297–8
testing material and specimen
preparation 298–9
testing procedures 300–2
testing programme 300
direct shear tests 286
extended Mohr–Coulomb failure
criterion 307–11
granite soil, loosely saturated and
unsaturated decomposed
shear behaviour 363–6, 368–74
soil type and specimen
preparation 358–61
testing program and procedures
361–2
wetting behaviour 374–9
loosely compacted isotropically
consolidated decomposed
volcanic soil (saturated) 335–6
initially unsaturated specimens
during saturation 339–41
isotropic compression 341
shear behaviour under undrained
conditions 342–4
loosely compacted isotropically
consolidated decomposed
volcanic soil (unsaturated)
355–7
isotropic compression of 345–6
shear behaviour under constant
water content 348–54
yielding behaviour 346–8
matric suction and 312
relationship between friction angle
(ϕ^b) and degree of saturation
(χ), 311–13
shear strength equation, alternate
solution to 325–6
shear strength function model
315–16
closed-form solutions 322–5
model comparison to example
data 316–22
soil suction effect on dilatancy
326–8
maximum dilatancy 334–5
soil type and test procedure
329–30

stress–dilatancy relationship
      333–4
test results   331–3
SWCC   313–15
triaxial tests on unsaturated soils
      280, 284
   axial stress application (stage two)
      281–2
   calibration of volume change
      293
   confining pressure application
      (stage one)   281
   consolidated drained (CD) test
      284
   consolidated undrained (CU) test
      284
   constant water (CW) content test
      284
   unconfined compression (UC) test
      284
   unconsolidated undrained (UU)
      test   284
   volume changes measurement
      287–93
testing   41
   axis-translation and osmotic
      techniques, comparison
      between   294
   direct shear box   44, 286
   matric suction and   295
   program and procedure   336–9
   triaxial test   41, 280
   unsaturated soils   294
volume changes measurement system
      287–93
   calibration   293
   design   289–92
shear wave velocity
   and matric suction relationship with
      anisotropic shear stiffness of
      CDT   422–4
   measurement
      anisotropic shear stiffness
         of completely decomposed
         tuff (clayey silt)
         413–16
      using bender elements   416
shearing
   direct shear box   44–7, 286
   resistance   14
   triaxial   45
   see also under triaxial tests

shrinkage
   normal   268
   residual   268
   structural   268
silt
   clayey silt (anisotropic shear stiffness)
      bender elements arrangement
         413–16
      completely decomposed tuff
         (CDT)   417–18
      matric suction and   422–4
      net mean stress and degree of
         anisotropy   426
      proposed equations verification
         426–30
      shear modulus measurement   416
      shear wave velocity and   413–16,
         422–4
      soil suction and net mean stress
         effect on void ratio   421
      soil types   417
      suction and degree of fabric
         anisotropy   425–6
      void ratio effect on suction   421
      test specimen preparation   417,
         419
      testing apparatus and measuring
         devices   411
      testing programme and procedures
         419–21
      triaxial testing   411–12
      theoretical considerations
         408–11
   decomposed volcanic
      (state-dependent elasto-plastic
      modelling)
      saturated   463
      undrained tests on   463
      unsaturated   463, 465
      water content tests on   463, 465
   quartz (shear stiffness)   394
      compression and swelling
         characteristics under constant
         suctions   397
      isotropic compression   397–9
      material tested   396–7
      suction effect on elastic shear
         modulus of   394–9
      triaxial equipment and
         instrumentation   394

simulated rainfall
  effect on pore water pressures and
    soil suction
    gravimetric water content
      responses 491
    horizontal total stresses response
      to PWP or suction changes
      493
    in situ PWP profiles variations
      488–9
    in situ water content responses
      491–2
    piezometric level responses
      486–8
    soil suction or pore water pressure
      responses 484–6
    volumetric water content
      responses 490–1
  field instrumentation aspects 484
  field performance observations 484
    ground deformations response
      495
    horizontal displacements due to
      soil suction changes 495–7
    horizontal total stresses response
      493–5
    in situ PWP profiles variations
      488–9
    in situ water content profiles
      variations 491–2
    piezometric level responses
      486–8
    soil suction or pore water pressure
      responses 484–6
    soil swelling at shallow depths
      upon wetting 497–9
    volumetric water content
      responses 490
  infiltration, *see* rainfall infiltration;
    rainfall infiltration into
    unsaturated soil slope (3D
    numerical parametric study)
  soil swelling at shallow depths upon
    wetting 497–9
  suction responses
    horizontal displacements and
      495–7
    horizontal total stresses and 493
    pore water pressure responses
      484–6
  swelling at shallow depths upon
    wetting 490–9

see also slope engineering
single effective stress parameter 14,
  16–17
single point, multiple specimen test
  239
see also collapsible soils
single specimen test, collapsible soils
  and 238
single-valued effective stress equation
  14
slope engineering 473
  field instrumentation programme
    478–80
    artificial rainfall simulations
      483–4
    evaporation monitoring 482
    horizontal movements and surface
      heave monitoring 482
    horizontal total stresses
      monitoring 482
    rainfall intensity monitoring 482
    runoff monitoring 482
    soil suction and water content
      monitoring 478, 481
  field performance observations
    ground deformations responses
      495
    horizontal displacements due to
      soil suction changes 495–7
    horizontal total stresses response
      493–5
    in situ PWP profiles variations
      488–9
    in situ water content profiles
      variations 491–2
    piezometric level responses
      486–8
    soil suction or pore water pressure
      responses 484–6
    soil swelling at shallow depths
      upon wetting 497–9
    volumetric water content
      responses 490
  recompaction (unsaturated loose fill
    slopes) 559–60
    analysis procedures 560–2
    influence on pore water pressure
      distributions 563–6
    influence on slope stability
      566–7
    model parameters 562–3
SNWTP 473

soil nails in loose fill slopes subjected
   to rainfall infiltration effects
   575–6, 586
22 s prototype time deformation
   pattern   593
60 s prototype time deformation
   pattern   594–6
analysis programme and
   modelling procedure   586
deformation mechanism   592
facing structure models   581, 582
failure mechanism   590–2
finite difference analyses   576–85
formation of the loose fill slope
   586
global stability evaluation   588
initial stress distributions in loose
   fill slope   588–9
input parameters for soil models
   582–5
simulated rainfall infiltration
   effects   587
slope geometry and soil nail
   arrangement   576–7
soil models   578–85
soil nails and facing structure, role
   of   597–8
soil nails and soil–nail interface
   models   580–1
soil nails installation in loose fill
   slopes   587
SP models   578
stress distribution subjected to
   surface infiltration effects
   589–90
un-reinforced loose fill slopes
   589–92
face infiltration effects   589–90
soil profile and properties   475–8
test site   474–5
slope flows   184
slope of infinite length   184
slope stability
conjunctive modelling effects on
   599–600, 609
   surface–subsurface flow analysis
      program   604
   surface–subsurface flow model
      600–3
   numerical simulation   604
   pore water pressure distribution
      604–8

finite element mesh and analysis
   procedures
   impeding layer effect on   568–9
   surface cover effect on   568–9
impeding layer effect on   567–9,
   572–4
recompaction influence on   566–7
surface cover effect on   567–8,
   570–2
two-dimensional
   saturated–unsaturated seepage
   analysis program SEEP/W   604
see also slope stability analyses
slope stability analyses   502
conjunctive modelling effects on
   609
critical slip surface location   503,
   504
damming influence on   547
   finite element mesh and numerical
      analysis plan   548–9
   hydraulic boundary conditions
      549–50
   input parameters and procedures
      550
   numerical simulation procedures
      549–50
   pore water pressure distributions
      553–7
drying SDSWCC   537–8
   pore water pressure distributions
      542
   slope stability and   545
   transient seepage analyses
      539–40
factor of safety variation
   damming effect on   558
   wetting SWCC and SDSWCC
      effect on   558
General Limit Equilibrium (GLE)
   method   504–7
   force equilibrium and safety
      511–12
   inter-slice force function   512–14
   moment equilibrium and safety
      510–11
   normal force equation   509–10
   safety equation and   510–12
   shear force mobilized equation
      507–9
groundwater response to infiltration
   525–6

slope stability analyses (*Continued*)
  input parameters and analysis
    procedures for   542
  limit equilibrium methods (other
    than GLE)   514, 514
  pore water pressure distributions
    damming and   553–7
    drying SDSWCC and   542
    influence on stability   542–5
    wetting SDSWCC and   542,
      551–3
  rainfall patterns
    at sections XX' and YY'   527–9
    effect on groundwater response
      527–36
    preliminary study of   523–4
    prolonged rainfall effect   534–6
    rainfall conditions simulated
      521–3
    rainfall return period (or amount)
      effect   532–4
  SDSWCC influence on   545–6
    damming influence on pore water
      pressure distributions   553–7
    drying SDSWCC   537–40, 542,
      545
    finite element mesh and numerical
      analysis plan   548, 549
    numerical simulations   539–42
    pore water pressure distributions
      542–5, 551–7
    wetting SDSWCC   538, 540–2,
      545, 549–52, 558
    transient seepage analyses   540–2
  three-dimensional numerical
      parametric study (unsaturated
      soil slope)   515
    3D finite element mesh   517–18
    boundary conditions   519
    FEMWATER   516–17
    governing equations   516–17
    hydraulic gradient role   530–2
    initial steady-state conditions
      519–21
    prolonged rainfall effect   534–6
    rainfall conditions simulated
      521–3
    rainfall patterns   523–4, 527–36
    subsurface flow modelling
      516–23
  transient seepage analyses   539,
    540, 541

wetting SDSWCC   538
  hydraulic boundary conditions
    549–50
  influence on slope stability   545
  numerical simulation procedures
    549–50
  pore water pressure distributions
    and   542, 551–3
  slope stability factor of safety,
    variation of   558
  transient seepage analyses   540,
    541
small strain shear stiffness of sands
    382–5
  see also sands (shear stiffness
    measurement)
small tip tensiometer   82
  see also tensiometer
SNWTP, *see* South-to-North Water
    Transfer Project
soil heterogeneity, rainfall infiltration
    and   216–18
soil mass
  consolidation   9
  cracking   8
  desaturation   9
  desiccation   8
  drying   8
soil mechanics
  cavitation   24
  diffusion aspects   117
  drier climatic regions   12
  flow of water in   95
  four-phase materials   20
  man-made structures and   10
  permeability aspects   10
  pore water under tension aspects
    26
  saturated   6
  stress state variables   13
  suction
    defined   12
    range   31
  surface tension   22
  unsaturated   6
  see also flow of air; flow of water
soil nails in loose fill slopes subjected
    to rainfall infiltration effects
    575–6
  analysis programme and modelling
    procedure

Stage 1 (formation of the loose fill slope) 586
Stage 2 (soil nails installation in loose fill slopes) 587
Stage 3 (simulated rainfall infiltration effects) 587
Stage 4 (global stability evaluation) 588
finite difference analyses
  facing structure models 581–2
  slope geometry and soil nail arrangement 576–7
  soil models 578–80
  soil models, input parameters for 582–4
  soil nail system, input parameters for 584–5
  soil nails and soil–nail interface models 580–1
  SP models 578
initial stress distributions in loose fill slope 588–9
nailed loose fill slopes with and without facing structure
  22 s prototype time deformation pattern 593
  60 s prototype time deformation pattern 594–6
  soil nails and facing structure, role of 597–8
un-reinforced loose fill slopes
  deformation mechanism 592
  failure mechanism 590–1, 592
  stress distribution subjected to surface infiltration effects 589–90
see also slope stability; slope stability analyses
soil particles, density of 611
soil solids 176
soil specimen holders 171
soil–water
  characteristics 93
    SDSWCC, see state-dependent soil–water characteristic curve (SDSWCC)
    SWCC, see soil–water characteristic curve (SWCC)
  content, axis-translation technique and 36
  interface, equilibrium at 32–3

soil–water characteristic curve (SWCC) 38, 102
  centrifuge technique principle 169–70
  ceramic cylinders 172–3
  degree of saturation and 128
  flow in water and 121
  flow laws and 121
  gravimetric water content 128, 129
  hydraulic properties of unsaturated soils and 196
  hysteresis
    drying 140, 143
    main branches 142
    scanning curves 142
    wetting 140, 142–3
  mathematical forms of 136–40
  measurement methods for 130
    axis-translation technique 38–41
    conventional 130–36
    test procedures 133
  natural and recompacted soils, comparison between 163–4
  natural soils 159, 165
  normal stress influence on 161–3
  rainfall infiltration 194–6
  recompacted soils 154, 165
  shear strength of unsaturated soils and 313–15
  shear testing and 296
  soil specimen
    descriptions 148
    holders 171
    preparation 152
  stress state influence on
    natural soils 159
    recompacted soils 159
  terminology used for 129
  triaxial apparatus for 166, 167
  volume change and 159, 160, 161
  volumetric pressure plate extractor 150–53
    testing procedures using conventional 152–3
    testing procedures using modified 152
  volumetric water content 128, 129
  water permeability functions 122
    correlation between 143–7
    pF scale and 126
    range of untis of soil suction 127

soil–water characteristic curve (SWCC)
   (*Continued*)
   soil suction components and
      123–4
   soil suction unit and   126
   water content aspects   128–9
   wetting   551, 558
   see also state-dependent soil–water
      characteristic curve (SDSWCC)
solid phase   20
   see also three-phase system
sorption   140, 142
   see also drying; wetting
South-to-North Water Transfer Project
   (SNWTP)   473
   field instrumentation programme
      478–80
      artificial rainfall simulations
         483–4
      evaporation monitoring   482
      horizontal movements and surface
         heave monitoring   482
      horizontal total stresses
         monitoring   482
      rainfall intensity monitoring
         482
      runoff monitoring   482
      soil suction and water content
         monitoring   478, 481
   field performance observations
      ground deformations responses
         495
      horizontal displacements due to
         soil suction changes   495–7
      horizontal total stresses response
         493–5
      in situ PWP profiles variations
         488–9
      in situ water content profiles
         variations   491–2
      piezometric level responses
         486–8
      soil suction or pore water pressure
         responses   484–6
      soil swelling at shallow
         depths upon wetting
         497–9
      volumetric water content
         responses   490
   soil profile and properties
      475–8
   test site   474–5

SP model   578
   see also soil nails in loose fill slopes
      subjected to rainfall infiltration
      effects
specific water content parameters
   determination   456
Spencer methods   514
squeezing technique   83
   see also osmotic suction
stabilization
   lime   275
   mechanical   275
   moisture content   276–7
   passive   277
   procedures
      drill hole lime   276
      mixed in place and recompacted
         276
      pressure injected lime   276
   see also collapse; swelling; wetting
standard penetration tests
   (SPTs)   476
state-dependent elasto-plastic
   modelling
   coarse-grained gravelly sand (DG)
      undrained tests on   457–9
      water content tests on   459–61
   dilatancy parameters
      compression-dependent   454–5
      suction-dependent   451–4
   fine-grained sandy silt (DV)
      undrained tests on   463
      water content tests on   463, 465
   mathematical formulations
      basic assumptions   436–7
      critical state   446
      elastic strains   437–9
      elasto-plastic variation of specific
         water content   446
      elasto-plasticity   437
      flow rules   443–5
      hardening rules   445
      loading-collapse yield curve
         442–3
      plastic strains   440
      yield functions   440–1
   model parameters determination
      critical state parameters   450–1
      dilatancy parameters   451–5
      elastic parameters   447
      hardening parameters   455

isotropic compression parameters
448–9
specific water content parameters
456
water retention parameters 448
undrained tests on saturated
specimens of
coarse-grained gravelly sand (DG)
457–9
fine-grained sandy silt (DV) 463
water content tests on unsaturated
specimens of
coarse-grained gravelly sand (DG)
459–61
fine-grained sandy silt (DV) 463,
465
wetting tests on unsaturated
specimens of coarse-grained
gravelly sand (DG) 461–3
state-dependent soil–water
characteristic curve (SDSWCC)
20, 93, 150
centrifuge for 168–70
ceramic cylinders 172–3
drying 147
equipment 150–51
experimental program 149
flow in water and 123
measurement 38–41
natural soil specimen 152
normal stress influence on 161
recompacted soil specimen 152
soil specimens
descriptions 148
holders 171
natural soil 152
preparation 152
recompacted 152
testing procedures 152–3
triaxial apparatus for 166–7
volumetric pressure plate extractor
149–52
wetting 148, 551, 558
see also soil–water characteristic
curve (SWCC); state-dependent
soil–water characteristic curve
(slope stability aspects)
state-dependent soil–water
characteristic curve (slope stability
aspects) 545–6
damming influence on pore water
pressure distributions 553–7

drying SDSWCC 537–8
pore water pressure distributions
542
slope stability 545
transient seepage analyses
539–40
finite element mesh and numerical
analysis plan 548–9
numerical simulations
input parameters and analysis
procedures 542
transient seepage analyses
539–41
pore water pressure distributions
damming and 553–7
influence on stability 542–5
wetting SDSWCC and 542,
551–2
wetting SDSWCC 538
hydraulic boundary conditions
549–50
influence on slope stability 545
numerical simulation procedures
549–50
pore water pressure distributions
and 542, 551–2
transient seepage analyses 540,
541
static liquefaction 336
steady-state flows 175
air flow
one-dimensional 188–90
two-dimensional 191–2
analyses 175
flow in infinite slope 184
one-dimensional flow 178–,
188–90
parametric studies 202
two-dimensional flow 181, 191–2
formulation 181
through unsaturated soil element
182–3
water flow 175
heterogeneous, isotropic
steady-state seepage 176–7
permeability coefficient variation
176
see also transient flows
steady-state infiltration 203, 220
see also infiltration
stiffness, shear, see shear
stiffness

strains
  elastic 437–9
  plastic 17, 440
  volumetric 17
  see also stress; stress state
    variables
stress
  CDT
    anisotropic shear stiffness 421
    degree of anisotropy 426
    void ratio 421
  dilatancy relationship 333–4
  distribution
    loose fill slope 588–90
    subjected to surface infiltration
      effects 589–90
    un-reinforced loose fill slopes
      589–90
  horizontal (slope engineering
    aspects) 482
  path testing in triaxial apparatus
    41–6
  PWP response 493
  simulated rainfall and 493
  slope engineering aspects 482
  state influence on SWCC
    constant volume assumption
      verification 159–61
    normal stress 161–3
  suction changes and 421, 493
  see also shear stiffness; shear
    strength; stress state
    variables
stress state variables 13
  capillary effect and 17
  collapse on wetting 17
  effective stress 13, 16
  effective stress equation 15
  elastic body 13
  external stresses 17–18
  inter-particle forces and 17–18
  normal 17
  plastic strains 17
  pressure 13
  reduced stress variables 14
  saturated soils 13
  SDSWCC curve 20
  suction and 17–18
  tangential 17
  temperature 13
  unsaturated soils 14
  volume 13

stress–strain relationship (granite soils)
  saturated 363–5, 368
  unsaturated 371–2
structural facing 581
  see also slope engineering
structural shrinkage 268
structures
  deep foundations 278
  flexible 277
  rigid 278
  see also drying; expansive soils;
    swelling
subsequent infiltration rate 213–16,
  220
  combined effect of saturated
    permeability and 216–17
  influence on flows 213–16
  see also infiltration
subsoils, in situ suction measurements
  and 87
substitution, ground 274
subsurface flow modelling
  conjunctive surface–subsurface
    602–3
  three-dimensional (3D)
    finite element mesh 517–18
    boundary conditions 519
    governing equations 516–17
    initial steady-state conditions
      519–21
    rainfall conditions simulated
      521–23
  see also flows
suction
  axis-translation technique 36
  capillary model 32
  CDT
    anisotropic shear stiffness of CDT
      void ratio 421
    degree of fabric anisotropy
      425–6
  decomposed granite (silty sand)
    elastic shear modulus and suction
      399–408
    shear stiffness 402–8
  defined 12
  dilatancy and
    dependent parameters 451–4
    during shear, evolution of 331–3
    maximum dilatancy 334–5
    soil type and test procedure
      329–30

stress–dilatancy relationship
333–4
test results 331–3
elastic shear modulus and suction
decomposed granite (silty sand)
399–408
quartz silt 394–9
equalization, influence of 60–62
horizontal total stresses response to
493
humidity control techniques 54
in osmotic technique 296
in situ suction measurements 87
case study 87, 92
climate description 87
hysteresis effect and 90
moisture continuity and 89, 90
sites description and 87
subsoils and 87
matric, *see* matric suction
measuring devices 63
measuring methods 64, 85
filter paper method 67–8
high-suction tensiometer probe
83
in situ suction measurements 87
jet fill tensiometer 82–3
mercury manometer 79
porous blocks (gypsum blocks)
70–71
pressure transducer 80–82
psychrometer 64–5
small tip tensiometer 82
squeezing technique 83
suction plates and pressure plates
76, 77
tensiometers 78
thermal conductivity sensors 71
thermistor/transistor
psychrometers 65
thermocouple psychrometers 66
vacuum gauge 80
osmotic
flow of water in soils and 95
measuring devices 63
squeezing technique 83
pF scale 126
plates 77
pore water under tension aspects
26
quartz silt

elastic shear modulus and suction
394–9
shear stiffness 397
shear stiffness and
decomposed granite (silty sand)
and 402–8
quartz silt and 397
shear testing and 294
simulated rainfall and
horizontal displacements 495–7
horizontal total stresses response
493–5
in situ PWP profiles vairations
488–9
piezometric level responses
486–8
suction responses 484–6
slope engineering aspects
473, 481
stress state variables and 17–18
SWCC and soil suction
components 123–5
unit 126
swelling determination and soil
suction methods 255–6
theory 32
total 63
units
pascals 127
pF scale 126, 127
range of units 127
see also stress state variables
surcharge pressures, expansive soil and
270–1
surface cover 567–9, 570–2
see also impeding layer; slope
stability
surface flow submodel 600, 601
see also conjunctive modelling
effects on slope stability;
subsurface flow modelling
surface flux 8
surface flux boundary condition
saturated soil mechanics and 9
unsaturated soil mechanics and 9
surface heave, horizontal movements
and 482
see also slope engineering
surface tension 22–3
see also cavitation
SWCC, *see* soil–water characteristic
curve

swelling
  caused by wetting   227
  calcium montmorillonite   254
  clay soils   254
  determination
    oedometer methods   255, 257
    soil suction methods   255–6
  expansive soils   253
    foundations design   273
    potential swell determination
      255
  quartz silt   397
  at shallow depths upon wetting
    497–9
  simulated rainfall   497–9
  stress state variable and   17
  suctions and   397
  see also drying; wetting
swell–shrink paths
  volumetric and vertical deformation
    271–2
  water content-void ratio   257–8,
    263–8
    initial placement conditions
      influence   268–9
    soils used and testing procedures
      259–63
    surcharge pressures influence
      270–1
swell–shrink tests   253
  cyclic   261–3
  expansive soils   253–4
  water content-void ratio   257–8

tangential stress   17
  see also stress state variables
temperature   13
  see also stress state variables
tensile strength   25
  see also shear stiffness; shear
    strength
tensiometers   78
  drift condition defined   80–1
  high-suction tensiometer probe   83
  jet fill   82–3
  pressure transducer and   80–1
  small tip   82
  see also psychrometers; suction,
    measuring methods
tension, pore water under   26
thermal conductivity sensors   71–6
  calibration aspects   72

hysteresis loop   73
  see also psychrometers; tensiometers
thermistor/transistor psychrometers
  65
thermocouple psychrometers   66
three-dimensional subsurface flow
  modelling
  3D finite element mesh   517–18
  boundary conditions   519
  governing equations   516–17
  initial steady-state conditions
    519–21
  rainfall conditions simulated   521–3
  see also slope stability analyses
three-phase system   20
  see also four-phase materials
three-phase unsaturated soil system   20
torsional shear   402
  see also shear stiffness; shear
    strength
total stress analysis of soil slopes   502
  see also slope stability
total stress approach   494
  see also stress state variables
total suction, measuring devices   63
transducer, pressure   80–82
  see also psychrometers; tensiometers;
    thermal conductivity sensors
transgression   27
transient
  infiltration   203
  seepage analyses   539–41
    see also state-dependent soil–water
      characteristic curve
      (SDSWCC)
transient flows   174
  flow of water   192–3
  parametric studies   202
  two-dimensional   191, 192
  see also steady-state flows
triaxial apparatus
  for state-dependent SWCC   166–7
  stress path testing in   41–6
  unsaturated clayey silt soils   411–12
triaxial tests   280
  shear stiffness of quartz silt   394
  volcanic soils, loosely compacted
    336
  volume changes measurement
    287–93
    calibration   293
    design   289–92

unsaturated soils 280, 284
  axial stress application (stage two)
    281–2
  calibration of volume change
    293
  confining pressure application
    (stage one) 281
  consolidated drained (CD) test
    284
  consolidated undrained (CU) test
    284
  constant water (CW) content test
    284
  unconfined compression (UC) test
    284
  unconsolidated undrained (UU)
    test 284
  volume changes measurement
    system 287–93
  see also direct shear box test
two fluid phases 8
two-dimensional flow 181, 191–2
  formulation 182
  through unsaturated soil element
    182–3
  transient flow of water 192–3
  see also one-dimensional flow
two-phase zone 7

unconfined compression (UC) test
  284
unconsolidated conditions 281
unconsolidated undrained (UU) test
  284
undrained behaviour, isotropically
  consolidated unsaturated
  decomposed volcanic soil
  348–54
undrained shear tests 368
undrained tests 282
  granite soils 362, 368, 457–9
  on saturated specimens of
    coarse-grained gravelly sand (DG)
    457–9
  fine-grained sandy silt
    (DV) 463
un-reinforced loose fill slopes
  deformation mechanism 592
  failure mechanism 590–2
  stress distribution subjected to surface
    infiltration effects 589–90
  see also slope stability analyses

unsaturated soils
  in arid regions 10
  cavitation 24
  coarse-grained gravelly sand (DG)
    constant water content tests on
      459–1
    wetting tests on 461–3
  collapse on wetting 17
  Darcy's law for 99, 100
  diffusion aspects 117
  direct shear tests on 285
  effective stress equation 14, 16
  external stress 17
  fine-grained sandy silt (DV) 463,
    465
  flow of air aspects 111
  flow of water in soils 95
  hydraulic properties 196–7
  impeding layer influence on slope
    stability 567–9, 572–4
  liquid water forms in 18
  mechanics 6
    fluid phases and 8
    seepage analyses and 10
    surface flux boundary condition
      and 9
  one-dimensional infiltration in
    199–202
  permeability coefficient variation
    and 176
  pore water under tension in 26
  pore-fluid volume distribution in
    181
  rainfall infiltration mechanism in
    194–5
  recompaction of unsaturated loose
    fill slopes 559–60
    analysis procedures 560–2
    model parameters 562–3
    pore water pressure distributions
      563–6
    slope stability 566–7
  reduced stress variables 14
  shear behaviour 279
    granite soils 369–79
    volcanic soils 345
  slope stability, see slope engineering,
    slope stability; slope stability
    analyses
  stress state variables 13–14
  suctions 12
  surface tension 22

unsaturated soils (*Continued*)
  swelling pattern   17
  system
    four-phase   20
    three-phase   20
  three-phase system   20
  triaxial tests on   280–4, 287–93
  two-dimensional water flow   182
  types   10
    loess   11
    plastic clays   11
    residual   11
    saprolitic   11
  unsteady-state flow analyses   175
    see also steady-state flows
  upward flux   8–9
    evaporation   8
    evapotranspiration   8
  vacuum gauge   80
  vadose zone
    capillary fringe   7
    climate changes and   8
    two-phase   7
  see also saturated soils

vaporization, explosive   25
vaporous cavitation   25
vapour pressure, partial   12
  see also pore water pressure
velocity head   98
vertical deformation
  at equilibrium cycle   271
  expansive soils   271
  see also horizontal movements
virgin fills   248
  collapse behaviour   244–6
  collapsible soils and   240
  testing
    procedures   243–4
    program   241–2
  see also non-virgin fills
void ratio
  and confining pressure effect on
    sands
    degree of saturation   385–8
    shear modulus ratio   385–8
  porosity and   615
  stress and suction effect on
    anisotropic shear stiffness of
    CDT   421
volcanic soil, loosely compacted
  335–6

isotropically consolidated (saturated)
  335–6
initially unsaturated specimens
  during saturation   339–41
  isotropic compression   341
  shear behaviour under undrained
    conditions   342–4
isotropically consolidated
  (unsaturated)   355–7
  isotropic compression   345–6
  shear behaviour under constant
    water content   348–54
  yielding behaviour   346–8
testing program and procedure
  336–9
volume as stress state variable   13
volume–mass and permeability
  relationship   101
volumetric change   15
  at equilibrium cycle   271
  expansive soils   271
  granite soils (unsaturated)   369
  measurement in triaxial cell
    287–9
    calibration   293
    system design   289–92
  SWCC and   154
volumetric pressure plate extractor
  150–51
  conventional   152–3
  modified   153
volumetric water content   108, 616
  simulated rainfall and   490–1
  SWCC and   127–8
  water coefficient of permeability and
    106–7
  see also gravimetric water content
volumetric–axial strain relationships
  372

water
  phase   7, 20
    coefficient of permeability with
      respect to   100
    density   612
    driving potential for
    porosity   615
    retention parameters   168, 448
    storage capacity, infiltration analyses
      and   209–12
    table   7
    see also flow of water

water coefficient of permeability
   hydraulic properties of unsaturated
      soils 196–7
   matric suction and 106
   volumetric water content and
      107–9
water content
   changes 400
   elasto-plastic variation of 446
      see also state-dependent
         elasto-plastic modelling
   gradient 95
   gravimetric 128–9, 491, 616
   in situ 491–2
   parameters determination 456
   recompacted soils SWCC and
      154–7
   shear stiffness of decomposed
      granite (silty sand) and 399
   simulated rainfall and
      gravimetric water content 491
      in situ water content 491–2
      volumetric water content 490–1
   slope engineering aspects 473, 481
   small strain shear stiffness of sands
      382–5
   suction and (slope engineering
      aspects) 478, 481
   SWCC and 127
   tests on unsaturated specimens of
      coarse-grained gravelly sand (DG)
         459–61
      fine-grained sandy silt (DV) 463,
         465
   unsaturated volcanic soil shear
      behaviour 348
   volumetric 128–9, 490–1, 616
water content-void ratio swell–shrink
      paths of compacted expansive
      soils 257–8, 263–8
   initial placement conditions
      influence 268–9
   soils used and testing procedures
      259–63
   surcharge pressures influence
      270–1
water diffusion through
   air 117
   chemical 121
water flow, see flow of water

water permeability functions, SWCC
   and 122
pF scale 126
range of units of soil suction 127
soil suction
   components 123–4
   units 126
water content aspects 127, 128
wet-bulb 66
wetting 140, 142–3
   collapse caused by wetting 227
   hysteresis and 140
   pore water pressure distributions
      and 551–2
   recompacted soils SWCC and 157,
      158
   simulated rainfall 497–9
   state-dependent SWCC and 147,
      538
      hydraulic boundary conditions
         549–50
      influence on slope stability 545
      numerical simulation procedures
         549–50
      pore water pressure distributions
         and 542, 551–2
      transient seepage analyses 540,
         541
   SWCC and 165, 551, 558
   swelling
      caused by wetting 227
      shallow depths swelling upon
         wetting 497–9
      wetting-induced 16
   tests 16
      coarse-grained gravelly sand (DG)
         461–3
      granite soils 362
      transient seepage analyses 540, 541
      unsaturated granite soil specimens
         374
   see also drying
Whatman No. 42 68, 69

yield
   curve 442–3
   functions 440
yielding behaviour, isotropically
      consolidated unsaturated
      decomposed volcanic soil 346–8